# An Introduction to
# GEOGRAPHICAL INFORMATION SYSTEMS

Visit the *An Introduction to Geographical Information Systems*, Fourth Edition, Companion Website at **www.pearsoned.co.uk/heywood** to find valuable student learning material including:

- Activities and revision questions for self-study or discussion
- Multiple choice questions to test understanding
- Web links for further investigation
- Datasets offering practice opportunities
- Future of GIS Blog - an opportunity to share your ideas about what the future holds for GIS

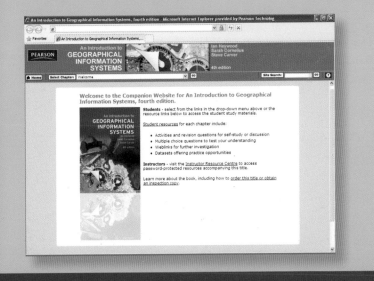

**PEARSON**

We work with leading authors to develop the
strongest educational materials in geography
bringing cutting-edge thinking and best learning
 practice to a global market.

Under a range of well-known imprints, including
Prentice Hall, we craft high-quality print and electronic
publications which help readers to understand and
apply their content, whether studying or at work.

To find out more about the complete range of our
publishing, please visit us on the World Wide Web at:
www.pearsoned.co.uk.

# An Introduction to
# GEOGRAPHICAL INFORMATION SYSTEMS

Ian Heywood
Sarah Cornelius
Steve Carver

Fourth edition

Prentice Hall
is an imprint of

**PEARSON**

Harlow, England • London • New York • Boston • San Francisco • Toronto • Sydney • Singapore • Hong Kong
Tokyo • Seoul • Taipei • New Delhi • Cape Town • Madrid • Mexico City • Amsterdam • Munich • Paris • Milan

**Pearson Education Limited**
Edinburgh Gate
Harlow
Essex CM20 2JE
England

and Associated Companies throughout the world

*Visit us on the World Wide Web at:*
www.pearsoned.co.uk

---

First published 1998
Second edition published 2002
Third edition published 2006
**Fourth edition published 2011**

ISBN: 978-0-273-72259-5

**British Library Cataloguing-in-Publication Data**
A catalogue record for this book is available from the British Library

**Library of Congress Cataloging-in-Publication Data**
Heywood, D. Ian.
   An introduction to geographical information systems / Ian Heywood, Sarah Cornelius, Steve Carver. -- 4th ed.
       p. cm.
   Includes bibliographical references and index.
   ISBN 978-0-273-72259-5 (pbk.)
   1. Geographic information systems. I. Cornelius, Sarah. II. Carver, Steve. III. Title.
   G70.212.H49 2011
   910.285--dc22
                                        2011002783

10 9 8 7 6 5 4 3 2
15 14 13 12

Typeset in 11/13pt Spectrum by 30
Printed and bound by Graficas Estella, Navarra, Spain

# Brief contents

 Part 1  Fundamentals of GIS

Part 2  Issues in GIS

# Contents

## Supporting resources

Visit **www.pearsoned.co.uk/heywood** to find valuable online resources

**Companion Website for students**

- Activities and revision questions for self-study or discussion
- Multiple choice questions to test understanding
- Web links for further investigation
- Hypothetical datasets offering practice opportunities in Esri- and MapInfo-readable formats and in Excel
- Future of GIS - an opportunity to share your ideas about what the future holds for GIS

**For instructors**

- Instructor's Manual including lecture outlines and practical exercises
- PowerPoint slides containing the artwork and images from the text

**Also: The Companion Website provides the following features:**

- Search tool to help locate specific items of content
- E-mail results and profile tools to send results of quizzes to instructors
- Online help and support to assist with website usage and troubleshooting

**For more information please contact your local Pearson Education sales representative or visit www.pearsoned.co.uk/heywood**

# List of figures

# Preface

## ■ HOW THIS BOOK IS WRITTEN

Your encounters with GIS to date may be similar to those of a Martian arriving on Earth and being faced with a motor car. Imagine a Martian coming to Earth and stumbling across a motor car show-room. Very soon he (or she) has heard of a 'car' and may even have seen a few glossy brochures. Perhaps you are in the same position. You have heard of the term GIS, maybe even seen one or two demonstrations or the paper output they produce.

Developing the analogy of the Martian and the car leads us to a dilemma. There are two approaches to explaining to the Martian what the car is and how it works. The first method is a bottom-up approach. This involves taking the car apart into its component pieces and explaining what each part does. Gradually, we put the pieces back together so that by the time we have reassembled the car we have a good appreciation of how the car works in theory. However, we may still have little idea about how to use it or what to do with it in practice.

The second method, the top-down approach, starts by providing several examples of what the car is used for and why. Perhaps we take it for a test run, and then explore how the different components of the car work together to produce an end result. If this approach is adopted, we may never be able to build a car engine, but we will have a clear appreciation of how, when, why and where a car could be used. In addition, if we explore the subject in suffi-cient technical detail we will know how to choose one car in preference to another or when to switch on the lights rather than the windscreen wipers.

We feel that the same two methods can be used to inform you about GIS. Since we believe you are reading this book not because you want to write your own GIS software, but because you wish to develop a better appreciation of GIS, the approach adopted is similar to the top-down method. We focus on the practical application of GIS technology and, where necessary and appropriate, take a more detailed look at how it works. In a book of this size it is impossible for us to explain and describe every

aspect of GIS. Therefore, we concentrate on those areas that enable you to make sense of the applica-tion of GIS, understand the theories upon which it is based and appreciate how to set up and imple-ment your own GIS project.

## ■ THE FOURTH EDITION

The fourth edition has been expanded to cover the latest advances in GIS including Web and mobile applications of GIS. The case studies, written by academics, professionals, teachers and research stu-dents to help illustrate the many ways in which GIS is being used around the world, have been updated and several new ones added to this fourth edition.

Many of the changes and corrections have been made in response to suggestions made by reviewers, and from lecturers and teachers using the book. Many, many thanks to all of you who commented on the first, second and third editions.

With rapid developments in GIS over recent years, the chapter on the future of GIS which was written for the earlier editions of the book had become outdated. Thus, one of the major changes for the fourth edition has been to restructure and rewrite Chapters 9 and 13. Much of what we pre-dicted might happen to GIS in earlier editions of the book is now reality and these events are cap-tured in an expanded Chapter 9, which looks at the development of GIS from a historical perspective. In Chapter 13 we offer comments on where GIS is now and on some of the likely future challenges. We also invite you to formulate your own predictions for the future of GIS and encourage you to share these with us via the website for the book.

## ■ HOW TO USE THIS BOOK

There are three ways to use this book:

1  Browse through the text, consider the chapter headings, and review the learning outcomes

at the start of each chapter. Supplemented by reading Chapter 1 this will give you a quick tour through the world of GIS and highlight some of the important issues and concepts. If you wish you can explore specific topics further using the boxes, case studies and the companion website.

2  A better way to approach the text is to read the book chapter by chapter. This will allow you to discover the ideas presented and develop an understanding at your own pace. To assist you we have included learning outcomes at the start of each chapter, reflection boxes within each chapter, and revision questions and pointers for further study at the end of each chapter. The reflection boxes can be found at the end of major sections of text to encourage you to pause and reflect on key issues and themes. They may ask you to think beyond the text itself and about the role of GIS in your studies, your work and the world around you. The revision questions are provided to encourage you to examine and revisit the major themes and issues introduced in each chapter and reinforce your learning. Further study sections at the end of each chapter have two parts; suggestions for further reading including full references and bibliographic details, and suggestions for activities you can do yourself either on the Web, on paper or using a GIS package if you have one available. A selection of weblinks is also provided at the end of each chapter to help you explore online materials and information resources.

3  The third way to use this book is as a reference source. You might wish to delve into those sections that you feel are relevant at a particular time. Perhaps you can read the appropriate chapter to supplement a course you are following. There is a comprehensive index and glossary at the back of the text and pointers to further sources of information. We have tried to reference only readily available published material. References to other additional information sources on the Internet can be found on the website for the book. These lists of further sources are by no means comprehensive, but are offered as a starting point. We suggest that you also consult library and online sources for more up-to-date materials.

## ■ BEFORE YOU START

You could not drive a car without an understanding of the road network or road signs. Similarly, a full understanding of GIS requires some computing background, particularly in topics like operating systems and file management. This book assumes basic familiarity with PC computing. Anyone who has any experience of word processing, spreadsheets, databases or mapping packages, should be able to apply that knowledge in a GIS context. The typical first-year undergraduate IT course offered in subject areas like geography, biology, business studies or geology is sufficient for you to cope with the ideas presented in this book. We assume that you are familiar with terms such as hardware and software, the Internet and the major components of a computer: for example, monitor, keyboard, hard disk drive, CD-ROM drive, processor and memory. We make no other assumptions – this book is written to be accessible to students of GIS from any professional or academic background: from archaeology, through biology, business studies, computing, demography, environmental management, forestry, geography, history … and on to zoology.

If you want to become a GIS expert, you will need to be comfortable with more advanced computing issues, and will have to expand your computing background to include skills in such areas as programming and networks. These issues are beyond the scope of this book, but we hope to provide you with a valuable head start and many pointers as to where to continue your journey.

A book cannot substitute for hands-on experience in a subject area as practical as GIS. Therefore, for a fuller appreciation of GIS we encourage you to enroll on a course that offers practical experience of GIS, or to find a system to use in your own time. There are a number of excellent GIS 'courses' now available electronically and reference has been made to these on the website for the book.

## ■ HOW THIS BOOK IS STRUCTURED

The book is organized into 13 chapters in two parts. The first part of the book deals with GIS theory and concepts. After an introductory scene-setting chapter, that also introduces four case studies that are used throughout the book, important topics like spatial data, database theory, analysis operations and output are all examined. In the second part of the book we have focused on a selection of GIS 'issues' – the development of GIS, data quality, organizational issues and project management. There are, of course, other issues we could have considered, but those selected reflect many of the key themes of current interest, and areas important for anyone undertaking practical work in GIS. The book ends with a postscript on the future of GIS which sets out a challenge for you to help us explore what the future for GIS might hold. Each chapter has been carefully structured to link together into a unified whole, but can be read in isolation. There are several common elements to each chapter to help you get the most from the book. These are:

- Introduction and list of learning outcomes.
- Boxes providing additional detail on theory and practice.
- Full colour maps, diagrams, tables and photographs.
- Illustrated case studies with further details of issues and applications in GIS.
- Reflection questions and activities.
- Conclusions including revision questions.
- Pointers to further study including references to additional reading and activities.
- Web links.

In summary, the chapters are:

## PART 1: FUNDAMENTALS OF GIS

### Chapter 1: What is GIS?

This chapter provides an overview of GIS. It examines what GIS is, what it can do and, in brief, how it works. The chapter starts by looking at the types of generic questions GIS can answer and expands on these with reference to a series of case studies which are then used throughout the rest of the book. GIS is then defined, and a range of issues and ideas associated with its use identified. Much of the material introduced in this chapter will be covered in more detail later in the book.

### Chapter 2: Spatial data

This chapter looks at the distinction between data and information and identifies the three main dimensions of data (temporal, thematic and spatial). The main characteristics of spatial data are described. A review of how the traditional map-making process shapes these characteristics is presented. The three basic spatial entity types (points, lines and areas), which form the basic building blocks of most GIS applications, are introduced. Maps and a range of other sources of spatial data are reviewed.

### Chapter 3: Spatial data modelling

How do you model spatial form in the computer? This chapter considers in detail how the basic spatial entities (points, lines and areas) can be represented using two different approaches: raster and vector. Two other entity types that allow the modelling of more complex spatial features (networks and surfaces) are introduced. Finally, modelling of three- and four-dimensional spatial data is reviewed.

### Chapter 4: Database management

This chapter introduces the methods available for handling attribute data in GIS. The need for formal methods for database management is discussed. The principles and implementation of a relational database model are considered in detail, since this model is the most frequently used in current GIS. Database options for large-scale users are presented, including the use of centralized and distributed database systems. Finally, a brief introduction to the object-oriented approach to database management is provided.

### Chapter 5: Data input and editing

This chapter gives an overview of the process of creating an integrated GIS. It takes us from source data through data encoding, to editing and on to manipulatory operations such as re-projection, transformation, and rubber sheeting. The chapter provides examples of how these processes are carried out, and highlights issues pertinent to the successful implementation of a GIS application.

## Chapter 6: Data analysis

Methods for making measurements and performing queries in GIS are introduced in this chapter. Proximity, neighbourhood and reclassification functions are outlined, then methods for integrating data using overlay functions explained. Interpolation techniques (used for the prediction of data at unknown locations) are introduced and the analysis of surfaces and networks considered. Finally, analysis of quantitative data is reviewed.

## Chapter 7: Analytical modelling in GIS

This chapter provides a summary of process models before considering how they can be implemented in GIS. These models are then approached from an applications perspective, and three examples are examined: physical process models; human process models and decision-making models. To conclude, the chapter considers some of the advantages and disadvantages of using GIS to construct spatial process models.

## Chapter 8: Output: from new maps to enhanced decisions

An understanding of the basic principles of map design is essential for the effective communication of information and ideas in map form. In addition, an understanding of the complexity of the map design process helps appreciation of the power of maps as a visualization tool. This chapter considers the advantages and disadvantages of cartographic and non-cartographic output. In the conclusion to this chapter there is a brief discussion of the role of GIS output in supporting decision making.

## ■ PART 2: ISSUES IN GIS

### Chapter 9: The development of computer methods for handling of spatial data

This chapter considers how GIS have developed to their current state. The methods of handling spatial data that were used before computers were available are examined. These give an insight into what we require computers to do, and how they can help (or hinder) existing practice. Computer methods for handling spatial data existed before GIS, so these are reviewed, then developments in GIS are discussed together with developments in a selection of complementary disciplines. The chapter does not attempt to present a comprehensive history of GIS but aims to give some context for the systems and concepts we work with today.

## Chapter 10: Data quality issues

The terms used for data errors and quality are explained at the beginning of this chapter, since the first step to solving problems is to be able to recognize and describe them. The remainder of the chapter outlines the types and sources of errors in GIS to help you identify and deal with problems at the appropriate stage of a GIS project. Techniques for modelling and managing errors are also considered.

## Chapter 11: Human and organizational issues

This chapter takes us away from relatively small-scale research-type applications of GIS, where one user can be in control from start to finish, and takes a look at some of the issues surrounding larger-scale commercial and business applications. In utilities, local government, commerce and consultancy, GIS must serve the needs of a wide range of users, and should fit seamlessly and effectively into the information-technology strategy and decision-making culture of the organization. In this chapter we examine the users of GIS and their needs; GIS education; how to justify investment in GIS; how to select and implement a system, and the organizational changes that may result.

## Chapter 12: GIS project design and management

How do we understand a problem for which a GIS solution is being sought? Two methods are introduced in this chapter: constructing a rich picture and a root definition. The method for constructing a GIS data model is then discussed. Here a distinction is made between the conceptual data model and its physical implementation in the computer. A closer look at the various project management approaches and techniques and tools available for the implementation of a GIS project follow. Potential implementation problems, tools for managing risk and tips for project evaluation are also considered. To conclude, a checklist is provided to help with the design and implementation of your own GIS project. A case study is used to illustrate the approach throughout the chapter.

### Chapter 13: The future of GIS

Who knows what the future for GIS will hold? So much has happened since the first edition of this book was published in 1998 and we revisit some of our earlier predictions for GIS in this chapter. Some of the challenges which remain for data, for technology and for applications are also considered. We provide some 'stories from the future' to stimulate your own thoughts on what the future for GIS might be, and invite you to contribute your own ideas via the website for the book.

### ■ IN SUMMARY

We hope that after you have read this book, you will have the knowledge and enthusiasm to start applying GIS in the context of your own course, discipline or organization. Whilst the text will not have taught you how to drive a specific GIS product, we hope that it will give you an appreciation of the concepts on which GIS are based, the methods they use and the applications to which they can be put. In addition, the book should give you an appreciation of some of the difficulties and considerations associated with setting up any GIS project. We hope you enjoy the book!

### Note

In common with Bonham-Carter (1995) we have found it convenient to use the abbreviation GIS in a number of ways:

- to refer to a single system (e.g. a GIS software package that illustrates this is...)
- to refer to several or all geographical information systems (e.g. there are many GIS that...)
- to refer to the field of geographical information systems (e.g. GIS has a long history), and
- to refer to the technology (e.g. the GIS answer would be...).

### Trade marks

Mention of commercial products does not constitute an endorsement, or, indeed, a refutation of these products. The products are simply referred to as illustrations of different approaches or concepts, and to help the reader relate to the practical world of GIS.

*Ian Heywood, Sarah Cornelius and Steve Carver*

# Guided tour

## Navigation and setting the scene

Learning outcomes **show what you'll gain from reading the chapter.**

Introduction **maps out the principal themes and content for the chapter.**

Conclusions **recall and highlight the key issues.**

## Reflection, testing and practice

End of chapter revision questions **help revisit and test the main themes.**

Reflection box **encourages you to ponder key issues or think further.**

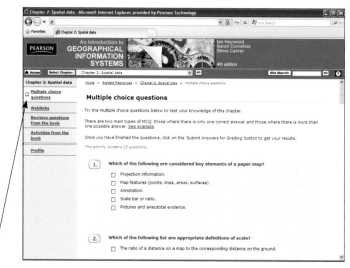

Datasets on the website linked to the Happy Valley case study **provide an opportunity for practice and application using simulated data.**

Multiple choice questions on the website **permit self-testing of basic understanding and help to build confidence.**

## Aiding your understanding

Theory boxes **help explain the thinking or technology behind a topic, issue or practice.**

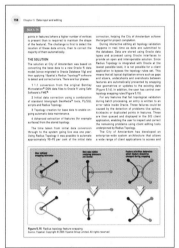

Visual presentation of material **shows GIS at work in exciting ways and helps demonstrate GIS in action.**

## Applying GIS in the real world

Practice boxes **give detailed examples of GIS being used in the real world.**

Multi-disciplinary case studies **show how GIS principles and technology are used in multiple ways, by different people, and across a range of industries.**

## Further study

End of chapter activities **and** revision questions **for self-study or discussion encourage the application and practice of knowledge and research beyond the textbook.**

Reading **and** Web Links **help you investigate key topics of further interest.**

# Case studies

| Case study | Location | | | | | | | | Sector | | | | | | | | | | |
|---|---|---|---|---|---|---|---|---|---|---|---|---|---|---|---|---|---|---|---|
| | Any | UK | Europe | Africa | Asia | Remote areas | Australia | North America | Tourism | Environmental planning | Commercial planning | Transport planning | Planning | Social/public sector | Business | Hazard assessment | Local government | Environmental Mgmt. | Urban planning |
| **1.1** Questions GIS could answer in Happy Valley — *Steve Carver, University of Leeds and Sarah Cornelius, University of Aberdeen* | ✗ | | | | | | | | ✓ | | | | | | | | | | |
| **1.2** Issues raised by the nuclear waste case study — *Steve Carver, University of Leeds and Sarah Cornelius, University of Aberdeen* | ✗ | | | | | | | | | ✓ | | | | | | | | | |
| **1.3** Issues raised by the Zdarske Vrchy case study — *Steve Carver, University of Leeds and Sarah Cornelius, University of Aberdeen* | | | ✗ | | | | | | | ✓ | | | | | | | | | |
| **1.4** Issues raised by the house-hunting case study — *Steve Carver, University of Leeds and Sarah Cornelius, University of Aberdeen* | ✗ | | | | | | | | | | ✓ | | | | | | | | |
| **2.6** Integrating data with GIS: • Mapping the UK's railways, *Peter Robson, Network Rail* • Solving crime in Salford City, *Paul Coward, Salford GIS Ltd* | | ✗ | | | | | | | | | | ✓ | | | | | | | |
| **2.8** High-resolution data solutions to complex modelling problems — *Stuart Lane, Durham University and Joe Holden, Leeds University* | ✗ | | | | | | | | | | | | | | | | | ✓ | |
| **3.8** Monitoring coastal change with GIS — *Jonathan Raper, City University, London* | ✗ | | | | | | | | | | | | | | | | | ✓ | |
| **4.4** Encroachments on public land in Accra, Ghana — *Isaac Bonsu Karikari, Millennium Development Authority, Ghana* | | | | ✗ | | | | | | | | | | ✓ | | | | | |
| **4.6** Delivering census interaction data: WICID — *John Stillwell, University of Leeds* | | ✗ | | | | | | | | | | | | ✓ | | | | | |
| **5.2** Business process outsourcing and the Indian GIS industry — *Ankur Das, India* | | | | | ✗ | | | | | | | | | | ✓ | | | | |
| **5.4** OpenStreetMap — *Tim Waters, Geospacial consultant, Leeds* | | ✗ | | | | | | | | | | | | | | | | | |
| **5.5** Terrestial laser scanning — *Jeff Warburton, University of Durham* | ✗ | | | | | | | | | | | | | | | | | | |
| **5.6** Fireguard: Using GIS and remote sensing for fire management — *Jim Hogg, University of Leeds* | | | ✗ | | | | | | | | | | | | | ✓ | | | |
| **5.10** Going Dutch with small scale topography — *Martin Gregory, 1Spatial* | | | ✗ | | | | | | | | | | | | | | | ✓ | |
| **5.11** Ordnance Survey (OS) data collection – from plan to MasterMap — *Paul Corcoran, Ordnance Survey* | | ✗ | | | | | | | | | | | | ✓ | | | | | |
| **5.12** An integrated GIS for Happy Valley — *Steve Carver, University of Leeds and Sarah Cornelius, University of Aberdeen* | ✗ | | | | | | | | ✓ | | | | | | | | | | |
| **6.1** Data analysis in Happy Valley — *Steve Carver, University of Leeds and Sarah Cornelius, University of Aberdeen* | ✗ | | | | | | | | ✓ | | | | | | | | | | |
| **6.3** Modelling remoteness in mountain areas — *Steve Carver and Steffen Fritz, University of Leeds* | | | | | | ✗ | | | | ✓ | | | | | | | | | |
| **6.5** Siting a nuclear waste repository — *Steve Carver, University of Leeds and Sarah Cornelius, University of Aberdeen* | ✗ | | | | | | | | ✓ | | | | | | | | | | |
| **6.6** Raster analysis of the nuclear waste siting example — *Steve Carver, University of Leeds and Sarah Cornelius, University of Aberdeen* | ✗ | | | | | | | | | ✓ | | | | | | | | | |

| Case study | | Location | | | | | | | | Sector | | | | | | | | | | | |
|---|---|---|---|---|---|---|---|---|---|---|---|---|---|---|---|---|---|---|---|---|---|
| | | Any | UK | Europe | Africa | Asia | Remote areas | Australia | North America | Tourism | Environmental | Environmental planning | Commercial planning | Transport planning | Planning | Social/public sector | Business | Hazard assessment | Local government | Environmental Mgmt. | Urban planning |
| 6.9 | Identifying mountains with GIS<br>*Jo Wood, City University, London* | ✗ | | | | | | | | | ✓ | | | | | | | | | | |
| 6.10 | Wind farm visibility in Cape Bridgewater, Australia<br>*Ian Bishop, University of Melbourne* | | | | | | | ✗ | | | ✓ | | | | | | | | | | |
| 7.1 | The development of a snow cover model for Happy Valley<br>*Steve Carver, University of Leeds and Sarah Cornelius, University of Aberdeen* | ✗ | | | | | | | | ✓ | | | | | | | | | | | |
| 7.3 | Using GIS to help manage air quality through estimating emissions of atmospheric pollutants<br>*Sarah Lindley, University of Manchester* | | ✗ | | | | | | | | | | | | | | | | | | ✓ | |
| 7.5 | GIS-enabled retail sales forecasting<br>*Tony Hernandez, Ryerson University* | | | | | | | | ✗ | | | | | | | | | ✓ | | | | |
| 7.8 | Fuzzy GIS and the fear of crime<br>*Andrew Evans, University of Leeds* | | ✗ | | | | | | | | | | | | | | ✓ | | | | | |
| 8.2 | Retail geovisualization: 3D and animation for retail decision support<br>*Tony Hernandez, Ryerson University* | ✗ | | | | | | | | | | | | | | | | ✓ | | | | |
| 8.4 | Virtual London<br>*Michael Batty, University College London* | | ✗ | | | | | | | | | | | | | | | | | | | ✓ |
| 8.5 | Happy Valley Internet GIS<br>*Steve Carver, University of Leeds and Sarah Cornelius, University of Aberdeen* | ✗ | | | | | | | | ✓ | | | | | | | | | | | |
| 8.7 | Public participation GIS<br>*Richard Kingston, University of Manchester* | | ✗ | | | | | | | | | | | | | | ✓ | | | | | |
| 8.8 | MapAction: a personal view<br>*Chris Ewing, MapAction* | ✗ | | | | | | | | | | | | | | | | ✓ | | | | |
| 9.1 | Developments in computer cartography<br>*Steve Carver, University of Leeds and Sarah Cornelius, University of Aberdeen* | ✗ | | | | | | | | | | | | | | | | | | | |
| 9.2 | CGIS: an early GIS<br>*Steve Carver, University of Leeds and Sarah Cornelius, University of Aberdeen* | | | | | | | | ✗ | | | | | | | | | | | | |
| 9.4 | CORINE: an international multi-agency GIS project<br>*Steve Carver, University of Leeds and Sarah Cornelius, University of Aberdeen* | | | ✗ | | | | | | | | | | | | | | | | | |
| 10.7 | Viewshed analysis in archaelogy<br>*Graham McElearney, University of Sheffield* | ✗ | | | | | | | | | ✓ | | | | | | | | | | |
| 10.8 | Uncertainty and terrain data in ice-sheet modelling<br>*Ross Purves and Felix Hebeler, University of Zurich* | | | | | | | | ✗ | | ✓ | | | | | | | | | | |
| 10.10 | Errors in the nuclear waste case study<br>*Steve Carver, University of Leeds and Sarah Cornelius, University of Aberdeen* | ✗ | | | | | | | | | | | | | | | | | | | |
| 11.6 | A problematic implementation<br>*Steve Carver, University of Leeds and Sarah Cornelius, University of Aberdeen* | | ✗ | | | | | | | | | | | | | | | | | | |
| 13.3 | The Culloden Battlefield Guide<br>*Steve Bone, Consultant* | | ✗ | | | | | | | | | | | | | | | | | | |

# Acknowledgements

During the production of four editions of this book many, many people have helped with support, advice and ideas both directly and indirectly. We have taken the opportunity here to thank some of those who have helped, but we recognize that we have omitted many others who deserve appreciation.

For help with the text of the first edition we should thank Steve Wise and an anonymous reviewer who made constructive comments on early drafts. Tony Hernandez, Chris Higgins, Derek Reeve, David Medyckyj-Scott and Tony Heywood are also thanked for reading drafts of various sections. Help with the diagrams, plates and cartoons and word processing came from Gustav Dobrinski, Bruce Carlisle, Simon Kenton and Tracey McKenna.

All of the case studies used in the book are the result of the work of teams of researchers and academics. Stan Openshaw and Martin Charlton were involved with the original radioactive waste siting study. Jim Petch, Eva Pauknerova, Ian Downey, Mike Beswick and Christine Warr worked on the Zdarske Vrchy study. The initial ideas for the House Hunting game came from a team that included James Oliver and Steve Tomlinson. This last case study was made available by the GeographyCAL initiative based at the University of Leicester following additional input and support from Roy Alexander, John McKewan and John Castleford. In particular, we would like to thank the many people who have given their time freely to write and update the case studies you see in this fourth edition. These people – friends, colleagues, associates and students who have worked with us on GIS projects – have allowed us to pass on the richness and variety of GIS applications across the globe. They are all introduced in the author bibliographies.

Many other individuals have supported our work on this book – Helen Carver, Thomas Blaschke and Manuela Bruckler deserve special mention. In addition, the Universities of Leeds, Aberdeen and Salzburg, the Manchester Metropolitan University and the Vrije Universiteit Amsterdam have allowed us to use their resources to help compile the materials.

We would like to thank the reviewers who have provided valuable feedback, advice and suggestions for this new edition. Our thanks go to:

Darius Bartlett, University College Cork, Ireland

Josie Wilson, Sheffield Hallam University

Dr Robert MacFarlane, Northumbria University, UK

Dr Duncan Whyatt, Lancaster University, UK

Steve Wise, University of Sheffield, UK

Thomas Balstroem, University of Copenhagen, Denmark

Anne Lucas, Geography Department, University of Bergen

Angela Cuthbert, Millersville University, USA

Tim Elkin PhD, Camosun College, Victoria, BC, Canada

Christine Earl, Carleton University, Canada

Dr Jane Law, University of Cambridge, UK

Dr Adrian Luckman, Swansea University, UK

Dr HA Oosterhoff, University of Groningen, The Netherlands

Bernd Etzelmuller, University of Oslo, Norway.

At Addison Wesley Longman, Prentice Hall and Pearson Education we have worked with many individuals. Particularly we should thank Vanessa Lawrence for help and advice at the start of the project, Sally Wilkinson, who took over during the middle stages, and Tina Cadle-Bowman, Matthew Smith, Shuet-Kei Cheung and Patrick Bonham, who worked with us on earlier editions of the book. For their work on the second edition thanks are due to Matthew Smith, Bridget Allen, Morten Funglevand, Paula Parish and Nicola Chilvers. For their support and innovative ideas during the preparation of this fourth edition we must thank Andrew Taylor, Janey Webb, Sarah Busby, Maggie Wells and Joe Vella. For the fourth edition a special thanks is due to Rufus Curnow who steered the book towards publication though many missed deadlines! In addition, we have been supported by a team of picture researchers and designers who have helped to create the striking product you see.

And lastly, we should thank all our students and colleagues. We started this project to help our students understand the mysteries of GIS. Some have been an inspiration, and gone on to exciting and rewarding careers in the GIS field. Others struggled with basic concepts and ideas, and helped us to realize just what we needed to take time to explain. Writing for distance learning students has taught us all a tremendous amount and our colleagues from the UNIGIS distance learning course deserve particular mention – Jim Petch, Derek Reeve, Josef Strobl, Nigel Trodd, Christine Warr, Steve Tomlinson and Bev Heyworth. We apologize to them unreservedly if they see ideas and examples inspired by their course materials within the book!

We hope that the experience we have gained between the initial idea for the book and publication of the fourth edition has helped to make this a useful and readable introduction to GIS.

*Ian Heywood, Sarah Cornelius and Steve Carver*

# Publisher's acknowledgements

The publisher would like to thank the following for their kind permission to reproduce their photographs and artwork:

Figure 1.1: Courtesy of author; Figure 1.2: Richard Armstrong; Figure 1.3: Courtesy of author (a), Getty Images: Jeff T. Green (b), British Nuclear Fuels Ltd (c, d); Figure 1.4: Press Association Images: Stefan Rousseau; Figure 1.5: Courtesy of author; Figure 1.7: NERC 2008 (a), Office for National Statistics (b), Ordnance Survey (c), Joint Nature Conservation Committee (d); Figures 1.8, 1.9, 1.10 and 1.13: Courtesy of author; Figure 1.14: Courtesy of Integraph Corp. (a), Courtesy of author (b), Fotolia.com: Maksym Dykla (c), Courtesy of Main Street GIS (d), Courtesy of Trimble (e); Figure 1.15: Courtesy of Esri: Copyright © 2011 Esri. All rights reserved (a,b); Figures 1.16, 1.17 and 1.18: Courtesy of author; Figures 1.19 and 1.20: United States Geological Survey; Figure 1.21 and 2.1: Courtesy of author; Figure 2.2: Crown Copyright (a); Alamy Images: Dominic Burke (b); Courtesy of author (c); Figure 2.3: Courtesy of author; Figure 2.4: Corbis: NASA (a), Torbay Council (b); Figure 2.5: TopFoto: Roger-Viollet; Figures 2.7, 2.8 and 2.9: Courtesy of author; Figure 2.10: Ordnance Survey: Reproduced by permission of Ordnance Survey on behalf of HMSO. Crown copyright 2011. All rights reserved. Ordnance Survey Licence number 100030901 (a–g); Figure 2.12: NASA (a), Paul B. Anderson: (b,d,e), Courtesy of John Savard (c); Figure 2.17: Ordnance Survey: Reproduced by permission of Ordnance Survey on behalf of HMSO. Crown copyright 2011. All rights reserved. Ordnance Survey Licence number 100030901 (a); Figure 2.18a: Courtesy of Peter Robson; Figure 2.18b: Computer Aided Development Corporation Ltd (MAPS software interface) and Ordnance Survey: Reproduced by permission of Ordnance Survey on behalf of HMSO. Crown copyright 2011. All rights reserved. Ordnance Survey Licence number 100030901 (map data); Figure 2.19: U.S. Department of Agriculture (a), © Bluesky International Limited (b), National Oceanic and Atmospheric Administration (c), United States Geological Survey (d); Figure 2.21: Camera Press Ltd: CNES Dist / SPOT Images / Explorer / Gamma (a), i-cubed LLC: i-cubed & USGS (b), Science Photo Library Ltd / Earth Satellite Corporation (c), Wetterzentrale, Germany (d), i-cubed & GeoEye (e); Figure 2.22: Precision Terrain Surveys Ltd (a, b), Courtesy of Dr Jonathan Carrivick, University of Leeds (c, d); Figures 2.23, 2.24 and 2.25: Courtesy of Stuart Lane; Figure 2.26: Alamy Images: Widstock (a), Press Association Images / AP (b); Figure 2.27: Courtesy of Magellan: (b, c); Figures 3.3 and 3.4: Courtesy of author; Figure 3.5: Courtesy of Esri: Copyright © Esri. All Rights Reserved; Figures 3.6 and 3.7: Courtesy of author; Figure 3.16: Ordnance Survey: Reproduced by permission of Ordnance Survey on behalf of HMSO. Crown copyright 2011. All rights reserved. Ordnance Survey Licence number 100030901; Figures 3.17 and 3.18: Courtesy of author; Figure 3.19: NASA / JPL / NIMA; Figure 3.20; NASA / JPL / NIMA (a, c), Courtesy of Environment Agency Science Group Technology (b); Figure 3.21: United States Geological Survey; Figure 3.23: Courtesy of author; Figure 3.25: United States Geological Survey; Figure 3.27: Copyright © Nexus (a), Courtesy of Steve Carver (b); Figure 3.30: Courtesy of author; Figure 3.31: Rockware, Inc. Graphics prepared using Rockworks, by Rockware Inc., Golden, Colorado, USA (a), Centre for Advanced Spatial Analysis (CASA), University College London (b); Figures 3.32 and 3.33: Jonathan Raper; Figure 4.1: Courtesy of Esri: ArcMap / ArcView / ArcInfo Graphical User interface is the intellectual property of Esri and is used herein with permission. Copyright © 2011 Esri. All rights reserved (a); Clark Labs, Clark University (b); Figure Figures 4.10 and 4.11: Taylor and Francis; Figures 4.14, 4.15 and 4.16: The Centre for Interaction Data Estimation and Research; Figure 5.2: Courtesy of author; Figure 5.3: Courtesy of Able Software Corporation; Figure 5.6: Courtesy of Esri: The ArcMap - ArcInfo graphical user interface is the intellectual property of Esri and is used herein with permission. Copyright © 2011 Esri. All rights reserved; Figure 5.7: Epson UK (a), Courtesy of Stefan Kühn (b), Courtesy of Colortrac / Smart LF Ci 40 Wide Format Scanner (c); Figure 5.8: United States Geological Survey; Figure 5.9: Courtesy of OpenStreetMap; Figure 5.10: Courtesy of Dr Jeff Warburton; Figures 5.11 and 5.12: James Hogg; Figures 5.14 and 5.15: 1Spatial Group Limited; Figure 5.16: Courtesy of author; Figure 5.17: Courtesy of Peter H. Dana; Figure 5.18: Courtesy of Esri: Copyright © 2011 Esri. All rights reserved (a–f); Figures 5.22 and 5.23: Ordnance Survey: Reproduced by permission of Ordnance Survey on behalf of HMSO. Crown copyright 2011. All rights reserved. Ordnance Survey Licence number 100030901; Figures 5.25, 6.2 and 6.3: Courtesy of author; Figures 6.11 and 6.12: United States Geological Survey; Figure 6.13: Centre for Advanced Spatial Analysis (CASA), University College London (a), Courtesy of Steve Carver (b); Figures 6.17, 6.19 and 6.23: Courtesy of Steve Carver; Figure 6.24: Courtesy of Horizon Simulation Ltd; Figures 6.25 and 6.26: Courtesy of author; Figure 6.27: Courtesy of Institute of Cartography, ETH Zurich (a, b), Courtesy of author (c); Figures 6.28 and 6.29: Courtesy of author; Figures 6.30 and 6.31: Ordnance Survey; Figure 6.33: Courtesy of author; Figures 6.34, 6.35 and 6.36: Courtesy of Ian Bishop; Figure 6.38: Courtesy of the R Project for Statistical Computing: © R Foundation, from www.r-project.org; Figure 7.1: Courtesy of author; Figure 7.5: Courtesy of Sarah Lindley; Figure 7.7: Courtesy of author; Figure 7.8; Courtesy of Tony Hernandez; Figure 7.11: Courtesy of Andrew Evans; Figure 7.12: Courtesy of author; Figure 7.13: Courtesy of author (a), Courtesy of SciLands GmbH (b); Figure 7.14: Courtesy of Sheelagh M. Coles; Figure 7.15: Clark Labs, Clark University; Figures 8.1 and 8.2: Courtesy of author; Figure 8.6: Benjamin David Hennig, University of Sheffield; Figure 8.7: Courtesy of Tony Hernandez; Figures 8.8 and 8.9: Courtesy of Esri: ArcMap / ArcView / ArcInfo Graphical User interface is the intellectual property of Esri and is used herein with permission. Copyright © 2011 Esri. All rights reserved. Figures 8.11 and 8.12: Courtesy of Michael Batty; Figure 8.13: Photo courtesy Google UK; Figure 8.14: Courtesy of Richard Kingston; Figure 8.15: Courtesy of MapAction, www.mapaction.org; Figure 8.16: Courtesy of author. Figure 9.2: Courtesy of author; Figure 9.3: Montgomery County Executive / Street Centerline Base Map, Montgomery County, Maryland, USA; Figure 9.4: Courtesy of Encom Technology Pty Ltd; Figure 10.2: Courtesy of author; Figure 10.5: Courtesy of Armin Gruen / Institute of Geodesy and Photogrammery; Figure 10.6: Courtesy of author; Figure 10.7: Courtesy of author (a, c, d, e, f), Ordnance Survey: Reproduced by permission of Ordnance Survey on behalf of HMSO. © Crown copyright 2011. All rights reserved. Ordnance Survey Licence number 100030901 (b); Figure 10.8: Courtesy of Ian Bishop (left), Photo courtesy Google UK (right); Figure 10.9: Clark Labs, Clark University; Figure 10.16: Courtesy of Dr Graham McElearney, University of Sheffield; Digimap download data © Crown Copyright Ordnance Survey. All rights reserved; Figures 10.18 and 10.19: Courtesy of author; Figures 10.20 and 10.21: Courtesy of Ross Purves and Felix Hebeler; Figures 10.22 and 10.23: Courtesy of author; Figure 13.1: Courtesy of Gordon Sell Public Relations; Figure 13.2: Photo courtesy Google UK; Figures 13.3 and 13.4: Courtesy of Steve Bone; Figure 13.5: Houston Transtar Consortium; Figure 13.6: Courtesy of Esri: Image Copyright © DigitalGlobe. Used with permission. The ArcMap / ArcInfor Graphical User Interface (GUI) is the intellectual property of Esri and is used herein with permission. Copyright © Esri; Figure 13.7: Courtesy of Esri: Copyright © 2005-2006 Data Transmission Network Corporation, d / b / a Meteorlogix. The ArcMap / ArcInfo Graphical User Interface (GUI) is the intellectual property of Esri and is used herein with permission. Copyright © Esri. (a), NASA (b), Sea Island Software (c); Figure 13.8: Digital Globe; Figure 13.9: Courtesy of Esri: Copyright © 2005 Esri, TANA. Used with permission; Figure 13.10: Digital Globe; Figure 3.11: Courtesy of author; Figure 13.12: NASA (a), Jonathan L. Bamber (b); Figure 13.13: Courtesy of the MECAA project (a), Courtesy of Landsat.org (b).

Every effort has been made to trace the copyright holders and we apologise in advance for any unintentional omissions. We would be pleased to insert the appropriate acknowledgement in any subsequent edition of this publication.

# Abbreviations and acronyms

| | |
|---|---|
| 1:1 | One-to-one relationship |
| 1:M | One-to-many relationship |
| 2.5D | Two-and-a-half-dimensional |
| 2D | Two-dimensional |
| 3D | Three-dimensional |
| 4D | Four-dimensional |
| AGI | Association for Geographical Information (UK) |
| AI | Artificial Intelligence |
| AM/FM | Automated Mapping/Facilities Management |
| BGS | British Geological Survey |
| BS | British Standard |
| BSU | Basic Spatial Unit |
| CAD | Computer-aided Design |
| CAL | Computer-aided Learning |
| CASE | Computer-assisted Software Engineering |
| CCTV | Closed-circuit Television |
| CD-ROM | Compact Disc – Read Only Memory |
| CEN | Comité Européen de Normalisation (European Standards Commission) |
| CGIS | Canadian Geographic Information System |
| CORINE | Coordinated Information on the European Environment |
| CoRWM | Committee on Radioactive Waste Management |
| CRT | Cathode Ray Tube |
| CUS | Census Use Study (USA) |
| DBMS | Database Management System |
| DEM | Digital Elevation Model |
| DIGEST | Digital Geographic Information Exchange Standards |
| DNF | Digital National Framework |
| DoE | Department of the Environment (UK) |
| DOS | Disk Operating System |
| dpi | Dots Per Inch |
| DSS | Decision Support System |
| DTM | Digital Terrain Model |
| DXF | Data Exchange Format |
| EAM | Entity Attribute Modelling |
| ED | Enumeration District (UK) |
| EDA | Exploratory Data Analysis |
| EDM | Electronic Distance Metering |
| ESMI | European Spatial Meta-information system |
| ESRC | Economic and Social Research Council (UK) |
| ESRI | Environmental Systems Research Institute |
| ETHICS | Effective Technical and Human Implementation of Computer-based Systems |
| EUROGI | European Umbrella Organization for Geographical Information |
| GAM | Geographical Analysis Machine |
| GBF-DIME | Geographical Base File, Dual Independent Map Encoding |
| GDF | Geographic Data File |
| GeoTIFF | Geographic Tagged Image Format File |
| GIS | Geographical Information System |
| GISc | Geographic Information Science |
| GLONASS | Global Orbiting Navigation Satellite System |
| GML | Geography Markup Language |
| GPS | Global Positioning Systems |
| GUI | Graphical User Interface |
| HTTP | Hypertext Transfer Protocol |
| IAEA | International Atomic Energy Authority |
| ID | Identifier |
| IS | Information System |
| ISO | International Standards Organization |
| IT | Information Technology |
| LAN | Local Area Network |
| LBS | Location-based Services |
| LCD | Liquid Crystal Display |
| LiDAR | Light Detection and Ranging |
| M:N | Many-to-many relationship |
| MAUP | Modifiable Areal Unit Problem |
| MCE | Multi-criteria Evaluation |
| MEGRIN | Multipurpose European Ground Related Information Network |
| MSS | Multispectral Scanner |
| NAVSTAR | Navigation System with Time and Ranging |
| NCDCDS | National Commitee on Digital Cartographic Data Standards (USA) |

| | |
|---|---|
| NCGIA | National Center for Geographic Information and Analysis (USA) |
| NDA | Nuclear Decommissioning Authority (UK) |
| NERC | Natural Environmental Research Council (UK) |
| NGO | Non-governmental Organization |
| NII | Nuclear Installations Inspectorate (UK) |
| NIMBY | Not In My Back Yard |
| NIREX | Nuclear Industry Radioactive Waste Executive (UK) |
| NGDC | National Geospatial Data Clearinghouse |
| NJUG | National Joint Utilities Group (UK) |
| NTF | National Transfer Format (UK) |
| NTF | Neutral Transfer Format (UK) |
| NTS | National Trust fo Scotland |
| NWW | North West Water plc |
| OA | Output Area (UK Census of Population) |
| OCR | Optical Character Recognition |
| OGC | Open GIS Consortium |
| OGIS | Open Geodata Interoperability Standards |
| OO | Object-oriented |
| OS | Ordnance Survey (UK) |
| OSM | Open Street Map |
| OSOCC | On-site Operations Co-ordination Centre |
| PC | Personal Computer |
| PDA | Personal Data Assistant |
| PDF | Portable Document Format |
| PNG | Portable Network Graphics |
| PERT | Program Evaluation and Review Technique |
| PHP | Hypertext Preprocessor |
| PLSS | Public Land Survey System (USA) |
| PPGIS | Public Participation GIS |
| PRINCE2 | Projects IN Controlled Environments |
| RAD | Rapid Application Development |
| REGIS | Regional Geographic Information Systems Project (Australia) |

| | |
|---|---|
| RINEX | Receiver Independent Exchange Format |
| RRL | Regional Research Laboratory (UK) |
| SDLC | Systems Development Life Cycle |
| SDSS | Spatial Decision Support System |
| SDTS | Spatial Data Transfer Standard |
| SLC | System Life Cycle |
| SPOT | Système pour l'Observation de la Terre |
| SQL | Standard Query Language (or Structured Query Language) |
| SSA | Soft Systems Analysis |
| SSADM | Structured Systems Analysis and Design |
| SWOT | Strengths, Weaknesses, Opportunities, Threats |
| TCP/IP | Transmission Control Protocol/ Internet Protocol |
| TIGER | Topologically Integrated Geographic Encoding Reference File (USA) |
| TIN | Triangulated Irregular Network |
| TLS | Terrestrial Laser Scanner |
| TM | Thematic Mapper |
| TV | Television |
| URL | Uniform Resource Locator |
| UK | United Kingdom |
| UN | United Nations |
| UNDAC | United Nations Disaster Assessment Team |
| URISA | Urban and Regional Information Systems Association (USA) |
| USA | United States of America |
| UTM | Univeral Transverse Mercator |
| VIP | Very Important Point |
| VR | Virtual Reality |
| VRML | Virtual Reality Modelling Language |
| WALTER | Terrestrial Database for Wales |
| WAP | Wireless Application Protocol |
| WAAS | Wide Area Augmentation System |
| WWW | World Wide Web |
| XML | Extensible Markup Language |
| ZVI | Zone of Visual Influence |

# About the authors

## Book authors

**Ian Heywood** is a freelance consultant and Director of the Ideas Academy Ltd., which specializes in helping organizations and teams use creative thinking combined with new technology to unlock and deliver new business opportunities. He is based in Scotland and has a particular interest in the convergence of GIS, GPS and Social Media. Ian was previously Director of Growing Business and Skills for Scottish Enterprise Grampian. He has undertaken research in GIS and taught at Universities in the UK, Austria and the Netherlands. Ian has a PhD from the University of Newcastle upon Tyne.

**Sarah Cornelius** is a Lecturer in the School of Education at the University of Aberdeen, UK, specializing in adult and online learning. She was previously an Associate Lecturer for the Open University and has taught GIS at Manchester Metropolitan University and the Free University of Amsterdam. She has also worked as an independent consultant developing education and training for private sector organizations using GIS.

**Steve Carver** is a Senior Lecturer in the Department of Geography at the University of Leeds, UK, where he teaches GIS at both undergraduate and postgraduate level. Steve is an active researcher and has undertaken field based GIS projects in the USA, Siberia, Greenland and the European Alps. Steve is also Director of the Wildland Research Institute and has a PhD in the siting of nuclear waste facilities.

## Case study authors

**Michael Batty** is Bartlett Professor of Planning at University College London, UK, where he directs the Centre for Advanced Spatial Analysis (CASA). His research is focused on new approaches to urban simulation modelling, with a strong focus on visualization. He is the Editor of Environment and Planning B. (See Box 8.4: Virtual London, p. 274.)

**Ian Bishop** is a Professorial Fellow in the Department of Infrastructure Engineering at the University of Melbourne. His research areas include environmental visualization, landscape assessment and decision support tools for land use change. He is currently working with Geographic Information Systems (GIS), visualization technology and agent-based models to improve the management of natural resources and to contribute to public participation in planning and design issues. (See Box 6.10: Wind farm visibility in Cape Bridgewater, Australia, p. 215.)

**Steve Bone** is a freelance consultant based in the Highlands of Scotland working in areas of web-analysis, heritage sustainability, sales and elearning. He started his business career in the logistics industry managing the distribution of oil field equipment across Europe. Following various moves within the industry he became a business adviser within the economic development agency for the North of Scotland in 1997. After a number of moves to manage various services within the agency Steve returned to private sector employment managing projects aimed at making the best use of technology in practical business scenarios. (See Box 13.3: The Culloden Battlefield Guide, p. 395.)

**Paul Corcoran** is a Geospatial Information Science Lecturer in the School of Natural and Built Environments at the University of South Australia. In 2007 he began a part time PhD exploring and expanding the use of geospatial information within the realms of Indigenous lands and waters management within the state of South Australia. Previously he worked for the Ordnance Survey, Great Britain's national mapping agency, where he undertook various surveying, GIS and management roles, predominantly in the North of England. (See Box 5.11: Ordnance Survey (OS) data collection – from Plan to MasterMap, p. 166.)

**Paul Coward** is a Geography graduate who has spent over twenty years working as a consultant in the GIS industry. He is one of the founding Directors of Salford GIS Ltd in the UK where he currently works. He has undertaken various roles including; systems developer, software trainer, systems and data auditor, project manager and strategic consultant. Paul has worked with over one hundred local authorities as well as numerous private sector and central government organisations. He has worked on contracts in Belfast, Spain, Abu Dhabi, and Copenhagen. (See Box 2.6: Integrating data with GIS: Solving crime the Salford City way, p. 54.)

**Ankur Das** is presently working towards a planning degree at the School of Planning, University of Cincinnati in the United States. Previously he worked in the GIS industry in India as a GIS Officer and a part time entrepreneur. He was an undergraduate student of Geography in India after which he pursued a postgraduate degree in Geographical Information Systems at the University of Leeds graduating in 2002. His research interests focus on satellite image processing for change detection; and environmental planning with new approaches for sustainable development. (Box 5.2: Business process outsourcing and the Indian GIS Industry, p. 142.)

**Andrew Evans** is a Lecturer in the School of Geography at the University of Leeds, UK, where he teaches Masters courses in GIS, AI and computer programming. His interests include building systems to enable online democracy and modelling social interactions, though he has sidelines in glaciology and Dark Age mythopoetic landscapes. (See Box 7.8: Fuzzy GIS and the fear of crime, p. 249.)

**Chris Ewing** is a volunteer for MapAction, a UK-based charity which uses GIS and GPS technology to help in disaster situations worldwide. If a natural disaster occurs the chances are that MapAction will be there providing mapping support. He has been involved with MapAction for four years and has attended natural disasters in Jamaica (Hurricane Dean) and Bolivia (severe flooding) and capacity-building missions in Lesotho and Niger. His skills and experience include information management, GIS, GPS, and disaster preparedness and response. Outside of MapAction, Chris now works for the Impact Forecasting team at Aon Benfield. (See Box 8.8: MapAction: a personal view, p. 282.)

**Steffen Fritz** works at the International Institute for Applied Systems Analysis (IIASA) where he is working on global land cover, land-use modeling, earth observation and benefit assessments. He studied Physics and Geography at the University of Tübingen and gained a Master of Science from the University of Durham in 1996. He undertook his PhD in the field of wild land research at Leeds University. Between 2001 and 2006 he worked at the Joint Research Centre of the European Commission in the field of global land cover mapping, monitoring agriculture and food security. (See Box 6.3: Modelling remoteness in mountain areas, p. 184.)

**Martin Gregory** is the Enterprise Solutions Sales Manager at 1Spatial. He has been with the company since 1989 when it was formerly known as Laser-Scan. During that period he's worked across many disciplines of Geographic Information including, data capture, customer services, project management and technical consultancy. He now has a commercial focus across multiple market sectors, developing and rolling out solutions to address spatial data management issues in order to provide efficiency benefits to the organisations concerned. (See Box 5.10: Going Dutch with small scale topography. p. 157.)

**Felix Hebeler** is an environmental consultant. He previously studied Geography and Biology (at the University of Giessen, Germany and Rhodes University, RSA) with an emphasis on environmental and ecological issues and a focus on modelling using Geographical Information Systems. His PhD investigated the modelling of uncertainties in digital elevation data and their influence on large-scale numerical models, such as those use for ice sheet modelling. (See Box 10.8: Uncertainty and terrain data in ice-sheet modelling, p. 333.)

**Tony Hernandez** is the Director of the Centre for Study of Commercial Activity, Eaton Chair in Retailing and Professor in the Department of Geography at Ryerson University, Toronto. Tony teaches courses at the undergraduate and graduate level in the area of business geomatics, with focus on the retail and service economy. His research interests include geographic information systems, spatial analysis, geovisualization and decision support. (See Box 7.5: GIS-enabled retail sales forecasting, p. 241 and Box 8.2: Retail geovisualization: 3D and animation for retail decision support, p. 267.)

**Jim Hogg** is the Managing Director of Graticule, a software engineering company specializing in bespoke software development, Software as a Service (SaaS) for business resource applications hosted on public and private clouds including Geographical Information Systems, social networking systems and an innovative rapid emergency response system using unmanned airborne vehicles and telecommunications. He was previously Lecturer in Physical Geography at the University of Leeds, UK, where he carried out research and teaching in surveying, remote sensing, digital image processing, geographical information systems. During this period he completed work on the EU FIREGUARD project. He has a degree in Geography from the University of Glasgow and a PhD on multispectral remote sensing in forestry from the University of Melbourne, Victoria, Australia. (See Box 5.6: Fireguard: Using GIS and remote sensing for fire management, p. 150.)

**Joseph Holden** is Chair of Physical Geography at the University of Leeds, UK, and is Head of water@ leeds. He leads research projects primarily in wetland environments but with more general interests in hillslope hydrology, geomorphology, biogeochemical cycling and climate monitoring. He is editor of the renowned textbook 'An introduction to physical geography and the environment' also published by Pearson Education and authored over 50 journal publications and another 30 published books, book chapters and reports. In 2007 he was awarded the Philip Leverhulme Prize and in 2011 the Gordon Warwick Medal for outstanding research achievements. He sits on a number of international grant and journal review panels. (See Box 2.8: High-resolution data solutions to complex modelling problems, p. 63.)

**Isaac Bonsu Karikari** is the Land Administration Project Manager with the Millennium Development Authority (implementing the Millennium Challenge Account Ghana Program). Prior to this, he was the Lands Commission's Team Leader of the World Bank's supported Land Administration Project in Ghana. He was also the Head of Geographic Information Systems (GIS), Training and Manpower Development Unit of the Commission. He is a Fellow of the Ghana Institution of Surveyors. He holds a PhD from the University of Leeds, UK, having obtained his MSc from the International Institute for Geo-information Science and Earth Observation – ITC, Netherlands, where he studied Geo-information Systems for Urban Applications. (See Box 4.4: Encroachments on public land in Accra, Ghana, p. 122.)

 **Richard Kingston** is a Lecturer in Spatial Planning and GIS in the School of Environment and Development at the University of Manchester, UK. He has been researching Public Participation in GIS (PPGIS) and Planning Support Systems for around 15 years. His research focus is on developing, testing and implementing Planning Support Systems for public participation in spatial planning. He has published widely on PPGIS in academic journals and books. Further details on his publications and current research can be found at: www.ppgis.manchester.ac.uk. (See Box 8.7: Public participation GIS, p. 280.)

 **Stuart Lane** is a Professor in the Department of Geography at the University of Durham where he specializes in quantitative approaches to the understanding of hydraulic and hydrological processes. In particular, he specializes in quantification of complex geomorphological surfaces, notably river systems, using remote sensing methods. He has worked in New Zealand, the Alps and the UK both on research and consultancy contracts. (See Box 2.8: High-resolution data solutions to complex modelling problems, p. 63.)

 **Sarah Lindley** is Senior Lecturer in Geographical Information Systems in the School of Environment and Development (Geography), University of Manchester, UK. Most of her work is multi-disciplinary and informed by over 15 years' experience in Geographical Information Science. She was a member of the UK Department of the Environment, Food and Rural Affairs (DEFRA) Air Quality Expert Group 2002-9 and continues to carry out research in the field of air pollution, where her work is associated with understanding spatial patterns in emissions and pollutant concentrations. Sarah is also involved in research looking at decision support for urban and regional sustainable development. Recent work has focused on climate change risk, vulnerability and adaptation, particularly in relation to urban heat. Information about recent projects and publications can be found at: www.manchester.ac.uk/research/sarah.lindley/personaldetails. (See Box 7.3: Using GIS to help manage air quality through estimating emissions of atmospheric pollutants, p. 232.)

 **Graham McElearney** is a Learning Technologist at The University of Sheffield's (UK) Corporate Information and Computing Services, where he advises academic staff across the institution on the use of e-learning. He is also an Honorary Lecturer at the University's Department of Archaeology, where he teaches the use of GIS. His interests include the use of GIS, multimedia and virtual reality in archaeological research, and the use of learning technology and e-learning in teaching archaeology. (See Box 10.7: Viewshed analysis in archaeology, p. 328.)

 **Ross Purves** is a Lecturer in the GIS Division of the Department of Geography at the University of Zurich, Switzerland. Prior to coming to Zurich, he worked in the Geography Department of the University of Edinburgh, UK. His research in GIScience covers a range of themes, from Geographic Information Retrieval to the environmental modelling of processes related to snow and ice and, in particular, issues relating to the uncertainty in modelling of such processes. (See Box 10.8: Uncertainty and terrain data in ice-sheet modelling, p. 333.)

**Jonathan Raper** is Head of the Department of Information Science and founder of the giCentre at City University, where he leads a group of 19 academics, researchers and postgraduates working on, among other things, location-based services, augmented reality, geographic data mining, geovisualization and terrain and cityscape modelling. Jonathan is a member of the Steering Committee of the Location and Timing Knowledge Transfer Network and Editor in Chief of the Journal of LBS, newly launched by Taylor and Francis. (See Box 3.8: Monitoring coastal change with GIS, p. 103.)

**Peter Robson** is a senior GI technical specialist at Network Rail. He has previously worked as an environmental consultant and for the GIS companies, ESRI and Laser-Scan (now known as 1Spatial). He has an MSc in GIS from the University of Leicester and a dual honours degree in Ecology and Geography from the University of Sheffield. (See Box 2.6: Integrating data with GIS: Managing the UK's railways, p. 54.)

**John Stillwell** is Professor of Migration and Regional Development in the School of Geography at the University of Leeds, UK, and Director of the Centre for Interaction Data Estimation and Research (CIDER) which is funded by the ESRC Census Programme. He has recently finished coordinating the ESRC's 'Understanding Population Trends and Processes' (UPTAP) programme. His research interests are primarily in the fields of migration and applied spatial analysis and he co-edits the journal of Applied Spatial Analysis and Policy. (See Box 4.6: Delivering census interaction data: WICID, p. 126.)

**Jeff Warburton** is a Reader in the Department of Geography at Durham University, UK. His research is concerned primarily with fluvial geomorphology, sediment budgets, shallow landslides, mountain sediment systems and geocryology. He also has an interest in hydraulic modelling, sediment transfer and the geomorphology of mountain streams and upland rivers. He has served on the Executive Committee of the British Society for Geomorphology, was Joint-Chair of the BGRG Working Group on Upland Sediment Budgets and is on the steering committee of the IAG SEDIBUD working group. In 2000 he was made a Millennium fellow for his work on 'Understanding the Landscape of the North Pennines' in the UK. (See Box 5.5: Terrestrial laser scanning, p. 148.)

**Tim Waters** is a long-time contributor to the open geodata project, OpenStreetMap (OSM). With the project, he has run mapping parties, provided hands-on training, has helped with data imports and frequently speaks about the project. He is a member of the Humanitarian OSM Team and is particularly interested in the benefits that OSM can give to developing nations. Other interests include historical maps and fuzzy GIS interfaces and he is active within the innovative Leeds Psychogeographical Institute. He has a MSc and is currently an online GIS developer for FortiusOne and GeoCommons. (See Box 5.4: OpenStreetMap, p. 147.)

**Jo Wood** is a Reader in Geographic Information at the giCentre, City University, London. He has been involved in teaching and research in GI Science since 1990. His research interests are in terrain modelling, visualization of surfaces, object-oriented modelling of GI and collaborative networks in GI. He is author of the widely used GIS LandSerf, the text Java Programming for Spatial Sciences and is course director of the Masters in Geographic Information (MGI). When not analysing terrain with a GIS, he can usually be found walking or cycling across it. (See Box 6.9: Identifying mountains with GIS, p. 211.)

# 1 PART

# Fundamentals of GIS

# CHAPTER 1

# What is GIS?

## Learning outcomes

By the end of this chapter you should be able to:

- Explain what GIS is

- Give examples of the applications of GIS

- Outline the characteristics of GIS

- Describe how the real world is represented in GIS

- Provide examples of the analysis GIS can perform

- Know where to find more information about GIS

# ■ INTRODUCTION

Every day you ask questions with a spatial component. Whether you are at work, studying or at leisure you probably ask spatial questions. Many of these questions you answer for yourself without reference to a map or a GIS, but both of these tools could help. GIS has particular value when you need to answer questions about location, patterns, trends and conditions such as those below:

■ *Location.* Where is the nearest bookshop? Where are stone age settlements located in Europe? Where are areas of forestry in which Norwegian Spruce trees can be found?

■ *Patterns.* Where do high concentrations of students live in this city? What is the flow of traffic along this motorway? What is the distribution of crime incidents in London?

■ *Trends.* How are patterns of retailing changing in response to the development of out-of-town superstores? Where have glaciers retreated in the European Alps? Where have changes to the population of polar bears occurred?

■ *Conditions.* Where can I find holiday accommodation that is within 1 km of a wind surfing beach and accessible by public transport? Where is there flat land within 500 m of a major highway? Where are there over 100,000 potential customers within a 5-mile radius of a railway station?

■ *Implications.* If I move to a new home in this location, how far will I be from the office, gym or coffee shop? If we build a new theme park here, what will be the effect on traffic flows? What would be the time saving if we delivered our parcels using this route, rather than an alternative?

This chapter provides an overview of GIS. It examines what GIS is, what it can do and, in brief, how it works. We begin with a look at the types of generic questions GIS can answer and expand on these with reference to case studies. GIS is then defined, and a range of issues and ideas associated with its use identified. Much of the material introduced in this chapter will be covered in more detail later in the book, and pointers to the appropriate sections are provided. If we return to the car analogy introduced in the Preface, this is where you find out what exactly a car is, why it is useful and what you need to know to make it work.

The generic questions that a GIS can help to answer can be summarized as:

■ Where are particular features found?

■ What geographical patterns exist?

■ Where have changes occurred over a given period?

■ Where do certain conditions apply?

■ What will the spatial implications be if an organization takes certain action?

Box 1.1 provides examples of problems that can be addressed by asking these types of questions in an imaginary ski resort called Happy Valley.

---

## BOX 1.1 Questions GIS could answer in Happy Valley

CASE STUDY

The Happy Valley GIS has been established to help the ski resort's managers improve the quality of the ski experience for visitors. The examples below are only a few of the situations in which asking a spatial question can help with the management of Happy Valley. You will find many other examples throughout the rest of this book.

**1** *Where are particular features found?* When skiers visit Happy Valley they need to know where all the visitor facilities are located. To help, the Happy Valley GIS is used to produce maps of the ski area. Visitors can also ask direct questions about the location of facilities using 'touch screen' computerized information points located in shops and cafes throughout the ski resort. These information points provide skiers with a customized map showing them how to find the facilities they require.

**2** *What geographical patterns exist?* Over the last few ski seasons there have been a number of accidents involving skiers. All these incidents

**BOX 1.1**

have been located and entered into the GIS. The Happy Valley management team is trying to establish whether there is any common theme or spatial pattern to the accidents. Do accidents of a certain type occur only on specific ski pistes, at certain points on a ski piste such as the lift stations, or at particular times of day? So far one accident black spot has been identified where an advanced ski run cuts across a slope used by beginners, just below a mountain restaurant.

**Figure 1.1** Facilities in Happy Valley ski resort

**3** *Where have changes occurred over a given time period?* In Happy Valley avalanches present a danger to skiers who wish to venture off the groomed ski pistes. The management team and the ski patrol use the GIS to build up a picture of snow cover throughout the area. This is done by regularly recording snow depth, surface temperature, snow water content and snow strength at a number of locations. A study of the geographical changes in these parameters helps the management team prepare avalanche forecasts for different locations in Happy Valley.

**4** *Where do certain conditions apply?* Every day, during the winter season, the Happy Valley management team provides information on which ski pistes are open. Since this depends on the snow cover, avalanche danger and wind strength, data on these

factors are regularly added to the GIS from reports provided by the ski patrols and local weather service. The warden can use this information and the GIS to help identify which runs should be opened or closed.

**5** *What will the spatial implications be if an organization takes certain action?* The access road to Happy Valley is now too narrow for the number of skiers visiting the area. A plan is being prepared for widening the road. However, any road-widening scheme will have impacts on a local nature reserve as well as surrounding farm land. The Happy Valley GIS is being used to establish the amount of land that is likely to be affected under different road-widening schemes.

Datasets and activities relating to the Happy Valley Case Study can be found online at **www.pearsoned. co.uk/heywood**

**Figure 1.2** Avalanche!
*Source:* Richard Armstrong

If you have a geographical background you may be asking what is new about these generic questions. Are these not the questions that geographers have been contemplating and answering for centuries? In part they are, though in many cases geographers and others using spatial data have been unable to find answers to their questions because of the volume of

data required and a lack of time and techniques available to process these data. The following examples of GIS applications are used to illustrate the capabilities of GIS as a tool for geographical analysis. All involve the manipulation of data in ways that would be difficult or impossible by hand, and each illustrates different issues associated with the application of GIS.

## Searching for sites

Searching for the optimum location to put something is a task performed by individuals and organizations on a regular basis. The task may be to find a site for a new retail outlet, a new oil terminal or a new airport. Sometimes the task is more demanding than others, involving searches through large numbers of maps and related documents.

Radioactive waste is waste material that contains levels of ionizing radiation considered harmful to health. It is produced by the nuclear industry, nuclear power generation, weapons manufacture, medical and research establishments (Figure 1.3). Radioactive waste is categorized according to the level of radiation emitted and the length of time for which it will be radioactive. Careful management is required, and ultimately governments and the nuclear industry have to find appropriate disposal options. Over the last 20 years finding a suitable site for the disposal of radioactive waste in the UK has become a sensitive and important issue. The Nuclear Decommissioning Authority (NDA, formerly known as NIREX – the Nuclear Industry Radioactive Waste Executive) is the UK organization with responsibility

(a)

(b)

(c)

(d)

**Figure 1.3** A nuclear power station (a) and nuclear waste containers (b, c and d)
*Sources*: (a) Courtesy of author; (b) Jeff T. Green / Getty Images; (c,d) British Nuclear Fuels Ltd

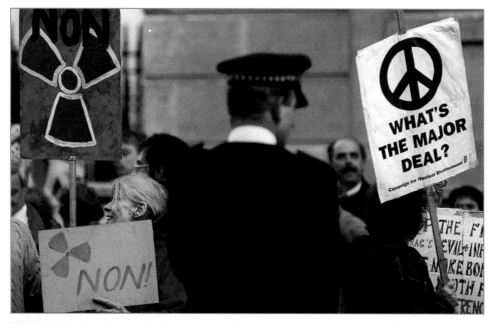

**Figure 1.4** Anti-nuclear protestors
*Source*: Stefan Rousseau/Press Association Images

for the identification of suitable radioactive waste disposal sites. Advice on what to do with the UK's radioactive waste is provided by the Committee on Radioactive Waste Management (CoRWM). The NDA has the task of interpreting current government radioactive waste policy and siting guidelines and presenting possible sites at public inquiries. One of the problems for NDA is the lack of comprehensive and coherent guidelines for the identification of suitable sites. Another is that radioactive waste is a strong political issue because nobody wants a disposal facility in their neighbourhood and protests against potential sites are common (Figure 1.4). However, NDA is expected to show that it has followed a rational procedure for site identification (Department of the Environment, 1985). Hydrology, population distribution and accessibility are examples of important siting factors, but how such factors should be interpreted is left up to NDA. Where, therefore, should NDA site a nuclear waste repository?

NIREX (the NDA's predecessor) used a pen-and-paper approach to sieve through large numbers of paper maps containing data about geology, land use, land ownership, protected areas, population and other relevant factors. Areas of interest were traced from these maps by hand, then the tracings were overlaid to identify areas where conditions overlapped. NIREX is not the only organization that has

searched for sites using these techniques. This was a standard approach employed in the siting of a wide range of activities including shopping centres, roads and offices (Figure 1.5). The method is time-consuming and means that it is impossible to perform the analysis for more than a few different siting criteria. The best sites are often missed. GIS techniques offer an alternative approach, allowing quick remodelling for slight changes in siting criteria, and produce results as maps eminently suitable for presentation at public inquiries.

Openshaw *et al.* (1989) first demonstrated the use of GIS for nuclear waste repository siting, and their method is summarized in Figure 1.6. They established a number of data layers, each containing data for a separate siting criterion (for example, geology, transport networks, nature conservation areas and population statistics). These were converted from paper to digital format by digitizing (this technique is explained in **Chapter 5**), or acquired from existing digital sources such as the UK Census of Population. These data layers were then processed so that they represented specific siting criteria. The geology layer was refined so that only those areas with suitable geology remained; the transport layer altered so that only those areas close to major routes were identified; and the nature conservation layer processed to show protected areas where no development is permitted. The population layer was analysed so that areas with high population densities were removed. The four data layers are shown in Figure 1.7.

**(a)** Step one

**(b)** Step two

**(c)** Step three

**Figure 1.5** Tracing paper overlay

GIS software was then used to combine these new data layers with additional layers of information representing other siting criteria. The final result was a map showing the locations where all the specified siting criteria were satisfied and, thus, a number of locations suitable for the siting of a nuclear waste repository. The advantage of using the GIS to perform this task was that the siting criteria could be altered and the procedure repeated with relative ease. Examples for several siting scenarios are shown in Figure 1.8. These illustrate how changes in siting criteria influence the geographical distribution of potential sites.

This example shows how a GIS approach allows comparative re-evaluation and testing of data and conditions. In this way the decision maker can evaluate options in a detailed and scientific manner. The work of Openshaw *et al.* (1989) also illustrates three other important issues associated with the use of GIS: the problem of errors in spatial data sets; the difficulty in establishing criteria for abstract spatial concepts; and the potential value of using GIS to communicate ideas (Box 1.2).

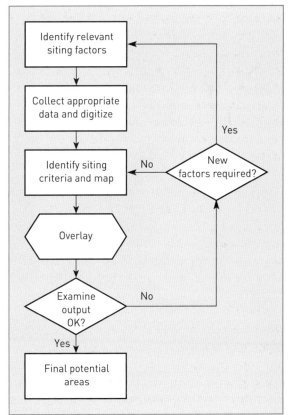

**Figure 1.6** Using GIS for siting a NDA waste site
*Source*: Adapted from Openshaw *et al.* (1989)

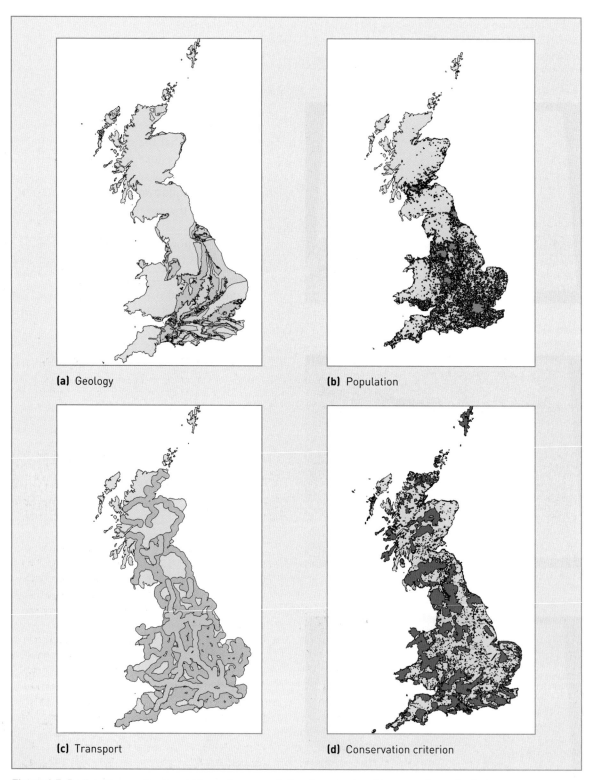

**(a)** Geology

**(b)** Population

**(c)** Transport

**(d)** Conservation criterion

**Figure 1.7** Radioactive waste case study: geology, population, transport and conservation criteria maps; *Sources*: (a) British Geological Survey. © NERC, IPR/71-39C reproduced by permission; (b) Office for National Statistics; (c) Ordnance Survey; (d) Joint Nature Conservation Committee, www.jncc.gov.uk

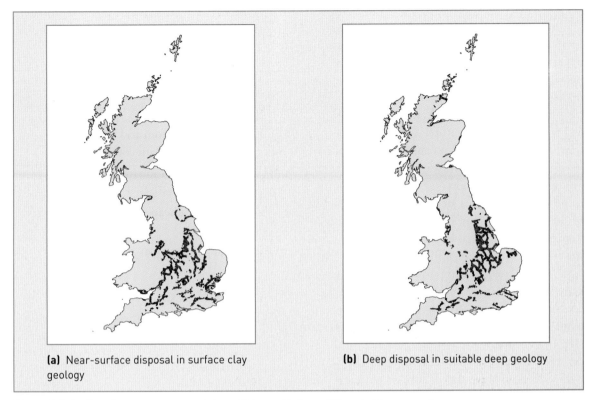

**(a)** Near-surface disposal in surface clay geology

**(b)** Deep disposal in suitable deep geology

**Figure 1.8** Radioactive waste case study (a and b): results from different siting scenarios

## BOX 1.2 Issues raised by the nuclear waste case study

CASE STUDY

- *Errors in source data*, such as those introduced during the conversion of data to digital form, may have a significant effect on the GIS site-searching process. Mistakes in capturing areas of appropriate geology from paper maps may lead to inappropriate waste repository sites being identified, because areas on the ground will have different geological properties from those recorded in the GIS. Errors in spatial data sets and the associated issues of data quality are discussed in detail in **Chapter 10**.

- The GIS site-searching process relies on the translation of *abstract* (or *'fuzzy'*) concepts such as 'near to' and 'far from' into precise conditions that can be mapped. This can be a problem. How do you create a map that shows all the geographical zones 'far from' a centre of population? The only method is to make an arbitrary decision about what sort of

distance (for example, 10, 20 or 30 km) 'far from' represents. In some cases rules may be applied to guide this process, but in others the numerical representation of a criterion may depend upon the preferences of the person responsible for choosing and implementing the criterion.

- *GIS output* can be used to inform public participation in the decision-making process. A series of maps could be used to illustrate why a particular geographical location has been identified as a suitable site for the disposal of radioactive waste. However, the issues raised above – data quality and the problems of creating spatial criteria for abstract concepts – suggest that output from GIS should be viewed with caution. Just because a map is computer-generated does not mean that the picture it presents is correct. More on this issue can be found in **Chapter 8**.

## Evaluating land use planning

Virtually every country in the world has areas of natural beauty and conservation value that are managed and protected in the public interest. Those managing these areas face the problem of balancing human activities (such as farming, industry and tourism) with the natural elements of the landscape (such as climate, flora and fauna) in order to maintain the special landscape character without exploitation or stagnation.

The protected area of Zdarske Vrchy, in the Bohemian–Moravian highlands of the Czech Republic (Figure 1.9), is an example of an area that has suffered as a result of ill-considered state control. Unregulated farming, tourism and industrial activities have placed the landscape under severe pressure. Czech scientists and environmental managers have relied on traditional mapping and statistical techniques to monitor, evaluate and predict the consequences of this exploitation. However, with change in the political administration of the country came scientists and managers who were looking not only into a policy of sustainable development for Zdarske Vrchy, but also at GIS as a tool to help with policy formulation (Petch *et al.*, 1995).

GIS is seen as a tool to bring together disparate data and information about the character and activities that take place in the Zdarske Vrchy region. Data from maps, aerial photographs, satellite images, ecological field projects, pollution monitoring programmes, socio-economic surveys and tourism studies have been mapped and overlaid to identify areas of compatibility and conflict.

**Figure 1.9** Zdarske Vrchy

In this context GIS permits scientists and managers in Zdarske Vrchy to interact with their data and ask questions such as:

- What will be the long-term consequences of continuing recreational activity for the landscape? (Figure 1.10)
- Where will damage from acid rain occur if a particular industrial plant continues to operate?
- Where is the best location for the re-introduction of certain bird species?
- Where should landscape conservation zones be established?

   (after Downey *et al.*, 1991, Downey *et al.*, 1992 and Petch *et al.*, 1995).

One application of GIS in Zdarske Vrchy has been to identify areas of the landscape for conservation. Traditionally, water storage in the region has relied on the use of natural water reservoirs such as peat wetlands and old river meanders (Petch *et al.*, 1995). Current land-use practices, in particular forestry and farming, are resulting in the gradual disappearance of these features. The consequences of these changes have been localized droughts and floods in areas further downstream. In turn, these changes have brought about a reduction in plant species and wildlife habitats. Managers in the Zdarske Vrchy region wanted to identify conservation zones to protect the remaining natural water reservoirs as well as to identify those areas where it may be possible to restore the water retention character of the landscape. To do this, they needed to establish the characteristics of the landscape that determine whether or not a particular location is likely to retain water. Specialists in hydrology, geology and ecology were consulted to identify a range of important criteria describing:

- the type of soil and its water retention ability;
- the character of the topography (for example, presence or absence of hollows or hills);
- the type of land use, as certain agricultural practices exploit the water retention capacity of the landscape; and
- the presence or absence of human-enhanced water drainage channels.

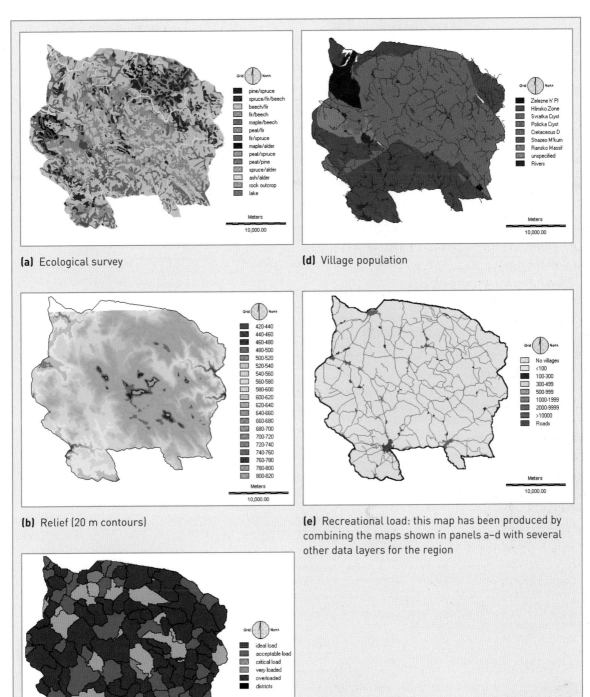

**(a)** Ecological survey

**(d)** Village population

**(b)** Relief (20 m contours)

**(e)** Recreational load: this map has been produced by combining the maps shown in panels a–d with several other data layers for the region

**(c)** Geology and hydrology

**Figure 1.10** GIS landscape assessment data layers: Zdarske Vrchy

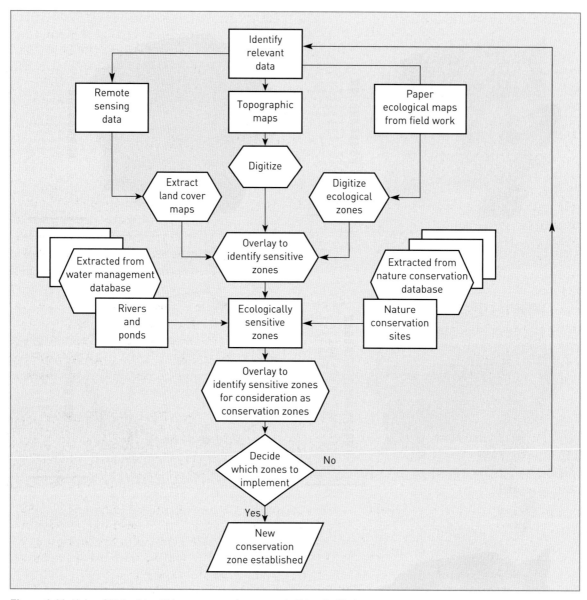

**Figure 1.11** Using GIS for identifying conservation zones in Zdarske Vrchy
*Source*: Adapted from Petch *et al.* (1995)

GIS professionals were then asked to find appropriate sources of spatial data that could be used to represent these criteria. A range of sources was identified including:

- paper maps (for soil type and geology);

- contour maps (for topography);

- ecological field maps (for drainage conditions); and

- remote sensing (for land use).

These data were acquired and entered into the GIS. The hydrologists, ecologists and geologists were then asked how each of the criteria (land use, topography, soil type and drainage) might interact to influence the retention capacity of the landscape at a particular location. First, the scientists used the GIS to look at the relationship of the criteria they had identified in areas where natural water reservoirs were still in existence. This involved adding more data to the GIS about the location of existing water retention zones. These data came from the regional water authorities as paper maps. The relationship between the geographical distribution of the various landscape characteristics at these locations was then used to develop a model. This model allowed the managers of the Zdarske Vrchy region to predict which other areas could be restored as natural water reservoirs. The next stage was to check which of these areas were located in existing conservation zones, as it was easier to change existing conservation regulations than to set up new conservation areas. Figure 1.11 summarizes the method used and shows how the GIS was used to integrate data from a range of different sources. Figure 1.11 also shows how these data were overlain with additional data about existing water retention zones and existing conservation areas to identify potential new conservation sites.

The Zdarske Vrchy project shows how GIS can be used to bring together data from a wide variety of sources to help address a range of environmental management problems. This use of GIS is not unique to environmental planning in Eastern Europe and is being practised by environmental managers all over the world. In addition, the Zdarske Vrchy project reveals a number of other important issues associated with the use of GIS. These include the problem of data sources being in different map projections, the value of GIS as a modelling tool and the role for GIS as a participatory problem-solving tool (Box 1.3).

## Finding a new home

At some stage in our lives, most of us will need to look for a new home. Perhaps because of a new job, or a change in family circumstances, our accommodation requirements will change and we will have to look for a new place to live. This can be a time-consuming and frustrating task. The requirements of individual family members need to be considered. Do they need to be close to schools, major roads or a railway station? Perhaps they would prefer to be in an area where insurance costs are lower. Maybe they want to be in an urban area to be close to shops and

## BOX 1.3 Issues raised by the Zdarske Vrchy case study

CASE STUDY

- The greatest problem associated with bringing data together for the creation of the Zdarske Vrchy GIS was deciding which *map projection* to adopt as the common frame of reference. Several different projection systems were used by the source maps. A projection system is the method of transformation of data about the surface of the earth on to a flat piece of paper. Because many methods exist which can be used to perform this task, maps drawn for different purposes (and maybe even for the same purpose but at different points in time) may use different projection systems. This does not present any problems as long as the maps are used independently. However, when the user wishes to overlay the data in a GIS, the result can be confusing. Features that exist at the same location on the ground may appear to lie at different geographical positions when viewed on the computer screen. This problem became apparent in the Zdarske Vrchy project when the road network, present on two of the maps, was compared. Map projections are explained in more detail in **Chapter 2**.

- The Zdarske Vrchy case study also shows how GIS can be used to create models of environmental processes with maps used as the building blocks for the model. The topic of *modelling and GIS* is returned to in **Chapters 6 and 7**.

- Bringing people together to search for a solution to a common problem is often difficult. Different specialists will have different ideas about the problem. For example, an ecologist might recommend one approach, an engineer a second and an economist a third. The Zdarske Vrchy project showed how, through the use of GIS, the common medium of the map could be used as a tool to help experts from different backgrounds exchange ideas and compare possible solutions. The idea that GIS can be used as a *participatory problem-solving tool* has also received considerable attention from the GIS research community (Carver et al., 1997) and often involves the use of web-based mapping tools. **Chapter 7** considers this topic in more detail.

their place of work. To find a new home acceptable to all the family a decision support system (software to help you make a decision) may be appropriate. GIS can act in this role.

Much of the data needed to help answer the questions posed above can be gathered and converted into a format for integration in GIS. Heywood *et al.* (1995) have successfully completed this exercise and created a house-hunting decision support system. To find a suitable home, participants were first required to decide which of a series of factors (insurance costs, proximity to schools, railways and roads, urban areas) were important in their decision making. These factors were allocated weights and scores reflecting their importance. Constraints, areas where a new home would not be suitable under any conditions, were also identified. Constraints excluded certain areas from the analysis altogether; for example, participants could decide that they did not wish to live within 500 m of a major road.

Once the weighting process had been completed, the data selected were combined in a GIS using a multi-criteria modelling technique. This technique will be explored in more detail in **Chapter 7**, but, in brief, the data layers were combined using weightings, so that the layer with the highest weight had the most influence on the result. The resulting maps were used to help target the house-hunting process. The method is summarized in Figures 1.12 and 1.13.

Locations of houses that were for sale were plotted over the top of the suitable areas identified, and ranked according to the number of criteria which they meet. If further details of these houses were available in computerized form, they were accessed by pointing at the map to find out: for instance, how many bedrooms they had, or whether they had a garden. To achieve this, the information on the map (locations of properties) was linked with a database of house features. This type of data is often referred to as 'attribute' data. Attribute data and database concepts are considered in detail in **Chapter 4**.

There are examples of software developed to perform similar tasks, including Wigwam in the UK (Anon., no date) and GeoData in the USA (Esri, 1995). These systems have been designed to make it possible for a home buyer to visit an estate agent, explain the type of house and neighbourhood they prefer, and come away with a map showing the locations of houses for sale which meet their requirements. These products bring help to the home buyer deciding where to look for a new home in an unfamiliar area. GIS in this context is a decision support system. Online applications such as upmystreet.com and Google Street View also allow users to search for properties and explore the characteristics of the area in which they are found.

The house-hunting example shows how GIS can be used to link databases with similar types of data. This improves the speed and efficiency with which an appropriate location can be found. In this respect, it is a similar application to the NDA case study discussed earlier. However, it differs from the NDA example in that a large part of the search process may be carried out using the attributes associated with a spatial feature. In the case of the house-hunting example, these may be the number of bedrooms a property has or its price. Heywood *et al.* (1995) raise other issues associated with the use of GIS as a decision support tool. These are the problem of different GIS software products giving different results, the problem of defining search criteria and the human constraints on the decision-making process (Box 1.4).

The examples above are not enough to illustrate the range of applications and problems that GIS can be used to address. Even when a problem cannot be solved entirely using GIS, there may be the potential for some GIS input to aid the decision-making process. Table 1.1 offers additional pointers to further examples of GIS use by local government, defence agencies, utility companies, commerce and business.

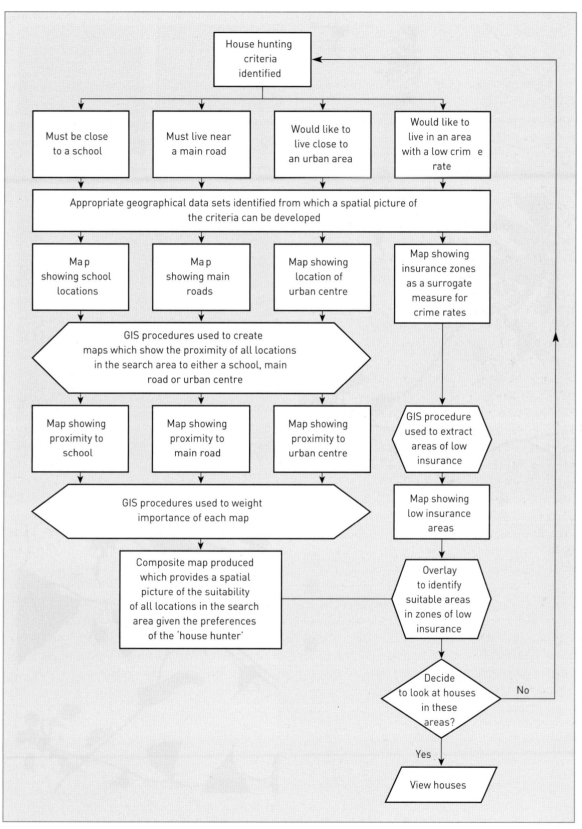

**Figure 1.12** Using GIS to assist in house hunting

**(a)** Railway constraint

**(b)** Countryside constraint

**(c)** Proximity to roads

**(d)** Proximity to school

**(e)** Combination of railway constraint and countryside constraint

**(f)** Combination of proximity and constraint maps [(c), (d) and (e)] with proximity to road used as the most important factor

LEGEND

— Motorway

— A-road

— U-road

- - - Railway

☐ Estuary

▨ Urban Area

Most suitable ▨

Least suitable ▨

Acceptable ☐

Not acceptable ■

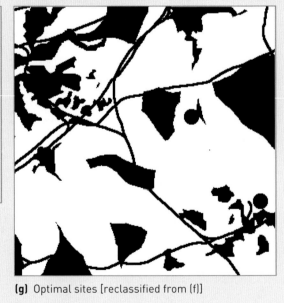

**(g)** Optimal sites [reclassified from (f)]

**Figure 1.13** The house-hunting case study
*Source*: GeographyCal®

## BOX 1.4 Issues raised by the house-hunting case study

- Heywood *et al.* (1995) compared the results they obtained using two GIS software products to identify appropriate neighbourhoods for one house buyer. The results were different. The differences were in part explained by small differences in the search methods used by each GIS. Different GIS will implement similar methods in slightly different ways, and there will also be variations due to the way data are stored in the GIS. Therefore, a clear understanding of the way *GIS software* works is crucial if you are to be able to understand and explain your results. **Chapters 2 to 8** expand on these issues.

- Heywood *et al.* (1995) also considered that there could be difficulties with results due to the way the problem was defined at the outset. For example, the limited selection of data suggested above may be available to help with the site selection, but in reality it may be that you wish to be in close proximity to a swimming pool and a medical centre. If data are not available for these factors, they cannot be included in the analysis. So, defining the problem, and identifying all relevant criteria, are crucial steps in the *design of GIS projects*. The result you obtain will be influenced by the questions you ask. If you do not ask the right questions, you will not get the right answer. Therefore, good project design is an essential component of using GIS. In **Chapter 12** we provide you with a methodology to help you plan your own GIS project.

- *Human factors* such as awareness and training also influence the effectiveness of GIS as a decision support system as they will help the user formulate appropriate questions. **Chapter 11** looks at these issues in more detail.

**TABLE 1.1 Application areas for GIS**

| Activity | Application | Activity | Application |
|---|---|---|---|
| Socio-economic/ government | Health | Utilities | Network management |
| | Local government | | Service provision |
| | Transport planning | | Telecommunications |
| | Service planning | | Emergency repairs |
| | Urban management | Environmental management | Landfill site selection |
| | International aid and development | | Mineral mapping potential |
| Defence agencies | Target site identification | | Pollution monitoring |
| | Tactical support planning | | Natural hazard assessment |
| | Mobile command modelling | | Disaster management and relief |
| | Intelligence data integration | | Resource management |
| | Homeland security and anti-terrorism | | Environmental impact assessment |
| Commerce and business | Market share analysis | | |
| | Insurance | | |
| | Fleet management | | |
| | Direct marketing | | |
| | Target marketing | | |
| | Retail site location | | |

## REFLECTION BOX

- Make a note of all the questions you have asked or heard recently that have a spatial component. Can you classify them into questions about location, patterns, trends, conditions and implications?
- In the house-hunting example GIS has been used to try to improve the method we use when searching for a new home. Would you consider using a

GIS to help you find a new home? Explain the reasons for your answer.
- Think about the impact of GIS on the case studies. Has the use of GIS been beneficial? Have there been any problems? To what extent might GIS be regarded as a unifying technology which spans many different disciplines in each instance?

## DEFINING GIS

There have been so many attempts to define GIS that it is difficult to select one definitive definition. Maguire (1991) offers a list of 11 different definitions. This variety can be explained, as Pickles (1995) suggests, by the fact that any definition of GIS will depend on who is giving it, and their background and viewpoint. Pickles also considers that definitions of GIS are likely to change quickly as technology and applications develop further. Some of the shorter definitions give an idea of what a GIS is, albeit in a superficial way. For example, Rhind (1989: 28) proposes that GIS is 'a computer system that can hold and use data describing places on the Earth's surface'. Fuller definitions give more idea of what GIS can do, as well as what they are. Those provided by Burrough (1986: 6): 'a set of tools for collecting, storing, retrieving at will, transforming, and displaying spatial data from the real world for a particular set of purposes', and the Department of the Environment (1987: 132): 'a system for capturing, storing, checking, integrating, manipulating, analyzing and displaying data which are spatially referenced to the Earth', fall into this category.

In general, the definitions of GIS cover three main components. They reveal that GIS is a *computer system*. This implies more than just a series of computer boxes sitting on a desk, but includes hardware (the physical parts of the computer itself and associated peripherals – plotters and printers), software (the computer programs that run on the computer) and appropriate procedures (or techniques and orders for task implementation). They also tell us that GIS uses *spatially referenced* or *geographical data*, and

that GIS carries out various *management and analysis tasks* on these data, including their input and output. The Department of the Environment (1987) lists the capabilities that a 'well-designed GIS' should be able to provide:

1. Quick and easy access to large volumes of data.
2. The ability to:
   - select detail by area or theme;
   - link or merge one data set with another;
   - analyse spatial characteristics of data;
   - search for particular characteristics or features in an area;
   - update data quickly and cheaply; and
   - model data and assess alternatives.
3. Output capabilities (maps, graphs, address lists and summary statistics) tailored to meet particular needs.

In short, GIS can be used to add value to spatial data. By allowing data to be organized and viewed efficiently, by integrating them with other data, by analysis and by the creation of new data that can be operated on in turn, GIS creates useful information to help decision making. A GIS can, as was alluded to in the house-hunting case study, be described as a form of spatial decision support system.

Some authors consider that there are important elements of a GIS in addition to those common to the definitions above. For example, Burrough (1986) suggests that GIS have three main elements: 'computer hardware, application software modules, and a proper organizational context'. Others, such as Maguire (1989), stress that data are the most important part of GIS. In practice, none of the main elements (the computer system, data or processing tools) will function as a GIS in isolation, so all might be considered of equal importance. However, it is perhaps the nature of the data used, and the attention given to the processing and interpretation of these data, that should lie at the centre of any definition of GIS.

GIS draws on concepts and ideas from many different disciplines. The term 'Geographic Information Science' has been adopted to refer to the science behind the systems. Geographic Information Science draws on disciplines as diverse as cartography, cognitive science, computer science, engineering, environmental sciences, geodesy, landscape architecture, law, photogrammetry, public policy, remote sensing, statistics and surveying. Geographic Information Science involves the study of the fundamental issues arising from the creation, handling, storage and use of geographic information (Longley *et al.*, 2005), but it also examines the impacts of GIS on individuals and society and the influences of society on GIS (Goodchild, 1997). Mark (2003) considers the definition and development of Geographic Information Science in detail.

Goodchild (1997) offers a useful summary of key concepts that help with the definition of GIS:

- Geographical information is information about places on the Earth's surface.
- Geographic information technologies include global positioning systems (GPS), remote sensing and geographic information systems.
- Geographical information systems are both computer systems and software.
- GIS can have many different manifestations.
- GIS is used for a great variety of applications.
- Geographic Information Science is the science behind GIS technology.

## ■ COMPONENTS OF A GIS

There is almost as much debate over the components of a GIS as there is about its definition. At the simplest level, a GIS can be viewed as a software package, the components being the various tools used to enter, manipulate, analyse and output data.

At the other extreme, the components of a GIS include: the *computer system* (hardware and operating system), the *software, spatial data, data management and analysis procedures* and the people to operate the GIS. In addition, a GIS cannot operate in isolation from an application area, which has its own tradition of ideas and procedures. It is this more comprehensive perspective that is adopted here.

### Computer systems and software

GIS run on the whole spectrum of computer systems ranging from portable personal computers (PCs) to multi-user supercomputers, and are programmed in a wide variety of software languages. Systems are available that use dedicated and expensive workstations, with monitors and digitizing tables built in; that run on bottom-of-the-range PCs or notebooks; and that run on portable Personal Data Assistants (PDAs), tablet PCs or handheld GIS/GPS devices (Figure 1.14). In all cases, there are a number of elements that are essential for effective GIS operation. These include (after Burrough, 1986):

- the presence of a processor with sufficient power to run the software;
- sufficient memory for the storage of large volumes of data;
- a good quality, high-resolution colour graphics screen; and
- data input and output devices (for example, digitizers, scanners, keyboard, printers and plotters).

Likewise, there are a number of essential software elements that must allow the user to input, store, manage, transform, analyse and output data. Discussion of these issues follows in **Chapters 4 to 8**. However, although GIS generally fit all these requirements, their on-screen appearance (user interface) may be very different. Some systems still require instructions to be typed at a command line, while others have 'point and click' menus operated using a mouse. Examples of popular GIS interfaces are shown in Figure 1.15. The type of interface individual users find easier to operate is largely a matter of personal preference and experience.

**(a)** Dedicated GIS workstation

**(b)** Desktop GIS

**(c)** GIS on a mobile

**(d)** GIS on tablet device

**(e)** GIS on a hand-held field computer

**Figure 1.14** GIS workstations
*Sources*: (a) Courtesy of Integraph; (b) Courtesy of author; (c) Maksym Dykla/Fotolia.com; (d) Courtesy of MainStreetGIS;
(e) Courtesy of Trimble

**(a)** Command line

**(b)** GUI

**Figure 1.15** The GIS interface
*Source*: Screenshots show Esri Graphical User Interface (GUI)
ArcMap / ArcView / ArcInfo Graphical User Interface is the
intellectual property of Esri and is used herein with permission.
Copyright © 2011 Esri. All rights reserved

Unlike the issue of software functionality there is limited discussion of hardware and interface technology in this book. This is because we consider these technologies to be changing so rapidly that any discussion would soon be out of date. If the reader is interested in the latest technical advances in GIS hardware and interfaces, then web addresses

are provided for a number of leading GIS developers and product directories on the book website (www.pearsoned.co.uk/heywood).

## Spatial data

All GIS software has been designed to handle spatial data (also referred to as geographical data). Spatial data are characterized by information about position, connections with other features and details of non-spatial characteristics (Burrough, 1986; Department of the Environment, 1987). For example, spatial data about one of Happy Valley's weather stations (Figure 1.16a) may include:

- latitude and longitude as a geographical reference. This reference can be used to deduce relationships with nearby features of interest. If the latitude and longitude of a weather station are known, the relative position of other weather stations can be deduced, along with proximity to ski slopes and avalanche areas;

- connection details such as which service roads, lifts and ski trails would allow the meteorologist access to the weather station;

- non-spatial (or attribute) data: for instance, details of the amount of snowfall, temperature, wind speed and direction.

In a similar way spatial data about a ski piste (Figure 1.16b) may include:

- a series of spatial references to describe position;

- details of other runs that cross or join the ski piste;

- attribute data such as the number of skiers using the piste and its standard of difficulty.

The spatial referencing of spatial data is important and should be considered at the outset of any GIS project. If an inappropriate referencing system is used, this may restrict future use of the GIS (Openshaw, 1990). The challenge is to adopt a flexible, robust and lasting referencing system, since a GIS may be intended to last many years. A full discussion of spatial referencing can be found in **Chapter 2**.

The traditional method of representing the geographic space occupied by spatial data is as a series of thematic layers. Consider, for example, traditional cartographic maps that may be available for an individual area. There may be a map for geology, one for soils and a topographic map showing cultural and environmental features on the surface. Computer models of space frequently use a similar approach. For example, the house-hunting GIS example discussed earlier contained layers of data including insurance, transport, schools, and urban–rural land use. This was the first method of modelling space to be developed. This method, known as the 'layer-based approach', is still used by most GIS.

**Figure 1.16** Weather station (top) and ski piste (bottom) in Happy Valley

An alternative method of representing reality in a computer is to consider that space is populated by discrete 'objects' (Goodchild, 1995). Goodchild (1995) presents the example of a utility company that needs to map and manage a vast array of telegraph poles, connection boxes and cables. Each of these may be regarded as a discrete object, and there is empty space between the objects. This method is known as the 'object-oriented approach'. **Chapter 3** looks at both these approaches to modelling spatial data in more detail and considers their advantages and disadvantages.

Spatial data, represented as either layers or objects, must be simplified before they can be stored in the computer. A common way of doing this is to break down all geographic features into three basic entity types (an entity is a component or building block used to help data organization). These are points, lines and areas (Figure 1.17). If we return to the example of Happy Valley, points may be used to represent the location of features such as restaurants, lift pylons or rescue stations. Lines can be used to represent features such as roads, rivers and ski lifts. Area features are used to represent geographical zones, which may be observable in the real world (such as the Happy Valley car park) or may be artificial constructs (such as administrative areas). Points, lines and areas can all be used to represent surfaces. For example, spot heights or contour lines can be used to create a surface model of the Happy Valley landscape. In addition, points (representing junctions) and lines (representing roads) can be used to create a network model of Happy Valley's roads. These representations of real-world phenomena are normally held in a GIS according to one of two models – *raster* (sometimes referred to as *grid* or *tesseral*) or *vector* (Dale and McLaughlin, 1988; Peuquet, 1990). These two different approaches are compared in Figure 1.18. The raster model is particularly applicable where remotely sensed images are used (since the data are collected in this format) and is considered the most appropriate choice for modelling continuous geographic phenomena such as snow depth. The vector model is more appropriate for mapping discrete geographic entities such as road and river networks and administrative boundaries. Several examples of raster and vector data layers are given in Figure 1.18 for you to compare. Modelling the geography of the real world in the computer is considered further in **Chapter 3**.

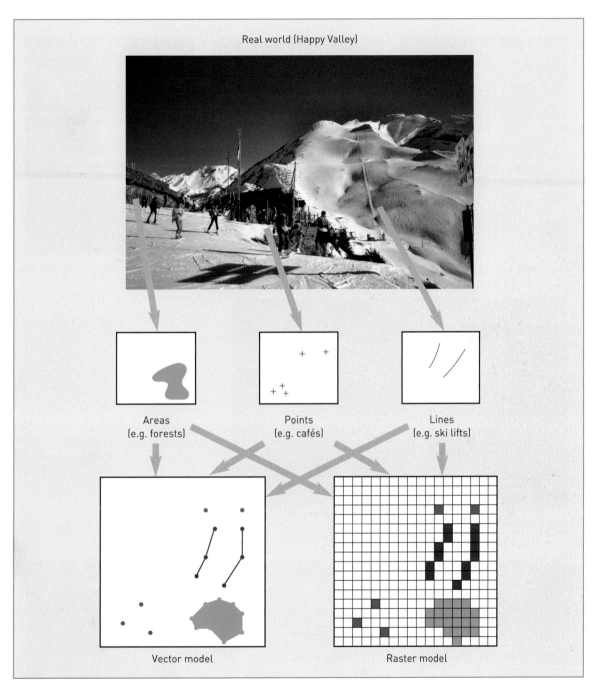

**Figure 1.17** Points, lines and areas

**Data management and analysis procedures**
The functions that a GIS should be able to perform include data input, storage, management, transformation, analysis and output. *Data input* is the process of converting data from its existing form to one that can be used by the GIS (Aronoff, 1991). It is the procedure of encoding data into a computer-readable form and writing the data to the GIS database.

This process should include verification procedures to check that the data are correct and transformation procedures to allow data from different sources to be used. GIS need to handle two types of data – graphical data and non-spatial attribute data. The graphical data describe the spatial characteristics of the real-world feature being modelled. For example, the hotels in Happy Valley may be described

(a) Raster digital elevation model  (b) Raster land cover data  (c) Satellite image

(d) Digital colour aerial photograph  (e) Vector contours and roads

(f) Vector soil polygons  (g) Vector census polygon boundaries  (h) Vector land cover polygons

**Figure 1.18** Examples of raster (a–d) and vector GIS data layers (e–h)

by a series of points. In some cases, particularly when area and line features are used to model real-world features, the graphical data may include information about the linkages between them. For example, if the boundary of an area feature such as a car park is also a snow fence that prevents skiers from overshooting the nursery slopes, this information may be stored with the graphical data. Non-spatial attribute data describe what the features represent. They tell the computer what a particular set of entities represents (for instance, a set of points may represent hotels). In addition, further non-spatial attribute data may be stored which provide extra information about the hotels (standard, number of rooms and restaurant facilities).

Data input and updating are frequently the most expensive and time-consuming part of any GIS project and their importance and complexity should never be underestimated. Approximately 80 per cent of the duration of many large-scale GIS projects is concerned with data input and management. Aronoff (1991) estimates that the construction of a large database could cost five to ten times more than the GIS software and hardware. Data input methods are discussed in **Chapter 5**.

The *data management* functions necessary in any GIS facilitate the storage, organization and retrieval of data using a database management system (DBMS). A DBMS is a set of computer programs for organizing information, at the core of which will be a database. Database applications that have no GIS component include management of payrolls, bibliographies, and airline booking systems. In the same way that DBMS organize these different types of data they can be used to handle both the graphical and non-graphical elements of spatial data. An ideal GIS DBMS should provide support for multiple users and multiple databases, allow efficient updating, minimize repeated (or redundant) information and allow data independence, security and integrity (Smith *et al.*, 1987). Relational databases, flat files and other database models used by GIS will be discussed in more detail in **Chapter 4**.

It is the ability of GIS to *transform* spatial data – for example, from one entity type (points, lines and areas) to another – and to perform *spatial analysis*, that distinguishes GIS from other types of information systems.

Transformation is the process of changing the representation of a single entity, or a whole set of data. In GIS, transformation may involve changing the projection of a map layer or the correction of systematic errors resulting from digitizing. In addition, it may be necessary to convert data held as rasters to vectors or vice versa.

Aronoff (1991) classifies GIS analysis procedures into three types (Figure 1.19):

(a) Soil map display     (b) Reselection of specific soil types     (c) Simple soil erosion model predictions

**Figure 1.19** Mapping, querying and modelling soil information
*Source*: Courtesy of United States Geological Survey

1 Those used for storage and retrieval. For example, presentation capabilities may allow the display of a soil map of the area of interest.

2 Constrained queries that allow the user to look at patterns in their data. Using queries, only erodible soils could be selected for viewing or further analysis.

3 Modelling procedures, or functions, for the prediction of what data might be at a different time and place. Predictions could be made about which soils would be highly vulnerable to erosion in high winds or during flooding, or the type of soil present in an unmapped area.

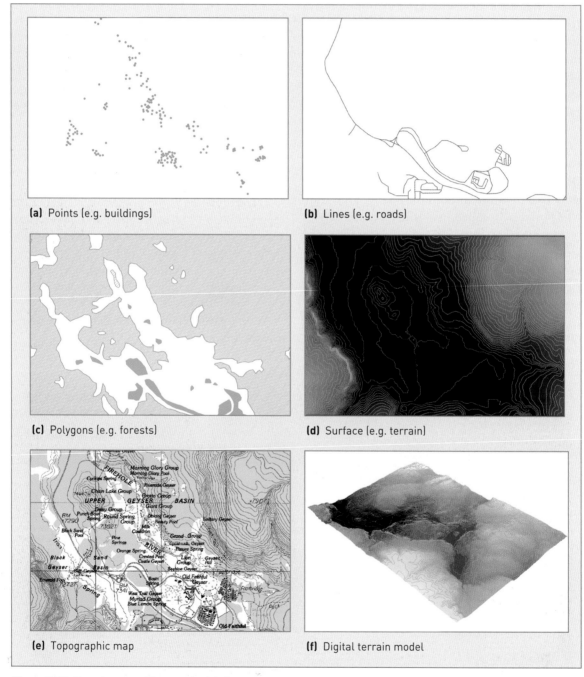

**(a)** Points (e.g. buildings)

**(b)** Lines (e.g. roads)

**(c)** Polygons (e.g. forests)

**(d)** Surface (e.g. terrain)

**(e)** Topographic map

**(f)** Digital terrain model

**Figure 1.20** From topographic maps to data layers
*Source*: Courtesy of United States Geological Survey

Transformation and analysis procedures can also be classified based on the amount of data analysed. Data in GIS are normally held in a series of layers. For instance, a 1:50,000 topographic map might be digitized to create a series of layers – one layer for road data, one for buildings, one for recreational interest (parking, picnic sites and youth hostels) and additional layers for soils and population data. Data layers normally contain data of only one entity type: that is, point, or line, or area data. This is illustrated in Figure 1.20. Analysis can be carried out either on one layer at a time, or on two or more layers in combination. The techniques available in GIS for manipulating and correcting data in preparation for analysis, and the analysis methods available, are discussed in greater depth in **Chapters 5, 6 and 7.**

The form of *data output* used will depend on cost constraints, the audience to whom the results are directed and the output facilities available. A local government agency may produce simple tables, graphs and maps for the communication of important points to politicians, whilst professional map makers may produce detailed plots for publication. In other cases, data may be output in digital form for transfer to another software package for statistical analysis, desktop publishing or further analysis. However, most GIS output is in the form of maps. These may be displayed on-screen for immediate communication to individuals or small groups,

photographed, stored digitally or plotted to produce permanent hard copy. Most GIS provide the ability to design screen formats and forms for plotting and these help to ensure that all maps have titles, keys, north arrows and scales, just as with traditional cartographic output. The facilities available for map design can be very extensive, incorporating myriad of different colours, symbols and line styles, and it is often possible to design additional symbols for your own use. This can make map design very time-consuming, but effective. The audience for any GIS product is an important consideration when designing output, but generally it is best to keep products clear and simple. Options for data output and communication of results from GIS are discussed further in **Chapter 8**.

## People and GIS

Most definitions of GIS focus on the hardware, software, data and analysis components. However, no GIS exists in isolation from the organizational context, and there must always be people to plan, implement and operate the system as well as make decisions based on the output. GIS projects range from small research applications where one user is responsible for design and implementation and output, to international corporate distributed systems, where teams of staff interact with the GIS in many different ways (Figure 1.21). In most

**Figure 1.21** People are a key component of GIS

organizations the introduction of GIS is an important event, a major change bringing with it the need for internal restructuring, retraining of staff and improved information flows. Research has been undertaken to highlight the factors that promote successful GIS and it has been suggested that in certain business sectors, innovative flexible organizations with adequate resources and straightforward applications are more likely to succeed (Campbell and Masser, 1995). However, not all GIS are successful. There is evidence that many systems fail, and more are under-used (Cornelius and Medyckyj-Scott, 1991). So the issues surrounding how to choose a system and how to implement it successfully require examination. These topics are covered in **Chapter 11**.

## REVISION QUESTIONS

- What is GIS? Write your own definition of GIS and consider why GIS can be difficult to define.

- GIS is routinely used in applications such as siting new industrial developments. What are the advantages of using GIS in this context?

- What type of questions could GIS help environmental managers address?

- What are the components of GIS?

- Explain the following terms and phrases:
  1 Spatial data    3 Spatial referencing
  2 Attribute data  4 Spatial entities

- What is the difference between a GIS and geographical information science?

## REFLECTION BOX

- Look again at Table 1.1. Can you add some applications for GIS in areas and activities in which you are interested or involved?
- If you have a GIS of your own, or access to one at work or college, make a list of all its components. Does it

match the definitions and component descriptions given above? How could it be improved?
- Try to summarize the components of a GIS in a diagram or table.

## CONCLUSIONS

GIS technology is now well established and, as we will see in **Chapter 9**, has been in use since the 1960s. Some of the work cited in this and subsequent chapters may have been written several decades ago – this is an indication of the maturity of GIS. The growth in application areas and products in recent years has helped GIS to become an accepted tool for the management and analysis of spatial data. This trend is set to continue as computer technology continues to improve with smaller, faster and more powerful devices, and as more data become available in digital formats directly compatible with GIS. In addition, the striking advances in related tech-

nologies used in surveying and field data collection, visualization and database management will also influence the development of GIS. Further comments on the future of GIS can be found in **Chapter 13**.

There have been some notable failures in GIS. Sometimes data difficulties or other technical problems have set back system developments and applications; however, there are also human and organizational problems at the root of GIS failures. Before we can begin to appreciate these fully, to ensure that our GIS applications are successful, it is important to have a good understanding of what a GIS can do and the data it works with (**Chapter 2**).

## FURTHER STUDY – ACTIVITIES

■ Visit online GIS tutors and glossaries. What definitions of GIS do they give and how do they differ?

■ Think about how GIS is used or could be used in your own organization or field of interest. How does it affect the way you work? If you are a student, how do you use GIS for your studies or in your everyday life?

■ Use the Web to research the problem of nuclear waste disposal. How many websites can you find that deal with siting issues and mention GIS?

■ Imagine you are looking for a new home in your area. Use online estate agent searches to find your ideal property. What aspects of the search engines you use are based on GIS principles? Repeat the exercise for an area you are unfamiliar with, for example in a different country. How does it help you as someone who is new to the area?

## FURTHER STUDY – READING

There are many sources of further reading that complement the material presented in this chapter. Recent editions of other widely used textbooks include Longley *et al.* (2010); DeMers (2008) and DiBiase (2010) – the last of these is an innovative online text. Longley *et al.* (2005) is a useful abridged version of a comprehensive reference work by Longley *et al.* (1999) which considers technical, theoretical and applied GIS issues.

For an overview of GIS from an applied perspective there are a range of texts available. Martin (1995) looks at GIS with a focus on socio-economic applications in a UK setting; while Grimshaw (1999) considers GIS from an information management perspective and offers some useful business case studies. Burrough and McDonnell (1998) and Bonham-Carter (1995) consider applications in land management and the geosciences respectively. Both of these books offer general introductory chapters, then considerable discussion of GIS data models and analysis methods that will be covered in later chapters of this text. Harder (1997), Mitchell (1997), Davis (1999), Bedford (2004), Giusti de Perez and Perez (2008) and Meehan (2007)

also cover applications of GIS at an introductory level in a range of application areas, whilst DeMers (2009) offers a GIS book in the 'For Dummies' series which is helpful for newcomers to the field.

The edited volume by Pickles (1995) provides an alternative perspective on GIS. This includes several extensive and in-depth critiques of GIS from various writers.

For up-to-date accounts of applications, technological developments and GIS issues magazines and journals should be consulted. These include *GeoWorld*, *GeoConnexions* and *GIS Development*. More academic journals include the *International Journal of Geographical Information Science*, *Geo-spatial Information Science*, *Geoinformatics and Transactions in GIS*. Journals and magazines from other disciplines (for example, surveying, computer-aided design (CAD) and computing) also contain introductory articles on GIS from time to time.

Bedford M (2004) *GIS for Water Management in Europe*. Esri Press, Redlands

Bonham-Carter G F (1995) *Geographic Information Systems for Geoscientists: Modelling with GIS*. Pergamon Press, New York

Burrough P A, McDonnell R (1998) *Principles of Geographical Information Systems*. Oxford University Press, Oxford

Davis D (1999) *GIS for Everyone*. Esri Press, Redlands

DeMers M N (2008) *Fundamentals of Geographic Information Systems*. 4th edn. Wiley, New York

DeMers M N (2009) *GIS for Dummies*. Wiley, New York

DiBiase D (2010) *Nature of Geographic Information*. Open Geospatial Textbook available at https://www.e-education.psu.edu/natureofgeoinfo/node/1672

Giusti de Perez R, Perez R (2008) *Analyzing Urban Poverty: GIS for the developing world*. Esri Press, Redlands

Grimshaw D J (1999) *Bringing Geographical Information Systems into Business*. 2nd edn. Longman, London

Harder C (1997) *ArcView GIS means Business: Geographic Information System Solutions for Business*. Esri Press, Redlands

Longley P A, Goodchild M F, Maguire D J, Rhind D W (eds) (1999) *Geographical Information Systems: Principles, Techniques, Management and Applications*. John Wiley, New York

Longley P A, Goodchild M R, Maguire D J, Rhind D W (eds) (2005) *Geographic Information Systems: Principles, Techniques, Management and Applications.* 2nd edn. Abridged version. John Wiley, New York

Longley P A, Goodchild M F, Maguire D J, Rhind D W (2010) *Geographic Information Systems and Science.* 3rd edn. John Wiley, Chichester

Martin D (1995) *Geographical Information Systems and their Socio-economic Applications.* 2nd edn. Routledge, London

Meehan B (2007) *Empowering Electric and Gas Utilities with GIS.* Esri Press, Redlands

Mitchell A (1997) *Zeroing In: Geographic Information Systems at Work in the Community.* Esri Press, Redlands

Pickles J (ed) (1995) *Ground Truth: The Social Implications of Geographic Information Systems.* Guilford Press, New York

## WEB LINKS

### Introduction to GIS:

- GIS.com: www.gis.com/content/what-gis

- The GIS Lounge: gislounge.com/what-is-gis/

- GIS Development online tutorial: www.gisdevelopment.net/tutorials/index.htm

- Ordnance Survey GIS Files: www.ordnancesurvey.co.uk/oswebsite/gisfiles/

- USGS GIS Poster: erg.usgs.gov/isb/pubs/gis_poster/

### GIS dictionaries and glossaries:

- Esri GIS Dictionary: support.esri.com/index.cfm?fa=knowledgebase.gisDictionary.gateway

- University of California Berkeley Dictionary of abbreviations and acronyms: www.lib.berkeley.edu/EART/abbrev.html

### Case study related materials:

- Nuclear Waste World Network: www.nuclearwaste.com/

- Radwaste: www.radwaste.org/

- NDA: www.NDA.co.uk

- Sun Valley Idaho, example mountain resort: www.sunvalley.com/

- Upmystreet local area information: upmystreet.com

Visit our website **www.pearsoned.co.uk/heywood** for further web links, together with self-assessment questions, activities, data sets for practice opportunities and other interactive resources.

# CHAPTER 2

# Spatial data

## Learning outcomes

By the end of this chapter you should be able to:

- Explain the difference between data and information

- Describe the main characteristics of spatial data

- Give examples of map projections and explain why they are important

- Provide details of different methods of spatial referencing

- Define topology

- Explain the thematic characteristics of spatial data

- List the main sources of spatial data

- Explain why data standards are an important issue in GIS

# ■ INTRODUCTION

All Geographical Information Systems are computer representations of some aspect of the real world. It would be impossible to represent all the features in which you are interested in a computer, so GIS present a simplified view of the world. For example, if you are interested in transport networks in order to find shortest routes for pizza delivery it might be impossible or impractical to input every detail of a road network (such as surface type and condition, signage, road markings, verge type). A simplified view, perhaps including road category, speed limit and known congestion points might be all that is necessary. Being able to develop and understand a simplified view of the world is an important first step for your own GIS projects.

The case studies introduced in **Chapter 1** also illustrate the use of a simplified view of the world. The view used in each case contains only the data the GIS designer considers necessary to solve a particular problem. Thus, the GIS for Zdarske Vrchy did not contain data on the location of houses for sale, nor did the house-hunting GIS contain data on the distribution of flora and fauna. The simplified view of the real world adopted by GIS is often termed a 'model'. A model is 'a synthesis of data' (Haggett and Chorley, 1967) which is used as a 'means of "getting to grips" with systems whose spatial scale or complexity might otherwise put them beyond our mental grasp' (Hardisty *et al.*, 1993). GIS is used to help build models where it would be impossible to synthesize the data by any other means. Models also contain our ideas about how or why elements of the real world interact in a particular way. Therefore, a GIS populated with data and ideas about how these data interact is a spatial model. Haggett and Chorley (1967) point out that a spatial model places emphasis on reasoning about the real world by means of translation in space. This is exactly the reason why GIS is used to solve geographical problems. However, in order to model the real world in a GIS, it has to be reduced to a series of abstract features or basic spatial entities (such as points, lines and areas, see Box 2.2) and this is not without problems. The reductionist nature of GIS has been severely criticized in some circles as being too simplistic for modelling complex human

systems (Pickles, 1995). This issue is covered in more detail in **Chapter 10**.

Before looking at how spatial models are constructed using a GIS it is necessary to consider the character of the spatial data they use as their raw material. In the context of our car analogy, this is where we try to understand the fuel a car requires to make it run. First, however, it is necessary to review our understanding of the term 'data' and take a closer look at the distinction between data and information.

Data are observations we make from monitoring the real world. Data are collected as facts or evidence that may be processed to give them meaning and turn them into information. There is a clear distinction between data and information, although the two terms are often used interchangeably. To help appreciate the distinction it is perhaps easiest to think of data as raw numbers, such as those you might see listed in a table. Field notes made by the Happy Valley ski patrol containing snow depth measurements, the printout from the Happy Valley automatic weather station or a table of responses from a survey of skiers are examples. All you see are numbers that have no particular meaning. To make the numbers useful you need to add context. For somebody else to interpret your tables or lists of figures they would need to know to what the data refer and which scale or unit of measurement has been used for recording the data. With these details the data become information. Information is data with meaning and context added (Hanold, 1972).

There are a wide variety of data sources, though all data fall into one of two categories: primary or secondary. Counts of skiers using a particular ski run are an example of primary data collected through first-hand observation. Secondary data will have been collected by another individual or organization: for example, consumer surveys of customers buying ski equipment. Many secondary data sources are published and include maps, population census details and meteorological data (Griffith and Amrhein, 1991). Box 2.1 provides examples of primary and secondary data sources used in Happy Valley.

All primary and secondary data have three modes or dimensions: temporal, thematic and spatial. For all data it should be possible to identify each of these three modes. For example, for data about an

## BOX 2.1  Sources of primary and secondary data used in Happy Valley

### PRIMARY DATA SOURCES

■ *Daily snow pack data* collected by the Happy Valley ski patrols. These data are used to help make decisions about which runs to open and which to close.

■ *Number of lift passes* purchased each day. These data are used to monitor the demand for skiing on different days of the week.

■ *The number of skiers* using a specific lift on a particular day. The automatic turnstiles at the entry points to all the lifts collect these data. They are used to monitor lift usage.

■ *The number of avalanches* recorded by the ski patrols. These data are used to help predict avalanche risk.

(a) Collection of snow pack data

(b) Ski lift pass

(c) Skiers on lift/tow

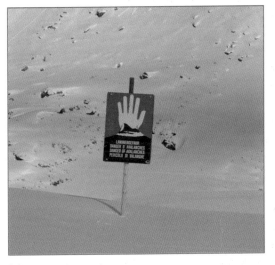

(d) Avalanche track

**Figure 2.1**  Primary data sources for Happy Valley

## BOX 2.1

### SECONDARY DATA SOURCES

- Published *meteorological maps* for the Happy Valley area. These data are used to assist with avalanche forecasting.
- *Local topographic maps.* Back-country ski trail maps are prepared using local topographic maps.

- National and regional *lifestyle* data derived from market research surveys are used to estimate the demand for skiing and target the marketing of Happy Valley.
- Datasets and activities relating to the Happy Valley Case Study can be found online at **www.pearsoned. co.uk/heywood.**

**(a)** Published meteorological maps

**(c)** Lifestyle/geodemographic map

**(b)** Piste map

**Figure 2.2** Secondary data sources for Happy Valley
*Sources*: (a) Published by the Met Office, (top) © Crown copyright 2006, (bottom) © EUMETSAT / Met Office 2006, used with permission; (b) Alamy / Dominic Burke

avalanche incident that took place in Three Pines Valley on 14 February 2002, the three modes are:

- *temporal* – 15:30 hrs 14 February 2002;
- *thematic* – wet slab avalanche triggered by two off-piste skiers; and
- *spatial* – Three Pines Valley, south-facing slope.

The temporal dimension provides a record of when the data were collected and the thematic dimension describes the character of the real-world feature to which the data refer. Additional thematic data for the avalanche incident might relate to the size and consequences of the avalanche. In GIS the thematic data are often referred to as non-spatial or attribute data. These are illustrated in the avalanche incident report map in Figure 2.3.

The spatial dimension of data can be regarded as the values, character strings or symbols that convey to the user information about the location of the feature being observed. In the case of the avalanche on 14 February 2002 we know that the incident occurred on a south-facing slope in Three Pines Valley. In this case, the spatial reference used is a textual description that would only be of use

to those who are familiar with the area. However, because GIS have no 'local knowledge', all spatial data used in GIS must be given a mathematical spatial reference. One of the most common is a map co-ordinate. Here, a co-ordinate pair (x,y) is used to locate the position of a feature on a uniform grid placed on a map. Spatial referencing is considered in more detail later in this chapter.

It is common to find the term 'temporal data' used to describe data organized and analysed according to time, thematic data used for data organized and analysed by theme, and spatial data for data organized and analysed by location. However, even though one dimension may be used to organize data, the other dimensions will still be present.

GIS places great emphasis on the use of the spatial dimension for turning data into information, which, in turn, assists our understanding of geographic phenomena. Therefore, we consider next the characteristics of spatial data in detail and examine how the map metaphor has shaped these characteristics. This is followed by a review of the thematic dimension of spatial data and discussion of a range of sources of spatial data, including surveys, aerial photographs, satellite images and field data sources.

## MAPS AND THEIR INFLUENCE ON THE CHARACTER OF SPATIAL DATA

The traditional method for storing, analysing and presenting spatial data is the map. The map is of fundamental importance in GIS as a source of data, a structure for storing data and a device for analysis and display. Perhaps more important, maps have shaped the way most of us think about space in two dimensions. Therefore, understanding maps and how they are produced is an essential starting point for exploring the characteristics of spatial data.

Maps take many different forms and come with a range of different scales. Examples vary from simple sketch maps, such as those used to show colleagues and friends how to get to a party, to the more complex topographic and thematic maps that can be found in national atlases.

It is common to make a distinction between thematic and topographic maps. Thematic maps show data relating to a particular theme or topic, such as

**Happy Valley Avalanche Incident Report**

Time/date: 15:30hrs 14th February 2002
Detail: Wet slab triggered by two off-piste skiers
Location: Three Pines Valley, south-facing slope
Avalanche extent shown in red

**Figure 2.3** Avalanche incident report, 14 February 2002

soil, geology, geomorphology, land use, population or transport. Topographic maps contain a diverse set of data on different themes. Thus, land use, relief and cultural features may all appear on the same topographic map. Unwin (1981) argues that the topographic map is simply a 'composite of many different kinds of maps'. Topographic and thematic maps are shown in Figure 2.4.

Even though there are many different types of maps, the mapping process is of a general nature. During this process the cartographer must (after Robinson *et al.*, 1995):

- establish the *purpose* the map is to serve;
- define the *scale* at which the map is to be produced;
- select the *features* (spatial entities) from the real world which must be portrayed on the map;
- choose a method for the *representation* of these features (points, lines and areas);
- *generalize* these features for representation in two dimensions;
- adopt a *map projection* for placing these features onto a flat piece of paper;
- apply a *spatial referencing system* to locate these features relative to each other; and
- *annotate* the map with keys, legends and text to facilitate use of the map.

When a cartographer produces a map, an underlying geometric structure is created which allows the user to describe the relationships between features. It will be clear, for instance, that islands lie within lakes, that fields are adjacent and that if you follow a particular road you will reach a certain destination. This structure, known as *topology*, is based on the geometric relationships of objects (see **Chapter 3**). Following the map-making process outlined above helps the cartographer shape the character of the final map. The key characteristics which result from this process are described in more detail below.

## Purpose

All maps, and other sources of spatial data, are generated with a purpose in mind. In most cases that purpose is to turn data into information that will be

**(a)** Topographic map

**(b)** Thematic map

**Figure 2.4** Examples of (a) Topographic map and (b) Thematic map
*Sources*: (a) Corbis/NASA; (b) Courtesy of Torbay Council

communicated to a third party. Every year the managers of Happy Valley produce a map of the ski area for use by visitors. The map shows the location of ski trails, car parks, hotels, emergency shelters and ski lifts. Its purpose is to help visitors orient themselves and decide how to spend their time. Naturally, such a map can have a strong influence over the user. For example, visitors are unlikely to dine at a restaurant if they cannot find it on the map. However, there are restaurants in the area that are not shown on the official Happy Valley map. The ski company wishes to encourage visitors to use its facilities, not those owned by competitors. This simple example illustrates how purpose can influence the character and quality of a spatial data set. Clearly, you would not use this map on its own if you were trying to compile a data set of the restaurants in Happy Valley. However, you may not know that the map is incomplete.

In some cases maps have a single purpose. The propaganda maps produced by the Allies during the Second World War were designed to convince the general public that the war effort was going well (Monmonier, 1996). The true geography of Europe was distorted to emphasize the area occupied by the allied forces. This was effective in boosting the morale of the Allies, but produced maps of limited use in other circumstances. Similar maps were produced during the Cold War to exaggerate the threat from the USSR (see Figure 2.5). Other maps, such as the topographic maps produced by national mapping agencies, aim to meet the needs of a wide range of users, ranging from utility companies to outdoor enthusiasts. In this case, the maps will be more geographically accurate than the propaganda maps described above, but they will still contain generalized data to enable them to be of wide generic use. These generalizations will limit their use for certain applications. A utility company is unlikely to use national maps, on their own, to plan the detailed installation of a new set of electricity cables, because these maps do not contain details about the location of the existing cable network.

Whilst not strictly a spatial characteristic in its own right, the purpose for which a spatial data set has been created will influence the quality and spatial detail provided by the data set. An appreciation of the purpose behind the production of a data set

**Figure 2.5** Cold War propaganda map
*Source*: Topfoto / Roger-Viollet

is, therefore, an essential prerequisite for judging whether or not the data are appropriate for use in a particular situation. Linked closely to purpose is the idea of scale.

## Scale

Virtually all sources of spatial data, including maps, are smaller than the reality they represent (Monmonier, 1996; Keates, 1982). Scale gives an indication of how much smaller than reality a map is. Scale can be defined as the ratio of a distance on the map to the corresponding distance on the ground (Martin, 1995). An alternative definition is offered by Laurini and Thompson (1992) as the order of magnitude or level of generalization at which phenomena exist or are perceived or observed. Scale can be expressed in one of three ways: as a ratio scale, a verbal scale or a graphical scale (Figure 2.6).

| Ratio | 1:5000 | 1:1,000,000 |
|---|---|---|
| Verbal (nominal) | 1 cm represents 50 m | 1 cm represents 10 km |
| Graphical | 0   100   200 km | 0  10 20 30 40 km |

**Figure 2.6** Expressions of scale

Examples of ratio scales are 1:5000 and 1:5,000,000. At a scale of 1:5000 a 1 mm line on the map represents a 5000 mm line on the ground. In the same fashion a line of 1 m on the map represents a line of 5000 m on the ground; the units do not matter as long as they are the same. A verbal scale would express the scale in words, for example '1 cm represents 50 m'. Finally, a graphic scale (or scale bar) is usually drawn on the map to illustrate the distances represented visually. Graphic scales are frequently used on computer maps. They are useful where changes to the scale are implemented quickly and interactively by the user. In such cases, recalculating scale could be time-consuming, and the ratios produced (which may not be whole numbers) may be difficult to interpret. Redrawing a graphic scale in proportion to the map is relatively straightforward and simple to understand. It is often possible in GIS to specify the scale at which you require your maps using a ratio representation.

Standard topographic maps contain examples of verbal, ratio and graphical scales. It should be remembered that small-scale maps (for example, 1:250,000 or 1:1,000,000) are those that cover large areas. Conversely, large-scale maps (for example,

1:10,000 or 1:25,000) cover small areas and contain large amounts of detail. With some data used in GIS, such as aerial photographs or satellite imagery, the scale is not immediately obvious and may have to be calculated by the user. Scale is also important when using spatial entities (points, lines and areas) to represent generalized two-dimensional versions of real-world features.

## Spatial entities

Traditionally, maps have used symbols to represent real-world features. Examination of a map will reveal three basic symbol types: points, lines and areas (Monmonier, 1996). These were introduced in **Chapter 1** (Figure 1.17) and are the basic spatial entities. Each is a simple two-dimensional model that can be used to represent a feature in the real world. These simple models have been developed by cartographers to allow them to portray three-dimensional features in two dimensions on a piece of paper (Laurini and Thompson, 1992; Martin, 1995). Box 2.2 provides more details on the types of features that points, lines and areas can be used to represent.

---

## BOX 2.2 Basic spatial entities

THEORY

### POINTS

Points are used to represent features that are too small to be represented as areas at the scale of mapping being used. Examples are a postbox, a tree or a lamp post (see Figure 2.7). The data stored for a postbox will include geographic location and details of what the feature is. Latitude and longitude, or a co-ordinate reference, could be given together with details that explain that this is a postbox in current use. Of course, features that are represented by points are not fully described by a two-dimensional geographical reference. There is always a height component since the postbox is located at some height above sea level. If three dimensions are important to a GIS application this may also be recorded, usually by adding a $z$ value representing height to give an $(x,y,z)$ co-ordinate.

### LINES

Lines are used to represent features that are linear in nature: for example, roads, powerlines or rivers (see Figure 2.8). It can be difficult for a GIS user to decide when a feature should be represented by a line. Should a road be represented by a single line along its centre, or are two lines required, one for each side of the road?

A line is simply an ordered set of points. It is a string of $(x,y)$ co-ordinates joined together in order and usually connected with straight lines. Lines may be isolated, such as geological fault lines, or connected together in networks, such as road, pipeline or river networks. Networks are sometimes regarded as a separate data type but are really an extension of the line type. More will be said about networks and their analysis in later chapters. Like

BOX 2.2

**(a)** Postbox                    **(b)** Tree                    **(c)** Lamp post

**Figure 2.7** Real-world objects commonly stored as a point

**(a)** Road                    **(b)** River                    **(c)** Power line

**Figure 2.8** Real-world objects stored as lines

points, lines are in reality three-dimensional. For instance, a hydro-geologist may be interested in underground as well as surface drainage. Adding a *z* co-ordinate (representing depth or height) to the points making up the line representing a stream allows an accurate three-dimensional representation of the feature.

## AREAS

Areas are represented by a closed set of lines and are used to define features such as fields, buildings or lakes (see Figure 2.9). Area entities are often referred to as 'polygons'. As with line features,

some of these polygons exist on the ground, whilst others are imaginary. They are often used to represent area features that do not exist as physical features, such as school catchment zones or administrative areas.

Two types of polygons can be identified: island polygons and adjacent polygons. Island polygons occur in a variety of situations, not just in the case of real islands. For example, a woodland area may appear as an island within a field, or an industrial estate as an island within the boundary of an urban area. A special type of island polygon, often referred to as a 'nested' polygon, is created

BOX 2.2

**(a)** Field

**(b)** Building

**(c)** Lake

**Figure 2.9** Real-world objects commonly represented as an area

by contour lines. If you imagine a small conical hill represented by contour lines, this will be represented in polygon form as a set of concentric rings. Adjacent polygons are more common. Here, boundaries are shared between adjacent areas. Examples include fields, postcode areas and property boundaries.

A three-dimensional area is a surface. Surfaces can be used to represent topography or non-topographical variables such as pollutant levels or population densities. Some authors (for example Laurini and Thompson, 1992; Martin, 1995) consider surfaces to be a separate fourth entity type. This issue will be considered in more detail in **Chapter 3**.

The representation of real-world features using the point, line and area entity types appears relatively straightforward. However, the method chosen to represent a spatial feature will depend on the scale used. Consider the way cities are represented on maps of different scales. On a world map a point would be the most appropriate method of representation, given the number of cities to be included. However, at national and regional scales a point could provide an oversimplified view of the extent of the geographical area covered by a city. A point used here would tell us nothing about the relative size of

**(a)** OS 'Route' map 1:625,000

**(b)** OS 'Travel' map 1:250,000

**(c)** OS 'Tour' map 1:250,000

**(d)** OS 'Landranger' map 1:50,000

**(e)** OS 'Explorer' map 1:25,000

**(f)** OS 'Landplan' map 1:10,000

**(g)** OS 'Superplan' map 1:200 to 1:10,000

**Figure 2.10** Representing a city at different map scales
*Source*: Reproduced by permission of Ordnance Survey on behalf of HMSO.
© Crown Copyright 2011. All rights reserved. Ordnance Survey Licence number 100030901

cities, so it is more likely that the cartographer would choose to represent the cities using areas. At the local scale even the area spatial entity may be considered too simplistic and the cartographer may choose to build up a representation of the city using a mixture of point, line and area entities. Points may be used for the representation of features such as telephone boxes, areas for residential blocks and parks, and lines for road networks. This is illustrated in Figure 2.10 for the city of London and shows how choosing the appropriate entity to represent real world features is often surprisingly difficult.

## Generalization

All spatial data are a generalization or simplification of real-world features. In some instances generalization is needed because data are required at a particular scale. In other cases generalization is introduced by the limitations of the technical procedures used to produce data. The grain size of photographic film, or the resolution of a remote sensing device, will determine the level of detail discernible in the resulting air photo or satellite image. Generalization may also be introduced directly by human intervention in order to improve the clarity of an image or to enhance its major theme.

All the data sources used in GIS — aerial photographs, satellite images, census data and particularly maps — contain inherent generalizations. This simplification of detail is necessary in order to maintain clarity. If a cartographer wishes to depict the course of a river on a map, decisions need to be taken regarding the amount of detail to be included. These decisions are largely governed by the scale of the map — the level of detail shown is proportional to scale. At very small scales (for example, 1:20,000,000), a large river like the Mississippi in the USA may appear as a single blue line. This line will show the approximate course of the river from its source in the northern state of Minnesota to its exit into the Gulf of Mexico. All bends and turns in the

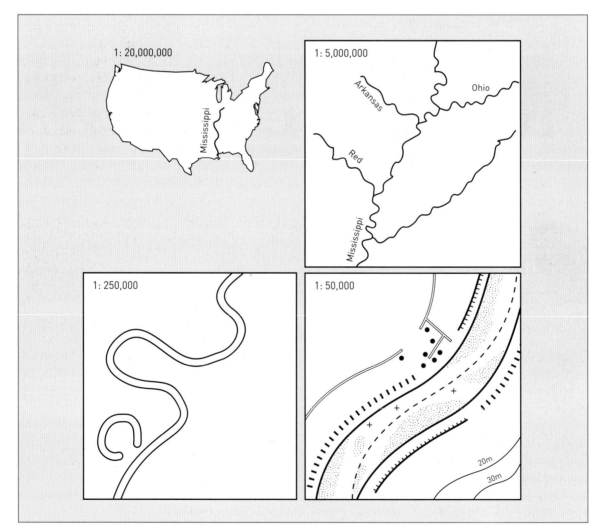

**Figure 2.11** Scale-related generalization

river will have been smoothed to create a simple, easy to understand map. At larger scales (for example, 1:5,000,000) it is possible to show something of the meandering nature of this great river and its network of tributaries: the Red, Arkansas, Missouri and Ohio. At even larger scales (for example, 1:250,000) it becomes possible to indicate width, river banks, small bends and meander cut-offs. At larger scales still (for example, 1:50,000) it may be possible to indicate depth and the positions of sandbanks and shoals that might be important for navigation purposes. As the scale increases, the cartographer has greater scope for including more detail. The relationship between scale and detail is referred to as *scale-related generalization* and is illustrated in Figure 2.11.

Decisions regarding what features to include on the final map and which to leave out also need to be made by the cartographer. If the cartographer were to include every single tributary of the Mississippi river network on the 1:20,000,000 scale map, the map would be covered by dense blue line work and impossible to read. For the sake of clarity, the cartographer has to be selective about drawing map features.

Another problem facing the cartographer is how to depict features in proportion to their size on the ground. If a river is drawn as a line 0.5 mm thick on a 1:1,000,000 scale map, this would imply that the river is 500 m wide. In reality it may be only 50 m wide. If the width of line features drawn on the map were determined rigidly by the map scale, then most features on small-scale maps could not be seen with the naked eye. For example, a 50 m wide river on a 1:1,000,000 scale map would have to be drawn using a line only 0.05 mm wide. Similarly, if a road running along the banks of the river were to be depicted accurately in this fashion, it would need to be drawn on top of the river on the map. In order to make the road distinguishable from the river, the cartographer has to displace the road to leave a gap between it and the river. To cope with these and other problems relating to the necessary generalization of map features, cartographers have adopted a broad code of practice relating to selection, simplification, displacement and smoothing. This is summarized in Box 2.3. Remember that most maps are just communication devices; a way of storing geographical information on paper and passing this information on to others. Maps have a long history, going back as far as the ancient civilizations and most were never originally intended to be used as a data source for GIS. A good understanding of the processes of cartographic generalization is therefore important if data from paper maps and other spatial data sources are to be used effectively within GIS.

# BOX 2.3 Cartographic generalization: code of practice

**PRACTICE**

1 *Selection*. First, the map feature for generalization is selected. If more than one source is available to the cartographer this may involve choosing the most appropriate representation of the feature or a blending of the two.

2 *Simplification*. Next, a decision will be taken to simplify the feature. For the example of the river this may involve the removal of some minor bends. The aim of generalization will usually be to simplify the image but maintain the overall trend and impression of the feature.

3 *Displacement*. If there are features that are located side by side in the real world, or that lie on top of one another, the cartographer may choose to displace them by a small degree so that they are both visible on the map image. This may have the effect of displacing a feature several hundred metres depending on the map scale used.

4 *Smoothing and enhancement*. If the source data from which a cartographer is working are very angular, because they have been collected from a series of sampling points, a smoothing technique may be used to apply shape and form to the feature. This will give a better representation.

*Source*: Adapted from Robinson *et al.* (1995)

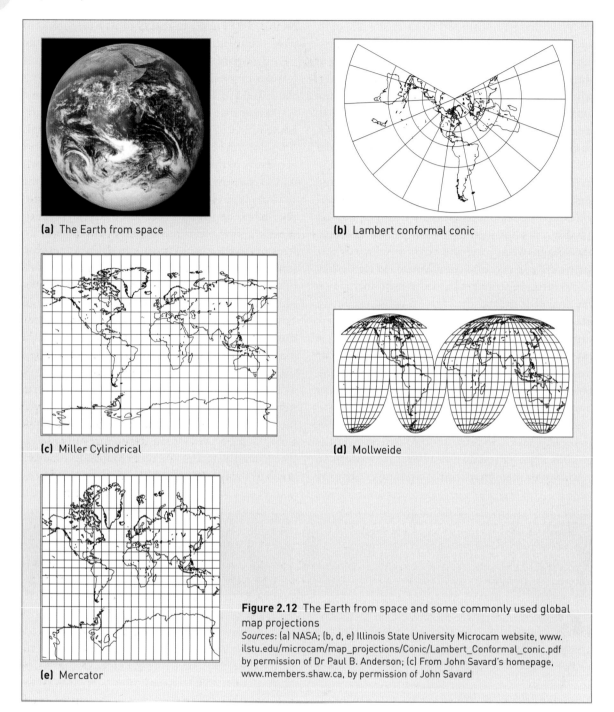

**(a)** The Earth from space

**(b)** Lambert conformal conic

**(c)** Miller Cylindrical

**(d)** Mollweide

**(e)** Mercator

**Figure 2.12** The Earth from space and some commonly used global map projections
*Sources*: (a) NASA; (b, d, e) Illinois State University Microcam website, www.ilstu.edu/microcam/map_projections/Conic/Lambert_Conformal_conic.pdf by permission of Dr Paul B. Anderson; (c) From John Savard's homepage, www.members.shaw.ca, by permission of John Savard

## Projections

For the GIS analyst to make use of simple spatial entities (points, lines and areas) it is necessary to locate them in two dimensions. The analyst, like the cartographer, must treat the world as a flat surface to achieve this. Whilst this is a gross generalization, for most purposes it works well. Moreover, current technology provides us with no other realistic

choice (though developments in three-dimensional modelling and virtual reality are changing this). The method by which the 'world is laid flat' is to use a map projection.

Map projections transfer the spherical Earth onto a two-dimensional surface. In doing so they approximate the true shape of the Earth. Figure 2.12 shows the Earth as seen from space and some commonly used projections depicting the surface of the Earth. The pattern of land masses and oceans are laid out in a familiar way. The process of transferring the spherical Earth onto a two-dimensional surface introduces errors into spatial data, the character of which will vary depending on the projection method chosen. Some projections will cause distance between spatial entities to be preserved whilst direction is distorted. In other cases, shape may be preserved at the expense of accurate area estimates. Figure 2.12 illustrates how different parts of the Earth are distorted to enable a fit onto a flat sheet of paper. One way to visualize the problem of representing a spherical world in two dimensions is to imagine a plastic beach ball overprinted with a map of the world showing lines of latitude and longitude. The inflated ball is a globe, with the countries in their correct locations, and shown as area entities with correct relative shapes and sizes. Imagine that you have to deflate the ball and lay it flat on a table whilst still displaying all the countries. The only way to do this is to cut the beach ball into pieces. In doing this you would find that the distances between countries will be altered and their shape distorted. The principle is the same with map projections.

If you imagine the beach ball has a hole at the 'north pole' large enough for a light bulb to be inserted, it is transformed into a light fitting. When the light is switched on, an image of the surface is projected onto the walls of the room. Careful examination of the images on the walls reveals that the centre of the image reflects the globe most accurately. It is on this simple concept that the whole range of map projections is based.

Today there are a wide range of map projections in use, and there were even more used in the past. Different map projections are used in different parts of the world for mapping different sized areas and for different applications. Think again of the globe as a light fitting. The picture of the Earth from our 'globe-light' will vary depending upon the shape of the room in which the light is placed.

In a circular room, assuming our globe is hanging from the north pole, there will be a continuous picture of the Earth. Countries nearest the equator will appear in their true relative geographical positions. The equator is the line of latitude nearest the wall and so represents the *line of true scale*, along which distances (and consequently the map scale) are not

distorted. However, at the top or bottom of the wall the location of the countries is distorted, with the distance between countries increased. Our view of the poles will be very distorted, or missing altogether. In fact, if the poles are included on our projected map, the points representing the north and south poles become so distorted as to be projected as a straight line equal in length to that of the equator. Of course, we also need to 'cut open' the image projected onto the walls so we can unroll it and lay it flat and this introduces the problem of discontinuity. When we 'step off' the map at its vertical edges we will reappear at the same latitude on the opposite edge.

In a square room with flat walls, only a part of the Earth's surface will be visible on any one wall. This will depend on the position of the light. The view will be of half the globe or less. Distortion will be similar to that in the circular room except that, since the single wall is straight and not curved, the image of the world will be distorted at all four edges of the wall and not just the top and bottom.

If the room is shaped like a tepee (with circular walls tapering towards a point at the apex of the structure) the line of true scale is no longer the equator, as in the circular room, but some line of latitude nearer the north pole that lies at a tangent to the sloping walls of the tepee. The exact line of latitude will depend on the angle of the tepee walls: the steeper the walls, the lower the latitude (nearer the equator); the shallower the walls, the higher the latitude (nearer the pole). Not all projections of this kind have the north pole uppermost; projections can have the south pole uppermost or may have the apex of the tepee centred over a different point altogether.

The circular room is equivalent to the family of *cylindrical* projections (which includes the Mercator projection), where the surface of the Earth is projected onto a cylinder that encompasses the globe (Figure 2.13a). This projection is very suitable for making maps of an area that have only a small extent in longitude. It has been chosen as the basic projection for use by the Ordnance Survey to map the UK. The transverse Mercator projection has the advantage of maintaining scale, shape, area and bearings for small areas. This explains why it has become a popular projection for mapping small areas of the globe. The single wall illustrates the *azimuthal* family of projections (Figure 2.13b) and the tepee is equivalent to the *conic* family (Figure 2.13c).

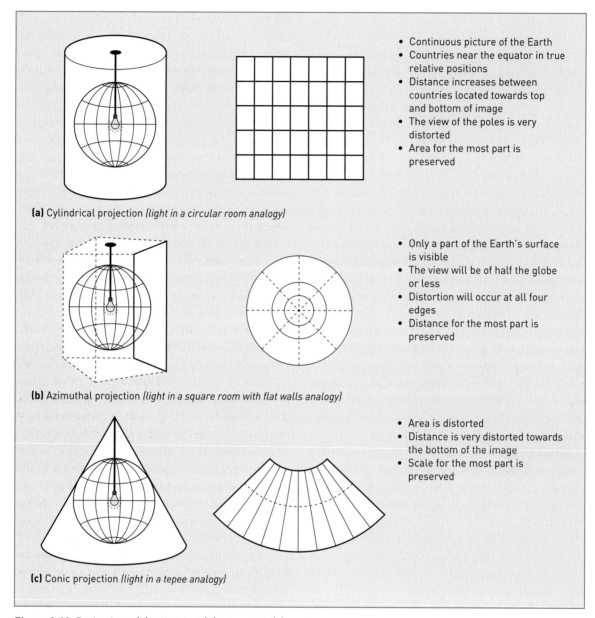

- Continuous picture of the Earth
- Countries near the equator in true relative positions
- Distance increases between countries located towards top and bottom of image
- The view of the poles is very distorted
- Area for the most part is preserved

(a) Cylindrical projection *(light in a circular room analogy)*

- Only a part of the Earth's surface is visible
- The view will be of half the globe or less
- Distortion will occur at all four edges
- Distance for the most part is preserved

(b) Azimuthal projection *(light in a square room with flat walls analogy)*

- Area is distorted
- Distance is very distorted towards the bottom of the image
- Scale for the most part is preserved

(c) Conic projection *(light in a tepee analogy)*

**Figure 2.13** Projections: (a) cylindrical;(b) azimuthal; (c) conic

Many of the map-based spatial data sources used in GIS have a projection associated with them. To undertake meaningful analysis it is necessary to know something about the projections being used. The results of analyses will be affected in different ways by different map projections. If a GIS application requires the accurate calculation of areas, then using a projection that distorts areas is obviously not suitable. When using data at large scales (covering small areas) the effects may be slight, but at small scales (covering large areas) the effects can be substantial. Finally, since one of the functions of a GIS application is to allow the integration of data from different sources, the ability to alter projections is a fundamental ability of many GIS. There are hundreds of different map projections and some GIS seem to offer the capability to re-project data for most of these. Only the most common projections have been considered above.

## Spatial referencing

A referencing system is used to locate a feature on the Earth's surface or a two-dimensional representation of this surface such as a map. There are a number of characteristics that a referencing system should have. These include stability, the ability to show points, lines and areas, and the ability to measure length, size (area) and shape (Dale and McLaughlin, 1988). Several methods of spatial referencing exist, all of which can be grouped into three categories:

- geographic co-ordinate systems;
- rectangular co-ordinate systems; and
- non-co-ordinate systems.

The only true *geographic co-ordinates* are latitude and longitude. The location of any point on the Earth's surface can be defined by a reference using latitude and longitude. Lines of longitude (also known as *meridians*) start at one pole and radiate outwards until they converge at the opposite pole (Figure 2.14). Conceptually they can be thought of as semi-circles. If you slice a globe along two opposing lines of longitude you will always cut the globe in half. The arbitrary choice for a central line of longitude is that which runs through the Royal Observatory in Greenwich in England, and is hence known as the *Greenwich meridian* or the *prime meridian*. Lines of longitude are widest apart at the equator and closest together at the poles. The relative distance between lines of longitude where they intersect lines of latitude (or parallels) is always equal. However, the real distance will vary depending on the line of latitude that is intersected. For example, the distance between the lines of longitude intersecting the same parallel will increase towards the equator, with the maximum distance existing at the equator itself.

Lines of latitude lie at right angles to lines of longitude and run parallel to one another. Each line of latitude represents a circle running round the globe. Each circle will have a different circumference and area depending on where it lies relative to the two poles. The circle with the greatest circumference is the equator (or central parallel) and lies equidistant from the two poles. At the two poles the lines of latitude are represented by a single point – the pole.

Using lines of latitude and longitude any point on the Earth's surface can be located by a reference given in degrees and minutes. For example, the city of Moscow represented as a point can be given a geographical co-ordinate reference using latitude and longitude of 55 degrees 45 minutes north and 36 degrees 0 minutes east (55° 45'N 36° 0'E). The first set of numbers, 55° 45'N, represents latitude. The N informs us that Moscow can be found north of the equator. The second set of numbers, 36° 0'E, tells us that Moscow lies to the east of the prime meridian. Therefore, the N and E together give the quarter of the globe in which Moscow is located (Figure 2.15a). The line of latitude on which Moscow lies is given by the degrees and minutes of this latitude away from the equator (Figure 2.15b). Finally, the line of longitude on which Moscow lies must be identified. Figure 2.15c shows how this angle is calculated based on relative distance from the prime meridian. Adopting this approach, all features on the surface of the Earth can be located relative to one another and the distance between them calculated. The shortest distance between two points on the Earth's surface is known as the *great circle distance*.

The latitude and longitude referencing system assumes that the Earth is a perfect sphere. Unfortunately this is not correct. The Earth is actually an oblate spheroid somewhat like an orange with flatter poles and outward bulges in equatorial regions. To complicate matters further, the surface of the Earth is far from smooth and regular, as you

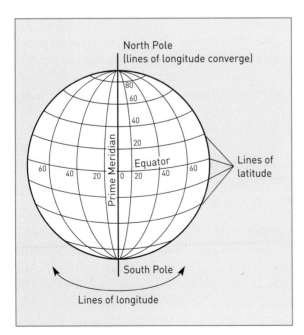

**Figure 2.14** Latitude and longitude

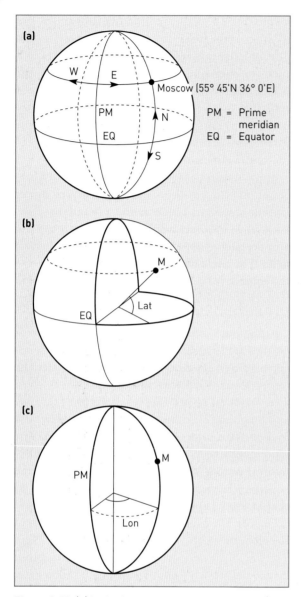

**Figure 2.15** (a) Latitude and longitude of Moscow; (b) calculating the latitude; (c) calculating the longitude

on the Earth's surface. The individual triangles are also of the same size and shape. The flexible nature of a triangular mesh means that it can be moulded to fit the slight bumps and blemishes that form the true surface of the Earth. The use of triangles to model the surface of the Earth will be examined in more detail in **Chapter 3** as it is an important concept in GIS.

At present, most of the spatial data available for use in GIS exist in two-dimensional form. In order to make use of these data a referencing system that uses *rectangular co-ordinates* is required. To obtain these, a map graticule, or grid, is placed on top of the map. This graticule is obtained by projecting the lines of latitude or longitude from our representation of the world as a globe onto a flat surface using a map projection. The lines of latitude and longitude become the grid lines on a flat map. As already pointed out, the problem is that when you project from a sphere onto a flat surface the image becomes distorted. When small areas are being studied, there will be only minor distortions in the layout of the grid. However, when large areas of the globe are projected onto a flat surface, the grid will tear and stretch. Therefore, all rectangular co-ordinate systems are designed to allow the mapping of specific geographical regions. A good example of a rectangular co-ordinate system is the UK Ordnance Survey's National Grid (Box 2.4). Another example is the Universal Transverse Mercator (UTM) plane grid system. This system uses the transverse Mercator projection and divides the Earth into 60 vertical zones that are 6 degrees of longitude wide, avoiding the poles. The system has been adopted by many organizations for remote sensing, topographic mapping and natural resource inventory (DeMers, 2005).

*Non-co-ordinate* systems provide spatial references using a descriptive code rather than a co-ordinate. Postal codes, widely used throughout the world, are an example. Some postal codes are fully numeric, such as the American ZIP codes, whilst others are alphanumeric, as in the case of the UK postcode. All have the same basic purpose: to increase the efficiency of mail sorting and delivery rather than to be an effective spatial referencing system for GIS users. There are several advantages to such systems: they are important to the postal services and are therefore maintained and updated, and they offer

will appreciate if you have visited any mountainous areas. At small scales these minor blemishes and imperfections in shape can be ignored. However, when dealing with large-scale maps of a small portion of the Earth's surface it is essential to make local corrections for these factors.

The Quaternary Triangular Mesh referencing system (Goodchild and Yang, 1989) tries to deal with irregularities in the Earth's surface. It replaces lines of latitude and longitude with a mesh of regular-shaped triangles. The advantage of this referencing system is that each triangle occupies the same area

## BOX 2.4  Ordnance Survey National Grid system

**PRACTICE**

The Ordnance Survey National Grid is a rectangular grid system based on the transverse Mercator projection (Figure 2.16). The grid is 700 × 1300 km covering all of Great Britain from the Scilly Isles to Shetland. This is divided into 500 km squares, which are then divided into twenty-five 100 km squares. Each 100 km square is identified by two letters. The first refers to the 500 km square and the second to the

**Figure 2.16** The Ordnance Survey National Grid system
*Source*: Adapted from Harley (1975)

100 km square. Each 100 km square is further divided into one hundred 10 km squares (10 km × 10 km), and the 10 km squares are divided into one hundred 1 km squares (1 km × 1 km). Grid references are commonly given as six figures prefixed by the letters denoting the 100 km square. An example could be SE 366 923. Here, the 'SE' denotes the 100 km square that has its origin 400 km east and 400 km north of the origin of the grid. The '366' and the '923' are the easting and northing recorded to the nearest 100 m.

coverage of all areas where people reside and work. Providing that individual codes do not refer to single addresses, they also provide a degree of confidentiality for data released using this as a referencing system. Box 2.5 provides more details on the UK postcode system.

In the western United States another non-co-ordinate referencing system is often used. This is known as the Public Land Survey System (PLSS). Here, there has been a recursive sub-division of the land into quarter sections. By knowing which section you are in, you can reference yourself to the Earth's surface (DeMers, 2005). Other non-co-ordinate referencing systems in use are based on administrative areas: for example, the units used for aggregation and presentation of population census data in different countries. For referencing within smaller areas, unique feature references may be used: for instance, the property reference numbers used by a local authority, or the pipeline references used by a utility company.

All spatial referencing systems have problems associated with them. Some are specific to the referencing system, such as the updating problems with postcodes or the difficulties caused by geographical co-ordinates with respect to map projections. However, some of the problems stem from the nature of the spatial entities that require referencing:

- Spatial entities may be mobile. Animals, cars and people move; therefore any spatial reference they are tagged with will only represent their known location at a particular time.

- Spatial entities may change. Rivers meander, roads can be relocated and policy areas redefined.

- The same object may be referenced in different ways. A house may be represented and referenced as both a point and an area on maps of different scales.

An additional problem for the GIS user is the large number of different spatial referencing systems in use. Choosing an appropriate referencing system can be difficult, and it will frequently be necessary to

## BOX 2.5 UK postcode system

In the UK the postcode system was developed about 25 years ago by the Royal Mail to help post sorting and delivery. Each code has two parts – the outward code and the inward code. The postcode system is hierarchical. The first one or two letters refer to a postcode area; these are followed by subsequent numbers and letters subdividing this into districts, sectors and unit postcodes (Department of the Environment, 1987) (Figure 2.17). The system is further complicated by the existence of single-user postcodes for business users and addresses which receive more than

15 items of mail per day (Raper et al., 1992). The system is very widely used in application areas such as health, marketing and education because of its ease of collection and widespread use for address-based data. However, as with any other postal code system there are problems:

- For entities without an address, a postcode system is useless. Entities without addresses include rivers, trees, fields and phone boxes.
- The spatial units – postal areas, districts and units – were designed to help mail delivery and bear no relationship to other spatial units commonly used

**BOX 2.5**

**(a)**                                    **(b)**

**Figure 2.17** The UK postcode system
*Source*: (a) Reproduced by permission of Ordnance Survey on behalf of HMSO. © Crown Copyright 2011. All rights reserved. Ordnance Survey Licence number 100030901

by those handling spatial information. However, in the UK there is a link to the Ordnance Survey grid reference and census enumeration districts.

■ Changes occur to postcodes. In the UK there is a three-month update cycle and approximately 18,000 changes are made each year. Changes may be corrections, due to the construction or demolition of properties, or to the movement of large users (who may be eligible to keep their postcode if they move within the same sector).

■ Some buildings have more than one postcode. Office blocks containing different companies, or blocks of flats where there are separate entrances or letter-boxes, may have several postcodes.

■ When comparing and plotting population distribution maps, it should be remembered that unit postcodes, which cover approximately 15 houses, will represent very small areas in urban environments, but may be huge in rural areas.

*Sources*: Adapted from Dale and McLaughlin (1988); Raper *et al.* (1992)

integrate data collected using different referencing systems. This can be problematic. For some referencing systems, however, links have been developed to allow the integration and conversion of data. A good example of this is the linkage between the UK's postcode system and the Ordnance Survey National Grid. By extrapolating the start of a postcode to the corner of a 100 m grid square, a grid reference has been allocated to each postcode.

## Topology

In GIS, *topology* is the term used to describe the geometric characteristics of objects which do not change under transformations such as stretching or bending and are independent of any co-ordinate system (Bernhardsen, 1999). The topological characteristics of an object are also independent of scale of measurement (Chrisman, 2002). Topology, as it relates to spatial data, consists of three elements: adjacency, containment and connectivity (Burrough, 1986).

Adjacency and containment describe the geometric relationships that exist between area features. Areas can be described as being 'adjacent' when they share a common boundary. For example, the ski slopes and car parks in Happy Valley may be adjacent. Containment is an extension of the adjacency theme and describes area features that may be wholly contained within another area feature such as an island within a lake. Connectivity is a geometric property used to describe the linkages between line features. Roads are usually connected together to form a road network through which traffic can flow.

An understanding of the geometric relationships between spatial entities is important for analysis and integration in GIS. Without knowledge of how entities are geometrically related to each other, it is impossible to answer questions such as 'What is the shortest route from A to B?' or 'How many ski slopes lie within or are next to zones of high avalanche risk?'

## ■ THEMATIC CHARACTERISTICS OF SPATIAL DATA

Attributes are the non-spatial data associated with point, line and area entities. Attributes are the characteristics of an entity (Laurini and Thompson, 1992). For example, the attribute data associated with spatial entities used in the Happy Valley GIS might tell us that:

- a point represents a hotel;
- a line represents a ski lift; and
- an area represents a forest.

Each spatial entity may have more than one attribute associated with it. For example, a point representing the hotel may have a number of other attributes: the number of rooms; the standard of accommodation; the name and address of the owner. Attributes give additional information about the character of the entities. They also allow certain GIS operations to be performed where it is the characteristics of the entities that are under scrutiny. Questions such as 'Where are all the hotels

with fewer than 20 bedrooms of luxury standard?' require the analysis of attribute data associated with the point entities used to represent the location of hotels (see **Chapter 4**).

The character of attribute data themselves can influence the utility of data sets in GIS analysis. One characteristic which is of considerable importance is the scale of measurement used to record and report the data. For example, every year the managers of Happy Valley complete a table for a ski resort guide. For this table they provide the name of the ski area, its ranking (1st, 2nd, 3rd, 4th most popular), its average winter temperature and the size of the ski area. Each item of data uses a different scale of measurement. The names given to these scales are nominal, ordinal, interval and ratio. Table 2.1 shows each of these scales in relation to the data collated for the ski resort guide. Each scale of measurement dictates how the data can be used.

**TABLE 2.1 Scales of measurement**

| Data | Unit of measurement | Scale |
| --- | --- | --- |
| Resort name | Text | nominal |
| Resort ranking | Value | ordinal |
| Average winter | °C | interval temperature |
| Size of ski area | m$^2$ | ratio |

On a *nominal* scale numbers are used to establish identity. In Happy Valley, numbers on a nominal scale include telephone numbers or ski pass codes. These numbers cannot be processed in a mathematical sense, since they do not represent order or relative value. Adding, subtracting or dividing numbers on a nominal scale will not produce a useful result. Adding together two phone numbers is possible, but the answer is meaningless.

The numbers in an *ordinal* scale establish order. Location in a ski lift queue is an example. In Happy Valley, the ordinal scale is used to publish the top 10 cafés and ski runs based on the number of people using them each week. Using an ordinal scale you can obtain an impression of the order of numbers, but no information about relative sizes. The most popular ski run (ranked 1) is not necessarily twice as popular as the ski run which is ranked second.

## REFLECTION BOX

- Pick a GIS project or application in which you are involved or one that you have read about. Identify all the data sources that are used in the project and categorize them into primary and secondary sources. Create a summary similar to that provided for Happy Valley in Box 2.1.
- Use a GIS or online mapping tool to display a map for an area of interest and experiment with the scale functions. You may be able to change the scale by zooming in and out, and you may be able to specify a particular scale for display. Look for differences between small- and large-scale maps

– for example, in the generalization of features or the level of detail shown. Does the representation of any features change (for example from point to area features) as you change the scale?
- Visit some of the websites suggested at the end of the chapter which allow you to investigate map projections further. Try to identify a projection which maintains shape, one which maintains distances and one which maintains area. For what type of GIS applications might your three examples be useful?

---

Arithmetic operations, whilst possible on ordinal data, will again give meaningless results.

On an *interval* scale the difference between numbers is meaningful but the scale does not have a real origin. Temperatures, in degrees Celsius, are a good example. On a temperature scale it is possible to say that there is a 10-degree difference between a thermometer that records a value of 10 degrees and one that records a value of 20 degrees. Thus, differences can be calculated. However, it would be incorrect to say that 20 degrees is twice as warm as 10 degrees, because zero degrees on the Celsius scale is not a true zero. There is still a temperature when the thermometer reads zero! Negative numbers are also possible on an interval scale.

On a *ratio* scale measurements can have an absolute or real zero, and the difference between the numbers is significant. Snow depth is an example. It is impossible to have a negative value for snow depth. Something is also known about relationships between data, for example a snow pack that is 3 m deep is twice as deep as one that is 1.5 m deep.

One of the problems with the scales of measurement used for the collection of attribute data is that the distinction between the various scales is not always obvious. Many data used in GIS are nominal or ordinal. It is important to take care when using these data in an analytical context. If the scale of measurement is not known, or the GIS user is unaware of what scale has been used or what operations can be carried out on data

from that scale, the GIS is unlikely to indicate when impossible or meaningless operations have been carried out. To a computer numbers are all the same and will be treated in the same ways. So, ranked scores for city sizes may be added together. Two different soil types could have a numerical code to tag them to the appropriate area in a GIS. If clay soils have the value 2 and sandy soils 3 on a nominal scale, multiplying them together to give soil class 6 would be a meaningless operation. On the other hand, population and area (both on a ratio scale) can be divided to give population density, or elevation at one point (interval scale) may be subtracted from elevation at another point to give difference in elevation.

## ■ OTHER SOURCES OF SPATIAL DATA

So far in this chapter we have considered the characteristics of spatial data and their thematic dimension. To do this we have drawn heavily on the map metaphor. However, there are a number of other sources of spatial data, including census and survey data, aerial photographs, satellite images and global positioning systems, which have additional special characteristics. These are reviewed below. In addition GIS applications may draw on other business-specific data sets. Box 2.6 provides brief details of a GIS application which requires the integration of a wide range of data sets, including video.

# BOX 2.6 Integrating data with GIS

**CASE STUDY**

## Mapping the UK's railways, Peter Robson

Network Rail is responsible for the maintenance and renewal of the railway in the United Kingdom. The company is one of the largest landowners in the country and required an IT solution to electronically map its property assets. Moreover, it needs to understand how land is used in specific areas to strategically develop its estate to provide funds for reinvestment into the operational railway. Historically, maps and plans were hand-drawn and, despite a gradual move to a variety of Geographical Information Systems (GIS) and computer-aided design (CAD) systems, there was no single company-wide solution. In order to improve efficiency and allow company-wide access to valuable data, a GIS, underpinned by topographic data acquired through a service level agreement with Ordnance Survey, was developed and delivered. This system has revolutionized the way the company gathers and interrogates information on its estate.

The GI portal (Figure 2.18a) is based on Esri's ArcGIS product. It provides access to digital mapping of the entire UK rail network and holds details of property ownership and status. It is the first system of its kind within the rail industry and is regularly used by close to 6000 employees from all areas of the business. It is one of the largest and most extensive web-based GIS in use in the UK transport sector.
The award-winning system contains both business-specific geographic data and data sets drawn from third parties. It includes data on operational and commercial property, land ownership and some engineering data sets. In 2003 Omnicom's National Video Survey (NVS) was integrated into the system and delivers video and positional track data on a large part of the rail network. The company has also acquired national coverage of aerial imagery, which will provide an extra dimension when it is integrated into Marlin GIS.

**Figure 2.18a** The GI portal GIS at Network Rail

Like any IT system, the GI portal is continually evolving to respond to the needs of the business. However, Network Rail has already witnessed significant business benefits since its deployment, including considerable cost savings, improved customer services and a more unified approach to property development. The system has provided tangible business benefits to Network Rail which, as an organization, fully appreciates the value of installing a GIS solution.

### FURTHER INFORMATION

www.networkrail.co.uk
www.geoconnexion.com/magazine/uk/article.asp?ID=2148

## Solving crime the Salford City way, Paul Coward

The MAPS (Multi-Agency Problem Solving) GIS integrates information on closed-circuit TV coverage with other data to assist in the prevention and detection of crime in the City of Salford, England.

The MAPS system was conceived when the Crime and Disorder Partnership at Salford City Council realised that they were unable to answer police and public questions as to whether locations of crime had CCTV coverage, let alone find the actual footage of incidents taking place.

The MAPS GIS was developed to record the location of the Council's CCTV Cameras and their 'field of view' (the areas covered by the camera). Camera locations were surveyed, and detailed maps produced to show the areas visible from each camera.

At the start of the project only the location of the 240 CCTV cameras, owned by Salford City Council, were included in MAPS. However, as soon as MAPS was launched and access provided to partner organisations including; Greater Manchester Police, the Transport Police, the University of Salford, the Manchester Fire Service; and the Primary Care Trust, it was clear that the potential of the system was far greater than initially thought. Within three months the number of cameras within the system had almost

tripled to 660 with additional camera locations being provided by partner organizations and through the Police Community Support Officers approaching local firms with CCTV cameras on their premises.

Eighteen partner organizations now have access to the information provided by MAPS with over eighty users trained to use the web-based system. Additional spatial data, to assist in the prevention and detection of crime, has now been added to MAPS. This includes information on; the location of licensed premises, property ownership, bus stops, bus routes, petrol stations, dispensing chemists, second hand goods shops; and scrap yards.

Since the rollout of MAPS the system has been widely used in support of crime prevention and detection including the planning of surveillance operations. The use of MAPS is now an integral part of the daily workflow for volume crime officers who can now quickly check whether the crime scenes have CCTV coverage without needing to visit the location of the camera. Perhaps the greatest benefit of MAPS is that Partners can now work together to track 'rolling' crimes which move between the cameras of the different agencies, something which would have been impossible before MAPS was established.

**Figure 2.18b** The MAPS system
*Sources*: MAPS interface courtesy of Computer Aided Development Corporation; Map data reproduced by permission of Ordinance Survey

**FURTHER INFORMATION**
www.cadcorp.com/pdf/ PA-CCTV_Image.pdf
*Sources*: Peter Robson, Network Rail and Paul Coward, Salford GIS Ltd

## Administrative, survey and census data

Administrative data, surveys and census data are collections of related information. They may be spatial in character if each item in the collection has a spatial reference that allows its location on the surface of the Earth to be identified.

■  *Administrative data*: these include data on births, deaths and marriages or details of those claiming

social security benefits which are collected routinely or on a one-off basis. They may provide useful data on topics which were not the primary reason for data collection. For example, data on benefits claimants can indicate levels of unemployment or sickness absence from work.

■  *Survey data*: a national coverage survey may be carried out on a sample of the population and ask respondents a series of detailed questions focusing on a particular topic: for example, health, occupation or housing. Surveys are also conducted for smaller geographical areas or for specific business or research purposes: for example surveys of shopping habits or food consumption.

■ *Census data*: a modern census collects demographic and household information about an entire population, unlike ancient censuses which were taken for taxation purposes or to determine the number of males who could be conscripted into the army. Knowing the population total may be useful in its own right, but details of how people are distributed across a country is of additional interest.

The utility of administrative, survey and census data in GIS applications depends on the detail of spatial referencing in the data source. Table 2.2 provides some examples of spatially referenced administrative and survey data collected at a national level. Administrative data are often released and aggregated to the same geographies as are used for the census (see below and Box 2.7). Nationally representative surveys may not include local-level geographies, but these sources often allow regional results to be mapped and underpin complex statistical models which provide explanations to complement detailed mapped distributions. Compared with censuses and administrative data, surveys tend to have only large-area regional geographies but contain comprehensive personal information.

Population census data normally have some element of spatial referencing. Most population censuses use a hierarchical series of spatial units to publish data. Two sets of spatial units are associated with the UK census with the smallest being the enumeration districts (EDs) used for data collection and Output Areas (OAs) used for reporting data (see Box 2.7). In the UK census data are not usually released in map form, but in tables for the spatial areas you request. However, since a spatial reference is attached which links the data to the areal units of collection, the data are immediately useful for spatial studies.

**TABLE 2.2 Examples of spatially referenced administrative and survey data collected in the UK**

| Administrative data | Government Survey Data |
|---|---|
| Births and deaths | Labour Force Survey |
| Marriages and divorce | General Household Survey |
| Source: www.statistics.gov.uk/ | British Crime Survey |
| | British Social Attitudes Survey |
| Unemployment benefits claimants | Health Survey for England |
| Housing benefits | National Food Survey |
| Disability and carers' allowances | National Travel Survey |
| Source: www.dwp.gov.uk/ | Source: www.esds.ac.uk/Government/ |

# BOX 2.7 Population Census

PRACTICE

Census data describe the socio-demographic state of a whole country from national level down to local area. No other data source provides such comprehensive spatial coverage of the human landscape. For example, the UK Census of Population provides a 'snapshot' of the distribution, size, structure and character of the people of Great Britain and Northern Ireland on a particular day once every 10 years (Openshaw, 1995). The data collected are used by national government for the allocation of billions of pounds of public expenditure and policy analysis. Census data are also very valuable commercially since they are essential ingredients in marketing analysis and retail modelling.

→

**BOX 2.7**

The UK Census of Population is a relatively simple questionnaire survey administered across Great Britain and Northern Ireland once every 10 years, and an estimated 25 million households were involved in the 2011 census. The census is administered separately in England and Wales, Scotland and Northern Ireland, but most of the questions asked and statistics published are common to all these countries, with just minor variations for country specific items.

The UK Census covers a wide range of subjects that describe the characteristics of the population of Britain and Northern Ireland. These include demography, households, families, housing, ethnicity, birthplace, migration, illness, economic status, occupation, industry, workplace, car ownership and mode of transport to work

Most censuses use a hierarchical series of spatial units to publish data, often based on the administrative geography existing at the time of the census. In the UK the smallest geographical units of data dissemination are Output Areas (OAs), a geography designed specially for the census to provide maximum geographic detail whilst still maintaining respondent anonymity (Martin, 1995). For the 2001 census OAs in England and Wales were based on postcodes and fitted within the boundaries of 2003 statistical wards and parishes. If a postcode crossed an electoral ward/division (or parish) boundary, it was split between two or more OAs. The recommended size for an OA was 125 households to ensure that smaller wards and parishes are incorporated into larger OAs and anonymity maintained. For the 2001 census data just over 175,000 OAs covered England and Wales (www.statistics.gov.uk/geography/census_geog.asp). Census data are also made available for a variety of other geographical areas and spatial scales. These include:

- *Administrative geographies*: Nation, region local government districts, electoral wards.
- *Electoral areas*: European constituencies, parliamentary constituencies, electoral wards.
- *Postal geographies*: postal areas, postal districts, postal sectors.

- *Small area statistical geographies*: Output Areas, Super Output Areas.

Most of the UK's census data are released in tables with pre-defined cross-tabulations of variables. Each record in a table will correspond to a particular area and will include an alphanumeric code which can be used to 'join' the data to GIS vector polygons so that choropleth maps of outputs can be displayed.

In the USA the Census has run since 1790 and has an additional role. The United States Constitution requires a census every 10 years to determine how many seats each state will have in the US House of Representatives. Census data are used to assist the apportionment (i.e. the distribution) of the 435 seats in the House of Representatives amongst the states. Census counts are also used to assist the redistricting (the redrawing of political districts) within each state after apportionment.

The most recent US Census, undertaken on 1 April 2010, was a short-form-only census asking questions about sex, age, race and Hispanic origin. A longer version, which was formerly received by one in every six households and covered topics such as education, employment, ancestry, disability and type of heating fuel, has been replaced by the American Community Census (ACS) which will be conducted on an annual basis to provide more frequent demographic, housing, social, and economic data. Online mapping tools were used to provide daily updates of the 2010 Census mail participation rates for local areas, and the National Historical Geographic Information System (www.nhgis.org) provides free access to aggregated census data and GIS compatible boundary files.

Other countries which have conducted censuses of population and housing since 2000 include Australia, New Zealand, Canada, Hong Kong, India, Singapore and Sri Lanka.

*Sources*: Paul Norman, School of Geography, University of Leeds; www.geog.leeds.ac.uk/projects/census; ESRC Census Development Programme www.census.ac.uk/; UK Office of National Statistics www.ons.gov.uk/census/index.html; Population Reference Bureau http://www.prb.org; and U.S. Census Bureau, www.census.gov

## Aerial photographs

Aerial photography was the first method of terrestrial remote sensing. It is the capturing of images from a position above the Earth's surface, or without contact with the object of interest (Curran, 1989; Mather, 1991). Unlike a map, which is a model of the Earth's surface and contains only a selection of data, an aerial photograph is a 'snapshot' of the Earth at a particular instant in time. As such it contains a mass of data and it is necessary to carry out some form of interpretation to make effective use of the information portrayed. Aerial photographs may be used in GIS as a background for other data, to give those data spatial context and to aid interpretation. Alternatively, the user may abstract information on land use, vegetation type, moisture or heat levels or other aspects of the landscape from the photograph. Aerial photographs are particularly useful for monitoring change, since repeated photographs of the same area are relatively inexpensive. For example, Gunn *et al.* (1994) have monitored

**(a)** Infrared vertical aerial photograph

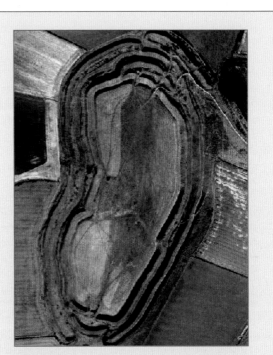

**(b)** Vertical colour aerial photographs showing archaeological remains

**(c)** Oblique colour aerial photograph

**(d)** Vertical black and white aerial photograph

**Figure 2.19** Aerial photographs
*Sources*: (a) U.S. Department of Agriculture; (b) Copyright © Bluesky International Limited used by permission; (c) National Oceanic and Atmospheric Administration (http://boulder.noaa.gov/gifs/aerial.jpg); (d) United States Geological Survey

changes in land use, particularly peat extraction, in County Fermanagh, Northern Ireland, from a time series of photographs. Interpretation of a sequence of photographs may allow the dating of events such as major floods which cause changes to the landscape. Curran (1989) identifies six characteristics of aerial photographs that make them of immense value as a data source for GIS:

- wide availability;
- low cost (compared with other remotely sensed images);
- wide area views;
- time-freezing ability;
- high spectral and spatial resolution; and
- three-dimensional perspective.

Additionally, aerial photographs can be used to obtain data not available from other secondary sources, such as the location and extent of new housing estates, or the extent of forest fires. One characteristic of aerial photographs that constitutes a possible disadvantage is the fact that they do not provide spatially referenced data. Spatial referencing has to be added to features on the image by reference to other sources such as paper maps. Several different types of aerial photographs are available, from simple black and white, which may be used for a wide variety of purposes, to colour and thermal infrared for heat identification.

The angle at which the photograph was taken is important. A photograph is referred to as *vertical* if taken directly below the aeroplane, and *oblique* if taken at an angle. Oblique photographs generally cover larger areas and are cheaper than vertical photographs. Vertical photographs are, however, the most widely used for GIS applications. Figure 2.19 contains examples of black and white, colour, infrared, vertical and oblique aerial photographs.

Before any aerial photograph information can be used in a GIS a number of factors must be considered. The first of these is scale. Scale varies across an aerial photograph, owing to the distance of the camera from the ground (Figure 2.20). The scale will be constant only at the centre of the image, and the greater the flying height, the greater the scale difference between the centre and edges of the image. This will also affect the angle of view towards the edge of the image creating the effect whereby tall,

vertical features such as mountains, buildings and trees appear to lean away from the centre of the image or the 'nadir' (that point that is vertically beneath the camera). This is particularly noticeable in the latest generation of high-resolution digital aerial photography and especially in urban areas where the 'lean' of tall buildings can obscure the streets below. Second, factors that may influence interpretation need to be considered. These include time of day and time of year. On photographs taken in winter, long shadows may assist the identification of tall buildings and trees, but may obscure other features on the image. Conversely, in summer, when trees are in full leaf, features that may be visible from the air in winter will be obscured.

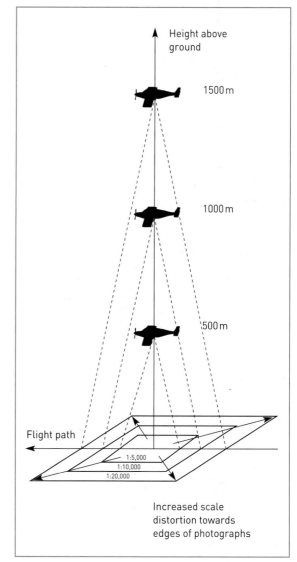

**Figure 2.20** Varying scale on aerial photograph

Aerial photography has been successfully used in archaeological surveys. Subtle undulations, indicating the presence of archaeological features such as the foundations of old buildings and the outlines of ancient roads and field systems just below the surface, can stand out in sharp contrast on winter or late evening images. Photographs taken during drought periods may reveal the detail of subsurface features as alternating patterns of green and dry vegetation caused by variations in soil depth.

Aerial photographs represent a versatile, relatively inexpensive and detailed data source for many GIS applications. For example, local government bodies may organize aerial coverage of their

**(a)** SPOT

**(b)** Landsat TM

**(c)** MSS

**(d)** Meteosat

**(e)** IKONOS satellite imagery

**Figure 2.21** Examples of satellite imagery
*Sources*: (a) CNES Dist/SPOT Image/Explorer/Gamma/Camera Press Ltd; (b) i-cubed: LLC and USGS; (c) Science Photo Library / Earth Satellite Corporation; (d) Wetterzentrale, Germany; (e) i-cubed: LLC and GeoEye

districts to monitor changes in the extent of quarrying or building development. At a larger scale, photographs can be used to provide data on drainage or vegetation conditions within individual fields or parcels that could not be obtained from conventional topographic maps (Curran, 1989).

## Satellite images

Satellite images are collected by sensors on board a satellite and then relayed to Earth as a series of electronic signals, which are processed by computer to produce an image. These data can be processed in a variety of ways, each giving a different digital version of the image.

There are large numbers of satellites orbiting the Earth continuously, collecting data and returning them to ground stations all over the world. Some satellites are stationary with respect to the Earth (geostationary), for example Meteosat, which produces images centred over Africa along the Greenwich meridian (Curran, 1989). Others orbit the Earth to provide full coverage over a period of a few days. Some of the well-known satellites, Landsat and SPOT, for example, operate in this way. Landsat offers repeat coverage of any area on a 16-day cycle (Mather, 1991). Figure 2.21 shows examples of images from earth observation satellites.

Most Earth observation satellites use 'passive' sensors that detect radiation from the sun that is reflected from the Earth's surface. Sensors may operate across different parts of the electromagnetic spectrum, not only those portions visible to the human eye. The multispectral scanner (MSS) on board Landsat simultaneously detects radiation in four different wavebands: near infrared, red, green and blue (Curran, 1989). After processing, the images can be used to detect features not readily apparent to the naked eye, such as subtle changes in moisture content across a field, sediment dispersal in a lake or heat escaping from roofs in urban areas.

A smaller number of satellites use 'active' sensors that have their own on-board energy source and so do not rely on detecting radiation reflected from the surface of the Earth. Examples are radar-based sensors such SAR (Synthetic Aperture Radar). These have the advantage of being able to explore wavelengths not adequately provided for by the Sun, such as microwave, and are both able to work at night and penetrate cloud layers.

LiDAR (Light Detection and Ranging) is a remote sensing system that uses aircraft-mounted lasers to collect topographic data from low altitude. The lasers are capable of recording elevation measurements with a vertical precision of 15 cm at spatial resolutions of around 2 m. Measurements are spatially referenced using high-precision GPS. The technology creates a highly detailed digital elevation model (DEM) that closely matches every undulation in the landscape. Even change in ground-surface elevation detail caused by buildings and trees can be detected. This makes LiDAR extremely useful for large-scale mapping and engineering applications. The images shown in Figure 2.22 are derived from LiDAR data. Box 2.8 outlines the use of LiDAR data in a hydrological application.

Scanned images are stored as a collection of pixels, which have a value representing the amount of radiation received by the sensor from that portion of the Earth's surface (Burrough, 1986). The size of the pixels gives a measure of the resolution of the image. The smaller the pixels the higher the resolution. The Landsat Thematic Mapper collects data for pixels of size 30 m by 30 m. Much greater resolution is possible, say 1 m by 1 m, but this has in the past been restricted to military use. Recent changes in US legislation and the availability of Russian military satellite data have made access to very high-resolution data easier such as that provided by the QuickBird and IKONOS satellites. Resolution is an important spatial characteristic of remotely sensed data and determines its practical value. A Landsat Thematic Mapper image with a pixel size of 30 m by 30 m would be unsuitable for identifying individual houses but could be used to establish general patterns of urban and rural land use. Box 2.9 provides further discussion of resolution.

For GIS, remotely sensed data offers many advantages. First, images are always available in digital form, so transfer to a computer is not a problem. However, some processing is usually necessary to ensure integration with other data. Processing may be necessary to reduce data volumes, adjust resolution, change pixel shape or alter the projection of the data (Burrough, 1986). Second, there is the opportunity to process images or use different wavebands for the collection of data to highlight features

**(a)** Vertical urban image showing residential housing and roadway

**(b)** Vertical urban image showing industrial units, trees, cuttings and embankments

**(c)** Oblique image of Kverkfjoll volcano, Iceland

**(d)** Oblique image of Odenwinkelkees glacier, Austria

**Figure 2.22** LiDAR imagery: vertical urban (a, b); oblique (c, d)
*Sources*: (a, b) Precision Terrain Surveys (PTS); (c, d): Reproduced with permission of Dr Jonathan Carrivick

of particular interest: for example, water or vegetation. The repeated coverage of the Earth is a further advantage, allowing the monitoring of change at regular intervals, although for some types of imagery, cloud cover whilst the satellite passes overhead may prevent a useful image being obtained. Finally, the small scale of images provides data useful for regional studies, and applications have included mapping remote areas, geological surveys, land use monitoring and many others (see Curran,

1989; Maguire, 1989; Mather, 1991). Trotter (1991) considers the advantages of remotely sensed data for GIS applications in the area of natural resource management to be:

- low cost relative to other data sources;
- currency of images;
- accuracy;
- completeness of data; and
- uniform standards across an area of interest.

# BOX 2.8 High-resolution data solutions to complex modelling problems

CASE STUDY

## Stuart Lane and Joseph Holden

Land management can influence landscape response to rainfall events altering the flow of water in rivers and leading to changes in flood peaks. High-resolution data can help us understand these impacts and enable us to develop management tools that are spatially focused on key areas of the landscape to provide maximum benefits from economic investment. This case study outlines the use of LiDAR DEMs and accurate GPS surveys to create better connectivity maps to inform management decisions in an upland catchment in northern England.

### SCALE AND RESOLUTION

DEM data sets such as those derived by digitizing contour lines from paper maps have lacked the spatial resolution (both horizontal and vertical) to allow distinction between areas of hillslopes that are likely to be wet and those that are likely to be drier. Comparisons of the Topographic Index (see **Chapter 7**) calculated from 2 m resolution LiDAR data (see Box 2.9 for more details on resolution) with 16 m and 64 m DEMs (generated by resampling the LiDAR data) show remarkable differences and a much finer spatial network can be seen showing wetter and drier areas (Figure 2.23). The vertical accuracy of the LiDAR data has been measured at ±0.12 m based on GPS surveys using equipment and methods accurate to within ±0.03 m (Lane *et al.*, 2004). The darker areas of Figure 2.23 are those areas which are likely to have higher soil moisture because they have a larger drainage area feeding into them from upslope and/or they are likely to be on gentle gradient slopes rather than short steep slopes. This information is very important for hydrological models as the high resolution data can show which areas of the catchment are likely to be fully saturated (soil pores are all full of water and the water table is at the surface) at particular times during a rainfall event.

### APPLICATIONS

When the soil is saturated from a given point on a hillslope all the way to the stream channel it can be said to be hydrologically well connected. In this instance it is likely that overland flow will move quickly across the land surface into the stream and there will be a rapid response to any further rainfall

**(a)** 2 metre

**(b)** 16 metre

**(c)** 64 metre

**Figure 2.23** Topographic Index for (a) 2 metre; (b) 16 metre; (c) 64 metre resolution DEM

**BOX 2.8**

inputs. It is also likely that any surface pollutants (e.g. from fertilisers or animal droppings) can be washed into the river channel network. However, while a given point on a hillslope might be saturated, overland flow might not be directly connected to the stream network because there may be places on the same hillslope where the soil is not fully saturated and water flowing over the surface from upslope can infiltrate into the soil at that point. Therefore, any surface pollutants may be prevented from reaching the stream channel as the flow is 'disconnected'. Figure 2.24 shows an example of a hydrological model prediction based on the Topographic Index using high-resolution LiDAR data for a particular time during a heavy rainfall event. While a large proportion of the catchment is saturated (green), the blue areas on Figure 2.24 are those where there is direct connectivity with the stream channel. It is therefore these areas where careful management may be required to reduce grazing or fertiliser inputs. This sort of approach means that, instead of having a blanket management policy for a whole catchment, we can adopt spatially distributed management plans allowing some activities to take place on some parts of the landscape and to reduce or ban these activities from other more sensitive parts of the landscape. Using high-resolution data means that this approach can be done more fairly as it allows individual farmers to have mixed management on a field-by-field basis rather than banning one farmer from applying fertiliser and letting another farmer use it. This information can also help in the definition of buffer zones to protect water courses from diffuse pollution (Lane *et al.*, 2004). Although suitable models for mapping hydrological connectivity using DEMs have been in existence for some time (for example, the D8 algorithm, see **Chapter 7**), appropriately detailed data sources such as SAR and LiDAR have become available more recently.

Accurate mapping of hillslopes is also useful for determining the effects of human modifications to the drainage network (such as from artificial drainage) and in planning flood defences. These techniques have been used in Upper Wharfedale, England, to model the impacts of upland land drains. These drains, locally known as 'grips', are shallow ditches that were cut diagonally across the contour lines into deep peat soils during the twentieth century to aid drainage and improve grazing (Holden *et al.*, 2007). However, these grips have not been shown to improve the land for grazing or grouse and are more often associated with negative environmental impacts. The grips have an impact on the saturation by preventing flow from upslope flowing down the hill to the lower parts of the slope. This means the lower parts of the slope below the drains are drier than they would otherwise be (Figure 2.25). This is important because peats are large stores of carbon, but this carbon store is only maintained when they are kept close to saturation. Drained peats are known to release more carbon into the atmosphere and into local rivers, causing the water to be discoloured brown. High-resolution mapping of the grips in the landscape allows determination of which grips are having the greatest impact. In Figure 2.25 this is shown by the red colouration which indicates all areas where the peat will have a smaller length of slope draining into it because of the presence of grips that divert the flow away into the stream channel. Such maps help land managers

**Figure 2.24** Map of hydrological connectivity in Upper Wharfedale, northern England

Saturated, no connectivity
Saturated, with connectivity

**Figure 2.25** Area of catchment with Topographic Index affected by grips

**BOX 2.8**

target remediation strategies to store more carbon on the land by indicating which grips should be targeted for blocking and back-filling. Grip-blocking is very expensive and so techniques such as those outlined here are very useful and can allow managers to get maximum carbon and water quality benefits from the money they spend. Such work has also been coupled to hydrological models to show how grips have an impact on river flow and flood peaks.

**FURTHER INFORMATION**

Grip blocking: www.blanketbogswales.org
LiDAR: www.lidar.com
LiDAR mapping: www.sbgmaps.com/lidar.htm

*Source*: Stuart Lane, University of Durham and Joseph Holden, University of Leeds

## BOX 2.9  Resolution

THEORY

Resolution is defined as the size of the smallest recording unit (Laurini and Thompson, 1992) or the smallest size of feature that can be mapped or measured (Burrough, 1986). In the Zdarske Vrchy case study introduced in Chapter 1, the data are stored as raster layers, the size of each individual cell being 30 m × 30 m. In this case the resolution of each image is 30 m, since this is the lowest level to which the data can be described. In the case of mapping census variables or other socio-economic data collected within administrative boundaries, the Department of the Environment (1987) refers to Basic Spatial Unit (BSU) as the smallest spatial entity to which data are encoded. The BSUs should be constant across all the data used for a particular application so that the data collected are immediately comparable. In order to facilitate subsequent analysis it is necessary to choose the smallest unit possible when collecting the data. These BSUs can then be built up into any other unit by the process of aggregation. In the UK Census of Population the BSU for publicly accessible data is the enumeration district (ED), containing approximately 500 residents (150 in rural areas). EDs can easily be aggregated to form wards. Wards can then be aggregated into districts, and districts into counties. Disaggregation (the reverse process to aggregation) to areas smaller than the original BSU is fraught with difficulties and based on so many assumptions as to pose serious problems for data quality.

## Field data sources: surveying and GPS

There are several methods of collecting raw data in the field for direct input into a GIS. These are most often used when the required data do not exist in any other readily available format such as a map or satellite image. Traditional manual-surveying techniques using chains, plane tables, levels and theodolites are examples of direct field measurement, but the data collected need to be written down on paper first. Modern digital equivalents of these manual techniques have been adapted so that the data collected are stored in digital format ready for direct input into GIS. Examples include total stations (high-precision theodolites with electronic distance metering (EDM) and a data logger) and hand-held laser range finders (Figure 2.26).

A technique of field data collection which has found particular favour with GIS users is the use of satellite navigation systems or GPS (Global Positioning Systems). These are portable devices that use signals from GPS satellites to work out the exact location of the user on the Earth's surface in terms of $(x,y,z)$ co-ordinates using trigonometry (Figure 2.27). Position fixes are obtained quickly and accurately at the push of a button. The accuracy obtainable from GPS receivers ranges from 100 m to as little as a few millimetres depending on how they are used. Originally designed for real-time navigation purposes, most GPS receivers will store collected co-ordinates and associated attribute information in their internal memory so they can be downloaded directly into a GIS database.

**(a)** EDM

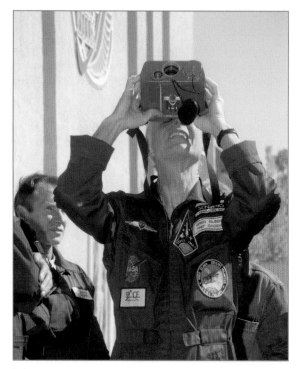

**(b)** Laser range finder in use

**Figure 2.26** Modern field survey equipment
*Sources*: (a) Alamy / Widstock; (b) Press Association Images/PA

**(a)** Satellite configuration

**(b)** GPS receiver

**(c)** GPS in use

**Figure 2.27** GPS receiver and satellite configuration
*Source*: (b, c) Courtesy of Magellan

The ability to walk or drive around collecting co-ordinate information at sample points in this manner has obvious appeal for those involved in field data collection for GIS projects. Box 2.10 gives further details on GPS.

In recent years a number of companies have developed high-resolution, automated laser range scanners for industrial and engineering applications. When coupled with high accuracy differential GPS, these terrestrial laser scanners (TLS) can be used to collect highly detailed terrain data with sub-metre accuracy that can processed to create a DEM and ultimately used in GIS. Apart from the obvious advantage of accuracy, multiple versions of these DEMs can be created over a period of time ranging from hours to years using repeat

## BOX 2.10  GPS basics

**THEORY**

GPS is a set of satellites and control systems that allow a specially designed GPS receiver to determine its location anywhere on Earth 24 hours a day (Barnard, 1992). Two main systems exist: the American NAVSTAR and the Russian GLONASS. A European system, GALILEO, is also due to come into service by 2014. The American system consists of 24 satellites orbiting the Earth in high-altitude orbits. These satellites have a 12-hour orbit time, and pass over control stations so that their orbits can be closely monitored and their positions precisely identified. Satellites and ground-based receivers transmit similarly coded radio signals, so that the time delay between transmission and receipt of the signals gives the distance between the satellite and the receiver. If a receiver can pick up signals from three or four satellites, trigonometry is used to calculate the location and height of the receiver.

A GPS user will see a position 'fix' displayed on their receiver. Until May 2000 all readings were affected by selective availability (SA), a deliberate error added to the signals by the US military. This has now been switched off and fixes of far greater accuracy can be obtained. One fix, obtained from a single receiver, will have an accuracy of about 25 m for 95 per cent of the time. More advanced data collection methods, including the averaging of fixes and the use of two receivers in parallel (differential GPS), can be used to obtain readings down to the sub-centimetre level. The US military still retain the ability to switch SA back on – for example, during times of crisis or conflicts such as 9/11 and the Iraq war, but the existence of GLONASS and GALILEO mean this will have a much reduced impact.

Differential GPS techniques require two receivers, one fixed at a known location (the base station) and the other at an unknown location (the roving receiver). If both receivers are set up in exactly the same manner and use the same satellites to compute their location, then the positional error recorded will be the same for both receivers. A highly accurate positional fix can therefore be obtained for the roving receiver by subtracting the positional error calculated for the base station.

Despite 24-hour global coverage, GPS use can be hampered by certain factors. These include problems where the path between the satellite and the receiver is obstructed by buildings, dense tree cover or steep terrain, and in polar regions where favourable satellite configurations are not always available. Receivers capable of using signals from the GPS, GLONASS, and shortly, the GALILEO constellations, will, to a certain extent, circumvent these problems by making sure the receiver is always in view of a minimum number of satellites. GPS signals can also be augmented by additional signals from geostationary satellites and groundstations. This provides the standard GPS system with regional real-time differential corrections for suitably enabled receivers giving approximately five-fold improvement in the spatial accuracy of position fixes. The Wide Area Augmentation System (WAAS) provides GPS signal corrections in the USA and similar systems are being developed in Asia and Europe.

GPS is finding a wide range of applications, varying from navigation (air, sea and land), to geomorphological mapping and urban surveying (see publications such as *GPS Solutions* or the *Journal of Global Positioning Systems* for up-to-date examples). Developments in portable computing and mobile communications have opened up a whole new area of application for GPS within consumer and personal electronics. Knowing exactly where you are in relation to a GIS database on your portable computer, or receiving location information via your WAP phone, can be of great value and is a highly saleable commodity. The use of this location technology as a form of GIS 'output' is known as *Location-based Services (LBS)* and is discussed further in **Chapter 5**.

surveys. This allows the GIS-based analysis of a range of physical processes and phenomena including mass movement (rockfall and landslides), flood plain mapping and river channel change, and glacial advance and retreat. Box 5.5 provides a case-study example of TLS being used to provide data on fluvial geomorphology in dynamic alpine river systems.

## GIS data standards

The number of formats available for GIS data is almost as large as the number of GIS packages on the market. This makes the sharing of data difficult and means that data created on one system is not always easily read by another system. This problem has been addressed in the past by including data conversion functions in GIS software. These conversion functions adopt commonly used exchange formats such as DXF and E00. As the range of data sources for GIS has increased, the need for widely applicable data standards to facilitate exchange of data has been recognized. Some of the standards in current use are listed in Table 2.3. More details of one of these standards, BS 7666, are provided in Box 2.11.

**TABLE 2.3  Current geographical information standards**

| | |
|---|---|
| BS 7666 | Spatial data sets for geographic referencing |
| CEN TC 287 | European norms for geographic information |
| DIGEST | Digital Geographic Information Exchange Standards |
| DNF | Digital National Framework |
| GDF | Geographic Data File |
| GeoTIFF | Geographic Tagged Image Format File |
| GML | Geography Markup Language |
| ISO 6709 | Standard representation of latitude, longitude and altitude |
| ISO 8211 | Specification for a data descriptive file for information interchange |
| ISO 15046 | Geographic information |
| NEN 1878 | Netherlands transfer standard for geographic information |
| NTF | Neutral Transfer Format |
| OGIS | Open Geodata Interoperability Specification |
| RINEX | Receiver Independent Exchange Format |
| SDTS | Spatial Data Transfer Standard |
| UGDCS | Utilities Geospatial Data Content Standard |

*Source*: Adapted from Harding and Wilkinson (1997)

# BOX 2.11  BS 7666

**PRACTICE**

British Standard 7666 specifies a nationally accepted, standard referencing method for land and property in the UK. The standard was developed by a multi-disciplinary working party that included representatives from, amongst others, local government, the Ordnance Survey, Her Majesty's Land Registry, the Royal Mail, the Forestry Commission and academia. The standard provides a common specification for the key elements of data sets of land and property in Great Britain. It assures the quality of land and property information in terms of content, accuracy and format.

BS 7666 has four parts. It includes a specification for:

■ a street gazetteer (BS 7666 Part 1);
■ a land and property gazetteer (BS 7666 Part 2);
■ addresses (BS 7666 Part 3);
■ a data set for recording Public Rights of Way.

Part 3 provides the specification for addresses. The specification provides a nationally consistent means of structuring address-based information. Use of the standard should simplify the exchange and aggregation of address-based and related data.

The standard specifies that an address must contain sufficient information to ensure uniqueness within Great Britain. The combination of a primary addressable object name and a secondary addressable object name achieves this. An address must also contain the name of at least one or more of the street, locality, town and administrative area data, so that it is unique.

In addition, a postcode is mandatory for a mailing address, although a postcode may not exist for non-postal addresses. A postal address is a routing instruction for Royal Mail staff that must contain the minimum information necessary to ensure secure

**BOX 2.11**

delivery. Its presentation and structure are specified in another international standard: ISO 11180.

The specification includes details of how locality name, town name, administrative area and postcode should be specified. There are also detailed text conventions, requiring for example the storage of text in upper-case format; no abbreviations except for ST (Saint) and KM (Kilometre); and no underlining of text. Table 2.4 provides examples of land and property identifiers acceptable under BS 7666.

TABLE 2.4 **Examples of land and property identifiers**

| Unique property reference number | Secondary addressable object name | Primary addressable object name | Unique street reference number | Street name |
|---|---|---|---|---|
| 000100001 | | Palace Deluxe | 00010001 | Pine Avenue |
| 000100002 | Caretaker's Flat | Palace Deluxe | 00010001 | Pine Avenue |
| 000100003 | Mountain View | 23 | 00010006 | High Street |
| 000100004 | Ski Lodge | 10 | 00010002 | Ski School Road |

*Source*: Adapted from www.housing.dtlr.gov.uk/research

The Open Geospatial Consortium (OGC), formed in 1994, is an international consortium of almost 400 companies, government agencies and Universities working to advance international standards for geospatial interoperability (OGC, 2010b). The OGC were responsible for proposing the Geography Markup Language (GML) as a GIS data standard. GML and early adopters of this standard are described in Box 2.12.

**BOX 2.12 GML**

PRACTICE

The Geography Markup Language (GML) is a non-proprietary computer language designed specifically for the transfer of spatial data over the Internet.

GML is based on XML (eXtensible Markup Language), the standard language of the Internet, and allows the exchange of spatial information and the construction of distributed spatial relationships. GML has been proposed by the Open GIS Consortium (OGC) as a universal spatial data standard. GML is likely to become very widely used because it is:

■ Internet friendly;
■ not tied to any proprietary GIS;
■ specifically designed for feature-based spatial data;
■ open to use by anyone; and
■ compatible with industry-wide IT standards.

It is also likely to set the standard for the delivery of spatial information content to PDA and WAP devices, and so form an important component of mobile and location-based (LBS) GIS technologies.

Amongst the organizations adopting GML is the Ordnance Survey (OS), the national mapping agency for the UK. The OS will deliver DNF (Digital National Framework) data in GML. DNF is a version of the OS's large-scale topographic database that will eventually encompass all types of spatial data and all data scales. In the DNF nearly 230,000 tiles of large-scale topographic data have been merged into a single, seamless topologically structured point, line and topographic database containing information on buildings, boundaries, roads, railways, water and other topographic features. Each feature in the DNF is assigned a unique 16-digit identifier that allows it to be unambiguously referenced and associated with other features.

By adopting GML, the OS is making the DNF accessible to more software systems and users than would be possible using any other single data standard.

*Sources*: Holland (2001); GISNews (2001)

## REFLECTION BOX

- How reliable do you think census data are? Try to list some of the problems that might be faced when collecting population census data in your own country. Give some examples of GIS applications or projects in which census data might be used.
- Use the Web to find the following (or think of some examples of your own):
  - an aerial photograph of Manchester, England;
  - a satellite image of the Great Wall of China;
  - an aerial photograph for an archaeological application; and
  - a satellite image for an agricultural application. What sensors have been used for the collection of these images? What are their characteristics? How might they be used in a GIS project?
- Think about how GPS could be used in a GIS project of your own. What data could GPS be used to collect? How would the data be collected? What might be the problems of using GPS data?

## CONCLUSIONS

In this chapter we have looked at the distinction between data and information, identified the three main dimensions of data (temporal, thematic and spatial) and looked in detail at how different spatial data sources portray the spatial dimension. The main characteristics of spatial data have been identified and a review of how the traditional map-making process has shaped these characteristics has been presented. In addition, we have considered a range of other sources of spatial data. The discussion has shown that any source of spatial data may be influenced by some, or all, of the following factors:

- the purpose for which they have been collated;
- the scale at which they have been created;
- the resolution at which they have been captured;
- the projection which has been used to map them;
- the spatial referencing system used as a locational framework;
- the nature of the spatial entities used to represent real-world features;
- the generality with which these entities have been modelled; and
- the topological structure used to represent the relationship between entities.

In some data sources, one factor will dominate; in others it will be the interplay of factors that gives the data their character. Appreciating the main characteristics of spatial data is important because these characteristics will determine how the data can be used in building a GIS model. For example, data collected at different resolutions should only be integrated and analysed at the resolution of the coarsest data set. In the Zdarske Vrchy case study the 30 m by 30 m resolution of the land use map generated from TM satellite data dictated the resolution of the database for analysis.

Therefore, GIS models are only as good a representation of the real world as the spatial data used to construct them. Understanding the main characteristics of spatial data is an important first step in evaluating its usefulness for GIS. The next step is to understand how these data can be stored in a form suitable for use in the computer, as this will also influence the quality of the GIS model.

## REVISION QUESTIONS

- Explain the difference between data and information.

- What are the three basic spatial entities and how are these used to portray geographical features on paper maps and in GIS?

- Explain the differences between geographic and rectangular co-ordinate systems. What are their relative advantages and disadvantages?

- Explain what is meant by adjacency, containment and connectivity.

- Why is a knowledge of the different scales of measurement important in GIS?

- Explain the importance of map projections for users of GIS.

- Describe the characteristics of three sources of spatial data.

- Using examples, outline the importance of standards for spatial data.

## FURTHER STUDY – ACTIVITIES

- For a project you are involved in, list your data sources. Review each one and identify any issues about scale, entity definition, generalization, projections, spatial referencing and topology that you think might be relevant.

- Compare large- and small-scale maps of the same area. Select a small area and note the differences in how common features are represented on the two maps. How is this controlled by generalization?

- Use a world atlas to compare global and regional map projections.

- Use the national statistics sites in the list of websites below to find an up-to-date population figure for New Zealand and the Netherlands. How is the current population total calculated?

- Calculate spatial references for your home or office using latitude and longitude and the local grid co-ordinate system.

- Use the Web to find aerial photographs and satellite imagery for your home town or an area with which you are familiar. Does the resolution of the image allow you to discern familiar locations or features? Can you notice the effects of edge distortion, time of day/year and so on in your image?

- Use the Web to find and compare satellite images captured by different sensors.

## FURTHER STUDY – READING

Gatrell (1991) provides a good, thought-provoking introduction to the concepts of space and geographical data that is a good starting point for anyone coming to GIS from a non-geographical background. Robinson *et al.* (1995) and Keates (1982) provide comprehensive coverage of the subject of cartography from a more conventional viewpoint. Monmonier's book *How to Lie with Maps* (1996) offers a comprehensive and very readable introduction to the potential pitfalls of displaying data in map form. The discussion is just as applicable to maps on the computer screen as those on paper. Subjects such as scale, projections and generalization are covered in detail. Illiffe and Lott (2008) and Maher (2010) offer practical texts which include detailed information on map projections. A good discussion on UK spatial referencing can be found in Dale and McLaughlin (1988); DeMers (2008) provides a comparable review for the USA. The Chorley Report (Department of the Environment, 1987) provides brief details of postcodes and recommendations for use of spatial referencing in the UK. Raper *et al.* (1992) discuss the whole issue of UK postcodes in considerable depth and provide examples of address formats and postcode systems in a number of other countries including Austria, Germany, the Netherlands, Spain, Sweden and the USA. A similar discussion on the US ZIP code, though not in the same depth, can be found in DeMers (2008).

Comprehensive coverage of the principles and applications of remote sensing can be found in Curran (1989), Clayton (1995), Gibson and Power (2000) or Campbell (2007). Curran (1989) contains a particularly useful chapter on aerial photography that discusses the characteristics and interpretation of aerial photographs. A good introduction to GPS and its importance for GIS can be found in Kennedy (1996). Seegar (1999) offers the basic principles of geodesy relevant to GPS. Up-to-date information on GPS can be found in publications such as *GPS World*.

Campbell J B (2007) *Introduction to Remote Sensing*. 3rd edn. Taylor and Francis, London

Clayton K (1995) The land from space. In: O'Riordan T (ed) *Environmental Science for Environmental Management*. 2nd edn. Longman, London, pp. 198–222

Curran P (1989) *Principles of Remote Sensing*. Longman, London

Dale P F, McLaughlin J D (1988) *Land Information Management: An Introduction with Special Reference to Cadastral Problems in Third World Countries*. Clarendon Press, Oxford

DeMers M N (2008) *Fundamentals of Geographic Information Systems*. 4th edn. Wiley, New York

Department of the Environment (1987) *Handling Geographic Information*. Report of the Committee of Enquiry chaired by Lord Chorley. HMSO, London

Gatrell A C (1991) Concepts of space and geographical data. In: Maguire D J, Goodchild M F, Rhind D W (eds) *Geographical Information Systems: Principles and Applications*. Longman, London, vol. 1, pp. 119–43

Gibson P, Power C H (2000) *Introducing Remote Sensing: Digital Image Processing and Applications*. Routledge, London

Illiffe J, Lott R (2008) *Datums and Map Projections for GIS, Remote Sensing and Surveying*. 2nd edn. Whittles Publishing, New York

Keates J S (1982) *Understanding Maps*. Longman, London

Kennedy M (1996) *The Global Positioning System and GIS: An Introduction*. Ann Arbor Press, Ann Arbor, Michigan

Maher M (2010) *Lining up data in ArcGIS. A Guide to Map Projections*. Esri Press, Redlands

Monmonier M (1996) *How to Lie with Maps*. University of Chicago Press, Chicago

Raper J F, Rhind D W, Shepherd J W (1992) *Postcodes: the New Geography*. Longman, Harlow, UK

Robinson A H, Morrison J L, Muehrecke P C, Kimerling A J, Guptill S C (1995) *Elements of Cartography*. 6th edn. Wiley, New York

Seegar H (1999) Spatial referencing and coordinate systems. In: Longley P A, Goodchild M F, Maguire D J, Rhind D W (eds) *Geographical Information Systems*. Wiley, New York, vol. 1, pp. 427–36

## WEB LINKS

### National Mapping Agencies and other GIS data sources:

- United States Geological Survey: www.usgs.gov

- Ordnance Survey, UK: www.ordnancesurvey. co.uk

- Centre for Topographic Information, Canada: maps.nrcan.gc.ca/

- Geoscience, Australia: www.ga.gov.au/

- Eurographics, European National Mapping Agencies: www.eurogeographics.org

- GeoCommunity GIS Data Depot: data.geocomm.com/

### Maps and map projections:

- David Rumsey collection of historical maps (can be viewed in various ways, including Second Life): www.davidrumsey.com/

- Cartographical map projections: www. progonos.com/furuti/MapProj/Normal/TOC/ cartTOC.html

- Gallery of map projections: www.csiss.org/ map-projections/index.html

### Census and survey data:

- United States Census Bureau: www.census.gov/

- National Statistics, UK: www.statistics.gov.uk/

- Statistics Netherlands: www.cbs.nl/en-GB/ default.htm

- Australian Bureau of Statistics: www.abs.gov.au/

- Statistics New Zealand: www.stats.govt.nz/

- Population Reference Bureau Datafinder (world population database with online mapping): www.prb.org/DataFinder.aspx

### GPS:

- GPS World: www.gpsworld.com/gpsworld/

- UK GPS Information: www.ordnancesurvey. co.uk/oswebsite/gps/

- Garmin, What is GPS? www.garmin.com/ aboutGPS/

### Data standards:

- Federal Geographic Data Committee: www.fgdc.gov/

- Open Geospatial Consortium: www.opengeospatial.org/

Visit our website **www.pearsoned.co.uk/heywood** for further web links, together with self-assessment questions, activities, data sets for practice opportunities and other interactive resources.

# CHAPTER 3

# Spatial data modelling

# Learning outcomes

By the end of this chapter you should be able to:

■ Provide a definition of 'spatial data model'

■ Explain how spatial entities are used to create a data model

■ Distinguish between rasters and vectors

■ Describe a spatial data structure

■ Explain what topology is and how is it stored in the computer

■ List the advantages and disadvantages of different spatial data models

■ Discuss the way in which time and the third dimension are handled in GIS

## ■ INTRODUCTION

The previous chapter introduced the concept that a GIS, populated with data and ideas about how these data interact, is a model of reality. This model is created to help us develop our understanding of geographical problems. To appreciate this idea of a model, try to imagine how you might explain what a car is and what it can do to someone who has never seen or heard of a car. To do this you would have to develop and communicate your own model of the car. You might do this in two parts: first explaining the physical form of the car – for example, its shape, engine components, number of wheels and colour; then how it works – what happens to the fuel, how the engine components interact, how the brakes slow the wheels and so on. A GIS model can also be split into two parts: a model of spatial form and a model of spatial processes. The model of spatial form represents the structure and distribution of features in geographical space. In a model of spatial processes, the interaction between these features is considered. More details of process models are presented in **Chapter 7**. This chapter, and those which follow (particularly **Chapter 6**), focus on the representation and analysis of geographical features using models of form.

Building a model of the world for your GIS is a key stage in any GIS project. For example, imagine that you are a retail analyst who wants to create a GIS to help with decisions about a new store location. You have collected together maps, population statistics, aerial photographs and postcoded customer information. You need to get them into your GIS (see **Chapter 5**), but before you do that you must make some decisions about how the data should be modelled in the GIS. To do this you go through a series of stages of data abstraction (Peuquet, 1984) By applying this abstraction process you move from the position of observing the geographical complexities of the real world to one of simulating them in the computer. For any GIS project this process involves:

1  Identifying the spatial features from the real world that are of interest in the context of an application and choosing how to represent them in a conceptual model. In the Happy Valley GIS, hotel features have been selected for inclusion and will be represented by point entities.

2  Representing the conceptual model by an appropriate spatial data model. This involves choosing between one of two approaches: raster or vector. In many cases the GIS software used may dictate this choice.

3  Selecting an appropriate spatial data structure to store the model within the computer. The spatial data structure is the physical way in which entities are coded for the purpose of storage and manipulation.

Figure 3.1 provides an overview of the stages involved in creating a GIS data model. At each stage in the model-building process we move further away from the physical representation of a feature in reality and closer to its abstract representation in the computer. This chapter examines the stages in the modelling process as shown in Figure 3.1. First, the definition of entities for representation in the computer is considered. Then, the different spatial data models and structures available are outlined. Finally, the modelling of more complex features and the difficulties of including the third and fourth dimensions in a GIS model are discussed.

## ■ ENTITY DEFINITION

In order to define entities for inclusion in a computer model features of interest need to be identified, then a method for their representation in the computer must be established. As suggested in **Chapter 1**, all geographical phenomena can be represented in two dimensions by three main entity types: points, lines and areas. Figure 3.2 shows how a spatial data model could be constructed for the Happy Valley ski area using points, lines and areas. Points are used to represent hotels, lines to represent ski lifts and areas to represent forests. Figure 3.2 also introduces two additional spatial entities: networks and surfaces. These are an extension of the area and line concepts and are introduced in Box 3.1.

There are a number of problems associated with simplifying the complexities of the real world into five basic, two-dimensional entity types. These include the dynamic nature of the real world, the identification of discrete and continuous features, and the scale at which a particular problem needs addressing.

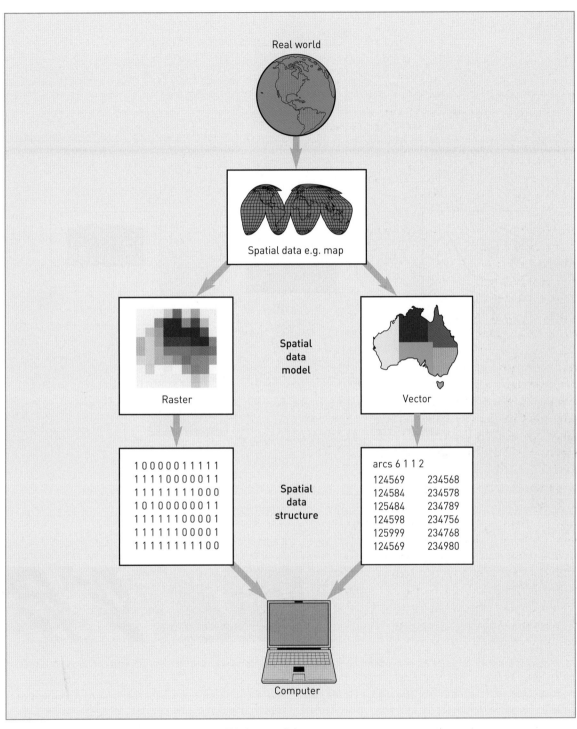

**Figure 3.1** Stages involved in constructing a GIS data model

The real world is not static. Forests grow and are felled; rivers flood and change their course; and cities expand and decline. The dynamic nature of the world poses two problems for the entity-definition phase of a GIS project. The first is how to select the entity type that provides the most appropriate representation for the feature being modelled. Is it best to represent a forest as a collection of points (representing the location of individual trees), or as an area (the boundary of which defines the territory covered by the forest)? The second problem is how to represent changes over time. A forest, originally

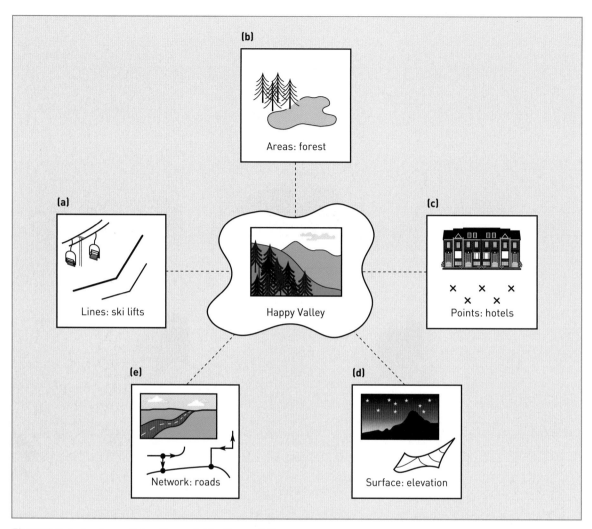

**Figure 3.2** A simple spatial entity model for Happy Valley

## BOX 3.1 Surfaces and networks

THEORY

A surface entity is used to represent continuous features or phenomena. For these features there is a measurement or value at every location, as is the case with elevation, temperature and population density as in Figure 3.3. In Happy Valley, elevation and snow depth are represented by surfaces. Both elevation and snow depth vary continuously over space – a value for each can be recorded at any location. This makes representation by a surface entity appropriate (Figure 3.4). The continuous nature of surface entities distinguishes them from other entity types (points, lines, areas and networks) which are discrete, that is either present or absent at a particular location.

A network is a series of interconnecting lines along which there is a flow of data, objects or materials. There are several networks in Happy Valley. One is the road network, along which there is a flow of traffic to and from the ski areas. Another is the river, along which there is a flow of water. Others, not visible on the land surface, include the sewerage and telephone systems. Examples of network data are shown in Figures 3.5 and 3.6.

Data sets and activities relating to the Happy Valley Case Study can be found online at **www. pearsoned. co.uk/heywood**.

BOX 3.1

**(a)** Terrain

**(b)** Rainfall

**(c)** Population

**Figure 3.3** Examples of surface data

**Figure 3.4**
Elevation and
snow depth
contours in
Happy Valley

**Figure 3.6**
Roads, rivers
and sewage
networks in
Happy Valley

**(a)** Railways

**(b)** Rivers

**(c)** Roads

**Figure 3.5**
Examples of
network data in
New Zealand
*Source*: From the
*Digital Chart of the
World*, courtesy
of Esri. Copyright
© Esri. All rights
reserved.

represented as an area, may decline as trees die or are felled until it is only a dispersed group of trees that are better represented using points.

In addition, the nature of the physical features themselves may remain static, but our perception of their geographic structure may differ according to our application area. For visitors to find their way around Happy Valley a map of the road network, with roads represented as a series of interconnected lines, would be sufficient. However, for road maintenance engineers, who must repair the surface after the winter season, it may be more appropriate to represent the roads as areas or even as surfaces.

The definition of entity types for real-world features is also hampered by the fact that many real-world features simply do not fit into the categories of entities available. An area of natural woodland does not have a clear boundary as there is normally a transition zone where trees are interspersed with vegetation from a neighbouring habitat type. In this case, if we wish to represent the woodland by an area entity, where do we place the boundary? The question is avoided if the data are captured from a paper map where a boundary is clearly marked, as someone else will have made a decision about the location of the woodland boundary. But is this the true boundary? Vegetation to an ecologist may be a continuous feature (which could be represented by a surface), whereas vegetation to a forester is better represented as a series of discrete area entities (Figure 3.7).

Features with 'fuzzy' boundaries, such as the woodland, can create problems for the GIS designer and the definition of entities, and may have an impact on later analysis. Consider a grant that the Happy Valley ski company has received to pay for maintenance of the ski area. If the grant is awarded, *pro rata*, according to the area of maintained piste and GIS is used to calculate the area of piste, then how the boundary between maintained piste and off-piste is defined will clearly be of financial importance!

An example of the problem of scale and entity definition is illustrated by considering two ways in which the geographical location of the ski lifts might be modelled. First, imagine that the marketing department at Happy Valley has decided to create a GIS that will permit skiers to check the

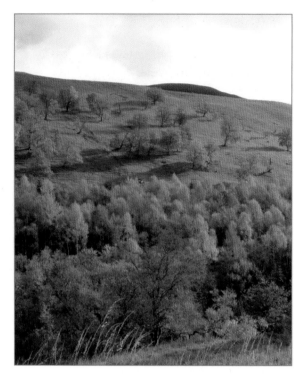

**Figure 3.7** Continuous canopy and open woodland: an example of an entity definition problem

status of ski lifts (whether they are open or closed) over the Internet. For the production of this digital lift map a simple set of line entities, to which information can be attached about the status of the lift (open or closed), is appropriate. The simple line entity gives the user the geographical reference and information they require to assess whether it is worth visiting Happy Valley. Second, consider the view of the ski lift network required by a lift engineer, who has to organize repairs to damaged lifts. This view might be quite different. Additional entities are required, perhaps including a set of point entities to mark the locations of the supporting pylons, and area entities to mark the embarkation and disembarkation zones.

Therefore, deciding which entity type should be used to model which real-world feature is not always straightforward. The way in which individuals represent a spatial feature in two dimensions will have a lot to do with how they conceptualize the feature. In turn, this will be related to their own experience and how they wish to use the entity they produce. An appreciation of this issue is central to the design and development of all GIS applications.

# SPATIAL DATA MODELS

Burrough (1986) recognizes that the human eye is highly efficient at recognizing shapes and forms, but the computer needs to be instructed exactly how spatial patterns should be handled and displayed. Computers require unambiguous instructions on how to turn data about spatial entities into graphical representations. This process is the second stage in designing and implementing a data model. At present there are two main ways in which computers can handle and display spatial entities. These are the raster and vector approaches.

The raster spatial data model is one of a family of spatial data models described as *tessellations*. In the raster world individual cells are used as the building blocks for creating images of point, line, area, network and surface entities. Figure 3.8 shows how

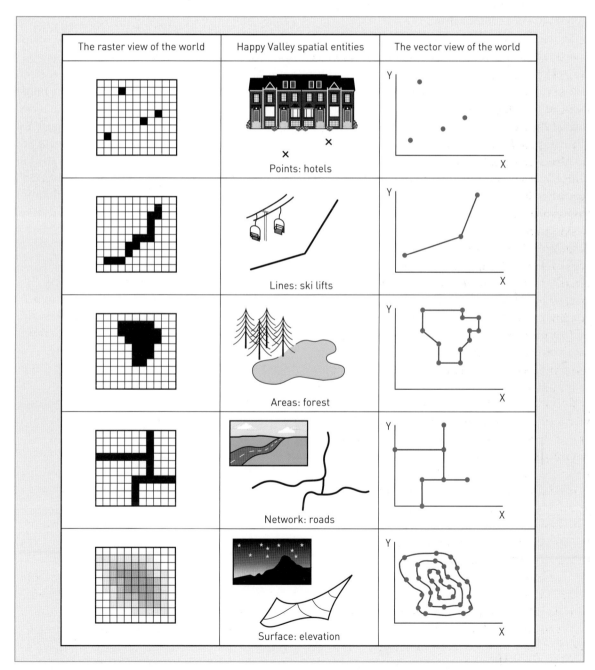

**Figure 3.8** Raster and vector spatial data

a range of different features from Happy Valley, represented by the five different entity types, can be modelled using the raster approach. Hotels are modelled by single, discrete cells; the ski lifts are modelled by linking cells into lines; the forest by grouping cells into blocks; and the road network by linking cells into networks. The relief of the area has been modelled by giving every cell in the raster image an altitude value. In Figure 3.8 the altitude values have been grouped and shaded to give the appearance of a contour map.

In the raster world the basic building block is the individual grid cell, and the shape and character of an entity is created by the grouping of cells. The size of the grid cell is very important as it influences how an entity appears. Figure 3.9 shows how the spatial character of the Happy Valley road network changes as the cell size of the raster is altered.

A vector spatial data model uses two-dimensional Cartesian (x,y) co-ordinates to store the shape of a spatial entity. In the vector world the point is the basic building block from which all spatial entities

are constructed. The simplest spatial entity, the point, is represented by a single (x,y) co-ordinate pair. Line and area entities are constructed by connecting a series of points into chains and polygons. Figure 3.8 shows how the vector model has been used to represent various features for the Happy Valley ski area. The more complex the shape of a line or area feature the greater the number of points required to represent it. Selecting the appropriate number of points to construct an entity is one of the major dilemmas when using the vector approach. If too few points are chosen the character, shape and spatial properties of the entity (for example, area, length, perimeter) will be compromised. If too many points are used, unnecessary duplicate information will be stored and this will be costly in terms of data capture and computer storage. Figure 3.9 shows how part of the Happy Valley road network is affected by altering the number of points used in its construction. Methods have been developed to automate the procedure for selecting the optimum number of points to represent a line or area feature.

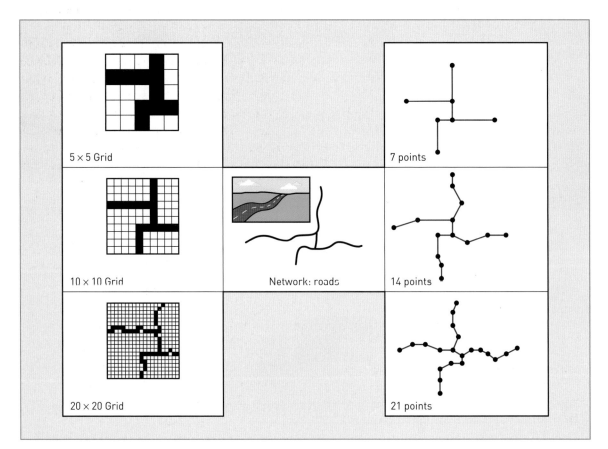

**Figure 3.9** Effect of changing resolution in the vector and raster worlds

The Douglas–Peucker algorithm is one example (see **Chapter 5**). This removes points that do not contribute to the overall shape of the line (Douglas and Peucker, 1973).

In the vector data model the representation of networks and surfaces is an extension of the approach used for storing line and area features. However, the method is more complex, and closely linked to the way the data are structured for computer encoding. Therefore, the approach used to store network and surface features is considered further in the following sections.

## ■ SPATIAL DATA STRUCTURES

Data structures provide the information that the computer requires to reconstruct the spatial data model in digital form. There are many different data structures in use in GIS. This diversity is one of the reasons why exchanging spatial data between different GIS software can be problematic. However, despite this diversity data structures can be classified according to whether they are used to structure raster or vector data.

### Raster data structures

In the raster world a range of different methods is used to encode a spatial entity for storage and representation in the computer. Figure 3.10 shows the most straightforward method of coding raster data.

The cells in each line of the image (Figure 3.10a) are mirrored by an equivalent row of numbers in the file structure (Figure 3.10c). The first line of the file tells the computer that the image consists of 10 rows and 10 columns and that the maximum cell value is 1. In this example, a value of 0 has been used to record cells where the entity is not present, and a value of 1 for cells where the entity is present (Figure 3.10b).

In a simple raster data structure, such as that illustrated in Figure 3.10, different spatial features must be stored as separate data layers. Thus, to store all five of the raster entities shown in Figure 3.8, five separate data files would be required, each representing a different layer of spatial data. However, if the entities do not occupy the same geographic location (or cells in the raster model), then it is possible to store them all in a single layer, with an entity code given to each cell. This code informs the user which entity is present in which cell. Figure 3.11 shows how different land uses can be coded in a single raster layer. The values 1, 2 and 3 have been used to classify the raster cells according to the land use present at a given location. The value 1 represents residential area; 2, forest; and 3, farmland.

One of the major problems with raster data sets is their size, because a value must be recorded and stored for each cell in an image. Thus, a complex image made up of a mosaic of different features (such as a soil map with 20 distinct classes) requires the same amount of storage space as a similar raster map showing the location of a single forest. To address this problem a range of data compaction

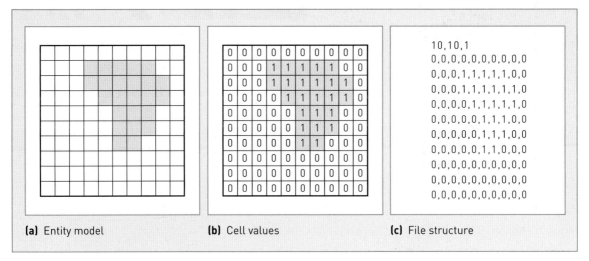

**(a)** Entity model          **(b)** Cell values          **(c)** File structure

**Figure 3.10** A simple raster data structure

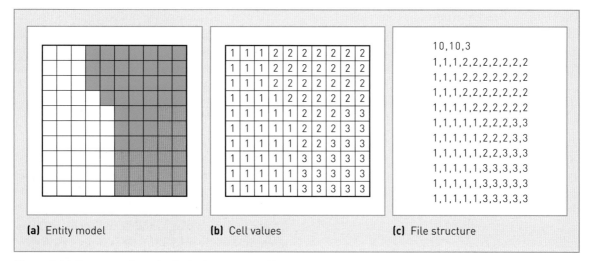

**(a)** Entity model        **(b)** Cell values        **(c)** File structure

**Figure 3.11** Feature coding of cells in the raster world

methods have been developed. These include run length encoding, block coding, chain coding and quadtree data structures (Box 3.2).

## Vector data structures

Figure 3.8 shows how the different entity types – points, lines and areas – can be defined by co-ordinate geometry. However, as with the raster spatial data model, there are many potential vector data structures that can be used to store the geometric representation of entities in the computer.

The simplest vector data structure that can be used to reproduce a geographical image in the computer is a file containing $(x,y)$ co-ordinate pairs that represent the location of individual point features (or the points used to construct lines or areas).

Figure 3.14a shows such a vector data structure for the Happy Valley car park. Note how a closed ring of co-ordinate pairs defines the boundary of the polygon.

The limitations of simple vector data structures start to emerge when more complex spatial entities are considered. For example, consider the

## BOX 3.2 Raster data compaction techniques

**THEORY**

1 *Run length encoding*. This technique reduces data volume on a row by row basis. It stores a single value where there are a number of cells of a given type in a group, rather than storing a value for each individual cell. Figure 3.12a shows a run length encoded version of the forest in Happy Valley. The first line in the file represents the dimensions of the matrix (10 × 10) and number of entities present (1). In the second and subsequent lines of the file, the first number in the pair (either 1 or 0 in this example) indicates the presence or absence of the forest. The second number indicates the number of cells referenced by the first. Therefore, the first pair of numbers at the start of the second line tell us that no entity is present in the first 10 cells of the first row of the image.

2 *Block coding*. This approach extends the run length encoding idea to two dimensions by using a series of square blocks to store data. Figure 3.12b shows how the simple raster map of the Happy Valley forest has been subdivided into a series of hierarchical square blocks. Ten data blocks are required to store data about the forest image. These are seven unit cells, two four-cell squares and one nine-cell square. Co-ordinates are required to locate the blocks in the raster matrix. In the example, the top left-hand cell in a block is used as the locational reference for the block.

3 *Chain coding*. The chain coding method of data reduction works by defining the boundary of the entity. The boundary is defined as a sequence of unit

**BOX 3.2**

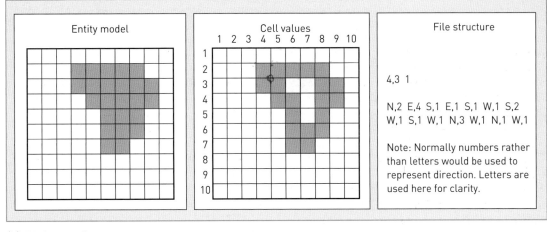

```
10,10,1
0,10
0,3    1,5    0,2
0,3    1,6    0,1
0,4    1,5    0,1
0,5    1,3    0,2
0,5    1,3    0,2
0,5    1,2    0,3
0,10
0,10
0,10
```

**(a)** Run length encoding

**Entity model**

**Cell values**

**File structure**

| Block size | No. | Cell co-ordinates |
|---|---|---|
| 1 | 7 | 4,2 8,2 4,3 6,5 6,6 6,7 7,7 |
| 4 | 2 | 8,3 7,5 |
| 9 | 1 | 5,2 |

**(b)** Block encoding

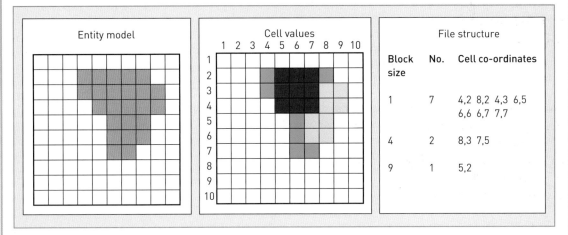

**Entity model**

**Cell values**

**File structure**

4,3  1

N,2  E,4  S,1  E,1  S,1  W,1  S,2
W,1  S,1  W,1  N,3  W,1  N,1  W,1

Note: Normally numbers rather than letters would be used to represent direction. Letters are used here for clarity.

**(c)** Chain encoding

**Figure 3.12** Raster data compaction techniques

**BOX 3.2**

cells starting from and returning to a given origin. The direction of travel around the boundary is usually given using a numbering system (for example 0 = North, 1 = East, 2 = South and 3 = West). Figure 3.12c shows how the boundary cells for the Happy Valley forest would be coded using this method. Here, the directions are given letters (N, S, E and W) to avoid misunderstanding. The first line in the file structure tells us that the chain coding started at cell 4,3 and there is only one chain. On the second line the first letter in each sequence represents the direction and the number of cells lying in this direction.

**4** *Quadtrees.* One of the advantages of the raster data model is that each cell can be subdivided into smaller cells of the same shape and orientation (Peuquet, 1990). This unique feature of the raster data model has produced a range of innovative data storage and data reduction methods that are based on regularly subdividing space. The most popular of these is the area or region quadtree. The area quadtree works on the principle of recursively subdividing the cells in a raster image into quads (or quarters). The subdivision process continues until each cell in the image can be classed as having the spatial entity either present or absent within the bounds of its geographical domain. The number of subdivisions required to represent an entity will be a trade-off between the complexity of the feature and the dimensions of the smallest grid cell. The quadtree principle is illustrated in Figure 3.13.

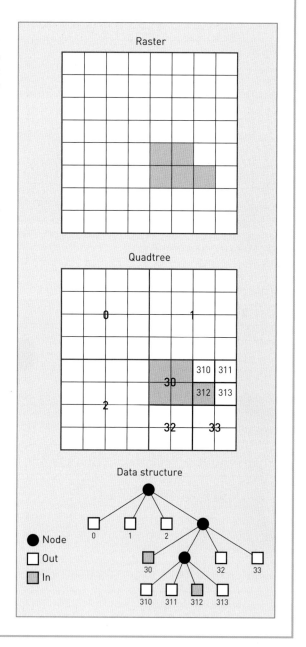

**Figure 3.13** The quadtree

Happy Valley car park divided into different parking zones (Figure 3.14b). The car park consists of a number of adjacent polygons. If the simple data structure, illustrated in Figure 3.14a, were used to capture this entity then the boundary line shared between adjacent polygons would be stored twice. This may not appear too much of a problem in the case of this example, but consider the implications for a map of the 50 states in the USA. The amount of duplicate data would be considerable.

This method can be improved by adjacent polygons sharing common co-ordinate pairs (points). To do this all points in the data structure must be numbered sequentially and contain an explicit reference which records which points are associated with which polygon. This is known as a *point dictionary* (Burrough, 1986). The data structure in Figure 3.14b shows how such an approach has been used to store data for the different zones in the Happy Valley car park.

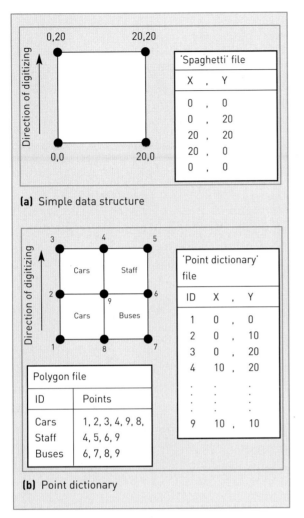

**(a)** Simple data structure

'Spaghetti' file

| X | , | Y |
|---|---|---|
| 0 | , | 0 |
| 0 | , | 20 |
| 20 | , | 20 |
| 20 | , | 0 |
| 0 | , | 0 |

Polygon file

| ID | Points |
|---|---|
| Cars | 1, 2, 3, 4, 9, 8, |
| Staff | 4, 5, 6, 9 |
| Buses | 6, 7, 8, 9 |

'Point dictionary' file

| ID | X | , | Y |
|---|---|---|---|
| 1 | 0 | , | 0 |
| 2 | 0 | , | 10 |
| 3 | 0 | , | 20 |
| 4 | 10 | , | 20 |
| . | . | | . |
| . | . | | . |
| . | . | | . |
| 9 | 10 | , | 10 |

**(b)** Point dictionary

**Figure 3.14** Data structures in the vector world

The Happy Valley road network illustrates a slightly different problem. The simple vector data structure illustrated in Figure 3.14a could be used to reproduce the network graphically without any duplication of data. However, it would not contain any information about the linkage between lines. Linkages would be implied only when the lines are displayed on the computer screen. In the same way, a series of polygons created using either the simple data structure described in Figure 3.14a or a point dictionary approach (Figure 3.14b) may appear connected on the screen when in fact the computer sees them as discrete entities, unaware of the presence of neighbouring polygons.

A further problem of area features is the island or hole situation. In Figure 3.15 a diamond-shaped area has been placed in the centre of the Happy Valley

car park to represent the information kiosk. This feature is contained wholly within the polygons classified as the car park. Whilst a simple vector file structure would recreate the image of the car park, it would not be able to inform the computer that the island polygon was 'contained' within the larger car park polygon.

For the representation of line networks, and adjacent and island polygons, a set of instructions is required which informs the computer where one polygon, or line, is with respect to its neighbours. Topological data structures contain this information. There are numerous ways of providing topological structure in a form that the computer can understand. The examples below have been selected to illustrate the basic principles underpinning topological data structures rather than describe the structures found in any one GIS environment.

Topology is concerned with connectivity between entities and not their physical shape. A useful way to help understand this idea is to visualize individual lines as pieces of spaghetti. Consider how you would create a model of the Happy Valley car park (Figure 3.15) using strands of spaghetti on a dinner plate. It is most likely that you would lay the various strands so that they overlapped. This would give the appearance of connectivity, at least whilst the spaghetti remained on the plate. If you dropped the plate it would be difficult to rebuild the image from the spaghetti that fell on the floor. Now imagine that the pieces of spaghetti are made of string rather than pasta and can be tied together to keep them in place. Now when the plate is dropped, the connectivity is maintained and it would be easier to reconstruct the model. No matter how you bend, stretch and twist the string, unless you break or cut one of the pieces, you cannot destroy the topology. It is this situation that computer programmers are striving to mirror when they create topological data structures for the storage of vector data. The challenge is to maintain topology with the minimum of data to minimize data volumes and processing requirements.

A point is the simplest spatial entity that can be represented in the vector world with topology. All a point requires for it to be topologically correct is a pointer, or geographical reference, which locates it with respect to other spatial entities. In order for a line entity to have topology it must consist of an

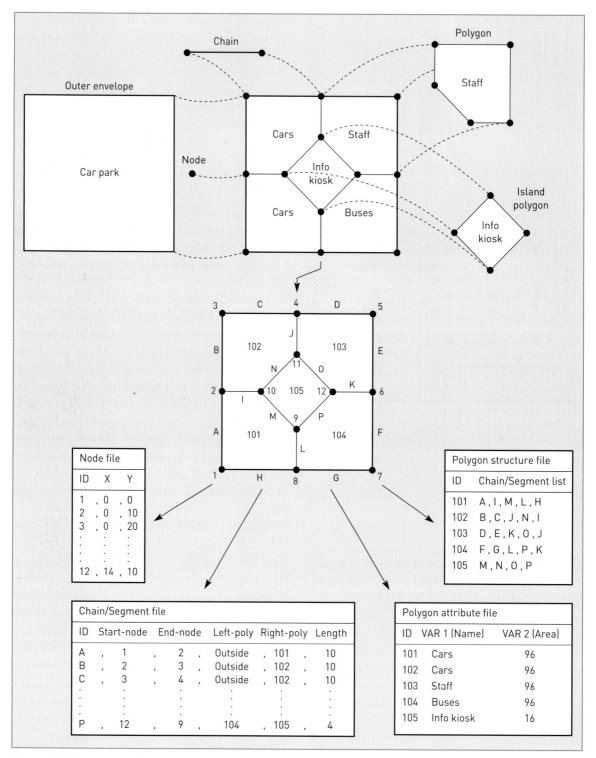

**Figure 3.15** Topological structuring of complex areas

ordered set of points (known as an arc, segment or chain) with defined start and end points (nodes). Knowledge of the start and end points gives line direction. For the creation of topologically correct area entities, data about the points and lines used in its construction, and how these connect to define the boundary in relation to adjacent areas, are required.

There is a considerable range of topological data structures in use by GIS. All the structures available try to ensure that:

- no node or line segment is duplicated;
- line segments and nodes can be referenced to more than one polygon;
- all polygons have unique identifiers; and
- island and hole polygons can be adequately represented.

Figure 3.15 shows one possible topological data structure for the vector representation of the Happy Valley car park. The creation of this structure for complex area features is carried out in a series of stages. Burrough (1986) identifies these stages as identifying a boundary network of arcs (the envelope polygon), checking polygons for closure and linking arcs into polygons. The area of polygons can then be calculated and unique identification numbers attached. This identifier would allow non-spatial information to be linked to a specific polygon.

This section has briefly considered a number of different approaches to the structuring of raster and vector data. It is important to remember that the examples presented here represent only a few of the data structures used by different GIS. There are many variations on these methods. Box 3.3 provides an example of how a spatial data model has been used to structure a commercial data set for use in GIS.

## BOX 3.3 Topologically structured data: MasterMap

**PRACTICE**

The UK Ordnance Survey's MasterMap is an example of a topologically structured data set in which each feature has a unique ID. It was developed in response to an increasing demand for spatially referenced topographic information, and for better structured data for use in GIS.

MasterMap is an object-oriented data set (see **Chapter 4**). It provides a series of data layers, each layer containing features representing real-world objects such as buildings, roads and postboxes. Each feature has a unique 16-digit identifier called a Topographic Identifier or TOID®. An example allocated to a building might be 0001100028007854. This TOID® remains stable throughout the life of a feature, so if a change is made, say the building is extended, the TOID® stays the same.

There are over 430 million features in the MasterMap database and 5000 updates are made every day. The data includes a layer with all 26 million postal addresses in England, Wales and Scotland referenced to less than 1 m. The MasterMap data have been widely adopted and used successfully in a range of projects.

The special features of MasterMap are:

- layers of data provide a seamless topographic database for the UK at scales of 1:1250 and 1:2500;
- the data are made up of uniquely referenced features created as points, lines or polygons;
- each feature has a unique 16 digit topographic identifier called a TOID®;
- MasterMap data can be supplied in a topologically structured format;
- users request a 'chunk' of the data which meets their needs and they can request data by theme or geographic area;
- metadata is included which contains additional detail about the features;
- the data are continually being updated and change-only updates are available to users.

Further information on OS MasterMap is available at www.ordnancesurveymastermap.com/.

**(a)** Topography

**(b)** Imagery

**(c)** ITN

**(d)** Address layer

**Figure 3.16** Ordnance Survey MasterMap
*Source*: Reproduced by permission of Ordnance Survey on behalf of HMSO. © Crown Copyright 2011. All rights reserved.
Ordnance Survey Licence number 100030901

## REFLECTION BOX

- Apply the steps in the abstraction process to a GIS project of your own choice:

    1 identify spatial features of interest and explain why you have selected them;

    2 choose an appropriate spatial data model and justify your choice; and

    3 choose an appropriate spatial data structure and give reasons for your choice.

- Entities that change and those that have fuzzy boundaries may be difficult to represent in a GIS model. Make a list of 10 features that may change, and 10 with fuzzy boundaries.

- Create a table summarizing the advantages and disadvantages of the raster and vector data models for a GIS application of your choice.

- Find accounts of two or three GIS applications in your field or area of interest. Try to establish the nature of the data model and data structure that has been used to represent the real world in each case.

## MODELLING SURFACES

The modelling of surface entities such as height, pollution and rainfall poses interesting problems in GIS. In this section we explore surface modelling by looking in detail at digital terrain models (DTMs). These are used as an example because of their wide application in GIS. Figure 3.17 shows some examples of surfaces in GIS.

The abbreviation DTM describes a digital data set which is used to model a topographic surface (a surface representing height data). To model a surface accurately it would be necessary to store an almost infinite number of observations. Since this is impossible, a surface model approximates a continuous surface using a finite number of observations. Thus, an appropriate number of observations must be selected, along with their geographical location.

**(a)** 3D contour

**(b)** Wire frame

**(c)** Shaded relief

**(d)** 3D DEM

**(e)** Air photo drape

**Figure 3.17** Example surface types

# BOX 3.4 Data sources for DTMs

Contours and spot heights from topographic maps are the most likely source of raw data for a DTM. These may be supplied in digital form for the express purpose of building a DTM, or they may need to be digitized from a paper map (see **Chapter 5**). The digital set of contours and spot heights represents an irregular set of (x,y,z) observations. The digital contours are line features representing lines on the terrain surface of equal height. These can be built into a set of nested polygons. The attributes of these polygons are height ranges as described by the bounding line features of which the polygons are made. Digital contour data are often supplemented by spot heights, which give the height at specific points between the contour lines. These are used to give the location and height of significant terrain features such as peaks, depressions or isolated knolls (Figure 3.18).

**Figure 3.18** Contours and spot heights

Height data may also be derived from the stereoscopic interpretation of paired images of the Earth's surface. The paired images may be either aerial photographs or satellite images that have been taken of the same patch of ground but from slightly different angles. The technique relies on being able to calculate elevation based on the

parallax displacement between the same point on both images. If done for many different points, a DTM may be built up for the area covered by the paired images. An example of this technique has been the introduction of detailed DTMs derived from high-resolution satellite imagery including SPOT and IKONOS (Figure 3.19).

**Figure 3.19** Stereoscopic satellite imagery and derived DTM data
*Source*: NASA/JPL/NIMA

Other kinds of remotely sensed data have been developed that directly measure elevation using radar or laser scanning sensors. Examples include Synthetic Aperture Radar (SAR), Light Detection and Ranging (LiDAR) and the Shuttle Radar Topography Mission (SRTM). These have resolutions varying from 90 or 30 metres for SRTM data to a few metres for SAR and LiDAR data. The spatial coverage of these products varies between almost global coverage for

**BOX 3.4**

SRTM data to selected small areas for LiDAR (Figure 3.20). One key advantage of these data is that they provide elevation data in the form of Digital Elevation Matrices (DEMs) that can be input directly to a GIS.

GPS receivers give accurate (x,y) co-ordinate locations and provide height information. They are not as accurate in the vertical plane as they are in the horizontal, though height data may be used in conjunction with (x,y) co-ordinate data to produce a DTM if the user is prepared to take readings all over the area of interest. It is more appropriate to use GPS readings for ground truthing of other data sets using high accuracy differential techniques.

Examples of data source selection criteria for DTMs are presented in Table 3.1.

**(a)** SRTM

**(b)** LiDAR

**(c)** SAR

**Figure 3.20** Examples of SRTM, LiDAR and SAR data
*Sources*: (a, c) NASA/JPL/NIMA; (b) Environment Agency Science Group Technology

**BOX 3.4**

**TABLE 3.1  DTM data source selection criteria**

| Data source | Capture method | DTM accuracy | Coverage |
|---|---|---|---|
| Maps | Digitizing or scanning | Low–moderate | All scales, but only where map coverage exists |
| Aerial photographs | Stereo plotters | High | Large areas |
| Satellite imagery | Stereo autocorrelation | Moderate | Large areas |
| Radar and laser scanning | Direct measurement | High–very high | All scales |
| GPS | Direct downloading from GPS receiver | High | Small project areas with moderate terrain |
| Ground survey | Direct downloading from field surveying equipment | Very high | Very small project areas, not feasible in rough terrain |

*Source*: Adapted from Stocks and Heywood (1994)

The 'resolution' of a DTM is determined by the frequency of observations used. DTMs are created from a series of either regularly or irregularly spaced (x,y,z) data points (where x and y are the horizontal co-ordinates and z is the vertical or height co-ordinate). DTMs may be derived from a number of data sources. These include contour and spot height information found on topographic maps, stereoscopic aerial photography, satellite images and field surveys (see Box 3.4).

## The raster approach to digital terrain modelling

There are differences between the way in which DTMs are constructed in raster-based and vector-based GIS. In raster GIS a DTM is simply a grid of height values in which each cell contains a single value representative of the height of the terrain that is covered by that cell (Figure 3.21). This type of simple raster DTM is perhaps more accurately referred to as a Digital Elevation Matrix (DEM) because it contains no information about the terrain surface being modelled other than the elevation or height above the datum. How accurately the terrain can be modelled using this method depends on the complexity of the terrain surface and the spacing (resolution) of the grid. If the terrain is complex, for example in a mountainous area like the European Alps or the North American Rocky Mountains, a fine grid will be required to represent it with any

degree of accuracy. If the terrain is relatively even, such as in the Netherlands or the Great Plains of the USA and Canada, fewer points will be needed to model it to the same degree of accuracy. Thus, a coarser-resolution DTM may be used.

Problems may occur with raster DTMs where the terrain is varied or complex. For example, if the area to be modelled includes an area of mountainous terrain to the north and a wide, gently sloping area to the south, choosing an appropriate resolution is difficult (Figure 3.21). A fine-resolution raster will accurately model the mountains in the west, but a high degree of data redundancy will be experienced in flat areas to the east. This could produce problems of large data storage requirements and long computing times. If you choose a low-resolution raster such as might be suitable for the flat eastern part of the area, the detail of the mountains to the west will be lost. The final choice will inevitably be a compromise. To counter this problem many GIS applications use a quadtree data structure or one of the raster data compression techniques described in Box 3.2 for storing raster DTM data.

## The vector approach to digital terrain modelling

In its simplest form a vector DTM mimics the raster version by using a regularly spaced set of spot heights to represent the terrain surface (Figure 3.22). A more advanced, more complex and more

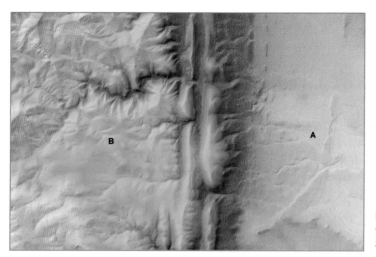

**Figure 3.21** Raster DTM (A) Simple terrain (B) Complex terrain
*Source*: United States Geological Survey

common form of vector DTM is the triangulated irregular network (TIN). In vector GIS a TIN is used to create a DTM from either regular or irregular height data. The TIN method joins the height observations together with straight lines to create a mosaic of irregular triangles. In the TIN model of a surface, the vertices of the triangles produced represent terrain features such as peaks, depressions and passes, and the edges represent ridges and valleys. The surfaces of individual triangles provide area, gradient (slope) and orientation (aspect). These values can be stored as TIN attributes or can be quickly calculated when the TIN is used in further analysis. Additional topographic features that do not necessarily describe lines of equal height can also be incorporated into the TIN model. These include breaks of slope such as ridges, troughs, and cliff edges/bases. Water features can also be incorporated into it to assist in the creation of an accurate DTM. Rivers and streams flow in valley bottoms and are therefore used to define trough lines, whilst lake and sea shores can be used to define flat areas

associated with surface water. An example of a TIN model is shown in Figure 3.23. A physical example is described in Box 3.5.

**(a)** DEM

**(b)** TIN for region of varying complexity

**Figure 3.23** Example DEM and TIN model for region of varying complexity

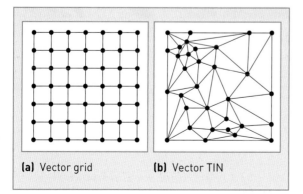

**(a)** Vector grid          **(b)** Vector TIN

**Figure 3.22** Digital terrain models

## BOX 3.5 Make your own TIN model

**PRACTICE**

A simple experiment you can try yourself is to construct a TIN model using a sheet of paper. Take a sheet of paper and crumple it into a loose ball. Now carefully open it out again so that it resembles a sheet of paper again but still has all the creases you have just put into it. This is an analogue of your TIN model. The creases are the ridges and valleys and the intersections of the creases are the peaks, depressions and passes. The areas of flat paper between the creases are the irregular triangles of the TIN model, which may be assigned area, slope and aspect values. Obviously this is not a perfect model as not all of the facets on your piece of paper will be triangles, but it does serve to illustrate the TIN principle.

Now hold the paper level with a light source, such as a window or desk lamp, to dramatize the effect. With a little imagination you can see a miniature terrain of peaks and valleys casting shadows in front of you. The tighter you crumple your paper the more complex the terrain you produce, and the more you try to flatten it out the lower the relief. This experiment also demonstrates the 'two and a half' dimensional nature of terrain models in GIS. You can see clearly that the model you have just created is only a surface with no depth.

---

The main advantage of the TIN model is efficiency of data storage, since only a minimum number of significant points need to be stored to reproduce a surface. In the previous section we considered the difficulties of choosing an appropriate resolution for a raster DTM in an area of mixed terrain. Since TINs are created from irregularly spaced points, more points are used to represent the mountainous areas and fewer to represent the flatter areas.

To achieve this efficiency in storage, TIN models use only 'surface significant' points to reproduce a terrain surface. These points are selected by the TIN model from the input data on the basis of their spatial relationship with their neighbours (Lee, 1991).

As in spatial interpolation (see **Chapter 6**), Tobler's law of geography (1976) regarding spatial autocorrelation comes into play. This states that points closer together are more likely to have similar characteristics (here in terms of their height) than those that are further apart. Consequently, if a height observation can be closely interpolated from its neighbours, then it is not considered to be 'surface significant' and is dropped from the TIN model. Those points that cannot be closely interpolated from their neighbours are considered to be 'surface significant' and are used as TIN vertices. Box 3.6 looks at four of the most common methods used to identify which points are surface significant.

## BOX 3.6 Methods for identifying surface significant points

**THEORY**

The four most common methods used to identify surface significant points in the construction of a TIN are as follows.

**1** *The skeleton method.* This method, developed by Fowler and Little (1979), is used to select surface significant points from a regular grid of height data using a 3 × 3 cell window. The window is passed over the data one cell at a time and the height value of the centre cell compared with its neighbours. A cell is labelled a peak if all the neighbouring cells in the window are lower; a pit (depression) if all the neighbouring cells are higher; and a pass if the heights of the neighbouring cells alternate lower and higher around the centre cell. Next, a 2 × 2 window is passed over the data. Cells in the 2 × 2 window that are highest are labelled as ridges and cells that are lowest are labelled as valleys. Peaks are then linked via ridges, and pits via valleys, to define the skeleton of the terrain. A line-thinning algorithm should then be used to identify the most crucial points from which to construct the TIN. This sounds simple enough but is a complex, computer-intensive, algorithm which is sensitive to noise in the data since it considers only the relative height differences between neighbouring cells.

2 *The filter or VIP.* The Very Important Point (VIP) method was developed by Chen and Guevara (1987) to identify points that define the character of a terrain. VIPs are determined by how well they can be approximated by their neighbours. If the predicted value of the central point is substantially different from its actual value then the point is considered to be a VIP and an essential component of the TIN. Points that are diametrically opposite the central point are connected by a straight line and the perpendicular distance of the central point in the third dimension calculated. This procedure is repeated for other diametrically opposed points and the average distance calculated. This average is then used as a measure of importance or significance of the point. The advantage of this method is that it is easy to compute and requires only one pass through the data. The number of points selected can also be specified by setting a 'significant threshold' at which points are considered to be VIPs.

3 *The hierarchy method.* This method, developed by DeFloriani *et al.* (1984), selects surface significant points by subdividing the total set of points into a series of successively smaller triangles in a manner similar to the quadtree data structure. The method evaluates all points located inside the first-level triangle by calculating the difference between their original elevations and the elevations interpolated from the enclosing triangle. This difference is used

as a measure of the error associated with the fit of the TIN model to each point at this first level of triangulation. The point with the largest error is identified and if this is greater than a predefined level of precision it is used as the node by which to further subdivide the first-level triangle into three smaller triangles. This process continues until all triangles in the TIN meet the pre-specified precision level. Advantages of this method include greater subdivision and smaller triangles in areas of more complex terrain and an efficient data structure allowing fast retrieval and access to information in the TIN. The main disadvantage is that the TIN inevitably consists of many long, thin triangles and the original first-level subdivision remains in the final TIN model. TIN models work better when the triangles are close to being equilateral (with angles of 60°) since this ensures that any point on the surface is as close as possible to a TIN vertex.

4 *The drop heuristic method.* This method works by starting with a full set of points and gradually discarding those of least importance. After each point is dropped a new TIN is constructed and the triangle containing the dropped point identified. The difference between the true elevation of the dropped point and the estimated elevation of the point based on the new TIN is calculated and a predefined tolerance level used to determine whether the point should remain in the TIN. The advantage of this technique is that it produces a TIN that best reflects the surface represented by the raw data, but it can be computationally very intensive.

The methods used to identify a set of surface significant points (except for the hierarchy method) do not help determine which points to connect to which. One of two methods is generally used to achieve the linking of TIN vertices. The first of these is known as distance ordering, where the distance between all pairs of points is calculated. These are sorted and the closest pair connected. The next closest pair are then connected if the resulting line does not cross any existing connections. This process is repeated until the TIN is complete. However, this method can produce too many thin triangles. The second method is Delaunay triangulation. Delaunay triangles are determined by a circle. Three points form the corners of a Delaunay triangle only when the circle that passes through them contains no other points.

## ■ MODELLING NETWORKS

A network is a set of interconnected linear features through which materials, goods and people are transported or along which communication of information is achieved. Network models in GIS are abstract representations of the components and characteristics of their real-world counterparts. They are essentially adaptations of the vector data model and for this reason raster GIS are generally not very good at network analysis. The vector network model is made up of the same arc (line segments) and node elements as any other vector data model but with the addition of special attributes as illustrated in Figure 3.24.

**Figure 3.24** Network data model

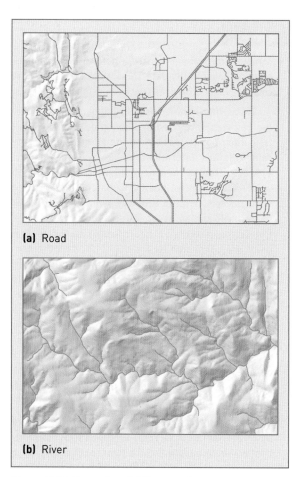

**Figure 3.25** Examples of GIS networks
*Source*: United States Geological Survey

In the network model the arcs become network links representing the roads, railways and air routes of transport networks; the power lines, cables and pipelines of the utilities networks; or the rivers and streams of hydrological systems (Figure 3.25). The nodes in turn become network nodes, stops and centres. Network nodes are simply the endpoints of network links and as such represent junctions in transport networks, confluences in stream networks, and switches and valves in utilities networks. Stops are locations on the network that may be visited during a journey. They may be stops on a bus route, pick-up and drop-off points on a delivery system, or sediment sources in a stream network. They are points where goods, people or resources are transferred to and from some form of transport system. Centres are discrete locations on a network at which there exists a resource supply or some form of attraction. Examples include shopping centres, airports, schools and hospitals. At a larger scale centres may be a whole city if the transport, resource or information networks for an entire country are being considered. Turns represent the transition from network link to network link at a network node, and therefore also represent the relationships between network links and greatly affect movement through the network system. For example, turns across oncoming traffic on a road network take longer than turns down slipways, whereas turns that go against the flow of traffic on one-way streets are prohibited altogether.

All the data regarding the characteristics of network links, nodes, stops, centres and turns are stored as attribute information in the vector model database. Two key characteristics of network features are *impedance* and *supply and demand*.

Impedance is the cost associated with traversing a network link, stopping, turning or visiting a centre. For example, link impedance may be the time it takes to travel from one node to another along a network link. If we use the example of a delivery van travelling along a city street, then the impedance value represents time, fuel used and the driver's pay. Factors influencing the impedance value will include traffic volume as determined by time of day and traffic control systems; direction (for instance, one-way streets); topography (more fuel is used going uphill); and weather (more fuel is used travelling into a strong headwind). Different links have different impedance values depending on local conditions. Turn impedance is also important and may be represented by the cost of making a particular turn. For example, when the delivery van reaches the end of the current link, the direction the driver chooses to turn may be strongly influenced by turn impedance. Stop impedance refers to the cost of making a stop and operates in a similar way to link and turn impedance. The delivery van driver may not be able to find a parking space and could get a parking ticket for stopping on double yellow lines. Impedance values are, therefore, very important in determining the outcome of route finding, allocation and spatial interaction operations. The various forms of impedance values that may be linked to a network are illustrated in Figure 3.26.

Link impedances:

→  One-way flows

⇄  Two-way flows

—  Narrow street – slow traffic flows

▬  Wide street – faster traffic flows

Turn impedances:

■  Junction controlled by traffic lights

▸■  No left turn (turn impedence) in direction of arrow

c  Traffic congestion

Stops and stop impedances:

D  Van depot (route start and end point)

S  Storage warehouse (point of supply) loading wait time

c  Customer location (point of demand)

P  Parking problems

**Figure 3.26** Link, turn and stop impedances affecting the journey of a delivery van

Supply and demand are equally important concepts in network analysis. Supply is the quantity of a resource available at a centre that is available to satisfy the demand associated with the links of a network. If we take the example of a hospital, then supply would be represented by the number of beds available. Demand, on the other hand, is the utilization of a resource by an entity that is associated with a network link or node. In the hospital case, demand is represented by the number of people requiring treatment living within the hospital's catchment area.

Correct topology and connectivity are extremely important for network analysis. Digital networks should be good topological representations of the real-world network they mimic. Correct geographical representation in network analysis is not so important, so long as key attributes such as impedance and distance are preserved. A classic example of this is the famous map of the London Underground system. This bears little resemblance to the real-world map of the underground system, which would be far too complex for underground users to follow. Instead, the map was redrawn as a linear cartogram to simplify the network, making it easily understood whilst at the same time maintaining relative distance and connectivity between all the stations on the network. Figure 3.27 demonstrates how this technique has been used to produce a map for the Tyne and Wear Metro. Figure 3.27a shows the network as a linear cartogram and Figure 3.27b the real geographic pattern of the network. Cartograms are covered in more detail in **Chapter 8**.

## ■ BUILDING COMPUTER WORLDS

This chapter has explored the way in which different spatial data models (raster and vector) and their respective data structures are used to store entities. In this section we will consider the methods used to construct computer worlds by grouping these entities together. At the present time we have two options: layers and objects.

The most common method of structuring the geography of the real world in the computer is to use a layered approach (Laurini and Thompson, 1992). This layered approach is illustrated in Figure

**(a)** Linear cartogram

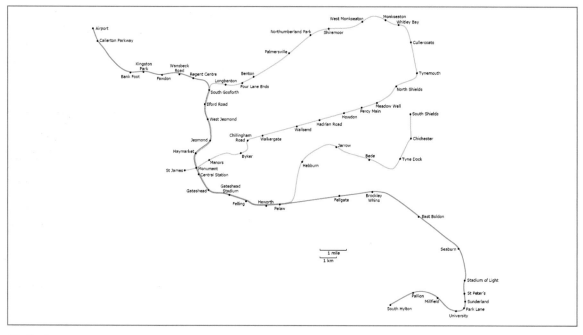

**(b)** Real space

**Figure 3.27** Two representations of the Tyne and Wear Metro Map
*Sources*: (a) © Nexus; (b) Steve Carver

3.28, in which Happy Valley has been dissected into a series of layers. Each layer is thematic and reflects either a particular use or a characteristic of the landscape. For example, layer one describes the general land use characteristics, layer two the soil type and layer three the hydrology. This layered approach to structuring spatial information has a long history characterized by the use of thematic maps to show different features for the same area. The map layer concept is helpful since it reminds us that many geographical features overlap with one another. For example, a village is built on top of soil and geology. The concept of breaking the geography of the real world down into a series of thematic layers was the process that was used to develop the first map overlay analysis (McHarg, 1969).

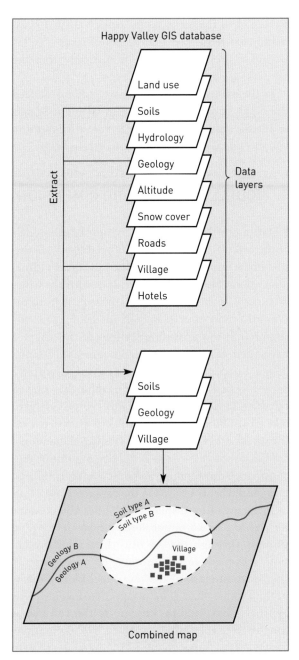

**Figure 3.28** The layer-based approach

data and retrieval and analysis. For example, a mosaic of individual map sheets may be used to provide geographical coverage of a country at a detailed scale.

Another approach to structuring geographical space views the real world as a set of individual objects and groups of objects. This approach draws on the methods of object-oriented (OO) programming (Worboys, 1995). At the conceptual level, the object-oriented approach is straightforward. Figure 3.29 shows how such an approach might be used to structure some of the information for Happy Valley. Notice that features are not divided into separate layers but grouped into classes and hierarchies of objects. This has particular advantages for modelling the geography of the real world, as it more accurately reflects the way we group spatial features using natural language (Frank and Mark, 1991). Adopting an object view of the world has many advantages over the more traditional layer-based approach. However, it does cause some problems associated with implementation into a workable GIS.

**Figure 3.29** The object-oriented approach

There are many different definitions of the object-oriented approach (Worboys *et al.*, 1991). However, all seem to have a similar aim: to bring the conceptual model of the modelling process closer to real-world things and events as opposed to the technicalities of computing (Reeve, 1997). Therefore, in theory at least, the OO approach to modelling should be easier to understand.

A logical extension to the layer concept is the use of tiles. This approach breaks down geographical space into a series of regular or irregular units that can be reassembled through the use of a co-ordinate system. The main purpose of the tiling concept is to assist with the storage of information in the computer. Spatial data sets are frequently large, and many GIS systems require that data are broken down into a set of logical units to assist with physical display of

Consider the design of a GIS to assist in the management and maintenance of Happy Valley. Happy Valley consists of a collection of objects such as ski lifts, car parks and hotels. Each of these has unique characteristics: geographical (size, shape and location); topological (the ski slopes are adjacent to the car parks); related to the activities which go on within objects (hotels may be used for holidaymakers, private parties or conferences); and related to the behaviour of objects in relation to one another (the lifts are used to get the skiers to the pistes). The OO approach tries to retain as much as possible of the character of these objects and their relationships to each other.

Let us assume that a traditional GIS model of Happy Valley has been produced using the concepts discussed earlier in this chapter. The GIS comprises several different data layers made up of point, line and area entities. These spatial entities have specific attributes that inform the user what they are. In addition, the spatial entities have some geometric and topological information stored about them so that, for instance, the car park knows where it is and which entities are next to it. These are all the data that are recorded and can be derived and used through the application of GIS. The OO approach suggests that we should stop thinking about these real-world features in an abstract manner, but instead think of them as objects, about which much more data can be stored. For example, because real-world objects are usually classified into groups, the OO model requests information about the classes to which an object belongs. This class membership can take one of two forms: that which reflects the real-world behaviour of the object; or that which provides information about how the object is to be treated in the computer world. Therefore, if we take the car park object as an example, we can indicate that our polygon is a car park and that it belongs to the class of objects known as ski resort.

In addition, entities may be grouped into classes according to their computer representation. Thus, car parks will also belong to the class of entity known to the computer as *areas*. Areas, in turn, are composed of another set of entities of the class *lines*. The beauty of defining things as objects and providing them with membership of a specific class is that it becomes much easier to perform a whole range of spatial and aspatial operations. Therefore, use of a command such as 'show all car parks' generates a picture of all the car parks in Happy Valley. A command that requested that a map of Happy Valley should be produced would automatically draw the car parks as part of Happy Valley.

In the discussion above we have only introduced the basics of an OO approach to GIS. The idea is revisited in **Chapter 4** and if you wish to know more you should take a closer look at Worboys (1995).

## ■ MODELLING THE THIRD DIMENSION

So far in our discussion of computer-based models of the real world we have largely avoided any discussion of the importance of the third and fourth (time) dimensions, although we have dealt with the basics of surface modelling. There is historical justification for this as most of the GIS modelling concepts explored in the preceding sections of this chapter have their origins in the techniques used for the design and construction of paper maps. **Chapter 2** demonstrated how the map metaphor has shaped the way GIS views the world. Thus, most GIS tend to take a two-dimensional or 'bird's eye' perspective on the world at a particular time. This two-dimensional (2D) snapshot view of the world has, to some extent, constrained our thinking about how spatial information should be structured in the computer.

In terms of its display capabilities the computer screen is a two-dimensional display device even though, through the use of orthographic projection, it is possible to simulate the appearance of the third dimension. However, technically this is nothing more than good perspective drawing. This approach has become known by the somewhat tongue-in-cheek term 'two and a half' dimensions (2.5D). Classic uses of the 2.5D approach are wire frame diagrams to show relief, an example of which is shown in Figure 3.30.

All the features we are trying to model have a third dimension. To produce systems capable of representing the complexities of the real world we need to portray the third dimension in more than a visual way. The representation of the third dimension of an entity can also help us model the form of an entity and associated spatial processes. For example, a three-dimensional (3D) scale model could be used for visualization of a proposed extension to the Happy Valley ski area. The use of the third dimension

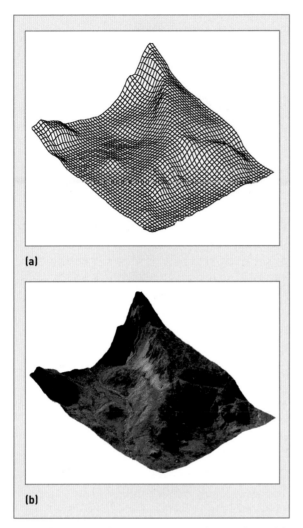

**(a)**

**(b)**

**Figure 3.30** A wire frame perspective of a terrain model (a) wire frame diagram for part of Snowdonia National Park, Wales (the grid interval is 20 m); (b) the wire frame model from part (a) draped with orthorectified aerial photography of the same area.

in this model is crucial to allow people to evaluate the potential impact the development will have on the landscape. Unfortunately, current GIS software products do not permit us to model the third dimension. Partly this is due to the technical difficulties of constructing full 3D models of geographical space. Those including full 3D topology are particularly difficult. Imagine, for example, constructing a 3D quadtree – an octree, as referred to by Raper and Kelk (1991) and Worboys (1995) – or providing full topology for a set of interconnecting 3D polygons (Figure 3.31). It would appear that until recently computer power has been insufficient to convince the major GIS software developers that investing in such data models

was worthwhile. However, as Raper and Kelk (1991) identify, work has been taking place in several sectors, notably the geoscientific, civil engineering and landscape architecture disciplines. Worboys (1995) describes how the simple raster grid cell can be extended into a 3D representation, known as a *voxel*. The problem with the voxel is that it is unable to record topological information. Worboys suggests that the solution to this problem lies in extending the vector approach by using constructive solid geometry to create 3D objects such as cubes, spheres and cylinders (Figure 3.31).

As Raper and Kelk (1991) pointed out, it is only a matter of time before we will see the emergence of 3D modelling as an integral part of the GIS toolbox. To some extent this has already begun to happen. For example, Jacobsen *et al.* (1993) described a new 3D data structure employed to improve modelling and understanding of the behaviour of glaciers, and Batty *et al.* (1998) describe the development of a Virtual London database based around photographic imagery, terrain models, building heights and geometry (see **Chapter 8**).

## ■ MODELLING THE FOURTH DIMENSION

A closer look at virtually any GIS database will reveal that the data layers represent no more than a collage of the state of various entities at a given time. This, in many ways, has much in common with a photograph as a GIS database is often a record of the state of entities or groups of entities at a particular time. If we are lucky, our GIS collage will be made up of entities captured at the same time. However, as the production of a GIS is often a long-term process, it is likely that it will include entities from different dates. Modelling time is made more complex since there are several different sorts of time that GIS developers need to consider (Langran, 1992): work practice time, database time and future time.

Work practice time is the temporal state of a GIS database used by many people. There may be several different versions of a GIS in existence at any one moment in time with several people (perhaps in different offices) working to update information. In a planning context the situation could arise where the same land parcel (area entity) is being edited by two people at the same time.

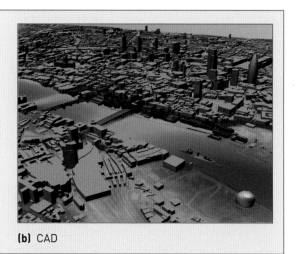

**(b)** CAD

**Figure 3.31** Examples of true 3D data structures
*Sources*: (a) Rockware Inc., with permission; (b) Centre for Advanced Spatial Analysis (CASA), University College London, with permission

Database time is the period for which the database is thought to be correct. Inevitably, this lags behind real time owing to the data capture process. Problems with database time occur if there is more than one version of the database in existence and no procedures to govern who can update it and when.

Many applications of GIS seek to model future time. Applications such as the forecasting of avalanches in Happy Valley require predictions to be made. This can be a complex problem since there may be several possible alternative scenarios. For example, a GIS used to model avalanches under different climatic conditions will yield several different future images that the GIS needs to handle.

The problems with developing a data structure that is flexible enough to handle the time elements of entities include *what* to include, *how frequently* to update the information and *when* to stop storing old information. In addition, the problem is complicated as an organization may need its GIS to be accessible by individuals at different geographic locations, and these individuals may make changes to the time status of an entity. To date, most of these problems are still within the research domain of GIS. Most users of GIS represent the different temporal states of a spatial entity either by creating a separate data layer for each time period or by recording the state of the entity at a given time as

**(a)** 3D geological

a separate field in the database (see **Chapter 4** and Box 3.7). An example of a GIS application in which the modelling of time has been addressed is provided in Box 3.8.

## BOX 3.7 Time and GIS

THEORY

Integrating the dimension of time into GIS presents challenges. There are two main reasons for this. First, it is still rare for the GIS analyst to be in a position where data about a spatial object are available for a continuous period (though this position is changing – see **Chapter 13**). Second, data models and structures that allow us to record, store and visualize information about an object in different temporal states are still in their infancy. This problem is bad enough when the geographic entity under investigation is fixed with respect to location, but is considerably more complex when the object is either mobile or changes its entity type through time.

Peuquet (1999) suggests that there are four types of temporal event. This provides an indication of the types of changes that may affect an entity:

- continuous – these events go on throughout some interval of time;
- majorative – these events go on most of the time;
- sporadic – these events occur some of the time; and
- unique – events that occur only once.

For a skier visiting Happy Valley, skiing would probably represent the majorative event during a single day, with visits to cafés and bars being sporadic events and injuring oneself in a fall being (hopefully) a unique event.

### HANDLING TIME IN GIS

In a raster or vector layer-based GIS, one option for handling time is to store multiple layers for the theme in which you are interested. For example, in the case of the skier a layer could be stored that shows the location of the skier at a particular time. As a number of layers build up for different times, each layer becomes a little like the individual frames in a movie. The problem with this approach is the need for a lot of duplicate data. One solution is to store only information that changes to reduce the data-storage requirements.

In a GIS using an object-oriented data model you can use a slightly different approach. The various elements and attributes which make up an object can each be given a time tag (Worboys, 1995). When the GIS analyst explores the database, the position of the skier changes as the 'time of view' is changed. The presence or absence of the skier is switched on or off at a given location in relation to the time tag. For example, if you choose to view a map of Happy Valley at 1630 hours, when the skier is in the bar, they are linked with the bar entity. When a map of 'building occupancy' is generated for this time the skier is aggregated into the appropriate count.

Much remains to be done before true temporal GIS can be realized (Peuquet, 1999), but a large part of the solution may lie in exploring how the temporal dimension of information has been dealt with by other disciplines.

## BOX 3.8 Monitoring coastal change with GIS

CASE STUDY

### Jonathan Raper

Since maps present a snapshot of all the real-world phenomena they are concerned with at the time of publication they are not designed to show change through time. As GIS have typically evolved from digital mapping systems they have inherited this limitation. This case study will explain the different situations that challenge GIS representation, and then outline an example solution to one of these challenges.

→

**BOX 3.8**

A static vector GIS with layers each containing geometry representing the state of real-world phenomena can be extended to handle changes in the phenomena. So, if you have a property GIS storing land parcels, it is possible to store multiple 'states' for each land parcel: for example, to deal with the sub-division of a land plot into two or more parts. Hornsby and Egenhofer (2000) presented a formal scheme for spatio-temporal knowledge representation, based on possible changes to phenomena, modelled as discrete objects at a high level of abstraction. In their scheme objects representing phenomena can be in the following states:

■ existing;
■ not existing with no history of a previous existence;
■ not existing but with a history of previous existence.

Changes from one state to another are defined as 'transitions', which in total allows nine combinations:

■ continue existence without history
■ create
■ recall
■ destroy
■ continue existence
■ eliminate
■ forget
■ reincarnate
■ continue non-existence with history.

Keeping track of this change involves some extensions to the static GIS data model known as 'temporal' GIS (Langran, 1992). Essentially, this involves keeping track of all the 'states' of each object in a database, as the creation of a new layer each time an object changes leads to an unacceptable expansion of layers. By creating two sets of tables to handle, first, object identity (with rows corresponding to objects known to the system) and second, all the change (with rows corresponding to change in objects that actually change) it is possible to store this kind of discrete change in well-defined phenomena. Systems like this have been implemented in commercial GIS such as the Esri Arcview Tracking Analyst, in which rows in a table store different states of an object and allow them to be visualized as they change through time.

These temporal GIS approaches are not suitable for other applications such as the representation of highly dynamic environmental phenomena which change at a fine temporal resolution (minutes to weeks) in

a continuous way – for example, in the migration patterns of animals or the continuous change of the tides in the coastal zone. In these application areas new GIS designs have been proposed, largely based on the object-oriented approach. The key design issues in object-oriented methods for handling spatio-temporal behaviour are illustrated by considering what approaches could be used to define spatio-temporal phenomena such as a moving herd of animals:

**1** You can use events to bound spatial objects in temporal duration so as to create spatio-temporally extended objects. Events in time are stored, which define different objects either side of an event, e.g. the location of the herd each week.

**2** You can make time a property of the objects so that objects have a spatial and temporal extent. For example, a line may represent the path followed by the herd of animals as they move around during a year.

Neither of these approaches has been implemented in a mainstream commercial GIS as they need designing specifically for the application envisaged. However, there are lots of implemented research systems (Raper, 2000).

Processes such as tidal flow or rainfall movement are almost impossible to represent in this design as there are no 'objects' to represent as part of the phenomena. In this case it is conventional to use a raster rather than vector approach, making each cell a size suitable for the application being modelled. This kind of dynamic, multi-dimensional process modelling aims to develop functional models of behaviour for these systems. Despite the fact that such processes surround us in society and the environment, there are few representations of this kind. It is likely that this shortfall derives both from computational limitations and from the lack of knowledge of the dynamic systems concerned.

An example of a representational problem faced by a real GIS application is a project to monitor coastal change on the east coast of England. Scolt Head is a barrier island consisting of a line of sand dunes backed by salt marshes, which forms part of the low-lying north Norfolk coast. It experiences a westerly longshore drift of sediment into a spit (sandy bar) at the western end (Far Point) reflecting the tidal circulation and the dominant north-east wave energy in this area. The dynamic form of these spits reflects a balance between wave and tide energy in the short term and relative sea-level change in the medium and long term. Hence, spits

**BOX 3.8**

**Figure 3.32** Far Point, Scolt Head Island, Norfolk (eastern England). Snapshot contour maps (1 m colour interval) of coastal spit landforms and their change through time (FPmonth/yearCL notation in labelling). Modelled area is 1 km across.

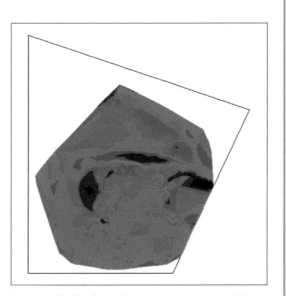

**Figure 3.33** Far Point, Scolt Head Island, Norfolk (eastern England). Difference map between intertidal terrain surface in March and September 1995 (blue is terrain lowering [erosion] and red is terrain elevation [deposition]). Modelled area is 1 km across.

and bars marginal to tidal channels may be both sensitive indicators of coastal change *and* reservoirs of sediment, whose budgets may be of considerable importance to the management of the adjacent coastline on an annual and decadal scale.

Representing these landforms and their change is complex. Using a total station surveying instrument data for sampling points at approximately 10 m horizontal resolution over 13 years of the rapidly changing spits on Scolt Head Island have produced over 20 separate maps of the terrain (Figure 3.32) at Far Point (Raper *et al.*, 1999) and over 20,000 data points describing terrain elevation and surface sedimentary composition. Change has been explored by comparing the surface elevation models at different times to see if the amounts lost and gained

from the surface can be correlated with wave energy inputs and sea-level rise (Figure 3.33). Looking at the areas of erosion and deposition and comparing their magnitude allows estimation of the changes produced by storms of particular magnitude.

The Scolt Head example shows how GIS can be used to model temporal change but also illustrates what still needs to be done. There is no way to automatically compare these surfaces in a GIS as they are based on (inevitably) different sample data points collected at different times as the landform itself is moving and changing through time. The GIS has no way to define spatio-temporal objects or events, and so contains no toolset to allow spatio-temporal analysis.

*Source*: Jonathan Raper, City University, London

**REFLECTION BOX**

- What data sources could be used to create a DTM for the following:
  - a 1 km × 1 km area for which an ecological survey is being undertaken;
  - a national survey of vegetation change;

- a visibility study of the area within 30 km radius of a proposed wind farm;
- a study of changes in the nature and extent of glaciers in the European Alps.

→

REFLECT

- Summarize the stages involved in creating a TIN model. Use a table or flowchart to help present your answer if you wish.
- What types of networks are common in your own discipline or field of interest? Could these be represented in a GIS? What problems might you encounter if you tried to represent them in a GIS?
- Try to apply Peuquet's classification of the types of temporal event to entities that you might like to use for a GIS project of your own. The types of changes that Peuquet (1999) identified were:
  - continuous
  - majorative
  - sporadic
  - unique.
- Speculate how these different types of temporal event might be handled by a GIS?

# CONCLUSIONS

This chapter has taken a detailed look at how to model spatial form in the computer. We have reviewed how the basic spatial entities (points, lines and areas) can be extended to include two other entity types (networks and surfaces) to permit the modelling of more complex spatial features. In addition, the use of the raster and vector approaches in spatial modelling have been examined. Finally, modelling of three- and four-dimensional data has been introduced.

Although few current GIS can handle three- and four-dimensional data, or even objects, most can now handle raster and vector data in two dimensions. It is useful at this point to remember that raster and vector data structures are two alternative methods for storing and representing geographical phenomena. As models they both have strengths and weaknesses in terms of:

- their data storage requirements;
- the faithfulness with which they reflect the character and location of the real-world features;
- the ability to change the scale of visual representation; and
- the analytical opportunities they present.

As a rule, raster data structures generally demand more data-storage than their vector counterparts. However, the more complex a spatial entity becomes, the closer the data volume requirements of the different data structuring techniques. During the 1970s and first half of the 1980s the data storage issue was at the forefront of the raster versus vector debate. However, with the increase in the processing power of the desktop PC and the falling price of data storage, the data volume debate is becoming less of an issue.

The world is not made up of cells, or co-ordinates joined by straight-line segments. Therefore, any data model (raster or vector) in GIS will only be an approximation of reality. However, in certain circumstances one model may give a better representation than another. For example, the road and river network in Happy Valley is better represented using a vector data model. On the other hand, where you have high spatial variability in the feature being modelled, a raster model may provide a better representation. Vegetation maps, where boundaries are difficult to identify and there is rapid and frequent change between vegetation types, may be better represented by a raster model.

The vector spatial data model handles changes in the scale of visual representation much more easily than its raster counterpart. This is because of the precise way in which data are recorded as a set of (x,y) co-ordinates. Changes of scale pose a problem in the raster world if a resolution is requested below the size of the cell specified at the outset of the project. Changes in scale in the raster world are often typified by the appearance of a blockier image. This gives rise to the oft quoted saying from Dana Tomlin 'Yes, raster is faster but raster is vaster, and vector just seems more correcter'.

There is a large distinction between the analytical capabilities of raster and vector GIS. Traditionally, vector spatial data models have been considered more appropriate for answering topological questions about containment, adjacency and connectivity. However, with the advent of more intelligent raster data structures such as the quadtree that contains information about the relationships between cells, this distinction is closing. **Chapter 6** looks in detail at analysis functions in the raster and vector world and compares and contrasts their suitability in the context of different applications.

**CONCLUSIONS**

Future developments are needed in spatial data models to ensure that our computer worlds accurately reflect the reality of geographical space. This is an area of GIS where it is likely that we will see much more complex systems capable of extending our computer worlds into the third and fourth dimensions (**Chapter 13**). One area for development is in the linkage between spatial and attribute data in GIS. The next chapter considers the structuring and storage of attribute data using a database approach.

## REVISION QUESTIONS

- Describe how the raster and vector approaches are used to construct point, line and area entities for representation in the computer.

- What are data structures? Outline their importance in GIS.

- What is a TIN? How are TINs constructed?

- What are surface significant points and how can they be defined?

- How can networks be modelled in GIS?

- Why is it difficult to model the third and fourth dimensions in GIS?

## FURTHER STUDY – ACTIVITIES

- Draw a simple polygon on a coarse grid (approximately 10 × 10 cells) and use different raster data encoding methods to represent this as a text file.

- Draw a more complex polygon on a piece of graph paper and encode this using the area or region quadtree method. Represent this as tree data structure on another piece of paper.

- Make a TIN model from a sheet of paper (see Box 3.5).

- 'Digitize' a simple polygon network using spaghetti and point dictionary methods.

- Select a small area of contours and spot heights from a topographic map. Try creating a TIN from these data using one of the methods described in Box 3.6.

- Use free surface modelling software (e.g. LandSerf – see web link below) to load and explore some surface data.

## FURTHER STUDY – READING

Peuquet (1990) provides an excellent description of the stages involved in moving from a conceptual spatial model of the real world through to its representation in the computer. Burrough (1986) and Aronoff (1991) provide good introductions to data modelling, particularly with regard to the distinction between the raster and vector data models. Wise (2002) offers more detail on raster, vector, surface and network models for GIS. Burrough (1986) also provides a good introduction to the TIN data model and extends the discussion to look in more detail at digital elevation models.

Laurini and Thompson (1992) take a look at the spatial data modelling process from the viewpoint of a computer scientist rather than a geographer. They provide a good introduction to the concept of object-oriented database design for handling spatial information. However, material about data modelling is dispersed throughout their book, so it may take you some time to find what you are looking for. If you want to dig deeper into spatial data structures, then Worboys and Duckham (2004) give a clear and readable account of a range of different approaches. Samet (1989) provides more advanced material, particularly on quadtree structures.

To explore the history and issues associated with the evolution of 3D GIS the paper by Raper and Kelk (1991) is a good introduction. Langran (1992) provides a comparable introduction to the issues surrounding the representation of time in the GIS environment.

Abdul-Rahman and Pilouk (2007) offer a comprehensive guide to modelling for 3D GIS.

Abdul-Rahman A, Pilouk M (2007) *Spatial Data Modelling for 3D GIS*. Springer-Verlag, Berlin

Aronoff S (1991) *Geographic Information Systems: A Management Perspective*. WDL Publications, Ottawa

Burrough P A (1986) *Principles of Geographical Information Systems for Land Resources Assessment*. Clarendon Press, Oxford

Langran G (1992) *Time in Geographical Information Systems*. Taylor and Francis, London

Laurini R, Thompson D (1992) *Fundamentals of Spatial Information Systems*. Academic Press, London

Peuquet D J (1990) A conceptual framework and comparison of spatial data models. In: Peuquet D J, Marble D F (eds) *Introductory Readings in Geographical Information Systems*. Taylor and Francis, London, pp. 209–14

Raper J F, Kelk B (1991) Three-dimensional GIS. In: Maguire D J, Goodchild M F, Rhind D W (eds) *Geographical Information Systems: Principles and Applications*. Longman, London, vol. 1, pp. 299–317

Samet H (1989) *Applications of Spatial Data Structures: Computer Graphics Image Processing and Other Areas*. Addison-Wesley, London

Wise S (2002) *GIS Basics*. Taylor and Francis, London

Worboys M F, Duckham M (2004) *GIS: A Computing Perspective*. 2nd edn. Taylor and Francis, London

## WEB LINKS

### GIS data structures:

- University of Melbourne, GIS Self Learning Tool – Raster concepts: www.sli.unimelb.edu.au/gisweb/GISModule/GIST_Raster.htm

- University of Melbourne, GIS Self Learning Tool – Vector concepts: www.sli.unimelb.edu.au/gisweb/GISModule/GIST_Vector.htm

### Modelling surfaces:

- Population surfaces: www.public.geog.soton.ac.uk/users/martindj/davehome/popmod.htm

- LandSerf: www.soi.city.ac.uk/~jwo/landserf/

### DTM data sources:

- SRTM (Shuttle Radar Topographic Mission) – Global Elevation Data: www2.jpl.nasa.gov/srtm/

- ASTGTM global DEM: lpdaac.usgs.gov/lpdaac/products/aster_products_table/routine/global_digital_elevation_model/v1/astgtm

- USGS DEMs: edc2.usgs.gov/geodata/index.php

Visit our website **www.pearsoned.co.uk/heywood** for further web links, together with self-assessment questions, activities, data sets for practice opportunities and other interactive resources.

# CHAPTER 4

# Database management

## Learning outcomes

By the end of this chapter you should be able to:

■ Explain the difference between databases and database management systems

■ Outline why databases are important in GIS

■ Give examples of the types of databases used with GIS

■ Explain how the relational database model works

■ Describe how to set up a relational database

■ Explain how databases are linked with GIS

■ List important considerations for GIS databases

# INTRODUCTION

Data about our world are produced continuously. They are being collected by remote sensing satellites; from automatic environmental monitoring equipment; during automated business transactions; and by individuals engaged in research and survey work. A large proportion of these data are managed in databases. In your own discipline there are undoubtedly databases under construction and in current use, holding data in text, numeric and multimedia formats. For example:

- An archaeologist might use the British and Irish Archaeological Bibliography to obtain information about research into Neolithic sites in Scotland.

- An architecture student might refer to an American database containing images of buildings and cities to help with a design project on the American Mid-West.

- A business manager might source information on economic outlook from a statistical database such as OECD.

- A soil scientist might consult the Soil Geographical Database of Europe to assist research into soil properties.

- A population research group may maintain a database of research projects on the Web, which users update themselves with details of new projects.

Many of our everyday activities also produce data that automatically find their way into databases. Imagine, for example, purchasing some new ski equipment from a major sports retailer. If you paid by credit card, debit card or cheque, the data in your bank's database would be updated. If the bar code on the item you purchased was scanned, information in the store's database would be updated and might trigger reordering from a supplier's database. Information from the guarantee card you complete and send back by post may trigger a marketing database to start sending targeted mail to you. If you used a store card to make your purchase, extra data about your buying habits are now available, and the store can also target mail to you. Four, or even more, databases may be updated as a result of one

purchase. Since each database holds your address, your postal code may be used as a spatial reference to link these new data to other data in a GIS.

The ski-buying example illustrates the amount of data which may be generated by one sales transaction. Scaling this up to consider the data generated by all the sales transactions in a day begins to give an idea of the amount of data that our society is generating. These data, if transformed into information, are a valuable resource, which can be traded in the same way that commodities were traded in the past. Market research companies now sell information about the characteristics of the population; environmental agencies sell information relating to nature conservation; and mapping agencies sell topographic information. We are in an age where the information resource will not run out; rather the problem is that we have too much of it. We need to be able to manage and share data and information efficiently to realize their value. One method of management is to use a database. Once the data are in a database you can list them in an order most appropriate for you, and undertake searches for specific information. For example, a bibliographic database can be searched using key words. The results can be sorted by date or author's surname. In many cases you can perform additional operations on the database. For example, using an online library database, it may be possible to reserve or renew books. Linking databases to GIS to provide additional spatial capabilities can further enhance the value of data and provide useful information.

Large organizations such as utilities, multinational companies, retailers and government agencies have adopted GIS for a whole range of applications, from automated mapping and facilities management (AM/FM) to decision support applications. An important issue for successful implementation of many of these new systems is their ability to integrate GIS with existing internal databases. Many of these existing databases (managing customer databases, inventories of parts or financial information) are relational, as the relational database approach to data management has permeated all sectors of the business world. These business databases may be data sources for GIS, or be required to integrate with GIS for decision support within the framework of a larger organizational 'information management strategy'. Issues of database management for large-scale, corporate applications of GIS are addressed in this chapter. The demands of large-scale applications are examined, along with the strengths and weaknesses of the current database approaches.

First, this chapter considers the conceptual and theoretical ideas behind databases in order to explain why they are used and how they work. Whichever type of GIS is being used, an understanding of databases is important. This chapter will

introduce the main types of database in current use and explain their characteristics, advantages and disadvantages. As the relational database is the most commonly used at present, the steps involved in establishing a relational database will be discussed in some detail.

## Terminology recap

In GIS, there are two types of data to be managed: spatial data and attribute data. An entity (point,

**Figure 4.1** Database facilities in GIS
*Sources*: (a) Screenshot shows the Esri Graphical User Interface (GUI). ArcMap/ArcView/ArcInfo Graphical User Interface is the intellectual property of Esri and is used herein with permission.Copyright © 2011 Esri all rights reserved; (b) IDRISI interface used by permission of Clark Labs, Clark University

line or area) has both spatial and attribute data to describe it (**Chapter 2**). Put simply, spatial data are the 'where things are' data and attribute data the 'what things are'. For example, a latitude and longitude reference gives the location of a point entity and to accompany this there would be attribute data about the nature of the real-world feature the point represents. If the point were a city in a global cities database, then the attribute information stored might include the city name, the country in which the city is situated and up-to-date population figures.

Conventionally, databases have stored only non-spatial entities; however, all entities in GIS are spatial. As a result the conventional database has been adapted to GIS in various ways. Some GIS are good at handling attribute data; some rely on links with conventional databases; and others have very limited database capabilities but good analysis facilities. Figure 4.1 shows some of the database facilities in some common GIS.

Databases offer more than just a method of handling the attributes of spatial entities; they can help to convert data into information with value. **Chapter 2** introduced the concept that data are raw facts that have little value without structure and context. A single number, for example '10', is data. With some explanation, this number becomes information. 'Ten degrees Celsius' is information, since now there is some meaning that can be interpreted by a user. Information results from the analysis or organization of data, and, in a database, data can be ordered, re-ordered, summarized and combined to provide information. A database could be used to sort a range of temperature values into order; to calculate maximum and minimum values or average temperature; show the temperature for a particular date; or to convert degrees Celsius to degrees Fahrenheit. With the additional mapping capabilities of a GIS, the locations of the points at which the temperatures have been monitored could be mapped, thus adding further value to the original data. Decision makers using GIS need information, not data, so a database offers one method of providing that information by organizing data into a more meaningful and usable form.

There are nearly as many definitions of a database as there are of GIS. Perhaps the simplest definition is that a database is a set of structured data. An organized filing cabinet is a database, as is a dictionary, telephone directory or address book. Thus, databases can be computer-based or manual.

## ■ WHY CHOOSE A DATABASE APPROACH?

Elsewhere in this book are examples of the difficulties faced when handling spatial data manually (see especially **Chapter 9**). Problems can also be encountered when processing attribute data manually. Imagine Happy Valley, the ski resort, as it might have been in the days before computerized databases. Different organizations and companies within the resort could be producing and using similar data for different purposes. There are ski schools, hotels and travel companies, all handling data such as clients' names and addresses and details of where they are staying. All these data are stored in different formats and separately by each organization. A box file may be used by one organization, a set of index cards by another and a third may have the data stored in the head of one of the employees! Each organization uses the data for different purposes: the ski school for booking lessons; the hotels for booking rooms; and the travel companies for arranging flight details and allocating hotels. The situation is illustrated schematically in Figure 4.2.

There is considerable duplication of data using traditional data management approaches. For example, a single visitor's address may be held three times, once by the travel company, then again by the hotel and ski school. It is likely that there will be errors made during the transcription and copying of an address, or even that different parts of the address will be held or omitted in each case. If a visitor changes their home address, they may remember to tell the travel company, but forget to tell the hotel or the ski school. Thus, different versions of the data could exist in the three companies. Errors are inevitable, and there is considerable effort involved in handling the data. Difficulties would be encountered if the organizations attempted to share their data. Even if the data were in computer files of one sort or another, there might be incompatibilities in data

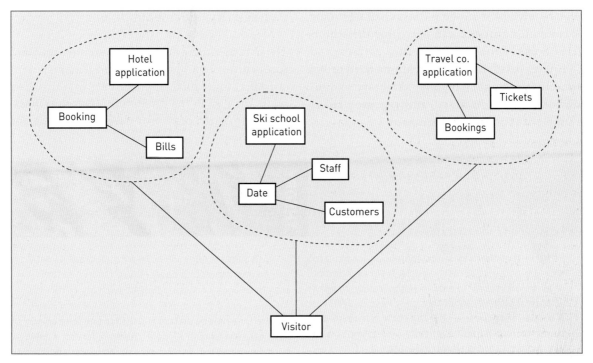

**Figure 4.2** The traditional approach to data handling. A different version of the visitor's details may be stored in each of the separate databases

formats which would prevent efficient data exchange.

Date (1986) summarizes the problems with the traditional approach to data management as:

- redundancy (the unnecessary repetition or duplication of data);

- high maintenance costs;

- long learning times and difficulties in moving from one system to another;

- the possibility that enhancements and improvements to individual files of data will be made in an *ad hoc* manner;

- data-sharing difficulties;

- a lack of security and standards; and

- the lack of coherent corporate views of data management.

In addition, the data storage mechanisms may be inflexible, creating difficulties in dealing with *ad hoc* 'one-off' queries. Oxborrow (1989) also identifies the problem of modelling the real world with traditional data management methods – many are simply not suitable. Data should be structured in such a way that they represent features from the real world and the relationships between them. Traditional methods

of storing data cannot represent these relationships, so a database approach is necessary.

## The database approach

A database is a simply a collection of related data. This can include non-computerized data such as those found in a telephone directory or address book. A generally accepted feature of data in a database is that they can be shared by different users (Stern and Stern, 1993). Data within a database are also organized to promote ease of access and prevent unnecessary duplication. In a paper-based telephone directory the entries are organized, usually alphabetically by surname, to allow ease of access to different users. However, there are problems with this paper-based database. It is not possible to search by first name, or using only part of an address. It is not possible to extract a list of all those people who live in a particular district. There is no security, as anyone who can read can access the information (the only security mechanism is to be ex-directory, then the data set is incomplete). It is not possible to adapt the telephone book for other uses, such as direct mailing. Updating requires expensive and time-consuming reprinting. A computer approach can overcome these problems. Some of the most often cited advantages of computer-based databases are summarized in Box 4.1. The electronic telephone directory enquiry service available in most countries provides an example of a computerized database.

## Database management systems

The data in a computer database are managed and accessed through a database management system (DBMS). Individual application programs will access the data in the database through the DBMS. For example, to book a new client's ski lessons, the

## BOX 4.1 Advantages of computer-based databases

**THEORY**

Imagine that there is a box containing index cards with the names, addresses and telephone numbers of all the ski school clients in Happy Valley, together with details of the ski lessons they have undertaken. A single index card is held for each individual, and cards are stored alphabetically by surname in the box. A typical card is shown in Figure 4.3.

If the data from the cards are transferred to a computer database, the following benefits could be achieved:

**1** *Different data access methods will be possible.* Data can be accessed by country of residence or lessons undertaken, not just surname of client as in the index box.

**2** *Data are stored independently of the application for which they will be used.* The database may have been established to assist the sending of bills to clients. However, the database could also be used to produce mailing lists for the promotion of special events, or for summarizing client data (where clients came from, number of beginners courses undertaken, number of one-off lessons) for annual reports.

**3** *Redundancy (the unnecessary duplication of data) will be minimized.* In a paper-based records system it may have been necessary to keep two sets of records for one client. For example, when one index card is full, another one may be started and the old one left behind. Alternatively, should a client change address, a new card may be added, and the old one not removed.

**4** *Access to data will be controlled and centralized.* A card index box kept on a desk is not very secure, and, unless it can be locked, can be viewed by anyone in the office. A computer database can have security built in – passwords can prevent access to all or part of the information, or to functions that will allow the updating or deletion of data. Allowing only one individual to access the raw data for updates and changes will also improve the reliability of the data.

**5** *A computer database is relatively easy to maintain and updating is possible.* The cards in the index box will soon become unreadable if clients' details change repeatedly. The computer version will prevent problems caused by unreadable text, and allow efficient updating.

**6** *Simple query systems and standardized query languages are available.* Queries, such as 'which clients have completed advanced training courses?', or 'how many Americans have taken training courses?' would be time-consuming with the card index box. A computerized system will offer simple query methods, or a standardized query language.

Datasets and activities relating to the Happy Valley Case Study can be found online at **www.pearsoned. co.uk/ heywood**.

*Sources*: Oxborrow (1989); Healey (1991)

| Surname | *Smith* |
| --- | --- |
| First name | *Jane* |
| Home Address | *12 High Street, Northton, Northtonshire, NN1 1NN* |
| Telephone No. | *0181 234 443*   *020 8222 442* |
| Lessons taken | *Beginners 1 week course( Jan 2001) Private lesson × 2 (March 2002)* |

**Figure 4.3** Card index record from ski school manual database

booking clerk will use an application produced using capabilities offered by the DBMS. This will instruct them to fill in a data entry form, which will then automatically update data in the database. The clerk will not need to interact with the database directly or understand how data are structured within the database. A conceptual view of the Happy Valley data as they might be organized for such a computer database, and the role of the DBMS, is shown in Figure 4.4 (adapted from Reeve, 1996).

There are many definitions of a DBMS. Dale and McLaughlin (1988) define a DBMS as a computer program to control the storage, retrieval and modification of data (in a database). Stern and Stern (1993) consider that a DBMS will allow users to join, manipulate or otherwise access the data in any number of database files. A DBMS must allow the definition of data and their attributes and relationships, as well as providing security and an interface between the end users and their applications and the data itself. From such definitions the functions of a DBMS can be summarized as:

- file handling and file management (for creating, modifying or deleting the database structure);

- adding, updating and deleting records;

- the extraction of information from data (sorting, summarizing and querying data);

- maintenance of data security and integrity (housekeeping, logs, backup); and

- application building.

The overall objective of a DBMS is to allow users to deal with data without needing to know how the data are physically stored and structured in the computer. To achieve this, DBMS usually comprise

software tools for structuring, relating and querying data; tools for the design of data entry and report forms; and application generators for the creation of customized applications.

A DBMS manages data that are organized using a database data model. This is analogous to the way in which spatial data are organized in a GIS according to a spatial data model (for example, raster or vector). Database data models for GIS are similar to those used for databases elsewhere.

## DATABASE DATA MODELS

There are a number of different database data models. Amongst those that have been used for attribute data in GIS are the hierarchical, network, relational, object-relational and object-oriented data models. Of these the relational data model has become the most widely used and will be considered in detail here, whilst the models based on objects are growing in popularity. Further details of the hierarchical and network models can be found in older GIS texts (Aronoff, 1991; Bernhardsen, 1999; DeMers, 2005).

### The Relational Database Model

At present the relational database model dominates GIS. Many GIS software packages link directly to commercial relational database packages, and others include their own custom-designed relational database software. Some GIS use a relational database to handle spatial as well as attribute data.

The relational data model is based on concepts proposed by Codd (1970). Data are organized in a series of two-dimensional tables, each of which contains records for one entity. These tables are linked by common data known as *keys*. Queries are possible on individual tables or on groups of tables. For the Happy Valley data, Figure 4.5 illustrates an example of one such table.

Each table in a relational database contains data for one entity. In the example in Figure 4.5 this entity is 'hotel'. The data are organized into rows and columns, with the columns containing the attributes of the entity. Each of the columns has a distinctive name, and each of the entries in a single column must be drawn from the same domain (where a domain may be all integer values, or dates

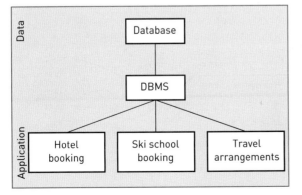

**Figure 4.4** The database approach to data handling

or text). Within a table, the order of the columns has no special significance. Other characteristics are listed by Reeve (1996). There can be only one entry per cell; each row must be distinctive (so that keys that use unique row entries are possible – in GIS location is often the key); and null values are possible where data values are not known.

The terminology of relational databases can be confusing, since different software vendors have adopted different terms for the same thing. Table 4.1 illustrates the relationship between relational database terminology and the traditional table, or simple computer file. Figure 4.6 applies this terminology to the table suggested for the Happy Valley database. A useful shorthand way of describing a table is using its 'intension'. For the table in Figure 4.5 this would be:

> HOTEL (Hotel ID, Name, Address, No. rooms, Standard).

The data in a relational database are stored as a set of base tables with the characteristics described above. Other tables are created as the database is queried and these represent virtual views. The table structure is extremely flexible and allows a wide variety of queries on the data. Queries are possible on one table at a time (for example, you might ask 'which hotels have more than 14 rooms?' or 'which hotels are luxury standard?'), or on more than one table by linking through key fields (for instance, 'which

**TABLE 4.1 Relational database terminology**

| Paper version | File version | RDBMS |
|---|---|---|
| Table | File | Relation |
| Row | Record/case | Tuple |
| Column | Field | Attribute |
| Number of columns | Number of fields | Degree |
| Number of rows | Number of cases | Cardinality |
| Unique ID | Primary key | Index |
| | Possible values | Domain |

*Source*: Adapted from Date (1986)

passengers originating from the UK are staying in luxury hotels?' or 'which ski lessons have pupils who are over 50 years of age?'). Queries generate further tables, but these new tables are not usually stored. There are few restrictions on the types of query possible.

With many relational databases querying is facilitated by menu systems and icons, or 'query by example' systems. Frequently, queries are built up of expressions based on relational algebra, using commands such as SELECT (to select a subset of rows), PROJECT (to select a subset of columns) or JOIN (to join tables based on key fields). SQL (standard query language) has been developed to facilitate

| Hotel ID | Name | Address | Number of rooms | Standard |
|---|---|---|---|---|
| 001 | Mountain View | 23 High Street | 15 | budget |
| 002 | Palace Deluxe | Pine Avenue | 12 | luxury |
| 003 | Ski Lodge | 10 Ski School Road | 40 | standard |

**Figure 4.5** Relational database table data for Happy Valley

**Figure 4.6** Database terminology applied to Happy Valley table

the querying of relational databases. The advantages of SQL for database users are its completeness, simplicity, pseudo English-language style and wide application. However, SQL has not really developed to handle geographical concepts such as 'near to', 'far from' or 'connected to'.

The availability of SQL is one of the advantages of the relational database model. Additionally, the model has a sound theoretical base in mathematics, and a simple logical data model that is easy to understand. The relational database model is more flexible than either of the previously used hierarchical or network models. However, the model will always produce some data redundancy and can be slow and difficult to implement. There are also problems with the handling of complex objects such as those found in GIS (and CAD and knowledge-based applications) as there is a limited range of data types, and difficulties with the handling of time. Seaborn (1995) considers that many of the limitations of relational databases in GIS stem from the fact that they were developed to handle simple business data, and not complex multi-dimensional

spatial data. However, they have been widely adopted and successfully used.

Relational databases are predominantly used for the handling of attribute data in GIS. For example, Esri's ArcGIS maintains an attribute table in relational database software, using a unique ID to link this to spatial data.

## ■ CREATING A DATABASE

Database design and implementation are guided by the relationships between the data to be stored in the database. The database design process is concerned with expressing these relationships, then implementation with setting up a new structure for these relationships within the chosen database software. The stages of database creation are summarized in Box 4.2.

The first stage in designing a database, data investigation, can result in a mass of unstructured information on information flows, relationships and possible entities. Thus, a key part of database

## BOX 4.2 The steps involved in database creation

**THEORY**

The steps involved in database creation, suggested by Oxborrow (1989) and Reeve (1996), are summarized in Table 4.2 and described below.

**TABLE 4.2 Steps in database creation**

| Oxborrow (1989) | Reeve (1996) |
| --- | --- |
| Data investigation | Needs analysis |
| Data modelling | Logical design |
| Database design | Physical design testing |
| Database implementation | Implementation |
| Database monitoring | Maintenance |

1 *Data investigation* is the 'fact finding' stage of data base creation. Here the task is to consider the type, quantity and qualities of data to be included in the database. The nature of entities and attributes is decided.

2 *Data modelling* is the process of forming a conceptual model of data by examining the relationships between entities and the characteristics of entities and attributes. This stage, like the data investigation stage, can be carried out independently of the software to be used.

3 *Database design* is the creation of a practical design for the database. This will depend on the database software being used, and its data model. This is the process of translating the logical design for the database (produced during the data modelling stage) into a design for the chosen DBMS. Field names, types and structure are decided. In practice, the design will be a compromise to fit the database design model with the chosen DBMS.

4 *Database implementation* is the procedure of populating the database with attribute data, and this is always followed by monitoring and upkeep, including fine tuning, modification and updating.

development is data modelling. There is a wide range of techniques available to assist this data modelling, including entity relationship modelling (Chen, 1976) and normalization.

There are four stages to entity relationship modelling (or entity attribute modelling, EAM): the identification of entities; the identification of relationships between entities; the identification of attributes of entities; and the derivation of tables from this. In a database for Happy Valley we may wish 'hotels', 'tour companies', 'ski schools' and 'visitors' to be regarded as entities. Each entity has distinctive characteristics and can usually be described by a noun. Its characteristics are the attributes (for example, a hotel will have a name, address, number of rooms and standard), and its domain is the set of possible values (for example, the standard may be budget, standard, business or luxury). The relationships between the entities can be described using verbs. Thus, a hotel *is located in* a resort; a visitor *stays* at a hotel, and a ski school *teaches* visitors. Three types of relationship are possible: one to one (one visitor stays at one hotel); one to many (one ski school teaches many visitors); or many to many (many tour companies use many hotels). These relationship can be expressed by the symbols: '1:1', '1:M' and 'M:N', and by using simple diagrams. Figure 4.7 illustrates a possible entity relationship model for the Happy Valley example:

The entity relationship model diagram is helpful in deciding what will be appropriate tables for a relational database. Where the relationship is 1:1, tables for each entity can be joined together or kept separate. Where the relationship is 1:M, two tables are needed with a key field to allow a relational join. Where the relationship is M:N, the tables should be separated. Where repeating fields occur, tables need to be broken down further to avoid redundancy. Figure 4.8 shows possible tables developed from the EAM model for Happy Valley.

Once the tables have been decided upon, the attributes needed should be identified. This is the process of developing intensions that was outlined earlier. The intensions for the Happy Valley example are:

HOTEL (<u>Hotel ID</u>, Name, other attributes ...)

TRAVELCO (<u>TravelCo ID</u>, TravelCo Name, other attributes ...)

SKISCHOOL (<u>SkiSchool ID</u>, SkiSchool Name, other attributes ...)

VISITOR (<u>Visitor ID</u>, Visitor Name,

Hotel ID, TravelCo ID, SkiSchool ID, other attributes ...)

LINK (<u>TravelCo ID</u>, SkiSchool ID)

The final result of entity relationship modelling is an EAM model diagram, a set of table definitions

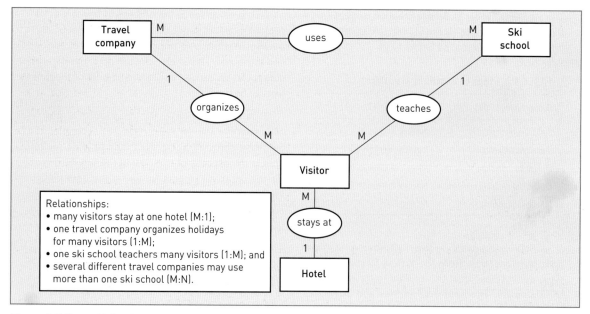

**Figure 4.7** Happy Valley EAM diagram

*Table 1: Hotel*

| Hotel ID | Name | Other attributes... |
|---|---|---|
| 001 | Mountain View | ..... |
| 002 | Palace Deluxe | ..... |
| 003 | Ski Lodge | ..... |
| ..... | ..... | ..... |

*Table 2: Travel company*

| Travel co. ID | Travel Co. Name | Other attributes... |
|---|---|---|
| T01 | Ski Tours | ..... |
| T02 | Snow Breaks | ..... |
| ..... | ..... | ..... |

*Table 3: Visitors*

| Visitor ID | Visitor Name | Hotel ID | Travel Co. ID | Ski School ID | Other attributes... |
|---|---|---|---|---|---|
| V001 | Smith J | 002 | T01 | S02 | ..... |
| V002 | Schmidt S | 001 | T02 | S02 | ..... |
| ..... | ..... | ..... | ..... | ..... | ..... |

*Table 4: Ski schools*

| Ski School ID | Ski School Name | Other attributes... |
|---|---|---|
| S01 | Snow Fun | ..... |
| S02 | Bert's Ski School | ..... |
| S03 | Ski School Professional | ..... |
| ..... | ..... | ..... |

*Table 5: Link table*

| Travel Co. ID | Ski School ID |
|---|---|
| T01 | S01 |
| T01 | S02 |
| T02 | S02 |
| ..... | ..... |

**Figure 4.8** Relational tables for Happy Valley database

and details of attributes (names, size and domain). The database can then be implemented. Another example of the use of entity relationship modelling to develop a GIS database is given by Healey (1991).

## Linking spatial and attribute data

The relationship between GIS and databases varies. For a simple raster GIS, where one cell in a layer of data contains a single value that represents the attributes of that cell, a database is not necessary. Here the attribute values are likely to be held in the same file as the data layer itself. However, there are few 'real' GIS like this, and those which exist are designed for analysis, rather than attribute data handling. An improvement on this approach is the ability to handle attribute values in a file separate from the raster image. Although this method also

lacks the flexibility of a true relational DBMS, it is possible to link the GIS software with proprietary relational DBMS to upgrade the capabilities. Most GIS, particularly vector-based systems, offer a hybrid approach (Batty, 1990; Maguire *et al.*, 1990; Cassettari, 1993). In this case spatial data are stored as part of the GIS data structure and attribute data are stored in a relational DBMS. This approach allows integration of existing databases with graphics by the allocation of a unique identifier to each feature in the GIS (Figure 4.9).

Finally, an alternative approach is an extended GIS, where all aspects of the spatial and attribute data are in a single DBMS. Seaborn (1995) considers these 'all-relational' GIS to have considerable potential, and cites examples of major organizations such as British Telecom, Electricité de France and New Zealand Lands which have adopted this approach.

**Figure 4.9** Linking spatial and attribute data in GIS

## REFLECTION BOX

- Are there any databases that you use for research or study purposes? Visit them again and identify what type of data they contain, what features they offer, and how they can be used.
- Produce a table that summarizes the advantages and disadvantages of manual and computer-based databases.

- Write your own definitions for the following terms. Check your own versions against those in the text, or an online dictionary.
  - Database
  - Database management system
  - Relational database

## ◼ GIS DATABASE APPLICATIONS

GIS databases cover a wide spectrum of applications, from those involving a single user with a PC and low-cost software working on small-scale research projects to huge corporate databases with data distributed over several sites, each with different computer systems and different users. Worboys (1995) offers a classification of potential database applications. This includes single-user small databases, corporate databases, office information systems, engineering databases, bibliographic databases, scientific databases, and image and multimedia databases, and has a separate category for geographic databases. Despite this separation of geographic databases, geographic data could be found in most of the other types of databases – single-user, corporate, office, image and multimedia. The differing nature of potential geographical applications leads to a range of issues for the GIS database user to consider.

For a single user, working on a PC database, issues and design considerations will be very different from those for a large, multi-user corporate database. For the single user, flexibility and ease of use may be important, whereas security, reliability (the probability of the system running at a given time), integrity, performance and concurrent access by different users, often using intranet or Internet systems, may be required by a large-scale, multi-user application. Many GIS applications are now large-scale corporate systems – for instance, in the gas, electricity, water and telecommunications industries. Large corporate GIS projects have special database demands. In some organizations GIS may be operated by a specialist department which controls access and manages the data. Increasingly, however, GIS are being integrated into overall information strategies, requiring the integration of GIS and general business data. Box 2.6 describes Network Rail's GIS – an example of a large-scale, multi-user application. There are two options for such databases, reflecting these differences: the centralized database system or the distributed database system.

In a centralized database system all the system components reside at a single computer or site (this includes data, DBMS software, storage and back-up facilities). The database can often be accessed remotely via terminals connected directly to the site. In Happy Valley, a centralized system may be available at the town hall, controlled and updated by council employees, and accessible to other users via terminals distributed throughout the town. The current trend is away from such central servers to databases located in different places but connected together by networks. These are distributed database systems.

Using a distributed database system the local user should believe that all the data they are using are located on their machine, even though some may be on the other side of the world. Using the Internet is an example – you work at your own computer, but access data stored on computers all around the world. It is not important where these other computers are, as long as you can obtain the data you need. Elmasri and Navathe (1994) describe such a distributed database system as one in which a collection of data that belongs logically to the same system is physically spread over the sites of a computer network. If the multinational company, SkiResorts Inc., were to buy the whole of Happy Valley (ski schools, hotels, tourist offices, etc.) it may decide to implement a distributed database for visitor monitoring and management. In this case, individual databases held by the ski schools and hotels could still be maintained by the originators but made accessible to other authorized users, including the multinational company's headquarters, via a communications network. Worboys (1995) considers that the distributed database approach is naturally suited to many GIS applications. One of the major advantages of a distributed database is that the reliability of the system is good since, if one of the database sites goes down, others may still be running. The approach also allows data sharing, whilst maintaining some measure of local control, and improved database performance may be experienced. In Happy Valley some of the problems of different users holding their own databases could be overcome by a distributed database approach. For example, data duplication and replications could be tracked, the consistency of copies of data could be maintained and recovery from individual site crashes would be possible. Box 4.3 summarizes some of the general issues that large corporate databases need to address.

## BOX 4.3 Issues for large corporate databases

PRACTICE

Large corporate database applications are characterized by the following needs:

**1** *The need for concurrent access and multi-user update.* This will ensure that when several users are active in the same part of the database, and could inadvertently update the same data simultaneously, they do so in an orderly manner, with the result they collectively intended.

**2** *The need to manage long transactions.* This is a database update composed of many smaller updates. The update takes place over a period of hours or days (not seconds, as in short transactions), most often by one user, but often by several users working in parallel, but independently. This situation might arise if major revisions to base map data were in progress.

**3** *The need for multiple views or different windows into the same database.* For example, a bank employee may be able to access and update customer information for any customer, but an individual customer can only access, but not update, their own records.

*Source*: Adapted from Seaborn (1995)

The specialist needs identified in Box 4.3 make large-scale GIS databases difficult to implement. There are other factors, not related to databases directly, which may restrict the evolution of corporate GIS databases. McLaren (1990) outlined some of these as different basic spatial units used for different data, data copyright and data ownership issues, large data volumes, formulation of projects at departmental level and institutional structures. Sanderson (2005) identifies other practical problems, all with financial implications for corporate users:

- expensive commercial products;

- high overheads (for example, high specification hardware may be needed and specialists may be employed to provide maintenance and support); and

- the need for end user training.

Many of these problems are not technical but human and organizational in nature. Such issues will be returned to in **Chapter 11**.

A variety of considerations will combine to ensure a successful database application. These are generally considered to be contemporaneous data, data as detailed as necessary for potential applications, data that are positionally accurate and internally accurate, up-to-date and maintained data, and data that are accessible to users. Databases should be usable by non-technical experts who do not need to know the principles of databases to add, edit, query or output data from the database.

The general problems with a database approach are complexity, cost, inefficiencies in processing and rigidity (Oxborrow, 1989). The complexity of DBMS means that training is required to design and maintain the database and applications. The costs include those of the software, development and design phases and maintenance and data storage. Processing inefficiencies may be caused by changes in user requirements. If changes require restructuring of the database, this may be difficult to implement. Finally, the data formats offered by DBMS are fixed. There are limited capabilities, in most DBMS, for the handling of long text strings, graphics and other types of data. All of these general problems apply to GIS databases, and there is an additional set of problems associated with data back-up, recovery, auditing, security, data integrity and concurrent update (Batty, 1990). Batty also considers that GIS are no different from other data management systems, and must face the challenges of sharing data between applications and handling data distributed across several computers.

Geographical data have three elements: space, theme and time (**Chapter 2**). Whilst we have discussed the handling of spatial data in **Chapter 3**, and thematic or attribute data in this chapter, temporal data have not been addressed in any detail. One of the limitations of a relational database – in fact, of most database systems available – is the inability to handle temporal data effectively. This is an issue that is being addressed by current research activities, such as for the study of encroachments in Ghana presented in Box 4.4.

Another area of current interest is the implementation of databases and GIS using the Internet. Box 4.5 gives an insight into how this might work. Box 4.6 describes the use of a relational databases in a web based application.

# BOX 4.4 Encroachments on public land in Accra, Ghana

**CASE STUDY**

### Isaac Bonsu Karikari

In Ghana Planning Schemes (PS) and Cadastral Plans (CP) are map-based tools used in strategic land-use planning. Unfortunately, these plans are often ignored by developers and encroachments (buildings erected at short notice without planning permission) frequently occur in planned government estates. Encroachments sometimes involve building on public as well as customary lands, with no regard for planning schemes prepared by Government for such areas. This is an issue that Ghana's land administration and planning systems have to face and is a problem with which GIS can help.

## BOX 4.4

The problem of encroachments is influenced by:

- the lack of adequate and accurate data for planning purposes;
- a serious lack of technical expertise in the collection, analysis and interpretation of data;
- perennial lack of adequate logistics and equipment to monitor encroachments; and
- the lack of cooperation between agencies mandated to oversee land administration and management.

This phenomenon has led to environmental problems, land disputes, conflicts and endless litigation in the law courts.

From a GIS perspective, the problem of land use planning and encroachments illustrates the difficulties faced by many GIS users when trying to collect, manage and maintain up-to-date databases of dynamic spatial objects. In other words, what might be the 'database truth' and what is the 'ground truth' are often two very different things. This is because the world is a dynamic and ever-changing place, and changes taking place on the ground often do so at a faster rate than the survey and monitoring process and are sometimes at odds with planned changes. Keeping the database truth and ground truth as close as possible is a task requiring a great deal of time and effort that covers the whole shelf-life of a project or database.

### CHOICE OF STUDY AREA

This case study outlines a project to monitor encroachments on public land in the East Legon Ambassadorial Estate in Accra, the capital of Ghana in west Africa. This is an area of the capital earmarked for the development of the estates of the embassies of foreign governments and so requires well-defined plans. The objectives of this case study were to:

- evaluate encroachments in the East Legon Ambassadorial Estates;
- record encroachments within the East Legon Ambassadorial Estates using a Global Positioning System (GPS) and Total Station equipment; and
- analyse and display the results in a Geographic Information System (GIS).

This project hoped to demonstrate that it was possible to collect accurate data of this type and overlay encroachment data on the available planning scheme and cadastral plan in a GIS in order to facilitate spatial analysis.

The area of study was chosen for three main reasons:

1 The area is heavily encroached and is a big concern for land administrators and planners as encroachments include unauthorized buildings and other structures within and around the estate and involve environmental degradation by the public through unauthorized sand and stone extraction/quarrying activities.

2 The area is the site for the main navigation beacon (VOR) for the capital's main airport that ought to be free from developments within 1000 feet *radius* of the VOR (with an additional 2000 feet radius buffer of restricted development).

3 The main aircraft approach lane to 'Runway 21' lies wholly within the area and as a result several additional navigational aids have been located in the vicinity. No structure should exceed the height of 85 feet and the wrong type of roofing material around the equipment could distort radiowaves and render equipment unreliable for aircraft operations.

### RESULTS

Encroachments in this area were surveyed and recorded by traversing between GPS control points to centimetre accuracy using Total Station surveying equipment. A total of 42 walled and un-walled premises were selected and recorded. The data collected in this manner were transferred to the GIS database and overlaid with the PS and CP maps along with a road layer. This enabled comparison to detect conflicts and new patterns that were expected to occur as a result of the different plans applied on the estate (see Figure 4.10).

Once the data was imported into the GIS it was noticed that the existing road network was at variance with the PS, but conformed quite closely to the CP. Recorded encroachments conformed more to the pattern of the CP than the PS, even though some roads within the CP had been built on. The PS has been totally disregarded by the developers. Further analysis revealed that all but seven of the 42 encroachments had been affected by the road buffer one way or another and therefore infringed building regulations, with five buildings completely blocking some roads. The southern sector was free from encroachments. The buffer zones around the VOR have been severely encroached leaving only a 500-foot zone of undeveloped land around this critical installation (Figure 4.11). Three particular buildings recorded pose a threat to the aviation industry and the Government is taking all

BOX 4.4

**Figure 4.10** Overlay of planning scheme, cadastral plan, existing roads and encroachments within the East Legon Ambassadorial Area
*Source*: From Karikari, I., Stillwell, J. and Carver, S. (2005) The application of GIS in the lands sector of a developing country: challenges facing land adminstrators in Ghana, *International Journal of Geographical Information Science*, Vol. 9 No. 3, Taylor & Francis Ltd, used by permission.

**Figure 4.11** Coded encroachments on the CP, showing also the boundary of the VOR (arrowed)
*Source*: From Karikari, I., Stillwell, J. and Carver, S. (2005) The application of GIS in the lands sector of a developing country: challenges facing land adminstrators in Ghana, *International Journal of Geographical Information Science*, Vol. 9 No. 3, Taylor & Francis Ltd, used by permission.

necessary steps to ensure maximum compliance with planning regulations in this area, meaning the buildings will ultimately need to be demolished.

## CONCLUSIONS

This pilot study revealed a number of important issues for this and similar projects:

**1** A number of core data sets are required to analyse the infringement of building regulations and their associated environmental conditions, which may in turn determine whether developments are in accordance with required planning regulations set by the planning authorities. These data sets include land-use, cadastral data/information, administrative boundaries and infrastructure data (roads, electricity pylons and sewers). Not all may be in digital format.

**2** There are difficulties involved in performing spatial analysis on conflicting or unreliable data sets for the same area such as the apparent mismatch between the PS and CP in Ghana.

**3** Planners need to work closely together to ensure plans are co-ordinated and enforceable. GIS has a key role to play in this process both in terms of data sharing and analysis.

The project concluded that the character of the neighbourhood had been seriously altered by unauthorized development making it impossible to go ahead with the planned estates for the foreign embassies. The exercise highlighted the following reasons behind the failure of the requisite agencies to use GIS as an analytical tool: confusion about what data set to use or lack of reliable data; lack of people with knowledge in Information Technology (IT) and GIS; and as a result of the above, lack of confidence to take advantage of the immense opportunities that IT and geotechnology now provide (Karikari *et al.*, 2005). Generic conclusions can be drawn about the need to keep databases up to date and ensure compatibility between plans, especially when dealing with dynamic objects in spatial databases such as urban expansion/land development, lest confusion and conflict become the inevitable result.

### FURTHER INFORMATION

This project was sponsored by the Ghana Forestry Commission through the DFID/FRR Ltd, UK via the Forest Sector Development Project (FSDP II).

*Source*: Isaac Bonsu Karikari, Millennium Development Authority. Ghana

## BOX 4.5 Web GIS

Most web-based GIS are client-server systems (Figure 4.12). Servers store data (possibly in a relational database management system) and provide a service to distributed clients. Clients use standard browser software to request this service or resources from the server. An Internet map server typically provides functions which allow, for instance, zooming or panning around map images.

Using a client server system a client requests a map image from a server. The Internet Map Server creates a map based on the request from the client as a graphics file in, for example, GIF or JPEG format.

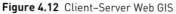

**Figure 4.12** Client–Server Web GIS

The server program transmits the image over the Internet back to the client's browser. The client and server systems communicate using a protocol – an established communication language such as HTTP (Hypertext Transfer Protocol).

If the user obtains a map and then decides to zoom in, another request would be sent to the server. The server would send a new version of the map image back to the client.

In some cases more of the processing can be done on the client side. In this case the Internet Map Server sends more complex data or structured data, for example in vector format, to the client. Users may have to obtain applications, plug-ins or Java applets for their client systems to allow them to process these data.

More complex networked systems include more than one server. Separate servers may be used as in the example in Figure 4.13. Here the clients have processing capabilities: data, features, maps and applications are all located on separate servers. The feature server may provide the results of a database query.

The applications of Web GIS are growing rapidly. They include:

- displaying static maps which users can pan or zoom whilst online;
- creating user-defined maps online which are in turn used to generate reports and new maps from data on a server;
- integrating users' local data with data from the Internet;

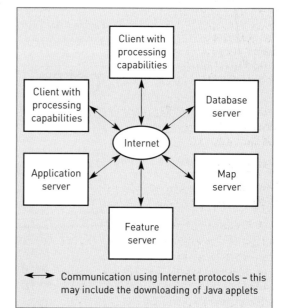

**Figure 4.13** A networked Web

- providing data that are kept secure at the server site;
- providing maps through high-speed intranets within organizations; and
- providing maps and data across the Internet to a global audience.

*Sources*: Harder (1998); Limp (1999); Longley *et al.* (2010)

## BOX 4.6 Delivering census interaction data: WICID

CASE STUDY

### John Stillwell

Without a population registration system in the United Kingdom, population censuses are very important sources of data for social science research and policy making. Although only a snapshot on one night every 10 years, the census is comprehensive, reliable and provides information for small areas across the whole country. The results of the census are available as a series of published or digital data sets (Rees *et al.*, 2002), one of which is the Origin–Destination Statistics containing information about flows of migrants and commuters between areas at different spatial scales. This case study introduces the Web-based Interface to Census Interaction Data (WICID) which provides registered academic users with access to these Interaction data.

WICID is the software system developed and managed by the *Centre for Interaction Data Estimation and Research (CIDER)*, formerly the Interaction Data Service (CIDS), a 'Data Support Unit' funded under the ESRC Census Programme 2006–2011. It has been constructed from a number of linked software components, including a web server (Apache), a database management system (PostgresSQL) and its add-on (PostGIS) that creates various geometric data types and also provides a set of functions to use those data types,

**Figure 4.14** The general query interface in WICID prior to selection
*Source*: The Centre for Interaction Data Estimation and Research

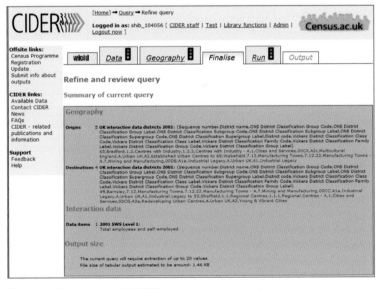

**Figure 4.15** Summary of WICID query
*Source*: The Centre for Interaction Data Estimation and Research

plus a widely used, general-purpose scripting language called PHP that is especially suited for Web development and can be embedded into

BOX 4.6

HTML. Use is also made of the library routines of a mapping package called Mapserver. These routines provide map-related functions for PHP to generate maps to support query selection. All the software components are freely available, permitting redistribution of the entire system (excluding data) without any licensing difficulties.

One of the fundamental aims of CIDER has been to create a user-friendly interface to these complex flow data sets in order to encourage their exploitation. Consequently, much effort has been directed at building a flexible, yet simple, query interface. Once logged into the system and running WICID, the user is presented with the 'General query interface' that requires the selection of *Data* and then *Geography* (Figure 4.14). Of course, the interaction data sets all require double geographies, and consequently the user has to decide which origin and destinations are required. One of the most innovative features of WICID is the facility to build geography selections in which the origins and destinations are drawn from sets of areas at different spatial scales and are not required to be the same.

There are various methods of selecting geographical areas: 'Quick selection' enables all areas at a certain scale to be selected; 'List selection' allows areas to be chosen from a list of all areas at each scale; 'Type-in-box' selection provides for one area to be selected at a time; and 'Copy selection' allows areas selected for origins to be copied into destinations or vice versa. The availability of web-mapping software in WICID means that there is a 'Map selection' option, enabling users to pick their origins and destinations from a map on screen. Similarly, there are a series of options allowing users to select variable counts in different ways. Once data variables, origins and destinations have been

**Figure 4.16** On-screen preview of flows extracted (WICID)
*Source*: The Centre for Interaction Data Estimation and Research

selected, the 'traffic lights' will turn green and the query is ready. Prior to extraction, users may finalize the query in some way; for instance to exclude all intra-area flow counts.

The query presented in Figure 4.15 illustrates the flexibility of the system by selecting origins (the 5 districts of West Yorkshire) that are different from destinations (the 4 districts of South Yorkshire). Information is provided on a number of origin and destination attributes including area classification codes. The data count in this instance is the number of employees and self-employed from the 2001 Census SWS.

Once the query has run and the data have been extracted, the flows extracted can be previewed on the screen (Figure 4.16) and downloaded using a preferred format with selected labelling. Additionally, WICID allows users to undertake some analysis of the data subset extracted. There is a suite of indicators that are computed for a selection of any five of the counts that have been selected in the query. These indicators include a set of general descriptive statistics and a measure of distance travelled as well as measures of connectivity, impact and intensity (Stillwell *et al.*, 2005).

CIDER has been in operation since 2006 and its web interface (WICID) has been modified to incorporate new sets of primary data from the 2001 Census. In addition, data from the 1981 and 1991 Censuses have been re-estimated for spatial units defined in 2001 and this allows more consistent comparison over time. In 2004 users in 169 academic institutions carried out 3246 sessions using WICID. However, use of the service increased significantly during the last quarter of 2004 and the first quarter of 2005 as the new 2001 Census interaction data sets became available.

Various functions within WICID have been developed to facilitate user access to data prior to extraction and data analysis following extraction. In particular, the map selection tool has been improved in response to feedback from users whose knowledge of the geographies of the UK is limited and who require some assistance in selecting sets of zones to build customized queries involving origins and/or destinations at different spatial scales.

Although WICID has been designed primarily with a view to facilitating access to and analysis of the census interaction data sets by researchers, it does provide a very useful example of the use of relational databases and metadata, as well as the software components required to build customized, Web-based databases with analytical and mapping facilities, planning support systems (Stillwell and

Geertman, 2003; Geertman and Stillwell, 2009) or decision support systems more generally.

CIDER is a service that promotes use of the interaction data sets. 'Easy access' and 'user-friendliness' are key principles in delivering the service. Users are at liberty to phone the help line in emergency situations or to email the developers with questions or for advice on using certain data sets or building particular queries. Further enhancements of WICID continue to be made to facilitate comparisons between censuses, to construct new geographies that users might find useful, to provide access to certain non-census interaction data sets, to enable users to create their own geographies and to extend the analytical facilities.

### FURTHER INFORMATION

Centre for Interaction Data Estimation and Research (CIDER): cider.census.ac.uk/
WICID Teaching resources: ahds.ac.uk/history/collections/chcc.htm
PostGIS: postgis.refractions.net
MapServer: mapserver.org/

The CIDER project (RES-348-25-0005) is funded under the ESRC Census Programme and the WICID software system has been developed by Oliver Duke-Williams.

*Source*: John Stillwell, University of Leeds

---

Real-time databases are an interesting area of current development. Managers often need to have access to real-time information about the location and state of their assets and the environment in which they are situated. Creating a database for this can be expensive, in terms of time and money. To ensure the value of data is retained, the database must be kept up to date. The advent of the Web and wireless networking have expanded the possibilities for making use of real-time databases that are updated automatically as events happen or as objects in the database change their state or location. For example, parcel delivery services such as UPS use dynamic real-time databases to allow customers to track the status and location of a package online. GPS-tracking of public transport networks and services such as buses and taxis give managers and customers real-time information on the current location and estimated time of arrival of a particular service.

## ■ DEVELOPMENTS IN DATABASES

There are problems with the relational approach to the management of spatial data. For example, spatial data do not naturally fit into tabular structures, and, as discussed earlier, the SQL query language does not have capabilities for spatial ideas and concepts. Early researchers suggested extensions to SQL to help with the handling of spatial data (Raper and Rhind, 1990) whilst Worboys (1995) suggested that work on 'extensible relational DBMS' would provide the way forward. MySQL is an example of a widely used, open-source relational database management system which has implemented spatial extensions, based on a specification developed by the Open Geospatial Consortium (OGC, 2010a). Karlsson (no date) outlines the extensions necessary to allow SQL-based relational databases such as My SQL to accommodate GIS objects and operations. Data types, operations, the ability to input and output GIS data and the indexing of spatial data all need to be addressed, together with issues of metadata and different co-ordinate systems.

Attention has also been focused on the development of object-oriented (OO) approaches to database design and there are some examples of OO database principles in current GIS.

The object-oriented database approach offers the opportunity to move away from the 'geometry-centric' data models that present the world as collections of points, lines and polygons. The fundamental aim of the OO model is to allow data modelling that is closer to real-world things and events (Longley *et al.*, 2010). Thus, the way events and entities are represented in a database should be closer to the way we think about them. The method for achieving this is to group together all the data describing a real-world entity, together with any operations that are appropriate to the entity, into an object. The key feature of an OO database is the power given to the user to specify both the structure of these objects and the operations that can be applied to them (Elmasri and Navathe, 1994). So, each entity is modelled as an object and can be represented by the simple formula:

Object = state + behaviour

The 'state' of an object is the set of values of its attributes and the 'behaviour' represents the methods of operating on it. The identity of an object is unique and does not change during its lifetime. Composite objects, which are made up of more than one other object, can be created. In addition, an object can belong to a superclass of objects, and then will inherit all the properties of this superclass. What does this mean in practice? Consider again the Happy Valley ski resort. One possible method for representing the entity 'hotel' was considered earlier during the discussion of the relational database model. An intension was proposed to describe a table for the entity 'hotel' in a relational database:

HOTEL (Hotel ID, Name, Address, No.rooms, Standard).

This tells us that the entity 'hotel' has attributes ID-number, name, address, number of rooms and standard. In the OO model, these attributes become the states and 'hotel' becomes a class of objects. In addition to the states of an object, information about its behaviour is stored. These are the operations that can be performed on the object. For example, it is possible for a hotel to open additional rooms, have its standard revised or be closed to visitors. So, for the object 'hotel':

Hotel = (Hotel ID, Name, Address, No.rooms, Standard) + (Open additional rooms, Change standard, Close to visitors)

In addition, the object 'hotel' may belong to a superclass of objects known as 'buildings'. All objects in this superclass will share similar properties – states and behaviours (for example, buildings all have an identity, size, shape and location, and they can be constructed, demolished or drawn on a map). Any object in the class 'hotel' will inherit the properties of the superclass 'building', thus the states and operations applicable to 'buildings' will also apply to 'hotels'. Therefore, a hotel has a size, shape, identity and location, and can be constructed, demolished or drawn on a map. The class of object 'hotel' has *inherited* the properties of the superclass 'buildings'. Thus, a *hierarchy* of objects is emerging. This hierarchy can work in the other direction too, and the object in the class 'hotel' may be *composite* – made up, for example, of subclasses 'restaurant', 'swimming pool' and 'garden'. This situation is illustrated in Table 4.3 and Figure 4.17.

In a GIS each class of object is stored in the form of a database table: each row represents an object and each column is a state. The methods that can be applied are attached to the objects when they are created in memory for use in an application (Longley *et al.*, 2010).

**TABLE 4.3 Definition of the object 'hotel' for the Happy Valley OO database**

| Object | State | Behaviour |
|--------|-------|-----------|
| Hotel | Name | Can be plotted on map |
| | Address | Can be added to database |
| | Number of bedrooms | Can be deleted from database |
| | Standard of accommodation | Standard can be upgraded or downgraded |
| | | Number of bedrooms can be increased or decreased |

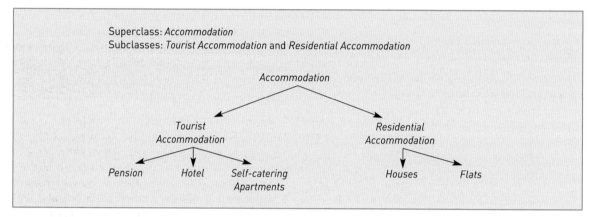

**Figure 4.17** Object hierarchy for Happy Valley OO database

The OO approach is possibly more appropriate for geographical data than the relational model, since it allows the modelling of complex, real-world objects, does not distinguish between spatial and attribute data and is appropriate for graphics operations. Longley *et al.* (2010) list the three particular features of object data models that make them particularly good for modelling geographic systems:

- *Encapsulation* – the packaging together of the description of state and behaviour in each object.

- *Inheritance* – the ability to re-use some or all of the characteristics of one object in another object.

- *Polymorphism* – the specific implementation of operations like draw, create or delete for each object.

However, Goodchild (1990) points out that there are problems with the approach, since many geographical ideas have implicit uncertainty and the spatial objects that we require our databases to model are often the products of interpretation or generalizations. Therefore, it can be difficult to represent the world as rigidly bounded objects. Other disadvantages are that the methods are still under research and development. For the user there are the additional problems that, to date, there is no standard data model, no clear theoretical base for the OO model and no standard query language.

## REFLECTION BOX

- Consider a GIS database created for a national utility company – perhaps an electricity supplier. What would you consider to be the most appropriate approach – centralized or distributed database? Why?
- Now imagine that you are one of the designers of the GIS database for the utility company. Using Box 4.3 to help you get started, make notes on some of

the issues that you will need to consider during the design of this database.
- Now imagine that you are a potential end user of the utility company database. Perhaps you record customer reports of faults and prepare instructions for field engineers to repair these. What are your database requirements?

## CONCLUSIONS

This chapter has introduced the methods available for the handling of attribute data in GIS. The need for formal methods for database management has been discussed, then the principles and implementation of a relational database model considered in detail, since this is the most frequently used in current GIS. Options for large-scale users have been presented, including the use of centralized and distributed database systems. Finally, a brief introduction to one of the trends in database management in GIS, the object-oriented approach, has been given.

An understanding of the database models for the management and handling of attribute data in GIS is essential for effective use of the systems available. It is important to understand how both spatial data and attribute data are structured within GIS to enable you to specify appropriate and achievable ques-

tions for your GIS to answer, and to implement these effectively. The way in which the spatial and attribute data are linked in an individual system is also important and an area where considerable differences are seen between systems. Returning to the car analogy, imagine that you have decided to buy a car but need to know what it can do. What is the car's top speed? How far can it travel on a full tank of fuel? How does the engine work, and how does the transmission/gear box make it move? How is the car structured, how effective is it and how complex is it to maintain? To keep the car running it is essential to know some of this technical information. Similarly, for a GIS user, a GIS will not be able to function unless the data are appropriately modelled and structured. But first we need to get some data into the GIS, or some fuel into the car!

## REVISION QUESTIONS

- What is the difference between a database and a database management system?
- Describe the main characteristics of the relational database model. Why have relational databases dominated in GIS?
- What are the main issues to be considered by users of large corporate GIS databases?
- Consider possible future directions and trends for GIS databases.
- List the main features of the object-oriented approach to databases.
- Why are Web-based applications of GIS databases becoming more common?

## FURTHER STUDY – ACTIVITIES

■ Draw up the tables you would expect to see in a simple relational database containing data for the following entities: student, course, teacher, department. Note which fields would be used as keys, and give examples of the types of queries that could be asked of the database.

■ Access an example of an online spatial database (see below for example web links). Use the database to retrieve some information of interest.

■ Use the Internet to explore the possibilities for parcel tracking with some of the major global delivery services. Next time you have cause to send a package using one of these services, track its progress online at regular intervals and note how the state and location of the packages changes in the database.

■ Access an online public transport vehicle tracking service. Watch how the state and location of services changes with time (see below for example web links).

## FURTHER STUDY – READING

A chapter that covers similar ground to this one, offering the development of a National Parks database as a useful example, is Healey (1991). Other texts that contain chapters introducing databases in GIS without too much technical detail, are Aronoff (1991), Burrough (1986) and Dale and McLaughlin (1988).

For a more technical introduction, from a computer science perspective, there is a huge range of texts on databases available. Some readable and useful examples are Oxborrow (1989) and Elmasri and Navathe (1994). Worboys and Duckham (2004) is an interesting book, considering all aspects of GIS from a computer science perspective, and this has a comprehensive section on databases.

Batty (1990) offers an article on exploiting relational database technology for GIS and Worboys (1999) reviews relational and object-oriented approaches. Further details of the object-oriented approach, including an example of an object data

model, can be found in Longley et al. (2010), whilst Worboys et al. (1991) and Egenhofer and Frank (2002) provide full explanations of the object-oriented approach. Corporate database issues have been considered by McLaren (1990).

Aronoff S (1991) *Geographic Information Systems: A Management Perspective*. WDL Publications, Ottawa

Batty P (1990) Exploiting relational database technology in GIS. *Mapping Awareness* 4 (6): 25–32

Burrough P A (1986) *Principles of Geographical Information Systems for Land Resources Assessment*. Clarendon Press, Oxford

Dale P F, McLaughlin J D (1988) *Land Information Management: An Introduction with Special Reference to Cadastral Problems in Third World Countries*. Clarendon Press, Oxford

Egenhofer M, Frank A (1992) Object-Oriented Modeling for GIS. *Journal of the Urban and Regional Information Systems Association* Vol. 4: 3–19

Elmasri R, Navathe S B (1994) *Fundamentals of Database Systems*. 2nd edn. Benjamin/Cummings, California

Healey R G (1991) Database management systems. In: Maguire D J, Goodchild M F, Rhind D W (eds) *Geographical Information Systems: Principles and Applications*. Longman, London, vol. 1, pp. 251–67

Longley P A, Goodchild M F, Maguire D J, Rhind D W (2010) *Geographical Information Systems and Science*. 3rd edn. Wiley, Chichester

McLaren R A (1990) Establishing a corporate GIS from component datasets – the database issues. *Mapping Awareness* 4 (2): 52–8

Oxborrow E P (1989) *Databases and Database Systems: Concepts and Issues*. 2nd edn. Chartwell Bratt, Sweden

Worboys M F (1999) Relational databases and beyond. In: Longley P A, Goodchild M F, Maguire D J, Rhind D W (eds) *Geographical Information Systems*. New York, Wiley, vol. 1, pp. 373–84

Worboys M F, Duckham M (2004) *GIS: A Computing Perspective*. 2nd edn. Taylor and Francis, London

Worboys M, Hearnshaw H, Maguire D (1991) Object-oriented data modeling for spatial databases. *International Journal of Geographical Information Systems* 4(4): 369–383

## WEB LINKS

### Definitions and concepts:

- Data models: unixspace.com/context/databases.html

### Database design:

- Unesco training materials on data modeling: ioc.unesco.org/oceanteacher/resourcekit/Module2/GIS/Module/Module_e/module_e2.html

### Databases with spatial extensions:

- MySQL: dev.mysql.com/tech-resources/articles/4.1/gis-with-mysql.html

- Oracle: www.oracle.com/technetwork/database/options/spatial/index.html

### Online databases:

- OECD: caliban.sourceoecd.org/vl=2215771/cl=97/nw=1/rpsv/home.htm

### Transport tracking:

- Nextbus: www.nextbus.com/predictor/newUserWelcome.shtml

- ACISK: www.acislive.com/pages/busnet.asp?SysID=13

Visit our website **www.pearsoned.co.uk/heywood** for further web links, together with self-assessment questions, activities, data sets for practice opportunities and other interactive resources.

# Data input
and editing

## Learning outcomes

By the end of this chapter you should be able to:

■ Explain the difference between analogue and digital data sources for GIS

■ Give examples of different methods of data encoding

■ Describe how paper maps are digitized

■ Explain how remotely sensed images are imported into GIS

■ Describe some of the problems that may be faced when encoding spatial data

■ Give examples of methods of data editing and conversion

■ Outline the process required to create an integrated GIS database

## ■ INTRODUCTION

Data encoding is the process of getting data into the computer. It is a process that is fundamental to almost every GIS project. For example:

■ An archaeologist may encode aerial photographs of ancient remains to integrate with newly collected field data.

■ A planner may digitize outlines of new buildings and plot these on existing topographical data.

■ An ecologist may add new remotely sensed data to a GIS to examine changes in habitats.

■ A historian may scan historical maps to create a virtual city from the past.

■ A utility company may encode changes in pipeline data to record changes and upgrades to their pipe network.

Being able to get data into GIS is clearly fundamental to how it works. A GIS without data can be likened to a car without fuel – without fuel you cannot go anywhere; without data a GIS will not be able to do any analysis or produce any output. However, this is perhaps where the similarity ends, as there is only one place to obtain fuel (a petrol/gas station) and only one method of putting fuel into a car (using a petrol/gas pump). Spatial data, on the other hand, can be obtained from many different sources (see **Chapter 2**), in different formats, and can be input to GIS using a number of different methods. Maps, which may come as paper sheets or digital files, may be input by digitizing, scanning or direct file transfer; aerial photographs may be scanned into a GIS; and satellite images may be downloaded from digital media. In addition, data can be directly input to GIS from field equipment such as GPS, or from sources of ready-prepared data from data 'retailers' or across the Internet.

Data would normally be entered into a GIS before being structured and used in analysis. However, since the characteristics of data and the way they are to be modelled influence data encoding methods, the characteristics of spatial and attribute data, and methods of representing and structuring them have been considered earlier in this book (see **Chapters 2**, **3** and **4**).

Once in a GIS, data almost always need to be corrected and manipulated to ensure that they can be structured according to the required data model. Problems that may have to be addressed at this stage of a GIS project include:

■ the reformatting of data (e.g. conversion of a postal or zip code to a grid reference);

■ the re-projection of data from different map sources to a common projection;

■ the generalization of complex data to provide a simpler data set; or

■ the matching and joining of adjacent map sheets once the data are in digital form.

This chapter looks in detail at the range of methods available to get data into a GIS. These include keyboard entry, digitizing, scanning and electronic data transfer. Then, methods of data editing and manipulation are reviewed, including re-projection, transformation and edge matching. The whole process of data encoding and editing is often called the 'data stream'. This is outlined in Figure 5.1 and used as a framework for the chapter.

Before further explanation of the stages in the data stream, it is necessary to make a distinction between analogue (non-digital) and digital sources of spatial data. Analogue data are normally in paper form, and include paper maps, tables of statistics and hard-copy (printed) aerial photographs. These data all need to be converted to digital form before use in a GIS; consequently, the data encoding and correction procedures are longer than those for digital data. Digital data are already in computer-readable formats and are supplied on CD-ROM or across a computer network or the Internet. Map data, aerial photographs, satellite imagery, data from databases and automatic data collection devices (such as data loggers and GPS) are all available in digital form. If data were all of the same type, format, scale and resolution, then data encoding and integration would be simple. However, since the characteristics of spatial data are as varied as their sources, the task is complicated. This variety has implications for the way data are encoded and manipulated to develop an integrated GIS database. Much effort has been made in recent years to develop universal GIS data standards and common data exchange formats (see **Chapter 2** for further details).

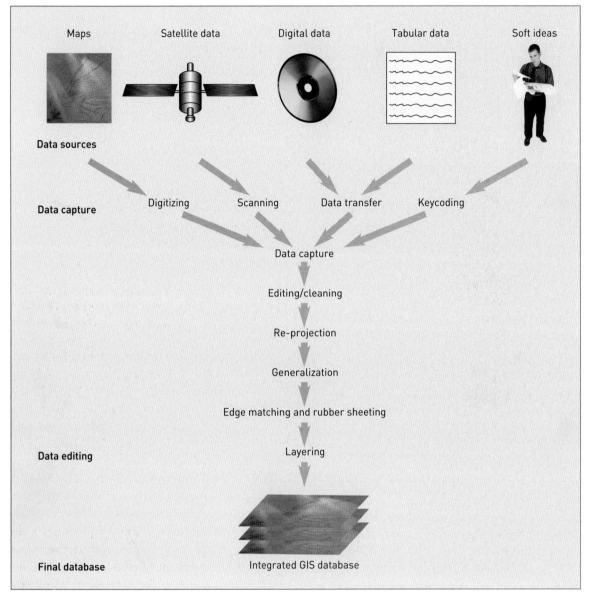

Maps   Satellite data   Digital data   Tabular data   Soft ideas

**Data sources**

**Data capture**   Digitizing   Scanning   Data transfer   Keycoding

Data capture

Editing/cleaning

Re-projection

Generalization

Edge matching and rubber sheeting

**Data editing**   Layering

**Final database**   Integrated GIS database

**Figure 5.1** The data stream

## ■ METHODS OF DATA INPUT

Data in analogue or digital form need to be encoded to be compatible with the GIS being used. This would be a relatively straightforward exercise if all GIS packages used the same spatial and attribute data models. However, there are many different GIS packages and many different approaches to the handling of spatial and attribute data. **Chapter 3** reviewed the two main data models in use in GIS (raster and vector), and **Chapter 4** considered the relational database model widely used for attribute data. While the general principles behind these models remain the same in different GIS packages, the implementation may be very different. Different too are the problems faced when transferring data from one package to another, or when preparing data for data analysis. The situation is being eased by the development of standards for spatial data encoding, a topic that will be returned to later in this chapter and in **Chapter 10**. Table 5.1 summarizes possible data encoding methods and gives an indication of their appropriateness for different data sources.

**TABLE 5.1 Possible encoding methods for different data sources**

| Data source | Analogue or digital source | Possible encoding methods | Examples |
| --- | --- | --- | --- |
| Tabular data | Analogue | ■ Keyboard entry<br>■ Text scanning | ■ Address lists of hotel guests<br>■ Tables of regional tourism statistics from official publications |
| Map data | Analogue | ■ Manual digitizing<br>■ Automatic digitizing<br>■ Scanning | ■ Historical maps of settlement and agriculture<br>■ Infrastructure and administrative maps |
| Aerial photographs | Analogue | ■ Manual digitizing<br>■ Automatic digitizing<br>■ Scanning | ■ Ski piste locations<br>■ Extent of spring floods |
| Field survey | Analogue | ■ Keyboard entry | ■ Piste condition |
| Tabular data | Digital | ■ Digital file transfer (with reformatting if necessary) | ■ National population census data<br>■ Data from meteorological station recording equipment |
| Map data | Digital | ■ Digital file transfer (with reformatting) | ■ Digital topographic data from national mapping agency<br>■ Digital height (DTM) data |
| Aerial photographs | Digital | ■ Digital file transfer (with reformatting) | ■ Background contextual data<br>■ Ski piste locations |
| GPS | Digital | ■ Digital file transfer (with reformatting) | ■ Piste lines |
| Satellite imagery | Digital | ■ Digital file transfer<br>■ Image processing and reformatting | ■ Land use data<br>■ Forest condition data |

All data in analogue form need to be converted to digital form before they can be input into GIS. Four methods are widely used: keyboard entry, manual digitizing, automatic digitizing and scanning. Keyboard entry may be appropriate for tabular data, or for small numbers of co-ordinate pairs read from a paper map source or pocket GPS. Digitizing is used for the encoding of paper maps and data from aerial photographs. Scanning represents a faster encoding method for these data sources, although the resulting digital data may require considerable processing before analysis is possible.

Digital data must be downloaded from their source media (pen drive, CD-ROM or the Internet) and may require reformatting to convert them to an appropriate format for the GIS being used. Reformatting or conversion may also be required after analogue data have been converted to digital form. For example, after scanning a paper map, the file produced by the scanning equipment may not be compatible with the GIS, so reformatting may be necessary. Keyboard entry, manual digitizing, automatic digitizing (including scanning) and digital conversion are covered in greater detail below.

## Keyboard entry

Keyboard entry, often referred to as key-coding, is the entry of data into a file at a computer terminal. This technique is used for attribute data that are only available on paper. If details of the hotels in Happy Valley were obtained from a tourist guide, both spatial data (the locations of the hotels – probably given as postal or zip codes) and attributes of the hotels (number of rooms, standard and full address) would be entered at a keyboard. For a small number of hotels keyboard entry is a manageable task, although typographical errors are very likely.

If there were hundreds of hotels to be coded, an alternative method would probably be sought. Text scanners and optical character recognition (OCR) software can be used to read in data automatically. Attribute data, once in a digital format, are linked to the relevant map features in the spatial database using identification codes. These are unique codes that are allocated to each point, line and area feature in the data set.

The co-ordinates of spatial entities can be encoded by keyboard entry, although this method is used only when co-ordinates are known and there are not too many of them. If the locations of the Happy Valley hotels were to be entered as co-ordinates then these could be read from a paper map and input at the keyboard. Where there are large numbers of co-ordinates and features to be encoded, it is more common to use manual or automatic digitizing.

## Manual digitizing

The most common method of encoding spatial features from paper maps is manual digitizing. It is an appropriate technique when selected features are required from a paper map. For example, the Happy Valley road network might be required from a topographical map of the area. Manual digitizing is also used for map encoding where it is important to reflect the topology of features, since information about the direction of line features can be included. It is also used for digitizing features of interest from hard-copy aerial photographs (for example, snow cover on a particular date could be encoded from aerial photographs).

**Figure 5.2** Digitizing table and PC workstation

Manual digitizing requires a digitizing table that is linked to a computer workstation (Figure 5.2). The digitizing table is essentially a large flat tablet, the surface of which is underlain by a very fine mesh of wires. Attached to the digitizer via a cable is a cursor that can be moved freely over the surface of the table. Buttons on the cursor allow the user to send instructions to the computer. The position of the cursor on the table is registered by reference to its position above the wire mesh.

Using a digitizer is quite straightforward. However, to achieve good results experience and awareness of some of the problems and limitations of digitizing are required. The general procedure for digitizing is summarized in Box 5.1.

## BOX 5.1 Using a manual digitizing table

PRACTICE

Before starting to digitize information from a paper map, care needs to be taken to ensure that the source document is in good condition and free from physical defects that could affect the digitizing process. Creases and folds can prevent the map from lying flat on the digitizer, while coffee stains, handwritten annotations and other blemishes can obscure detail. The procedure followed when digitizing a paper map using a manual digitizer has five stages:

1 *Registration*. The map to be digitized is fixed firmly to the table top with sticky tape. Five or more control points are identified (usually the four corners of the map sheet and one or more grid intersections in the middle). The geographic co-ordinates of the control points are noted and their locations digitized by positioning the crosshairs on the cursor exactly over them and pressing the 'digitize' button on the cursor. This sends the

co-ordinates of a point on the table to the computer and stores them in a file as 'digitizer co-ordinates'. These co-ordinates are the positions of the cursor cross-hairs relative to the wire mesh in the table. They are usually stored as decimal inches or centimetres from the bottom-left-hand corner of the table. Later, the geographic co-ordinates of the control points are used to transform all digitizer co-ordinates into geographic co-ordinates. Therefore, it is essential that the map is carefully registered on the digitizer table to ensure an accurate transformation of digitized features from digitizer to geographic co-ordinates. Once the map has been registered the user may begin digitizing the desired features from the map.

2 *Digitizing point features.* Point features – for example, spot heights, hotel locations or meteorological stations – are recorded as a single digitized point. A unique code number or identifier is added so that attribute information may be attached later. For instance, the hotel with ID number '1' would later be identified as 'Mountain View'.

3 *Digitizing line features.* Line features (such as roads or rivers) are digitized as a series of points that the software will join with straight line segments. In some GIS packages lines are referred to as arcs, and their start and end points as nodes. This gives rise to the term *arc–node topology*, used to describe a method of structuring line features (see **Chapter 3**). As with point features, a unique code number or identifier is added to each line during the digitizing process and attribute data is attached using this code. For example, when digitizing a road network, data describing road category, number of carriageways, surface type, date of construction and last resurfacing might be added.

4 *Digitizing area (polygon) features.* Area features or polygons – for example, forested areas or administrative boundaries – are digitized as a series of points linked together by line segments in the same way as line features. Here it is important that the start and end points join to form a complete area. Polygons can be digitized as a series of individual lines, which are later joined to form areas. In this case it is important that each line segment is digitized only once.

5 *Adding attribute information.* Attribute data may be added to digitized polygon features by linking them to a centroid (or seed point) in each polygon. These are either digitized manually (after digitizing the polygon boundaries) or created automatically once the polygons have been encoded. Using a unique identifier or code number, attribute data can then be linked to the centroids of appropriate polygons. In this way, the forest stand may have data relating to tree species, tree ages, tree numbers and timber volume attached to a point within the polygon.

Figure 5.3 shows an example of digitizing software.

**Figure 5.3** Digitizing software
*Source*: Courtesy of Able Software Corporation

Most manual digitizers may be used in one of two modes: point mode or stream mode (Figure 5.4). In point mode the user begins digitizing each line segment with a start node, records each change in direction of the line with a digitized point and finishes the segment with an end node. Thus, a straight line can be digitized with just two points, the start and end nodes. For more complex lines, a greater number of points are required between the start and end nodes. Smooth curves are problematic since they require an infinite number of points to record their true shape. In practice, the user must choose a sensible number of points to represent the curve (a form of user-controlled generalization). In addition, the digitizing equipment will have a minimum resolution governed by the distance between the wires in the digitizing table (typically about 0.1 mm). However, some digitizing packages allow the user to record smooth curves as mathematically defined splines or Bézier curves (Figure 5.5).

In stream mode the digitizer is set up to record points according to a stated time interval or on a distance basis (Jackson and Woodsford, 1991). Once the user has recorded the start of a line, the digitizer

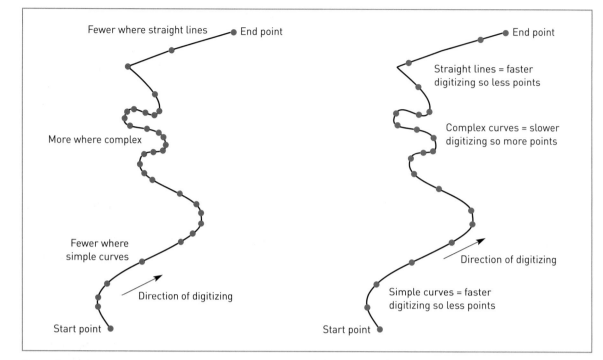

**Figure 5.4** Point and stream mode digitizing:
**(a)** *Point mode* – person digitizing decides where to place each individual point such as to most accurately represent the line within the accepted tolerances of the digitizer. Points are placed closer together where the line is most complex and where the line changes direction. Points are placed further apart where the line is less complex or made up of straight line segments.

**(b)** *Stream mode* – person digitizing decides on time or distance interval between the digitizing hardware registering each point as the person digitizing moves the cursor along the line. Points are placed closer together where the line is most complex only as the person digitizing slows the movement of the cursor down to more accurately follow the line. Points are placed further apart where the line is less complex or made of straight line segments allowing the person digitizing to move the cursor more quickly.

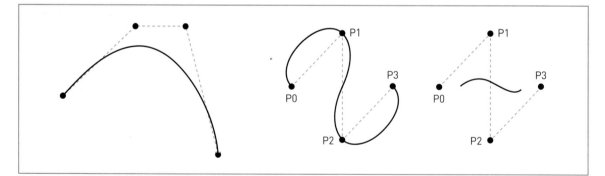

**Figure 5.5** Bézier curves and splines:
**(a)** *Bézier curves* are defined by four points; a start and end point (nodes) and two control points. When using Bézier curves to define curves when digitizing a curve between two points, the control points are used to mathematically determine the arc (path) of the curve on leaving the start point and on arriving at the end point. Bézier curves are used in many vector drawing/drafting packages such as Adobe Postscript.

**(b)** *Splines* are mathematically interpolated curves that pass through a finite number of control points (e.g. P0–P3). They can be exact (i.e. pass exactly through the control points) or approximating (i.e. pass approximately).

might be set to record a point automatically every 0.5 seconds and the user must move the cursor along the line to record its shape. An end node is required to stop the digitizer recording further points. The speed at which the cursor is moved along the line determines the number of points recorded. Thus, where the line is more complex and the cursor needs to be moved more slowly and with more care, a greater number of points will be recorded. Conversely, where the line is straight, the cursor can be moved more quickly and fewer points are recorded.

The choice between point mode and stream mode digitizing is largely a matter of personal preference. Stream mode digitizing requires more skill than point mode digitizing, and for an experienced user may be a faster method. Stream mode will usually generate more points, and hence larger files, than point mode.

The manual digitizing of paper maps is one of the main sources of positional error in GIS. The accuracy of data generated by this method of encoding is dependent on many factors, including the scale and resolution of the source map, and the quality of the equipment and software being used. Errors can be introduced during the digitizing process by incorrect registration of the map document on the digitizer table or 'hand-wobble'.

A shaky hand will produce differences between the line on the map and its digitized facsimile. Published estimates of the accuracy of manual digitizing range from as high as ± 0.8 mm (Dunn *et al.*, 1990) to as low as ± 0.054 mm (Bolstad *et al.*, 1990). As a rule an experienced digitizing technician should be able to encode data with an accuracy equal to the width of the line they are digitizing. Walsby (1995) considers that line characteristics have a significant influence on error creation, with the greatest number of errors being produced during the digitizing of small thin polygons and sinuous lines. In addition, she found that operators with limited cartographic experience who were also infrequent digitizers were most likely to create errors. Errors produced

during manual digitizing are covered in more detail in **Chapter 10**.

Many high-volume digitizing projects are outsourced by major GIS projects, often to specialist data suppliers in areas of the world such as the Indian sub-continent where a well-educated but relatively cheap labour force is available (see Box 5.2). Manual digitizing may also be used to digitize low volumes of data 'on demand' from scanned and geocorrected digital map images. Many GIS packages provide facilities for onscreen digitizing of point, line and polygon features using raster backdrop images as a guide (Figure 5.6). This technique can be employed by field engineers using hand-held devices and GPS to record the location of a feature or incident whilst they are on site.

## Automatic digitizing

Manual digitizing is a time-consuming and tedious process. If large numbers of complex maps need to be digitized it is worth considering alternative, although often more expensive, methods. Two automatic digitizing methods are considered here: *scanning* and *automatic line following*.

Scanning is the most commonly used method of automatic digitizing and is an appropriate method of data encoding when raster data are required, since this is the automatic output format from most scanning software. Thus, scanning may be used

**Figure 5.6** On-screen digitizing
*Source*: Screenshot shows Esri Graphical User Interface (GUI) ArcMap/ArcView/ArcInfo Graphical User Interface is the intellectual property of Esri and is used herein with permission. Copyright © 2011 Esri all rights reserved

# BOX 5.2 Business process outsourcing and the Indian GIS industry

**Ankur Das**

India is a developing country in which the IT industry has grown rapidly due to the presence of a relatively large pool of skilled labour and favourable governmental policies. GIS in India is at an early stage of implementation and growth. This case study examines the trends in data capture and input that are currently dominating the GIS industry in India and outlines the direction in which the industry is developing.

Although Chrisman (2002) predicted that advances in scanning technology would make manual digitizing obsolete, the presence of a large, highly skilled and IT-literate labour force in India has created a whole industry based on data capture and input. This industry services the GIS sector and other fields which require the conversion of paper documents to digital copy (for example, engineering, architecture and textiles). In addition to the availability of labour, the phenomena of business process outsourcing (BPO) has fuelled the growth of the IT sector, including the geospatial industry. The developed nations of North America and Europe have found a market in India where they can get work done with the same efficiency and quality that they would have achieved in their own country but at a much reduced cost. Growth of BPO has also been helped by the presence of an ever-growing, English-educated population which provides skilled and semi-skilled labour for the industry.

A handful of major companies operate at the top of the GIS market in India, all depending on work from the overseas market. Only a few companies cover the full spectrum of GIS work, and even in these companies the main component of their business comes from outsourcing from Europe and North America. About 5 per cent of business is from the domestic Indian market. Data entry work comprises the bulk of the activity of the GIS industry in India, and the industry is primarily dependent on overseas data entry contracts for its survival. There are two main components of the work:

- generation of new spatial data sets; and
- addition of new data layers and attributes to pre-existing data sets that are already in use.

The task of creating a spatial database by manual digitizing is one that demands many hours of monotonous but concentrated work and involvement.

A very small segment of the industry in Europe and North America are interested in such work and labour costs are prohibitive. In India the manual digitizing work is done quickly, accurately and efficiently. Digital data can be sent via the Internet and paper documents by courier to ensure rapid turnaround. Email communications permit easy supervision and feedback, allowing a company in India to produce a quality product that is acceptable to clients in a short time and at a competitive price.

Large data entry tasks require a large and moderately skilled workforce, and many Indian GIS data entry companies employ dozens of technicians working in digitizing workshops. These often operate shift working to ensure expensive equipment is kept in use 24 hours a day and that turn-around times are minimized. Multiple pass or double digitizing is commonly used to ensure high levels of data quality. Accuracy figures of 99.5 per cent are quoted by some companies. Quality-control procedures are rigorous and ensure that product quality is not compromised by the speed and price of the service.

Companies who use these services from the Indian industry may have their own GIS divisions with skilled manpower, but they prefer to outsource the work of data generation and editing. For example, a UK telecom service provider used BPO firms in India to update its customer database using GIS tools. Other examples of work completed in India include digitizing of flood hazard maps for US insurance companies, capture and coding of vegetation survey maps of the Netherlands and digitizing contours from US Quad sheets.

Some of the impacts of this reliance on data entry business are emerging in India. There has been stagnation in the quality of the available and emerging GIS skilled personnel. To meet project requirements, data entry companies often employ non-GIS people who undergo in-house training on the rudiments of the technology and then get to work. There is little scope for them to hone the skills and knowledge necessary to become a GIS specialist. This has had a tremendously negative effect on the employment prospects of the highly skilled people in GIS presently available in the country, however few in number they may be. Nonetheless, the data entry business is an important stepping stone to the next stage of development of the GIS industry in India and has generated employment for

→

**BOX 5.2**

many thousands of skilled and educated workers. Other issues that are influencing the development of GIS in India include the absence of a concrete mapping policy and the highly centralized nature of Indian decision-making. Academic users have been doing their best to raise the awareness for the acceptance of GIS technology. It remains to be seen how the development of the geospatial industry in India will evolve in the longer term and whether BPO operations will aid or retard India's development in this context.

**FURTHER INFORMATION**

Examples of Indian IT companies specializing in BPO operations including servicing the GI sector include:
Blue Matrix Software Consultancy: www.bluematrixit.com/
Riddhi Management Services: www.riddhi.org/
Micronetsolutions: www.micronetsolutions.itgo.com/
Prithvitech: www.prithvitech.com/
Trigeo Image Systems: www.trigeo.net/
Variac Systems: www.variacsystems.com/

*Source*: Ankur Das, India

to input a complete topographic map that will be used as a background raster data set for the over-plotting of vector infrastructure data such as pipelines or cables. In this case a raster background map is extremely useful as a contextual basis for the data of real interest.

A scanner is a piece of hardware for converting an analogue source document into digital raster format (Jackson and Woodsford, 1991). All scanners work by sampling the source document using transmitted or reflected light. Jackson and Woodsford (1991) provide a comprehensive review of scanning technology. The cheapest scanners are small flat-bed scanners – a common PC peripheral. High-quality and large-format scanners require the source document to be placed on a rotating drum, and a sensor moves along the axis of rotation.

Practical problems faced when scanning source documents include:

■ the possibility of optical distortion when using flat-bed scanners;
■ the automatic scanning of unwanted information (for example, hand-drawn annotations, folds in maps or coffee stains);
■ the selection of appropriate scanning tolerances to ensure important data are encoded, and background data ignored;
■ the format of files produced and the input of data to GIS software; and
■ the amount of editing required to produce data suitable for analysis.

The general procedure for using a scanner is described in Box 5.3.

**BOX 5.3  Using a scanner**    PRACTICE

There are three different types of scanner in widespread use (Figure 5.7):

■ flat-bed scanners;
■ rotating drum scanners; and
■ large-format feed scanners.

Large-format feed scanners are most suitable for capturing data for input to GIS as they are relatively cheap, quick and accurate. Flat-bed scanners, whilst being a common PC peripheral, are too small and inaccurate; and drum scanners, despite being very accurate, tend to be too slow and expensive.

All scanners work on the same principles (Hohl, 1998). A scanner has a light source, a background (or source document) and a lens. During scanning the absence or presence of light is detected as one of the three components moves past the other two.

The user should be aware of a number of issues when scanning maps for use in GIS:

**OUTPUT QUALITY**

If the output is to be used as a visual backdrop for other data in a GIS, then quality may not be a major issue. If, on the other hand, scanned

**BOX 5.3**

**(a)** Flat bed

**(b)** Drum

**(c)** Feed roller

**Figure 5.7** Types of scanner
*Sources*: (a) Epson (UK) Ltd used by permission;
(b) Stefan Kuhn, www.webkuehn.de;
(c) Colortrac, www.colortrac.com

output is to be used to produce vector data, the output must be as sharp and clear as possible. Quality can be influenced by the setting of a threshold above which all values are translated as white, and below which all values are black (Hohl, 1998). Setting brightness and contrast levels can also affect the quality of images obtained, as can gamma correction (a method that looks at a histogram of the image and places points strategically along the histogram to isolate data types). Filtering methods may be used to selectively remove noise from a document. This may be particularly useful when scanning old maps.

### RESOLUTION

This is the density of the raster image produced by the scanning process. The resolution of scanners is usually measured in dots per inch (dpi) as a linear measurement along the scan line. In general, choose a resolution that matches the data being captured. Commonly used guidelines are 200 dpi for text, 300 dpi for line maps and 400 dpi for high-quality orthophotos (Hohl, 1998). Interpolation may be used to increase the resolution of a scan. The values of pixels smaller than the scan head is physically capable of seeing are interpolated. This can create problems when scanning maps with sharp lines such as roads or contours as it tends to produce fuzzy rather than sharp lines in the scanned image.

### ACCURACY

The accuracy of the scanned image is important if the image is to be imported into a GIS database. The scanned document needs to be fit for its intended use in terms of its physical as well as its cartographic qualities. Remember that a scanner will not distinguish between coffee stains and real map data. The engineering tolerances of the scanner itself will also affect the accuracy of the output. For example, the rotating drum mechanism of drum scanners is much more accurate than the rollers employed to pull a map through a large-format feed scanner.

### VECTORIZATION

The output from scanned maps is often used to generate vector data. This process involves either automatic or interactive (user-controlled) raster to vector conversion. Problems occur with this process due to topographical effects at intersections, cartographic annotation such as contour heights or road names, generalization of features smaller than the resolution of the scanner and the subsequent coding of attributes (these are dealt with further in **Chapter 10**).

BOX 5.3

### GEOREFERENCING

Ultimately the output from a scanner needs to be correctly referenced according to the co-ordinate system used in the GIS. Normally this process is controlled using linear transformation from the row and column number of the scanned image to the chosen geographic co-ordinate system. This requires a series of fixed reference points on the image for which the geographic co-ordinates are known. Distortion across scanned images can create a problem in this transformation when using cheap flat-bed scanners.

The accuracy of scanned output data depends on the quality of the scanner, the quality of the image-processing software used to process the scanned data, and the quality (and complexity) of the source document.

The resolution of the scanner used affects the quality, and quantity, of output data. Cheaper flat-bed scanners have resolutions of 50–200 dpi whereas more expensive drum scanners use resolutions of 500–2500 dpi. The higher the resolution, the larger the volumes of data produced. This can place a heavy burden on computing power and storage space, and increase the time spent editing the scanned data to make it usable by a GIS.

Whilst digitizing tools may be provided within GIS software, scanning and image-processing software are seldom provided; therefore, users must consider the format of output files. If these are not compatible with the GIS being used, extra reformatting may be necessary. Image-processing software can be used to distinguish different features from a single scanned image. For example, data layers for roads, water and forestry may be produced if image-processing software is used to distinguish between map features on the basis of their colour.

Either ordinary 'off-the-shelf' maps or the colour separates used by publishers can be encoded using scanning (Figure 5.8). If colour separates are used the colour schemes used to distinguish features in a published map can be used to advantage. For example, on maps, contour lines are generally represented as orange or brown lines made up of the colours yellow and magenta. Large volumes of scanned data require substantial computer power, occupy sizeable file storage space, and need more editing to make the data usable in a GIS. Thus, the scanning and subsequent editing of contour data are made much easier.

Chrisman (1987) considers that scanning technology produces data to a higher quality than manual digitizing, and that the combination of scanned graphic and line detection algorithms controlled on-screen by the user will help to make manual digitizing obsolete. However, this has not happened yet, as the case study in Box 5.2 illustrates.

Another type of automatic digitizer is the automatic line follower. This encoding method might be appropriate where digital versions of clear, distinctive lines on a map are required (such as country boundaries on a world map, or clearly distinguished railways on a topographic map). The method mimics manual digitizing and uses a laser- and light-sensitive device to follow the lines on the map. Whereas scanners are raster devices, the automatic line follower is a vector device and produces output as (x,y) co-ordinate strings. The data produced by this method are suitable for vector GIS. Automatic line followers are not as common as scanners, largely due to their complexity. In addition, difficulties may be faced when digitizing features such as dashed or contour lines (contour lines are commonly broken to allow the insertion of height values). Considerable editing and checking of data may be necessary. However, automatic digitizing methods can be used effectively and are now widely employed by commercial data providers as they allow the production of data at a reasonable cost.

## Electronic data transfer

Given the difficulties and the time associated with keyboard encoding, manual digitizing and automatic digitizing, the prospect of using data already in digital form is appealing. If a digital copy of the data required is available in a form compatible with your GIS, the input of these data into your GIS is merely a question of electronic data transfer. However, it is more than likely that the data you require will be in a different digital format to that recognized by your GIS. Therefore, the process of

(a) Original

(b) Cyan

(c) Magenta

(d) Yellow

(e) Black

**Figure 5.8** Colour separates
*Source*: United States Geological Survey

digital data transfer often has to be followed by data conversion. During conversion the data are changed to an appropriate format for use in your GIS.

Spatial data may be collected in digital form and transferred from devices such as GPS receivers, total stations (electronic distance-metering theodolites), and data loggers attached to all manner of scientific monitoring equipment. All that may be required is wireless data transfer for a user to download the data to a file on their computer. In some cases it may be possible to output data from a collection device in a GIS format. For example, GPS receivers are available that will output data into the format required by a range of GIS packages, as well as into common spreadsheet and standard GPS formats (Table 5.2). GPS is being used as a physical replacement for digitizing data from paper maps. For example, community-based and end-user-driven programmes of 'digitizing' street networks such as OpenStreetMap have been initiated as a means of avoiding copyright charges on existing spatial databases (see Box 5.4). Recent advances in laser range finding have led

**TABLE 5.2 GPS data formats**

| Organization/Source | Format | Description | Web link |
|---|---|---|---|
| National Marine Electronics Association | NMEA | NMEA's standards for data communication between marine instruments (as used between a GPS and Autopilot, for example). | http://www.nmea.org/ |
| Garmin | PCX5 TRK WPT | Garmin's own standards for storing and transferring waypoint, route and track information. | http://www.garmin.com/ |
| DeLorme | GPL | DeLorme's GPS log format. | |
| GPXchange | GPX | Common format based on XML intended for interchange of GPS data between applications and Internet services. | http://www.gpxchange.com/ |
| RINEX | RINEX | Receiver INdependent EXchange Format | http://gps.wva.net/html.common/rinex.html |

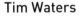

**BOX 5.4 OpenStreetMap**

CASE STUDY

### Tim Waters

OpenStreetMap (OSM) is a collaborative project aiming to create a free, editable world map (Figure 5.9). Raster maps, vector data and the raw database are available without charge. It is a volunteer run project, with over 100,000 registered users which is being driven, in part, by the availability of cheap GPS equipment, the Open Source movement and 'crowd sourcing' ( the process that allows anyone and everyone to contribute, in a similar manner to the popular Wikipedia project).

Data sources for OSM include public domain and donated data sets. Data are also collected by digitizing over suitable aerial imagery and out of copyright maps. However, the most common method of data collection is recording tracks using a GPS receiver. A GPS device stores its position at intervals and produces a track of the route that the operator takes. As the GPS receiver logs the position, the operator notes important features and items they think should be on the map. The operator may record waypoints, take photos, record spoken comments or draw sketch maps to help record additional attribute information.

The data are transferred from the GPS device onto a computer and the operator traces over the GPS tracks, digitizes new features and adds to or corrects existing information. Updated features and GPS tracks are uploaded to the OSM server, then become live for anyone else to see, download, use and edit.

→

**BOX 5.4**

The OSM data are composed of the following elements: nodes, ways and relations, and tags for attributes. At its simplest, a node is a point, and a way is a collection of nodes. A way can be thought of as a line. Creating a polygon by joining up a way makes an area. A relation is a group of features: for instance, a multipolygon (for example, an area with holes) or route (such as a bus route).

Attributes or tags are applied to nodes and ways and relations. Tags are value pairs, such as 'highway=primary' or 'amenity=doctors', and can be combined. There is no fixed schema for the use of tags and attributes; however, there are a number of agreed and recommended approaches.

**Figure 5.9** OpenStreetMap web interface
*Source*: OpenStreetMap (www.openstreetmap.org/). Copyright © OpenStreetmap and contributors, CC-BY-SA

**WEB LINK**

OpenStreet Map: www.openstreetmap.org/
*Source*: Tim Waters, Geospatial Consultant, Leeds

to new sources of terrain data using both airborne and ground-based systems. LiDAR is being used to create high-resolution terrain models using plane or helicopter mounted sensors (see **Chapters 2** and **3**). Fixed or mobile ground-based sensors or Terrestrial Laser Scanners (TLS) are also being used to create 3D terrain models and perspective views in both urban and rural locations (Box 5.5).

Electronic data transfer will also be necessary if data have been purchased from a data supplier,

# BOX 5.5 Terrestrial Laser Scanning

CASE STUDY

## Jeff Warburton

Terrestrial Laser Scanning (TLS) provides an effective means of capturing high resolution data very rapidly. The major benefit of TLS over conventional survey techniques is that it provides automated, high-speed data capture of complex surfaces in what are sometimes inaccessible environments. The technique is particularly valuable in studies of dynamic fluvial systems which are undergoing rapid landscape change. TLS:

- is very fast;
- can collect complex surfaces at high resolution (millimetre accuracy over a 100 m range);
- can be executed remotely (for example, away from a dangerous or hazardous environment); and
- provides a complete archive of the current scene.

The mass of 3D points recorded by the scanner is known as a *point cloud* (Figure 5.10). Point clouds generated are amenable to numerous 2D and 3D analyses and can be combined with ground photography to produce real-colour landscape models. However, care should be taken when implementing these surveys as the method contains pitfalls and limitations which need to be considered. For example, data voids may occur if scan stations are not carefully selected; scan artefacts can result from uneven point densities or differences in surface reflectivity (for instance, of rock, vegetation and water); and scanning cannot be undertaken in poor weather (for example, when there is too much atmospheric moisture). Furthermore, in fluvial applications the terrestrial laser scan pulses are not consistently returned from the water surface

**BOX 5.5**

(a)

(b)

(c)

**Figure 5.10** Terrestrial Laser Scanning (TLS) images of the Odenwinkelkees braidplain, Austria (July 2008). (a) Composite scan of the full study reach scanned at 0.25 m resolution. (b) Small area scan of a diagonal river bar at the head of the study reach (yellow box – scan resolution 0.1m). The images show 'point clouds' composed of thousands of individual scanned survey points. The black areas are the river channels which do not return a signal when scanned. (c) TLS in use.
*Source*: Courtesy of Jeff Warburton

of streams, rivers and ponded water – although in some situations this can be used to delineate the wetted extent and channel patterns. Heritage and Hetherington (2007) propose a protocol for the application of laser scanning in fluvial geomorphology. Their review demonstrates the limitations of the method and offers good practical advice for optimizing field results. Hetherington *et al.*, (2005) and Milan *et al.* (2007) have applied these techniques to assess fine sediment dynamics and channel morphological change in a proglacial river/glacier outwash plain in southern Switzerland. Using high resolution data (> 500 points per square metre) from a ten-day period early in the ablation season (June), the study successfully quantified a major episode of avulsion (rapid channel abandonment and reformation) and medial bar erosion, but also many transient lobe features and minor changes such as

bank accretion. Using the same data, Milan *et al.* (2007) demonstrated the precision of the method and determined the effects of survey frequency on the sediment budget estimates. It was found that daily surveys increased erosion volume estimates by 67 per cent and deposition estimates by 14 per cent compared with an eight-day survey interval.

### FURTHER READING

Heritage G, Hetherington D (2005) The use of high-resolution field laser scanning for mapping surface topography in fluvial systems. In Walling D E, Horowitz A J (eds) *Sediment Budgets 1*, IAHS Publication 291: 269–77

Heritage G, Hetherington D (2007) Towards a protocol for laser scanning in fluvial geomorphology. *Earth Surface Processes and Landforms.* 32 (1) 66–74

## BOX 5.5

Hetherington D, Heritage G, Milan D (2005) Daily fine sediment dynamics on an active Alpine glacier outwash plain. In Walling D E, Horowitz A J (eds) *Sediment Budgets 1*, IAHS Publication 291: 278–84

Milan D J, Heritage G L, Hetherington D (2007) Application of 3D laser scanner in the assessment of

erosion and deposition volumes and channel change in a proglacial river. *Earth Surface Processes and Landforms*. 32 (11), 1657–74

*Source*: Jeff Warburton, University of Durham

or obtained from another agency that originally encoded the data. For example, the UK Ordnance Survey and US Geological Survey both provide data for the construction of digital terrain models in pre-prepared digital formats. In addition, both agencies, in common with many other national mapping agencies, provide detailed topographic data in a range of digital formats. Remotely sensed data are normally provided in electronic form, since this is the easiest way to transport the large files representing individual 'scenes' from the various sensors (Box 5.6).

## BOX 5.6 Fireguard: using GIS and remote sensing for fire management

CASE STUDY

### Jim Hogg

The Fireguard project was funded by the European Commission under the Quality of Life Programme to evaluate the potential of high-resolution satellite imagery as a source of information for GIS-based forest management in fire prevention, pre-suppression and suppression in Mediterranean countries. This information is required in a timely and accurate manner, and in a format that is easy to handle and use for integration into strategic plans for fire suppression management. High-resolution, remote-sensing imagery integrated with GIS data is the key to providing this information. This case study shows how data from the QuickBird and IKONOS satellites have potential as a useful source of data for forest management.

### A BURNING ISSUE: WILDFIRES IN THE MEDITERRANEAN

Wildfires are one of the major threats to commercial forests and have a significant impact on the environment, public health and economies of rural areas. They cause ecological damage, destruction of property, and jobs loses in local communities. On average Europe experiences 60,000 wildfires annually. These fires destroy approximately 700,000 ha of forest, and fire-fighting and restoration cost an estimated 1.5 billion euros. The Mediterranean region is particularly susceptible to wildfires. Here approximately 500,000 ha of forest are burnt each year, and the last 25 years have seen a tripling in the area burnt.

The Mediterranean ecosystem is a heterogeneous and highly fragmented mosaic of grass, shrub, wood and forest lands. Vegetation cover is complex and highly variable, as are land use and terrain. As in most semi-arid climatic regions, the rate of fuel accumulation is high and the natural reduction through decomposition is low, so the region is extremely vulnerable to wildfires.

### THE CONTROLLING FACTORS

The three primary factors influencing wildfire behaviour are:

- vegetation (fuel properties);
- topography (slope, aspect and elevation); and
- meteorological or weather conditions (temperature, relative humidity, and wind speed and direction).

All three of these factors can be mapped and modelled using GIS and remote sensing. They all play crucial roles in all aspects of fire management strategy and planning, including fuel ignition, combustibility, flammability determination, fire behaviour prediction, fire hazard, risk, danger assessment, and in fire management systems. Weather is a major factor in a fuel's susceptibility to

BOX 5.6

ignition and flammability, and topography affects the direction and rate of spread of a fire. Fuel controls ignition, combustibility and flammability, and strongly influences fire intensity and spread. Fuel is, therefore, the most important factor for a fire to occur, as without combustible material there cannot be a fire.

Many mathematical models for predicting fire behaviour and determining fire danger are available for forest and fire management support. Examples include BEHAVE, FARSITE, FIRE! and CFFDRS. These models are used by forest and fire-fighting organizations for analysis of wildfire scenarios and suppression alternatives. They can help to predict the behaviour of active fires (for example, rate and direction of spread, fire intensity and perimeter, potential fire paths and corridors) and to plan pre-suppression and suppression strategies and measures (particularly prescribed burning and fuel-breaks). They are widely used for fire prevention planning and fuel modelling, as well as for general contingency planning purposes. The models also can be integrated into fire management decision support systems. All models require spatial information on vegetation cover characteristics (fuel), terrain and weather. Fuel information is by far the most important of the requirements and is perhaps the most difficult to source.

## THE PROBLEM

The biggest problem facing forest and fire service managers in fuel management for fire prevention and control is the high diversity and fragmentation of the Mediterranean plant community and the great variability it shows over short distances. The patchy, quilt-like distribution of vegetation arising from the discontinuous canopy closure that is characteristic of the Mediterranean landscape is a result, in part, of the private ownership of two-thirds of the region's land, with most holdings smaller than 5 ha. This makes it difficult to inventory and monitor the Mediterranean landscape on a regular and up-to-date basis using traditional field surveys. Information on fuel complexes (such as accumulation, stratification, distribution and dynamics) is required to assess potential fire behaviour and to be able to undertake effective proactive fire prevention and control measures.

## THE SOLUTION

New remote-sensing platforms such as IKONOS and QuickBird allow rapid sourcing of high-resolution multi-spectral and panchromatic satellite imagery.

Imagery can be sourced directly over the Internet at relatively low cost to provide immediate, up-to-the-minute coverage for areas of interest. Quick-view images are used to confirm image suitability (for example, location and cloud cover) before download. Imagery is available with various levels of pre-processing from standard imagery (for those with image-processing skills) to ortho-rectified GIS-ready imagery (for those users wanting to load the data directly into their GIS). In the Fireguard Project this type of imagery is being used to provide information on burnt areas (see Figure 5.11) and fuel complexes. The high resolution allows the identification of individual trees (see Figure 5.12). This level of definition provides the high level of detail required for assessment of forest fragmentation and fuel loads. The images provided by the QuickBird satellite have also helped with identification of woodland type (see Figure 5.12). Stereo IKONOS images are used to create high-resolution DEMs of the area. The display of the QuickBird images in a Digital Elevation Model (DEM) increases the capacity of visualization for the decision and settlement of intervention priorities.

**Figure 5.11** Infrared QuickBird imagery used for mapping burned areas

**Figure 5.12** Fire and fuel break in Pinus pinea stand as seen on the ground (left) and in panchromatic QuickBird image (right)

Panchromatic images, with 0.6m pixels, allow the project managers to operate the GIS at a very high scale (1:1150), more than enough to be used as a tool for decision support in forest management at the unit level. Pre-processed high-resolution satellite imagery associated with the increasing computation capacities make remote sensing solutions an essential tool for a wide range of forest resources management applications.

**WEB LINKS**
IKONOS images:
www.satimagingcorp.com/gallery-ikonos.html
www.infoterra-global.com/capabilities_spaceborne.php
www.space.com/news/ikonos_wtc_010912.html
Quickbird images:
www.digitalglobe.com/
www.ballaerospace.com/page.jsp?page=77

*Source*: Jim Hogg, University of Leeds

Thus, electronic data transfer is an appropriate method of data encoding where the data are already available in digital form (from a data collection device or another organization) in a format compatible with your GIS software. However, it may also be necessary to encode data using electronic transfer where the data are not in a format that is compatible with your GIS. To deal with this, you will need to transform or convert your data to an appropriate format. Most GIS software packages will allow you to convert data from a number of different formats; however, there may still be times when it is necessary to use additional software or even write your own

data transfer routine to alter formatting of data. A range of commonly used data transfer formats are presented in Table 5.3.

Finding out what data exist, how much they will cost, where they can be found and the format in which they are available are some of the most challenging stages in the development of a GIS application. Therefore, obtaining data from other sources requires users to address a range of important questions. These are outlined in Box 5.7.

As with so many aspects of GIS, Box 5.7 illustrates that there are not only technical issues to take into account when acquiring digital data from another source, but also human and organizational issues.

**TABLE 5.3 Common data transfer formats**

|  | Vector transfer formats | Raster transfer formats |
|---|---|---|
| Software or data specific | ESRI shapefile | IMG (ERDAS Imagine) |
|  | HPGL (Hewlett Packard) | Landsat |
|  |  | LiDAR |
| General | Postscript | Standard Raster Format |
|  | DXF (Digital Exchange Format) | Postscript |
|  | NTF (Neutral Transfer Format) | TIFF (Tagged Image Interchange format) |
|  | PDF (Portable Document Format) | PNG (Portable Network Graphics) |
|  |  | GIF (Graphic Interchange format) |
|  |  | JPEG (Joint Photographic Experts Group) |
|  |  | DEM (Digital Elevation Model) |
| Able to preserve topological information | DLG (Digital Line Graph) |  |
|  | TIGER (Topologically Integrated Geographic Encoding and Reference) |  |
|  | SDTS (Spatial Data Transfer Standard) |  |
|  | GML (Geography Markup Language) |  |

## BOX 5.7 Obtaining spatial data from other sources

Users must address a number of questions if they wish to obtain data in digital form from another source:

1 *What data are available?* There are few 'data hypermarkets' where you can go to browse, select and purchase spatial data. Instead, you usually have to rummage around in the data marketplace trying to find what you need at an affordable price. Advertisements for data in digital format can be found in trade magazines, data can be obtained from national mapping agencies and a range of data is available from organizations via the Internet. However, since the search process can be time-consuming, and rather 'hit or miss', databases of data sources have been created to help users locate appropriate data. These databases are known as *metadatabases*. Two European examples of meta-database projects, MEGRIN (Salge, 1996) and ESMI (Scholten, 1997), have also been designed to stimulate the data market and provide better access to data sources. The Internet is the nearest thing there is to a data hypermarket for GIS. Several organizations have set up data 'clearing houses' where you can browse for and purchase data online (e.g. GeoCommunity's GIS Data Depot). Some spatial data may be found 'hidden' within other databases. These can be extracted using techniques of data mining. Large databases are 'mined' using specialist software to identify hitherto unknown or unnoticed spatial patterns hiding within the complexities of the data. Data mining software tools use classification, clustering, statistical or topological association and numeric prediction methods to link aspatial data to spatial data. For example, data mining methods are often used by retail companies to spot spatial and other patterns in data gained from customer loyalty cards and electronic point of sale (EPOS) data.

2 *What will the data cost?* Data are very difficult to price, so whilst some digital data are expensive, others are freely available over the Internet. The pricing policy varies depending on the agency that collected the data in the first place, and this in turn may be affected by national legislation. Because of the possibility of updating digital data sets, many are bought on an annual licensing agreement, entitling the purchaser to new versions and corrections. The cost of digital data may be an inhibiting factor for some GIS applications. In extreme cases, users may duplicate data by redigitizing in order to avoid the costs of purchase. In addition to the cost of purchasing data there may be copyright charges and restrictions to be taken into account.

3 *On what media will the data be supplied?* Data may be available in a number of ways. These range from magnetic media (diskette or tape) and optical disks (CD-ROM) to network transfers across internal local area networks (LAN) or the Internet. Even these methods of data input are not without their problems. Magnetic media may be damaged during transit or by faults in the copying process, while networks are often subject to faults or interruptions. These may lead to data being lost or corrupted. Problems may also be experienced with out-of-date media. Older data sets may still be stored on half-inch magnetic tape, top-loading disk packs or even punched cards. Finding a suitable reader for these old media formats is becoming increasingly difficult.

4 *What format will the data be in – will standards be adhered to?* In recent years, a great deal of effort has been put into the development of national and international data standards to ensure data quality and to improve compatibility and transfer of data between systems. Harding and Wilkinson (1997) list 27 different national and international organizations responsible for the development of standards related to geographical information. In addition, the GIS software vendors have developed their own standards. As a result, software is becoming increasingly compatible and there is a plethora of standards, developed by vendors, users and national and international geographic information agencies. Data standards and their implications for data quality are discussed further in Chapter 10.

However, if data can be successfully obtained and input to GIS, considerable time may be saved when electronic data transfer is possible, and duplication of effort may be avoided. One of the main sources of off-the-shelf GIS data sets and remotely sensed imagery is the Web. Sourcing data from the Web brings its own unique set of problems. These are covered in detail in Box 5.8.

## BOX 5.8 Finding spatial data online

PRACTICE

### FINDING SPATIAL DATA ONLINE

There are several approaches to finding spatial data online.

1 *Use a search engine*. This can be a time-consuming task, as it is often difficult to focus a query precisely to find data that meets your needs.

2 *Find a list of suppliers*. Many universities and individual teachers have produced lists of spatial data sources to assist their own students finding information about their local area and country. It may be easier to search for these rather than specific data sets. However, bear in mind that these lists may be out of date.

3 *Search for an appropriate organization*. For example, if you are looking for geological data for Wisconsin you could try the United States Geological Survey's website.

4 *Use a Web-based information service* specifically designed to encourage the sharing of geospatial data. These services use metadata (information about data) to facilitate the searching process. This metadata may include when the data were collected, the geographical area it relates to, how it can be accessed and who to contact to obtain it. Examples include:

### GEOSPATIAL ONE-STOP

(www.geodata.gov)
A US government initiative designed to facilitate the sharing of spatial information between different levels of government. The service includes a portal that allows access to maps, data and other spatial services. It also provides information on future investments in data to prevent duplication of effort in data collection, and encourages development of standards and best practice.

### GEOCONNECTIONS DISCOVERY PORTAL

(http://geodiscover.cgdi.ca)
A Canadian initiative to provide tools to help users discover and access spatial data and services. It is part of the Canadian Geospatial Data Infrastructure (CGDI), and users can search metadata or user relevant search criteria such as place name.

### GIGATEWAY

(www.gigateway.org.uk)
A metadata search engine developed to increase awareness of and access to geographic information in the UK. Gigateway encourage as many different organizations as possible to share metadata through their service and have provided tools to help users create metadata to acceptable standards. Metadata are maintained using a standard used known as GEMINI (Geo-spatial Metadata Interoperability Initiative).

Other UK based examples include Magic (www.magic.gov.uk) which provides a one-stop portal to environment designations, Go-Geo (www.gogeo.ac.uk) which acts as a portal to spatial data for UK academics and Edina Digimap (www.edina.ac.uk/digimap/) which provides access to online maps and spatial data for the UK.

Wherever you find data there are questions that you may need to ask:

- In what format are the data provided – will I be able to enter them into my GIS?
- When were the data produced? Are they up to date, or timely for my project?
- What projection has been used?
- What scale has been used?
- Are there any restrictions on my use of the data?
- Who created the data? How reliable is it?

*Sources*: www.agi.org.uk; www.gigateway.org.uk; www.geodata.gov; www.geodiscover.cgdi.ca

## REFLECTION BOX

- Think about the main techniques available for data encoding.
  - Which do you think were more important 10 and 20 years ago?
  - Which are most important now?
  - Which do you think will be most important in 10 years' time?

- Imagine that you need to contact a data supplier in order to obtain some data for your GIS project. This could be a satellite data set, digital topographic data or geocoded address data – you decide. Draft an email message to the supplier containing questions you need to ask to ensure that when the data arrives you will be able to use it in your GIS software.

# ■ DATA EDITING

As a result of the problems encountered during data encoding, you cannot expect to input an error-free data set into your GIS. Data may include errors derived from the original source data, as well as errors that have been introduced during the encoding process. There may be errors in co-ordinate data as well as inaccuracies and uncertainty in attribute data. Data quality and data error will be considered in greater detail in **Chapter 10**. However, good practice in GIS involves continuous management of data quality, and it is normal at this stage in the data stream to make special provision for the identification and correction of errors. It is better to intercept errors before they contaminate the GIS database and go on to infect (propagate) the higher levels of information that are generated. The process is known as data editing or 'cleaning'. Data editing can be likened to the filter between the fuel tank and the engine that keeps the fuel clean and the engine running smoothly. Four topics are covered here: detection and correction of errors; reprojection, transformation and generalization; edge matching and rubber sheeting; and updating of spatial databases.

## Detecting and correcting errors

Errors in input data may derive from three main sources: errors in the source data; errors introduced during encoding; and errors propagated during data transfer and conversion. Errors in source data may be difficult to identify. For example, there may be subtle errors in a paper map source used for digitizing because of the methods used by particular surveyors, or there may be printing errors in paper-based records used as source data.

During encoding a range of errors can be introduced. During keyboard encoding it is easy for an operator to make a typing mistake; during digitizing an operator may encode the wrong line; and folds and stains can easily be scanned and mistaken for real geographical features. During data transfer, conversion of data between different formats required by different packages may lead to a loss of data.

Errors in attribute data are relatively easy to spot and may be identified using manual comparison with the original data. For example, if the operator notices that a hotel has been coded as a café, then the attribute database may be corrected accordingly. Various methods, in addition to manual comparison, exist for the correction of attribute errors. These are described in Box 5.9.

Errors in spatial data are often more difficult to identify and correct than errors in attribute data. These errors take many forms, depending on the data model being used (vector or raster) and the method of data capture. In our Happy Valley example, a spatial error may arise if a meteorological station has been located in the wrong place, if a forest polygon has been wrongly identified during image processing or if a railway line has been erroneously digitized as a road.

Examples of errors that may arise during encoding (especially during manual digitizing) are presented in Table 5.4. Figure 5.13 illustrates some of the errors that may be encountered in vector data. Chrisman (2002) suggests that certain types of error can help to identify other problems with encoded data. For example, in an area data layer 'dead-end nodes' might indicate missing lines, overshoots or undershoots. The user can look for these features to direct editing rather than having to examine the whole map. Most GIS packages will provide a suite of editing tools for the identification and removal of errors in vector data. Corrections can be done interactively by the operator 'on-screen', or automatically by the GIS software. However, visual comparison of the digitized data against the source document, either on paper or on the computer screen, is a good starting point. This will reveal obvious omissions, duplications and erroneous additions. Systematic errors such as overshoots in digitized lines can be corrected automatically by some digitizing software, and it is important for data to be absolutely correct if topology is to be created for a vector data set. This is covered further in **Chapter 10**. Automatic corrections can save many hours of work but need to be used with care as incorrectly specified tolerances may miss some errors or correct 'errors' that never existed in the first place (see **Chapter 10**). Box 5.10 provides an example of an application that makes use of automatic error correction and topology generation.

Errors will also be present in raster data. In common with vector data, missing entities and noise are particular problems. Data for some areas may be difficult to collect, owing to environmental or cultural obstacles. Trying to obtain aerial

## BOX 5.9 Methods of attribute data checking

PRACTICE

Several methods may be used to check for errors in the encoding of attribute data. These include:

**1** *Impossible values*. Simple checks for impossible data values can be made when the range of the data is known. Data values falling outside this range are obviously incorrect. For example, a negative rainfall measurement is impossible, as is a slope of 100 degrees.

**2** *Extreme values*. Extreme data values should be cross-checked against the source document to see if they are correct. An entry in the attribute database that says the Mountain View Hotel has 2000 rooms needs to be checked. It is more likely that this hotel has 200 rooms and that the error is the result of a typing mistake.

**3** *Internal consistency*. Checks can be made against summary statistics provided with source documents where data are derived from statistical tables. Totals and means for attribute data entered into the GIS should tally with the totals and means reported in the source document. If a discrepancy is found, then there must be an error somewhere in the attribute data.

**4** *Scattergrams*. If two or more variables in the attribute data are correlated, then errors can be identified using scattergrams. The two variables are plotted along the *x* and *y* axes of a graph and values that depart noticeably from the regression line are investigated. Examples of correlated variables from Happy Valley might be altitude and temperature, or the category of a hotel and the cost of accommodation.

**5** *Trend surfaces*. Trend surface analyses may be used to highlight points with values that depart markedly from the general trend of the data. This technique may be useful where a regional trend is known to exist. For example, in the case of Happy Valley most ski accidents occur on the nursery slopes and the general trend is for accidents to decrease as the ski piste becomes more difficult. Therefore, an advanced piste recording a high number of accidents reflects either an error in the data set or an area requiring investigation.

**TABLE 5.4 Common errors in spatial data**

| Error | Description |
| --- | --- |
| Missing entities | Missing points, lines or boundary segments |
| Duplicate entities | Points, lines or boundary segments that have been digitized twice |
| Mislocated entities | Points, lines or boundary segments digitized in the wrong place |
| Missing labels | Unidentified polygons |
| Duplicate labels | Two or more identification labels for the same polygon |
| Artefacts of digitizing | Undershoots, overshoots, wrongly placed nodes, loops and spikes |

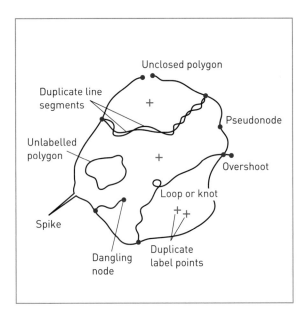

**Figure 5.13** Examples of spatial error in vector data

# BOX 5.10 Going Dutch with small scale topography

## Martin Gregory

The City of Amsterdam has brought its maps up to date with an innovative modernization programme. The City's GeoInformation Department (Geo en VastgoedInformatie) has redesiged and redefined its Small Scale Standard Topographical map, the 1:10,000 Kleinschalige BasisKaart (KBK). The aim of the redesign was to ensure that data are accurate, clean and aligned with national standards. This ensures the delivery of high-quality data to end customers, both online and offline. The City of Amsterdam identified topology as a key enabler to achieve this goal. This case study describes the challenges involved in modernization, the solution implemented and the benefits delivered.

## THE CHALLENGE

For many years paper maps were the most important end-product produced by the City of Amsterdam. However, now that the underlying source data are being provided for the general public and other government departments using a variety of media and scales, in both high-quality cadastral and topographic mapping, the topological (or geometrical) quality of digital data has become increasingly important.

The City of Amsterdam's original base data contained many typical topological problems: spikes, kickbacks, undershoots, overshoots, gaps and slivers (Figure 5.14). As well as removing these errors, it is vital to address less obvious visual problems such as exactly duplicated features and almost-duplicated

**Figure 5.14** Examples of original data problems and the corrected data after processing
*Source*: 1Spatial. Copyright © 2005 1Spatial Group Limited. All rights reserved

points in features (where a higher number of vertices is present than is required to maintain the shape of the feature). The challenge is first to detect the location of these data errors, then to correct the majority of them automatically.

## THE SOLUTION

The solution at City of Amsterdam was based on converting the base data to a new Oracle 9*i* data model (since migrated to Oracle Database 10g) and then applying 1Spatial's Radius Topology™ software to detect and correct errors. There are four phases:

**1** 1:1 conversion from the original Bentley Microstation® DGN data files to Oracle 9*i* using Safe Software's FME®.

**2** Initial data correction using a combination of standard Intergraph GeoMedia® tools, PL/SQL scripts and Radius Topology.

**3** Topology creation for base data to enable on-going automatic data maintenance.

**4** Advanced extraction of features (for example surfaces) from the stored topology.

The time taken from initial data conversion through to the system going live was one year. Using Radius Topology it was possible to automate approximately 90–95 per cent of the initial data

correction, helping the City of Amsterdam achieve the target for project completion.

During interactive editing all topology validation happens in real time as data are submitted to the database. Data are stored using Oracle data types and accessed using Oracle interfaces to provide an open and interoperable solution. Since Radius Topology is integrated with Oracle at the lowest possible level, it is not possible for a client application to bypass the topology rules set. This means that all typical digitization errors such as gaps and slivers, undershoots and overshoots between features are automatically prevented by snapping new geometries or updates to the existing data (Figure 5.14). In addition, the user has control over topology snapping rules (Figure 5.15).

For any features that fail topological validation during batch processing, an entry is written to an error table inside Oracle. These failures could be caused by the detection of problems like spikes, kickbacks or duplicated points in features. These are then queued and displayed in the GIS client application, enabling the user to inspect and correct the remaining problems using client editing tools underpinned by Radius Topology.

The City of Amsterdam has developed an enterprise-wide system architecture that allows a wide range of client applications to access and

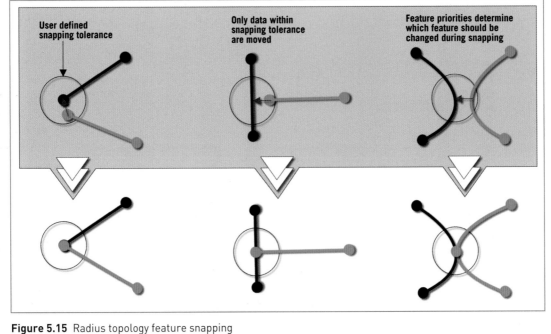

**Figure 5.15** Radius topology feature snapping
*Source:* 1Spatial. Copyright © 2005 1Spatial Group Limited. All rights reserved

**BOX 5.10**

update the spatial data. Each of these applications benefits from the transparent server-side topology validation provided by Radius Topology. Ensuring data quality within the interoperable environment is known as *Sustainable Interoperability.*

In summary, Radius Topology is being used at City of Amsterdam to:

- clean and create a topological structure for existing base data, and dynamically maintain data integrity for new feature updates;
- enable more efficient updating of shared topological structures by editing nodes and edges;
- display holes, overlapping elements, or other issues affecting data quality, using spatial queries;
- enable faster spatial querying of the database through topological searches;
- edit nodes and edges directly to help with the editing and splitting of area features;
- automatically maintain the quality of the base data as new updates are made; and
- extract cartographic features (for example contours or displayable faces).

## THE BENEFITS

The benefits of the approach taken by the City of Amsterdam include:

- *Time saving* – by using 1Spatial's Radius Topology it is estimated that the GeoInformation Department will be able to automate the majority of the initial data correction process. When compared with a purely manual correction workflow

this gives an estimated 80–85 per cent time saving.
- *Improved data quality* – the new technology automatically maintains existing data and validates any new data. This means that the City of Amsterdam's data are accurate and error-free at all times.
- *Improved efficiency* – after some initial training for end users on how to work with Radius Topology, all data are validated automatically at feature insertion, modification or deletion time. There is no need to run post-validation and correction processes.
- *Maximizing existing software investments* – Radius Topology is interoperable with mainstream GIS tools so users can continue to use the software tools they prefer and with which they are familiar. The data set produced is not only interoperable with GIS tools, but also with other processes in the municipality.

By using topology to manage data quality, the City of Amsterdam can rely on the integrity and accuracy of its data. With high-quality base data it will be able to offer enhanced data products to its customers and will be in a position to develop new data products by applying techniques such as map generalization.

## FURTHER INFORMATION

1Spatial: www.1Spatial.com
Amsterdam local government website: www.amsterdam.nl/

*Source:* Martin Gregory, 1Spatial

---

photographs of an area containing a major airport is one example, since it may be difficult to fly across the area using a flight path suitable for photography. Similarly, it may be difficult to get clear images of vegetation cover in an area during a rainy season using certain sensors or where snow cover obscures underlying vegetation patterns. Noise may be inadvertently added to the data, either when they were first collected or during processing. This noise often shows up as scattered pixels whose attributes do not conform to those of neighbouring pixels. For exa mple, an individual pixel representing water may be seen in a large area of forest. Whilst this may be correct, it could also be the result of noise and needs

to be checked. This form of error may be removed by filtering. Filtering is considered in this book as an analysis technique (**Chapter 6**) but, in brief, it involves passing a filter (a small grid of pixels specified by the user – often a 3 × 3 pixel square is used) over the noisy data set and recalculating the value of the central (target) pixel as a function of all the pixel values within the filter. This technique needs to be used with care as genuine features in the data can be lost if too large a filter is used. Figure 5.16 illustrates the effect of different-sized filters. Further details of errors that may be added to data during scanning and later processing are presented in **Chapter 10**.

**(a)** Original image

**(b)** 3 × 3 mean filter

**(c)** 5 × 5 mean filter

**(d)** 7 × 7 mean filter

**(e)** 3 × 3 minimum filter

**(f)** 3 × 3 maximum filter

**(g)** Sobel edge extraction filter

**Figure 5.16** Filtering noise from a raster data set

## Re-projection, transformation and generalization

Once spatial and attribute data have been encoded and edited, it may be necessary to process the data geometrically in order to provide a common framework of reference. **Chapter 2** showed how the projection system of the source document may affect the positioning of spatial co-ordinates. The scale and resolution of the source data are also important and need to be taken into account when combining data from a range of sources into a final integrated database. This section briefly considers the role of re-projection, transformation and generalization in the data stream.

Data derived from maps drawn on different projections will need to be converted to a common projection system before they can be combined or analysed. If not re-projected, data derived from a source map drawn using one projection will not plot in the same location as data derived from another source map using a different projection system. For example, if a coastline is digitized from a navigation chart drawn in the Mercator projection (cylindrical) and the internal census boundaries of the country are digitized from a map drawn using the Alber's Equal Area (conic) projection (as in the case of the US Bureau of Census), then the census boundaries along the coast will not plot directly on top of the coastline. In this case they will be offset and will need to be re-projected into a common projection system before being combined. Figure 5.17 shows another example.

Data derived from different sources may also be referenced using different co-ordinate systems. The grid systems used may have different origins, different units of measurement or different orientation. If so, it will be necessary to transform the co-ordinates of each of the input data sets onto a common grid system. This is quite easily done and involves linear mathematical transformations. Some of the other methods commonly used are:

- *Translation and scaling*. One data set may be referenced in 1-metre co-ordinates whilst another is referenced in 10-metre co-ordinates.

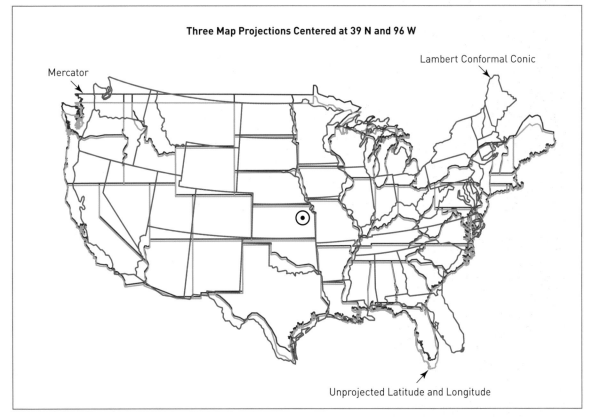

**Figure 5.17** Topological mismatch between data in different projections

*Source*: Courtesy of Peter H. Dana

If a common grid system of 1-metre co-ordinates is required, then this is a simply a case of multiplying the co-ordinates in the 10-metre data set by a factor of 10.

- *Creating a common origin.* If two data sets use the same co-ordinate resolution but do not share the same origin, then the origin of one of the data sets may be shifted in line with the other simply by adding the difference between the two origins (dx,dy) to its co-ordinates.

- *Rotation.* Map co-ordinates may be rotated using simple trigonometry to fit one or more data sets onto a grid of common orientation.

Data may be derived from maps of different scales. The accuracy of the output from a GIS analysis can only be as good as the worst input data. Thus, if source maps of widely differing scales are to be used together, data derived from larger-scale mapping should be generalized to be comparable with the data derived from smaller-scale maps. This will also save processing time and disk space by avoiding the storage of unnecessary detail. Data derived from large-scale sources can be generalized once they have been input to the GIS. Routines exist in most vector GIS packages for weeding out unnecessary points from digitized lines such that the basic shape of the line is preserved. The simplest techniques for generalization delete points along a line at a fixed interval (for example, every third point). These techniques have the disadvantage that the shape of features may not be preserved (Laurini and Thompson, 1992). Most other methods are based on the Douglas–Peucker algorithm (Douglas and Peucker, 1973). This involves the following stages:

1 Joining the start and end nodes of a line with a straight line.
2 Examining the perpendicular distance from this straight line to individual vertices along the digitized line.
3 Discarding points within a certain threshold distance of the straight line.
4 Moving the straight line to join the start node with the point on the digitized line that was the greatest distance away from the straight line.
5 Repeating the process until there are no points left which are closer than the threshold distance.

Chrisman (2002) presents a practical example of the effects of line thinning. From a line consisting of 83 points he repeatedly thins the border between the states of Maryland and West Virginia to a line with only five points. Chrisman also highlights the risk of generating topological errors during line thinning where lines are close together. An example of the results of repeated line thinning and fuzzy tolerances on topology and detail is shown in Figure 5.18 and Table 5.5.

When it is necessary to generalize raster data, the most common method employed is to aggregate or amalgamate cells with the same attribute values. This approach results in a loss of detail which is often very severe. A more sympathetic approach is to use a filtering algorithm. These have been outlined above and are covered in more detail in **Chapter 6**. If the main motivation for generalization is to save storage space, then, rather than resorting to one of the two techniques outlined above, it may be better to use an appropriate data compaction technique (see **Chapter 3**) as this will result in a volume reduction without any loss in detail.

TABLE 5.5 **Original coastline based on Digital Chart of the World 1:5,000,000 scale data (Figure 5.18)**

| Weed tolerance | | No. vertices | No. arcs | No. nodes | No. line segments | Total line length (km) | No. polygons | Total polygon area (million km²) |
|---|---|---|---|---|---|---|---|---|
| Original | 5.18a | 6,252 | 354 | 354 | 5,898 | 7,412 | 103 | 7,695 |
| 1,000 | 5.18b | 1,724 | 351 | 400 | 1,318 | 6,815 | 51 | 7,673 |
| 2,000 | 5.18c | 1,243 | 324 | 362 | 819 | 6,358 | 29 | 7,647 |
| 5,000 | 5.18d | 904 | 283 | 295 | 465 | 5,552 | 16 | 7,425 |
| 10,000 | 5.18e | 778 | 267 | 275 | 329 | 4,747 | 8 | 7,178 |
| 20,000 | 5.18f | 735 | 264 | 267 | 287 | 4,026 | 8 | 6,606 |

**(a)** Original

**(b)** Weed tolerance = 1.000

**(c)** Weed tolerance = 2,000

**(d)** Weed tolerance = 5,000

**(e)** Weed tolerance = 10,000

**(f)** Weed tolerance = 20,000

GENERALIZE
Tolerance

1st Trend Line

2nd Trend Line

3rd Trend Line

Resulting Arc

**(g)**

**Figure 5.18** The results of repeated line thinning
*Sources*: (a–f): From *The Digital Chart of the World*. Courtesy of Esri. Copyright © Esri. All rights reserved; (inset): From Esri, ArcGIS
online help system, courtesty of Esri. Copyright © 2011 Esri. All rights reserved

## Edge matching and rubber sheeting

When a study area extends across two or more map sheets, small differences or mismatches between adjacent map sheets may need to be resolved. Normally, each map sheet would be digitized separately and then the adjacent sheets joined after editing, re-projection, transformation and generalization. The joining process is known as *edge matching* and involves three basic steps. First, mismatches at sheet boundaries must be resolved. Commonly, lines and polygon boundaries that straddle the edges of adjacent map sheets do not meet up when the maps are joined together. These must be joined together to complete features and ensure topologically correct data. More serious problems can occur when classification methods vary between map sheets. For example, different soil surveyors may interpret the pattern and type of soils differently, leading to serious differences on adjacent map sheets. This may require quite radical reclassification and reinterpretation to attempt a smooth join between sheets. This problem may also be seen in maps derived from multiple satellite images. If the satellite images were taken at different times of the day and under different weather and seasonal conditions, then the classification of the composite image may produce artificial differences where images meet. These can be seen as clear straight lines at the sheet edges.

Second, for use as a vector data layer, topology must be rebuilt as new lines and polygons have been created from the segments that lie across map sheets. This process can be automated, but problems may occur due to the tolerances used. Too large a tolerance and small edge polygons may be lost; too small a tolerance and lines and polygon boundaries may remain unjoined. Finally, redundant map sheet boundary lines are deleted or dissolved. The process of edge matching is illustrated in Figure 5.19. Jackson and Woodsford (1991) note that, although some quasi-automatic scanning edge matching is available, in practice the presence of anomalies in the data produced can require considerable human input to the process.

Certain data sources may give rise to internal distortions within individual map sheets. This is especially true for data derived from aerial photography as the movement of the aircraft and distortion caused by the camera lens can cause internal inaccuracies in the location of features within the image. These inaccuracies may remain even after transformation and re-projection. These problems can be rectified through a process known as *rubber sheeting* (or *conflation*). Rubber sheeting involves stretching the map in various directions as if it were drawn on a rubber sheet. Objects on the map that are accurately placed are 'tacked down' and kept still whilst others that are in the wrong location or have the wrong shape are stretched to fit with the control points. These control points are fixed features that may be easily identified on the ground and on the image. Their true co-ordinates may be determined from a map covering the same area or from field observations using GPS. Distinctive buildings, road or stream intersections, peaks or coastal headlands may be useful control points. Figure 5.20 illustrates the process of rubber

**Figure 5.19** Edge matching

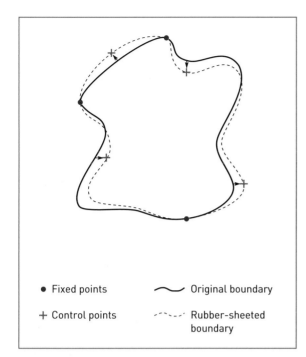

**Figure 5.20** Rubber sheeting

sheeting. This technique may also be used for re-projection where details of the base projection used in the source data are lacking. Difficulties associated with this technique include the lack of suitable control points and the processing time required for large and complex data sets. With too few control points the process of rubber sheeting is insufficiently controlled over much of the map sheet and may lead to unrealistic distortion in some areas.

## Geocoding address data

Geocoding is the process of converting an address into a point location (McDonnell and Kemp, 1998). Since addresses are an important component of many spatial data sets, geocoding techniques have wide applicability during the encoding and preparation of data for analysis.

Geocoding may be required to:

- turn hotel names and addresses collected during a questionnaire survey into a map of the distribution of holidaymakers in Happy Valley;
- locate an incident in a street where an emergency response vehicle is required, based on details provided over the telephone by a witness at the scene; and

- establish the home location of a store's credit card holders using their postcodes.

During geocoding the address itself, a postcode or another non-geographic descriptor (such as place name, property or parcel reference number) is used to determine the geographical co-ordinates of a location. UK postcodes can be geocoded with an Ordnance Survey grid reference. Several products are available that contain a single data record for each of the 1.6 million postcodes in the UK. In these files, each data record contains the OS Grid Reference (to 100 m resolution in England and Wales and to 10 m resolution in Scotland) and local government ward codes for the first address in each postcode. Many GIS software products can geocode US addresses, using the address, ZIP code or even place names.

Address matching is the process of geocoding street addresses to a street network. Locations are determined based on address ranges stored for each street segment.

Geocoding can be affected by the quality of data. Address data are frequently inconsistent: place names may be spelt incorrectly, addresses may be written in different formats and different abbreviations exist for words that appear frequently in addresses such as 'Street' and 'Avenue'. The use of standards for address data is particularly relevant to geocoding (see **Chapter 2**).

## Updating and maintaining spatial databases

A great deal of effort is put into creating spatial databases and so it makes good sense to keep important and hard-won data as up to date as possible. The world is a very dynamic place and things change, often rapidly and especially in urban areas where new buildings and roads are being built. This means that spatial data can go out of date and so needs regular updating. National mapping agencies such as the Ordnance Survey spend a lot of time and resources maintaining the currency of their national mapping products so that they are fit for use by their customers. Imagine working with some large-scale data of an urban area to plan the route of a new section of elevated highway only to find that a new multi-storey office building has been built on the vacant land that you had planned for the road to pass through. In this instance, using old and out-of-date map information would have cost you time and money (not to mention embarrassment!) as you would have to re-design your chosen route to avoid the building you didn't think was there. Keeping databases up to date avoids such problems and is a key aspect of ongoing data editing and maintenance. This process of updating the changes on the ground requires site visits by surveyors (often using portable computing devices and GPS) to collect information. The case study in Box 5.11 describes how, in the UK, the Ordnance Survey keep their mapping products up to date.

## BOX 5.11 Ordnance Survey (OS) data collection – from plan to MasterMap

CASE STUDY

### Paul Corcoran

The Ordnance Survey (OS), Britain's national mapping agency, is responsible for collecting and maintaining information about changes to the landscape of the British Isles. This information is used to populate OS's geospatial database, a digital model of the country that contains approximately 229,000 large-scale digital maps. Five thousand changes are made to the database every day, ranging from new buildings to changes in feature attribution. The OS's data collection business group has around 320 staff spread across Great Britain who use GIS and Global Positioning Systems (GPS) to enable this process to be efficiently and effectively undertaken.

The OS data-collection process follows closely the Longley et al. (2005) workflow illustrated in Figure 5.21. Two GIS systems are used – the Geospatial Management System 2 (GMS2 – for stages 1, 2 and 5) and the Portable Revision Integrated Survey Module (PRISM – for stages 3 and 4). The different stages of

this workflow are described in the context of a survey of a new footbridge across the River Aire in Leeds.

### STAGE 1 – PLANNING

GMS2 is a web-based GIS that uses a bespoke version of ESRI's ARC software and a spatially enabled Oracle relational database. GMS2 contains data from a variety of sources, including construction market intelligence companies and local authorities. The location of a new survey job is identified using the Ordnance Survey grid reference or a postcode. Once located, a job is created by digitizing a polygon indicating the extent of the site. When the polygon is complete the Oracle database is automatically accessed and attribute information about the site (for example, 'survey ready' data or 'survey type' required) is added.

Figure 5.22 shows the OS Landline data used as a backdrop, but GMS2 also has the ability to display other layers of information such as postcode polygons (red lines) and also the jobs created (here the polygon created for the River Aire footbridge can be seen in the centre).

### STAGE 2 – PREPARATION

Once the planning information is integrated into GMS2, queries can be executed to establish the jobs that need surveying. GMS2 queries are standard, but if more complex analysis is required, there is the facility to export the results to an external software package such as Microsoft Excel.

In our example a query has been executed and it has been established that the footbridge is ready for survey. The digital maps that cover the area are then identified by GMS2 and requested from OS Head Office in Southampton. To add the footbridge to the digital data, map reference SE3033SE needs to be extracted from Head Office and transferred onto the local Leeds server. In addition to the map, the surveyors

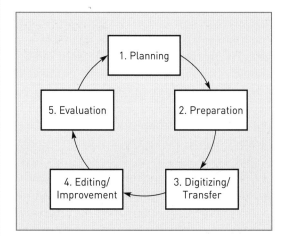

**Figure 5.21** Data collection workflow
*Source*: Adapted from Longley et al. (2005)

**Figure 5.22** River Aire footbridge on GMS2
*Source*: Reproduced by permission of Ordnance Survey on behalf of
HMSO. © Crown Copyright 2011. All rights reserved. Ordnance Survey
Licence number 100030901

are required to update other databases such
as Address Point. This involves matching new
addresses on the Postal Address File to the
location on the map.

## STAGE 3 – DIGITIZING AND TRANSFER

The map data extracted from Head Office
is transferred onto PRISM. This is another
bespoke software package that has
been developed by the OS and a private
software company. It enables data sets to
be combined, edited and maintained. Each
surveyor has a Pen Tablet computer with
the PRISM software loaded. In our example,
SE3033SE is transferred from the Leeds
server directly onto the Pen Tablet.

As Longley *et al.* (2005) confirm, the
next part of the process is the most time
consuming. Data are manually collected using
various surveying techniques. The footbridge could
be surveyed using traditional surveying instruments
but the most suitable method for this site would be to
use GPS.

PRISM is now ready to accept the new survey data.
The surveyor connects the GPS receiver, waits for the
appropriate number of satellites to be received and
then begins capturing the new data (Figure 5.24).

The resulting survey is shown in Figure 5.23. The
main data set in PRISM is topographic, constructed of
numerous layers that correspond to different feature

codes. Figure 5.23 shows these feature codes
in different colours (for example, buildings
are red and roads are green).

## STAGE 4 – EDITING AND IMPROVEMENT

PRISM has search facilities similar to GMS2,
together with data editing functions and the
ability to create queries. At the end of the
surveying various types of analysis can be
undertaken in PRISM. On the topographic
layer, analysis can be used to assess
the quality of the data – for example, the
number of under- or overshoots, whether
features are correctly coded or if the
topology is built properly. Drop-down
menus are used to create queries and
results appear as lists of errors that need
attention. Each feature on the map also has

**Figure 5.23** Updated digital data showing footbridge and layers
*Source*: Reproduced by permission of Ordnance Survey on behalf of
HMSO. © Crown Copyright 2011. All rights reserved. Ordnance Survey
Licence number 100030901

**Figure 5.24** Surveying the River Aire footbridge

attribute information attached which indicates when
the feature was created or altered and the survey
accuracy.

A crucial component in the data is the
Topographical Identifier (TOID®). This is an unique
identifier for each feature on the map (see Box 3.3).
The surveyor has to be aware of this throughout the
survey in order to maintain the lifecycle of a feature.
Customers attach their own data to TOIDs, therefore
TOIDs need to be maintained correctly so that
customers do not lose the link between the OS and
their data.

When the surveyor has finished the data capture
and is happy with the editing and improvement
functions, the data are then transferred back from
PRISM to the office server and eventually back to
Head Office.

### STAGE 5 – EVALUATION

GMS2 is updated to indicate that the footbridge has
been surveyed. The attribute data are amended. This
could take the form of archiving the job from GMS2
or amending it if the site needs a further visit in the
future (for example, to survey additional roads and
paths leading to the footbridge).

When both GIS systems have been successfully
updated, the new data are made available to
customers and the footbridge has finished its journey
from plan to MasterMap. The data collection cycle
can then begin again for another job.

### The future?

The next generation of Pen Tablets should help the
OS to develop a seamless database which would
benefit Central and Local Government as well as
other users. At the moment, surveyors are restricted
to extracting whole maps. When they do this, other
users in the OS are prevented from accessing the
information and carrying out their own processes.
The OS wants surveyors to extract only the specific
area of data they need thus leaving other information
available for others to use. Ultimately, this would
benefit customers, because if surveyors can order
parts of maps, then customers can be delivered
parts of maps: for instance, 'change only updates'.
This reduces the amount of 'digital traffic' and
makes the whole process cleaner and more efficient
for all users.

*Source*: Adapted from material provided by Paul Corcoran,
Ordnance Survey

## ■ TOWARDS AN INTEGRATED DATABASE

As outlined in **Chapter 3**, most GIS adopt a layer
view of the world. Each thematic layer in the database must be encoded, corrected and transformed
to create a GIS ready for analysis. To explore how
such an integrated database might be created in
practice, the Happy Valley case study is considered
in Box 5.12.

While Box 5.12 provides a hypothetical example
of the work involved in the creation of an integrated
GIS database, the variety and range of methods suggested is typical of real-world GIS applications.
For example, Petch *et al.* (1995) describe how over
30 different data sources to build the GIS database
and an equal number of data encoding and editing

techniques were necessary to create a GIS for environmental characterization in Zdarske Vrchy (Czech
Republic). A map of ecological landscape zones was
encoded by manual digitizing from a paper map (see
Figure 1.9a). This map was itself a generalization of
field surveys. A forest cover map was captured from
remotely sensed data (LandsatTM) updated from
more recent aerial photographs (Figure 5.25). The
DTM used was interpolated (see **Chapter 6**) from
contours which had been digitized from topographic maps. These data sets, and the others in the
GIS, were re-projected to a common Transverse
Mercator projection and transformed to a common
scale of 1:50,000. The Zdarske Vrchy project lasted
for four years in total, and the first two of these were
almost entirely devoted to the data encoding and
editing phase of the work.

# BOX 5.12 An integrated GIS for Happy Valley

The Happy Valley GIS team have a list of data requirements (Table 5.6). This includes details of the sources and uses of data and details of the entities, data models and attributes required. The data are from several different sources and require the use of a range of encoding and editing techniques. Some of the data are already in digital form whilst other data are from analogue sources.

**TABLE 5.6  Data in the Happy Valley GIS**

| Name of data layer | Source | Uses | Entity type | Data model | Attributes |
|---|---|---|---|---|---|
| Infrastructure | Scanned from 1:5000 scale plans | Detailed resort planning and management | Mixed | Raster | None |
| Hotels | Accommodation survey carried out by GIS team, postcodes used to identify point locations for hotels | Customer service planning and management | Point | Vector | Name, address, category (basic, standard or luxury), number of rooms |
| Ski schools | GIS team survey – point locations as for hotels | Customer service planning and management | Point | Vector | Ski school name, address, number of instructors |
| Meteorological | Downloaded from GPS readings | Skiing forecasts and conditions reporting, ski run planning and management | Point | Vector | Meteorological Station reference number, snow cover, precipitation, wind speed and direction, temperature and sunshine hours |
| Roads | Digitized from 1:25,000 paper maps | Resort planning, traffic management | Lines (network) | Vector | Road name and number |
| Ski pistes | Digitized from aerial photographs | Ski run planning and management | Lines | Vector | Grade of run (green, blue, red, black) |
| Ski resort boundary | Digitized from 1:25,000 paper map | Resort planning and management | Area | Vector | Name of resort |
| Topography | Downloaded from purchased 1:25,000 DTM | Ski run and resort management | Surface | Vector/Raster | Height |
| Land use | 10-m resolution SPOT image, ground-truthed with GPS readings | Vegetation surveys and monitoring, environmental assessment, ski run and resort planning/ management | Areas | Vector/Raster | Land use class (including agricultural, forestry, settlement, water) |

Survey data in analogue form (the ski school and hotel surveys) are encoded by manual keyboard encoding. The roads and resort boundary are manually digitized in-house, and the team experiment with line-following software to automatically digitize the contours from which to build a terrain model. The latter proves too difficult and expensive because of the availability of the hardware and cost of editing, so an external supplier of DTM data is sought and the digital data purchased at a scale compatible with the topographic map used for digitizing roads. An appropriate data format is documented in the licence agreement, which entitles the GIS team to any corrections to the data. Meteorological station locations are downloaded from a GPS receiver. After processing in the GPS software they are imported to the GIS. In this case attribute data are available as computer files downloaded from the automatic weather stations. However, only summary data are required in the GIS, so summary statistics are calculated and entered by keyboard encoding to an attribute database. Stereoscopic pairs of aerial photographs are interpreted to produce maps of the ski pistes. These are then encoded by manual digitizing. Finally, the land use data are purchased as a classified satellite image from the local remote-sensing data supplier. The team transfer the data electronically to their GIS, then check the data by undertaking some ground truthing of their own using GPS receivers to navigate to predefined sites.

After considerable effort, the Happy Valley GIS team now have a GIS database populated with data from a range of sources and in various formats and co-ordinate systems. There are vector data input by both manual and automatic digitizing; there are 10-metre resolution raster data sourced from satellite imagery; and there are a number of attribute files attached to vector point, line and area data. The task of the Happy Valley GIS team is to process these data so that they are free from errors (as far as is possible) and are in a common format and frame of spatial reference. The 1:25,000

**TABLE 5.7  Creating an integrated database for Happy Valley**

| Data set | Encoding and editing processes required |
| --- | --- |
| Survey data | ■ Keyboard encoding of co-ordinate data points into data file<br>■ Conversion of data file into GIS data layer<br>■ Design and implementation of attribute database<br>■ Linking of spatial and attribute data using unique identifiers |
| Meteorological stations | ■ GPS data downloaded into GPS software<br>■ Processed data transferred to GIS in compatible format<br>■ Point data layer created<br>■ Processing of attribute data in statistical package<br>■ Keyboard encoding of summary attribute data<br>■ Linking of spatial and attribute data using unique identifiers |
| Vector road data | ■ Manual digitizing<br>■ Automatic cleaning and topology generation<br>■ Manual on-screen editing to check errors missed and/or created by automatic procedure<br>■ Joining of map sheets and edge matching<br>■ Re-projection onto Lambert Conformal Conic projection<br>■ Creation of attribute database<br>■ Linking of spatial and attribute data |
| Ski pistes | ■ Photogrammetric interpretation of stereo pairs of aerial photographs<br>■ Manual digitizing of ski pistes<br>■ Automatic cleaning and topology generation<br>■ Manual on-screen editing to check errors missed and/or created by automatic procedures<br>■ Rubber sheeting<br>■ Creation of attribute database<br>■ Linking of spatial and attribute data |

data from which the DTM and valley infrastructure data originate are based on the Lambert Conformal Conic projection. This is chosen as the spatial frame of reference onto which all other data not based on this projection will be re-projected. The roads and boundary data were digitized from more than one map sheet. Thus, edge matching is required. Table 5.7 summarizes all the processes that were necessary for a selection of the data sets.

Datasets and activities relating to the Happy Valley Case Study can be found online at www.pearsoned.co.uk/heywood.

**Figure 5.25** Zdarske Vrchy case study: forest cover, created by reclassifying a LandstatTM scene for the area

REFLECTION BOX

■ Think about your own experiences of data entry and encoding, or the data entry and encoding used in an application with which you are familiar. What errors do you think may have been introduced as data were encoded and edited?

■ For an application or GIS project of your own choice, make a list of all the data sources you would like to include in your database. Identify all the encoding and editing processes that might be required and create a table with following headings:

| Data set | Source | Encoding method needed | Editing processes required |
| --- | --- | --- | --- |

## CONCLUSIONS

This chapter has given an overview of the process of creating an integrated GIS database using the data stream model (Figure 5.1) to guide the process. This has taken us from source data through data encoding, to editing and on to data manipulation operations such as re-projection, transformation and rubber sheeting. The chapter has given some insights into how some of these processes are carried out, and highlighted issues relevant to data encoding and editing which will be covered again elsewhere in the book.

Much of the effort required in an individual GIS project is concerned with data encoding and edit-

ing. A commonly quoted figure suggests that 50–80 per cent of GIS project time may be taken up with data encoding and editing. Whilst the amount of time and work involved is significant it is very important to ensure that the resulting database is of the best practical quality. The quality of the results of any analyses carried out on the GIS will ultimately depend on the quality of the input data and the procedures carried out in creating an integrated GIS database. This topic will be returned to in **Chapter 10** in a more in-depth discussion of data quality issues, but first the book continues with details of what can be done with the data now integrated in your GIS.

## REVISION QUESTIONS

- Describe the main phases of the data stream.

- How do source data type and nature affect data encoding?

- Describe the processes of point and stream mode digitizing.

- What are the main data sources for GIS?

- What are the main problems associated with using the web to find GIS data?

- Why is data editing important?

- What methods are available for detecting and rectifying errors in:
  (a) attribute data; and
  (b) spatial data?

- What are re-projection, transformation and generalization?

- What is geocoding?

## FURTHER STUDY – ACTIVITIES

- Use the web to try to find GIS data and remotely sensed imagery for your home town or region.

- Create a flowchart representing the data stream for a GIS project with which you are involved (if you are not currently working on a GIS project, create a hypothetical example or use one of the examples in the book).

- Manually digitize a sample set of simple features (points, lines and areas) from a topographic map by writing down their easting and northing (x,y or longitude, latitude) co-ordinates and entering them into a flat file using keyboard entry. Import the co-ordinates into a GIS (use a spreadsheet such as Excel if you don't have a GIS) and plot them on screen.

- Choose a stretch of coastline or river and try to identify the minimum number of points and their locations that can be used to represent its shape. If you can find the same coastline or river on maps of different scales, repeat this exercise with these maps and compare different versions to see the effect of scale.

## FURTHER STUDY – READING

There are a number of GIS texts that cover data encoding and editing issues. Hohl (1998) devotes a whole book to the subject of GIS data conversion, while DeMers (2005) offers two relevant chapters – on data input and data storage and editing. Jackson and Woodsford (1991) present an excellent, if a little dated, review of hardware and software for data capture. Chrisman (2002) offers a reflective review of techniques and quality issues associated with data encoding in a chapter on 'Representation'. Chapter 5 of Martin (1996) and Openshaw (1995) contain overviews of aspects of data encoding issues relevant to population census data. For reference to some of the important research work in data encoding you need to seek out articles in journals such as the *International Journal of Geographical Information Systems, Cartographica and Cartography*, and papers in conference proceedings. For example, Chrisman (1987) and Walsby (1995) address issues of efficient digitizing, and the now famous line-thinning algorithm is presented in Douglas and Peucker (1973). Miller and Han (2001) contains a series of chapters concerning data mining for GIS.

Chrisman N R (1987) Efficient digitizing through the combination of appropriate hardware and software for error detection and editing. *International Journal of Geographical Information Systems* 1: 265–77

Chrisman N R (2002) *Exploring Geographic Information Systems*. 2nd edn. Wiley, New York

DeMers M N (2005) *Fundamentals of Geographic Information Systems*. 3rd edn. Wiley, New York

Douglas D H, Peucker T K (1973) Algorithms for the reduction of the number of points required to represent a digitized line or its caricature. *Canadian Cartographer* 10 (4): 110–12

Hohl P (ed) (1998) *GIS Data Conversion: Strategies, Techniques, and Management*. Onward Press, Santa Fe

Jackson M J, Woodsford P A (1991) GIS data capture hardware and software. In: Maguire D J, Goodchild M F, Rhind D W (eds) *Geographical Information Systems: Principles and Applications*. Longman, London, vol. 1, pp. 239–49

Martin D (1996) *Geographical Information Systems and their Socio-economic Applications*. 2nd edn. Routledge, London

Miller H J, Han J (2001) *Geographic data mining and knowledge discovery*. Taylor and Francis, London

Openshaw S (ed) (1995) *The 1991 Census User's Handbook*. Longman Geoinformation, London

Walsby J (1995) The causes and effects of manual digitizing on error creation in data input to GIS. In: Fisher P (ed) *Innovations in GIS 2*. Taylor and Francis, London, pp. 113–24

## WEB LINKS

### Further information on data sources and input:

- The GIS Data Primer – Data Sources: www.innovativegis.com/basis/ primer/sources. html

- GIS and Scanning Technology: www.gisdevelopment.net/ technology/gis/ techgi0002b.htm

- OpenStreetMap: www.openstreetmap.org/

### Finding spatial data:

- US National Spatial Data Intrastructure Geospatial Onestop: www.geodata.gov/

- ESRI Geography Network: www.geographynetwork.com/

- Go-Geo! (service for UK academic community): www.gogeo.ac.uk/

- Digital chart of the world: www.maproom. psu.edu/dcw/

Visit our website **http://www.pearsoned.co.uk/heywood** for further web links, together with self-assessment questions, activities, data sets for practice opportunities and other interactive resources.

# 6

# Data analysis

# Learning outcomes

By the end of this chapter you should be able to:

■ List the data analysis operations available in GIS

■ Explain how features are measured in GIS

■ Describe methods of querying spatial and attribute data

■ Provide examples of network analysis

■ Outline methods of surface analysis

■ Describe the differences between analysis carried out on raster and vector data

■ Predict some of the problems that might be faced when conducting data analysis with GIS

## ■ INTRODUCTION

**Chapters 1 to 5** have introduced some of the fundamental concepts behind GIS. The nature of the spatial data used has been reviewed, and the encoding and structuring of these data to produce a computer representation of the real world outlined. Continuing the car analogy, we might consider that you now know how a car works, how it is built and what fuel to use. Now you are ready to take to the open road and use the car to get from one place to another. This chapter begins a journey in GIS, which takes the user from data to information and ultimately to decision-making. It covers some of the options in GIS for data analysis.

For many users data analysis is the most interesting and rewarding part of a GIS project. This is where they can start to find answers to some of their questions, and use GIS to help develop new questions for research. The analysis they undertake with GIS may lead to new information that will inform decision-making. For example, end users of a web-based mapping application may be able to answer simple queries about the location of places of interest, or to identify shortest routes between features using mobile GIS platforms. As a result they can make decisions about where to go next, and how to get there. In a more research oriented application 'what if' questions might be developed and investigated. For example, presentation of the results of research into complex planning issues may inform decision-making that will affect the development of whole communities and regions. **Chapter 1**

presented a list of types of questions that GIS can address. With data analysis we begin to be able to provide information to help answer some of these. Table 6.1 considers how these questions might be asked in the context of two typical GIS applications.

There is a wide range of functions for data analysis available in most GIS packages, including measurement techniques, attribute queries, proximity analysis, overlay operations and the analysis of models of surfaces and networks. To set these in an applied context, Box 6.1 considers what sort of data analysis might be required in the imaginary Happy Valley GIS. The questions presented in Box 6.1 will be returned to throughout the chapter and suggestions for methods of answering them will be presented.

Whilst Box 6.1 gives a substantial list of questions that the Happy Valley GIS could be used to answer, the list is by no means comprehensive. So too with this chapter – although a large number of data analysis functions will be examined, by no means all can be introduced and explained in a chapter of this size. Indeed, whole books have been devoted to the topic of spatial analysis in GIS (Fotheringham and Rogerson, 1994; Chou, 1996; Mitchell, 1999; Fischer and Getis, 2010). The aims of this chapter are to introduce the range of data analysis functions available, provide examples of how they might be applied and outline the problems faced in using them. The focus is on the practical application of data analysis functions in GIS, rather than a full explanation of the algorithms and concepts behind them. Reference is made to sources of further information throughout, and there are many suggestions for additional reading included at the end of the chapter.

**TABLE 6.1  Examples of questions for GIS analysis**

|  | Wind farm siting application | Retail store location application |
|---|---|---|
| **Location** | Where is the proposed wind farm location? | Where is the region in which the store will be located? |
| **Patterns** | What is the pattern of land use in this area? | What are the flows of traffic along major routes in this area? |
| **Trends** | How has land use changed over the last 20 years? | Which towns in the area have experienced the greatest population grown in recent years? |
| **Conditions** | Where is there an area of privately owned land of a suitable size, with adequate access from an existing road? | Where is there an area of commercially zoned land, close to major transport routes and with potential catchment of over 10,000 customers? |
| **Implications** | If we located the wind farm here, from where would it be visible? | If we locate a store here, what would be the impact on our closest existing store? |

## BOX 6.1 Data analysis in Happy Valley

The Happy Valley GIS team has set up its system. Table 5.6 in **Chapter 5** gave details of the data layers in the GIS. Using data analysis functions, the team hope to answer questions about the tourist facilities in the valley and to identify a possible location for a new ski piste. Some of the questions they will address are:

- Which is the longest ski piste in Happy Valley?
- What is the total area of forestry in the valley?
- How many luxury hotels are there in the valley?
- Where are the luxury hotels with more than 20 bedrooms?
- Where are the areas of forestry land in Happy Valley?
- Which hotels are within 200 m of a main road?
- In which land use types are the meteorological stations located?

- Which roads should be used for a scenic forest drive?
- What is the predominant land use type in the Happy Valley resort?
- Where could a new ski piste be located?
- What is the predicted snowfall for the new ski piste?
- What would be the quickest route by road from the centre of the resort to this new ski piste?
- What are the slope and aspect of the new ski piste?
- From where will the new ski piste be visible?

Data sets and activities relating to the Happy Valley Case Study can be found online at **www.pearsoned. co.uk/heywood**.

The chapter begins by introducing methods for measurement and queries in GIS. Proximity, neighbourhood and reclassification functions are outlined, then methods for integrating data using overlay functions are explained. Interpolation techniques (used for the prediction of data at unknown locations) are introduced, and since height values are frequently predicted, analysis of terrain and other surfaces follows. Finally, analysis of network data is considered. This chapter sets the scene for **Chapter 7** (Analytical modelling in GIS), which takes GIS through the next step, from a system capable of analysing models of spatial form to a system capable of building models of spatial process. There is also a strong link with **Chapter 12** on project management, where issues such as how to plan and prepare for successful analysis are covered.

An important first step to understanding spatial data analysis in GIS is developing a sound knowledge of the terminology used. Finding standard terms is sometimes difficult since different GIS software packages often use different words to describe the same function, and individuals with backgrounds in different fields tend to prefer particular terms. The terminology used in this chapter is summarized in Table 6.2.

## ■ MEASUREMENTS IN GIS – LENGTHS, PERIMETERS AND AREAS

Calculating lengths, perimeters and areas is a common application of GIS. Measuring the length of a ski piste from a digital map is a relatively straightforward task. However, it is possible that different measurements can be obtained depending on the type of GIS used (raster or vector) and the method of measurement employed. It is important to remember that all measurements from a GIS will be an approximation, since vector data are made up of straight line segments (even lines which appear as curves on the screen are stored as a collection of short straight line segments), and all raster entities are approximated using a grid cell representation.

In a raster GIS there is more than one answer to the question 'what is the distance between A and B?' where A and B are two ends of a straight line. The answer varies according to the measurement method used. Normally the shortest path, or Euclidean distance, is calculated by drawing a straight line between the end points of a line, and creating a right-angled triangle so that Pythagorean geometry can be used (Figure 6.1a). In this case the distance AB is calculated using the formula:

$$AB^2 = AC^2 + CB^2$$

**TABLE 6.2  Data analysis terminology**

| Term | Definition |
| --- | --- |
| Entity | An individual point, line or area in a GIS database |
| Attribute | Data about an entity. In a vector GIS attributes are stored in a database. For example, the street name for a line entity that represents a road may be stored. In a raster GIS the value of a cell in the raster grid is a numerical code used to represent the attribute present. For example, code '1' may be used for a motorway, '2' for a main road and '3' for a minor road. Further attributes of an entity (for instance street name) may be stored in a database linked to the raster image |
| Feature | An object in the real world to be encoded in a GIS database |
| Data layer | A data set for the area of interest in a GIS. For example, the Happy Valley GIS contains data layers on nine themes, including roads, ski runs, hotels and land use. Data layers in a GIS normally contain data of only one entity type (that is, points, lines or areas). It would be unusual for a data layer to contain, for instance, roads and hotels; these would be stored as separate thematic data layers |
| Image | A data layer in a raster GIS. It should be remembered that each cell in a raster image will contain a single value that is a key to the attribute present there |
| Cell | An individual pixel in a raster image |
| Function or operation | A data analysis procedure performed by a GIS |
| Algorithm | The computer implementation of a sequence of actions designed to solve a problem |

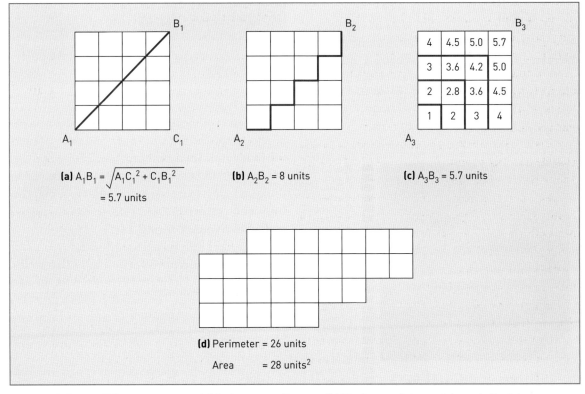

**Figure 6.1** Raster GIS measurements: (a) Pythagorean distance; (b) Manhattan distance; (c) proximity distances; (d) perimeter and area

Therefore:

$$AB = \sqrt{AC^2 + CB^2}$$

Alternatively, a 'Manhattan' distance can be calculated. This is the distance along raster cell sides from one point to the other. The name for this method comes from the way in which you would get across a city, like Manhattan, consisting of dense 'blocks' of buildings. As it is impossible to pass diagonally through a block, you have to traverse the sides (Figure 6.1b). A third method of calculating distance in a raster GIS uses 'proximity' (Berry, 1993). In this method, concentric equidistant zones are established around the start point A (Figure 6.1c). The resulting image shows the shortest straight line distance from every point on the map (including end point B) to the location of interest (A). Thus, the distance from A to B can be ascertained. Some authors (Berry, 1987; Tomlin, 1990) use the term 'spread' for functions which act in this way to reflect the way the distance surface grows outwards from the feature of interest (Figure 6.2).

To obtain a perimeter measurement in a raster GIS, the number of cell sides that make up the boundary of a feature is multiplied by the known resolution of the raster grid. For area calculations, the number of cells a feature occupies is multiplied by the known area of an individual grid cell (Figure 6.1d). The area of forestry in Happy Valley (from a classified satellite image in a raster GIS) could be obtained in this way (see Figure 6.3).

**Figure 6.3** Calculating lengths and areas from the Happy Valley GIS

Area and perimeter calculations in raster data can be affected by the cell size, origin and orientation of the raster grid. These problems are discussed further in **Chapter 10**, and they can be partly avoided by using cell sizes smaller than the minimum mapping unit, and consistent use of north–south aligned grids with common origins.

In a vector GIS distances are measured using Pythagoras's theorem to obtain the Euclidean distance (Figure 6.4). Geometry is also used to calculate perimeters and areas. Perimeters are built up of the sum of straight line lengths, and areas are calculated by totalling the areas of simple geometric shapes formed by subdividing the feature of interest (Figure 6.4). In vector GIS, length, perimeter and area data can be stored as attributes in a database, so these need to be calculated only once and then are permanently saved.

**Figure 6.2** House-hunting case study: distance from office calculated using proximity method

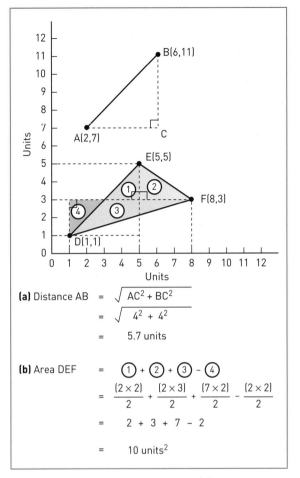

**(a)** Distance AB $= \sqrt{AC^2 + BC^2}$

$\qquad\qquad\quad\;\; = \sqrt{4^2 + 4^2}$

$\qquad\qquad\quad\;\; = \quad$ 5.7 units

**(b)** Area DEF $= \;①\; + \;②\; + \;③\; - \;④$

$\qquad\qquad\; = \dfrac{(2 \times 2)}{2} + \dfrac{(2 \times 3)}{2} + \dfrac{(7 \times 2)}{2} - \dfrac{(2 \times 2)}{2}$

$\qquad\qquad\; = \quad 2 + 3 + 7 - 2$

$\qquad\qquad\; = \quad 10 \; units^2$

**Figure 6.4** Vector GIS measurements: (a) distance; (b) area

Ski pistes in Happy Valley are likely to bend and weave their way through the trees. To find the total length of an individual piste it would be necessary, therefore, to sum the lengths of several straight line segments of forest edge that approximate the piste (Figure 6.3).

### ■ QUERIES

Performing queries on a GIS database to retrieve data is an essential part of most GIS projects. Queries offer a method of data retrieval, and can be performed on data that are part of the GIS database, or on new data produced as a result of data analysis. Queries are useful at all stages of GIS analysis for checking the quality of data and the results obtained. For example, a query may be used if a data point representing a hotel is found to lie erroneously in the sea after data encoding. A query may establish that the address of the hotel had been wrongly entered into a database, resulting in the allocation of an incorrect spatial reference. Alternatively, queries may be used after analysis has been conducted. For instance, following extensive searches using GIS for a suitable hotel to meet a particular client's requirements (perhaps one with a small number of bedrooms, located a short distance from the ski slopes, and accessible by public transport), a query may be used to obtain the name and address of the hotel to allow a booking to be made.

There are two general types of query that can be performed with GIS: spatial and aspatial. Aspatial queries are questions about the attributes of features. 'How many luxury hotels are there?' is an aspatial query since neither the question nor the answer involves analysis of the spatial component of data. This query could be performed by database software alone. In Box 6.1 the question 'Where are the luxury hotels in Happy Valley?' was asked. Since this requires information about 'where' it is a spatial query. The location of the hotels will be reported and could be presented in map form.

The method of specifying queries in a GIS can have a highly interactive flavour. Users may interrogate a map on the computer screen or browse through databases with the help of prompts and query builders. A user may point at a hotel on the computer screen and click to obtain the answer to 'What is the name of *this* hotel?'. Queries can be made more complex by combination with questions about distances, areas and perimeters, particularly in a vector GIS, where these data are stored as attributes in a database. This allows questions such as 'Where is the longest ski run?' to be answered in one step, where a raster GIS might require two – one to calculate the lengths of all the ski runs, and the second to identify the longest.

Individual queries can be combined to identify entities in a database that satisfy two or more spatial and aspatial criteria, for example 'Where are the luxury hotels which have more than 20 bedrooms?'. Boolean operators are often used to combine queries of this nature. These use AND, NOT, OR and XOR, operations that are also used for the combination of different data sets by overlay (Box 6.2).

# BOX 6.2 Boolean operators

**THEORY**

Boolean expressions used in GIS are AND, OR, NOT and XOR. These are explained best with the help of Venn diagrams, where each circle in the diagram represents the set of data meeting a specific criterion (Figure 6.5). In the diagrams, A is the set of hotels that are in the 'luxury' category, and B is the set of hotels that have more than 20 bedrooms.

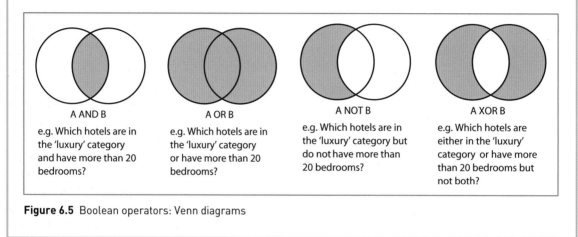

**A AND B**

e.g. Which hotels are in the 'luxury' category and have more than 20 bedrooms?

**A OR B**

e.g. Which hotels are in the 'luxury' category or have more than 20 bedrooms?

**A NOT B**

e.g. Which hotels are in the 'luxury' category but do not have more than 20 bedrooms?

**A XOR B**

e.g. Which hotels are either in the 'luxury' category or have more than 20 bedrooms but not both?

**Figure 6.5** Boolean operators: Venn diagrams

Boolean operators need to be used with care since 'Where are the hotels which are in the category "luxury" AND have more than 20 bedrooms?' will yield a different answer from the question 'Where are the hotels which are in the category "luxury" OR have more than 20 bedrooms?' The second query will probably identify more hotels.

## ■ RECLASSIFICATION

Reclassification is an important variation on the query idea in GIS, and can be used in place of a query in raster GIS. Consider the raster land use image referred to in Table 5.6 and shown in Figure 6.3. If we wished to ask 'Where are all the areas of forestry?' an answer could be obtained using a query or by reclassifying the image. Reclassification would result in a new image (Figure 6.3). For example, if cells representing forestry in the original image had a value of 10, a set of rules for the reclassification could be:

Cells with values = forestry (value 10) should take the new value of 1

Cells with values ≠ forestry should take the new value of 0

Reclassification would result in a new image with all areas of forestry coded with value 1, and all areas that were not forest coded with value 0. This is a Boolean image. Reclassification of this nature, producing a two-code image from a complex original image, is a very useful technique because it allows the resulting image (containing only values 1 and 0) to be used in further analysis.

Other reclassifications are also possible. In the Zdarske Vrchy case study reclassification was used to allocate a new value to different land use classes based on their ecological importance (Table 6.3). Forestry and agricultural land were considered to have a relatively high conservation value, whilst water and settlement had lower conservation value. A new image was produced after implementation of the following set of rules:

Cells with original values = 10 should take the new value of 5

Cells with original values = 11 should take the new value of 2

Cells with original values = 12 should take the new value of 1

Cells with original values = 13 should take the new value of 4

**TABLE 6.3 Reclassification values for the land use data layer**

| Land use | Original value | New value after reclassification: Boolean example | New value after reclassification: Weighting example |
|---|---|---|---|
| Forestry | 10 | 1 | 5 |
| Water | 11 | 0 | 2 |
| Settlement | 12 | 0 | 1 |
| Agricultural land | 13 | 0 | 4 |

The resulting map helped the ecologist to identify areas of high conservation value. However, the new classes (1, 2, 4 and 5) are still simply labels (ordinal values) and care needs to be taken to ensure appropriate further analysis of the image.

## ■ BUFFERING AND NEIGHBOURHOOD FUNCTIONS

There is a range of functions available in GIS that allow a spatial entity to influence its neighbours, or the neighbours to influence the character of an entity. The most common example is buffering, the creation of a zone of interest around an entity. Other neighbourhood functions include data filtering. This involves the recalculation of cells in a raster image based on the characteristics of neighbours.

The question 'Which hotels are within 200 m of a main road?' could be approached in a number of ways. One option would be, first, to produce a buffer zone identifying all land up to 200 m from the main roads; and second, to find out which hotels fall within this buffer zone using a point-in-polygon overlay (see later in this chapter). Then a query would be used to find the names of the hotels. An alternative approach would be to measure the distance from each hotel to a main road, then identify those which are less than 200 m away. This example illustrates that in most GIS data analysis there is more than one method of achieving an answer to your question. The trick is to find the most efficient method, and the most appropriate analysis.

Buffering, as already stated, is used to identify a zone of interest around an entity, or set of entities. If a point is buffered a circular zone is created. Buffering lines and areas creates new areas (Figure 6.6).

**(a)** Point

**(b)** Line

**(c)** Area

**Figure 6.6** Buffer zones around (a) point, (b) line and (c) area features

Buffering is very simple conceptually but a complex computational operation. Creating buffer zones around point features is the easiest operation; a circle of the required radius is simply drawn around each point. Creating buffer zones around line and area features is more complicated. Some GIS do this by placing a circle of the required radius at one end of the line or area boundary to be buffered. This circle is then moved along the length of the segment (Figure 6.6). The path that the edge of the circle tangential to the line makes is used to define the boundary of the buffer zone.

Figure 6.6 illustrates only the most basic set of buffer operations as there are many variations on these. For example, buffer zones may be of fixed or varying width according to feature attributes (Figure 6.7). When analysing a road network, wide buffer zones could be attached to motorways and narrower buffer zones to minor roads to reflect traffic densities. In the radioactive waste case study (introduced in **Chapter 1**) local accessibility of potential disposal sites was defined using buffering techniques. Buffer zones of varying widths were drawn around roads and railway lines. Motorways were buffered at 3 km (these being the main carriers of volume traffic) and primary routes at 1.5 km. Railways were also buffered at 3 km (Figure 6.8). Sites lying within any of these buffer zones were potentially feasible waste disposal sites as they were sufficiently accessible (Openshaw *et al.*, 1989).

**Figure 6.8** Radioactive waste case study: 3 km buffer zones (blue) around the rail network (red)

Buffering has a whole series of uses when combined with other data layers. In the radioactive waste example the buffer zones were used as part of a process to identify the land use, population totals and conservation status of accessible land (Openshaw *et al.*, 1989). Andersson (1987) used buffering of bus stop data with population data to identify the best locations for bus stops. A buffer zone was calculated for each potential bus stop, using a value that reflected the distance an individual was prepared to walk to reach the stop. The population density within this buffer zone was then calculated. A set of bus stops was identified which maximized the overall population catchment areas. This type of operation relies on multiple layer overlays. These are considered in greater depth later in this chapter.

Whilst buffer zones are often created with the use of one command or option in vector GIS, a different approach is used in many raster GIS. Here proximity is calculated. This method was outlined earlier (Figure 6.1c) and will result in a new raster data layer where the attribute of each cell is a measure of distance. Figure 6.9 shows an example of a proximity map produced for the hotels in Happy Valley, subsequently reclassed to create a buffer zone of 125 m for each of the hotels.

Other operations in raster GIS where the values of individual cells are altered on the basis of adjacency are called neighbourhood functions. Filtering is one example used for the processing of remotely sensed imagery. Filtering will change the value of a cell based on the attributes of neighbouring cells. The filter is defined as a group of cells around a target cell. The size and shape of the filter are determined by the operator. Common filter shapes are squares and circles, and the dimensions of the filter determine

**Figure 6.7** House-hunting case study: distance from office adjusted for road network

**(a)** distance surface

**(b)** 125 m buffer zones

**Figure 6.9** Proximity map for hotels in Happy Valley

the number of neighbouring cells used in the filtering process. The filter is passed across the raster data set and used to recalculate the value of the target cell that lies at its centre. The new value assigned to the target cell is calculated using one of a number of algorithms. Examples include the maximum cell value within the filter and the most frequent value. These and other filtering techniques are shown in Figure 6.10. Filtering operations might be used in the preparation of the forestry data layer in the Happy Valley GIS. The raster data obtained from a classified satellite image may require filtering to 'smooth' noisy data caused by high spatial variability in vegetation cover or problems with the data collection device.

In the Zdarske Vrchy case study a filtering algorithm was used to create a forest edge index for the region. A 6 cell × 6 cell filter was passed over a forest edge map derived from satellite imagery. A new value for the central cell was calculated, based on the number of cells in the window that contained forest edge. The forest edge index was then used to identify suitable habitats for the reintroduction of the black grouse. For this species of bird the edge of the forest is an important habitat (Downey *et al.*, 1992).

| 6 | 3 | 3 | 2 | 1 | 2 | 2 |
| 5 | 3 | 3 | 2 | 1 | 1 | 2 |
| 4 | 3 | 2 | 2 | 2 | 1 | 1 |
| 3 | 2 | 2 | 2 | 1 | 1 | 1 |
| 2 | 1 | 2 | 1 | 1 | 1 | 1 |
| 1 | 1 | 1 | 1 | 0 | 0 | 0 |
|   | a | b | c | d | e | f |

Example: Forest data in Happy Valley GIS
- Applying 3 3 square filter to recalculate value for cell c4:

minimum filter  c4 = 1
maximum filter  c4 = 3
mean filter       c4 = 1.89
modal filter      c4 = 2
(most frequently occurring class)
diversity filter  c4 = 3
(number of different classes)

**Figure 6.10** Raster GIS filter operations

A combination of distance and neighbourhood operations can be used to perform some quite complex distance or proximity calculations that take into account not only horizontal linear distance but also the effects of vertical distance or slope (for example, climbing or descending a hill). Other cost factors that can be accounted for include the effect of wind speed or resistance, trafficability, load carried or other push/pull factors. Cost factors such as slope and wind speed are not the same in all directions, so proximity models need to take this into account. When the factors controlling relative distance are the same in all directions, proximity models are said to be 'isotropic' (e.g. simple buffering or linear distance surfaces). When the factors controlling relative distance are not the same in all directions, promixity models are said to be 'anisotropic'. Many GIS systems now make provision for anisotropic proximity modelling. Box 6.3 describes the use of GIS for modelling remoteness in mountainous terrain for landscape assessment studies.

---

## BOX 6.3 Modelling remoteness in mountain areas

**CASE STUDY**

### Steve Carver and Steffen Fritz

Remoteness as a geographical measure of distance is not easy to measure, nor indeed describe. Remoteness may be defined geographically as the linear distance from one point in space to another point in space. This makes an assumption of equal speed and cost of travel in all directions, which, of course, is not possible in the real world where there are differences in transport networks, terrain and other variables. Another, more useful, definition of remoteness is the time taken to travel between the origin and destination. Using time of travel as a measure of remoteness requires that a number of geographical factors other than simple linear distance are considered. These include cost or push (benefit) factors that influence the ease of travel in a particular direction and the route chosen. For example, there are several geographical factors that are likely to influence pedestrian off-road access. These include terrain features such as the type of ground cover, presence of tracks and paths that make the going easier; prevailing weather conditions; barrier features such as lakes, cliffs, uncrossable rivers, fenced and private land; and hydrological conditions, including ground wetness, stream levels and snow cover. All these variables may be included in a spatial model to estimate remoteness. It is necessary to distinguish between definitions of remoteness for both mechanized and non-mechanized modes of transport, since different cost/push factors assume very different levels of significance within each category. This case study demonstrates the use of GIS-based models to map physical remoteness for roadless areas, where remoteness is defined as the time taken to walk from the nearest road taking into account the physical characteristics of the landscape.

Mapping physical remoteness is important given the rise in outdoor recreation seen during the last 40 years and substantial increases predicted for the next few decades (Ewert and Hollenhorst, 1997). New techniques are needed to help manage the landscape resource and, in particular, those areas beyond mechanized access, to prevent overuse and degradation. Remoteness maps are an essential element in the evaluation of wild areas and in the development of policy aimed at their protection.

### THE ROLE OF REMOTENESS IN LANDSCAPE CHARACTER

Remoteness plays a significant role in determining people's perceptions of landscape character particularly in mountain areas. Many people associate remoteness with wilderness, sometimes with scant regard for land use and landscape histories. Remoteness is, from this perspective, best referred to as a *perceived variable* in that, whilst it may have little or no biophysical effect beyond its geographical influence on the history of human use of the landscape, it does have a marked effect on how people feel about the landscape setting. Being able to accurately map remoteness by taking the associated geographical variables into account could prove to be of great value to studies of wild land and landscape character (Carver *et al.*, 2002). It is possible to estimate for any location within a roadless area the time it would take to walk to that location from a road or car park. By calculating the time taken to walk to all points in a roadless area to the nearest access point it is possible to generate a remoteness surface for the whole area of interest. This can be used to help define policy boundaries, zone recreation areas, map landscape character and even wilderness quality indices.

## MAPPING ACCESSIBILITY AND REMOTENESS USING GIS

Simple GIS-based access models often start by assuming equal ease of travel in all directions. When considering accessibility from point, line or area origin features, the simplest solution is to draw buffer zones or use simple linear distance (an isotropic surface). Some GIS models use cost and push factors to calculate accessibility surfaces not constrained to transport networks (for example Dana Tomlin's MAP software). In the context of off-road pedestrian travel, terrain variables such as gradient may be applied to give better estimates of access times. By incorporating relevant cost/push factors into pedestrian specific access models within a GIS, better estimates of remoteness for a given roadless area may be obtained. Naismith's Rule (1892) has long been used as a rule-of-thumb guide to estimate the time taken to walk from A to B over mountainous terrain. The basic rule states that a walker can maintain a speed of 5 km/h on level ground, but half an hour needs to be added for every 300 m of ascent. Several refinements have been made to Naismith's Rule over the years including those proposed by Langmuir (1984) who includes corrections for going down (minus 10 minutes per 300 m descent for slopes between 5° and 12°) and up (plus 10 minutes per 300 m descent for slopes greater than 12°). At a local level it is possible to develop a model that takes into account a range of topographic variables influencing on-foot travel times. This can be

achieved by incorporating Naismith's Rule and more recent corrections into a GIS model.

Using Naismith's Rule it is possible to calculate the time taken to traverse a set of cells in a digital elevation model (DEM) by taking gradient and slope direction relative to direction of travel into account. Accessibility from different directions relative to the same point in the landscape should be considered, along with the shortest path or access time. Using this approach it is possible to design a model that calculates the time taken to walk from single or multiple origin points to any destination on the terrain surface. Due to the fact that it is unknown which route a walker will take, the model considers only the quickest possible path.

The model described here integrates Naismith's Rule with Dijkstra's shortest path algorithm (Aho *et al.*, 1974; Dracos, 1998). Dijkstra's algorithm works by considering the relative costs of moving through each of the cells in a matrix. Costs are represented impedance values in the cell matrix. In order to implement Naismith's Rule within Dijkstra's algorithm, four different cost matrices were used. These include a height matrix, a distance matrix, a trace matrix (which marks all the cells that have been dealt with), and a results matrix (the values in which are changed during the analysis process). Barrier features such as lakes and cliffs are included in the matrix as NoData values and are ignored. This process has been automated within Esri's ArcGIS and uses custom C code. For a detailed description of the implementation of the algorithm see Fritz and Carver (2000). A similar result can also be achieved using the PATHDISTANCE command in the ArcGIS or the cost distance module in Idrisi. Figure 6.11 shows

**Figure 6.11** DEM of the Rocky Mountains, Colorado
*Source*: United States Geological Survey

**(a)** Remoteness

**(b)** Linear distance

**(c)** Residuals

**Figure 6.12** Results of remoteness model compared to simple linear distance
*Source*: United States Geological Survey

## BOX 6.3

a DEM of part of the Colorado Rocky Mountains overlaid with roads, rivers and lakes. Figure 6.12a shows the resulting remoteness surface from the Naismith's modified Dijkstra's algorithm applied to the DEM in Figure 6.11 to map remoteness from all roads surrounding the Rocky Mountains. Some of the geographical factors affecting off-road access times can be incorporated into the algorithm by modifying the distance matrix according to additional cost/push factors. These include ground cover, underfoot conditions, presence of tracks and footpaths, prevailing weather conditions and the effects of barrier features such as lakes, crags,

uncrossable rivers and private land. Remoteness maps can be compared with a simple linear distance surface (Figure 6.12b) to show how terrain and other geographical features make a significant difference to mapped off-road access times. It is clear that the greater the relief and the more barrier features that are present, the greater the divergence between linear distance maps and the remoteness mapping techniques described here. The degree of difference between the simple linear distance surface and the results from the Naismith's/Dijkstra's model can be represented as residuals. These are shown in Figure 6.12c.

**(a)** MapTube interface

**(b)** The Cairngorms

**Figure 6.13** Wildland quality in the Cairngorm National Park
*Source*: (a) MapTube interface courtesy of Centre for Advanced Spatial Analysis [CASA]

### POTENTIAL APPLICATIONS OF REMOTENESS MAPPING

Remoteness mapping may have several applications within the planning field. In particular, remoteness maps may be useful in route planning for new or upgraded footpaths, tracks or roads, evaluating new hydroelectric/wind energy schemes and the effect of associated access roads, and assessing the impacts of new recreational developments. In this context, information on remoteness may be useful in helping consider planning applications for new developments. Using the techniques described above it would be possible for any proposed development to create 'before' and 'after' remoteness maps as part of the associated environmental assessment. A further application area for remoteness mapping is in the zoning of protected areas for management and decision support purposes. With the increasing numbers of people participating in some form of outdoor recreation it is becoming necessary to employ zoning techniques within Britain's national parks and other protected areas. Remoteness surfaces based on the geographical factors described here may be useful in assisting park managers to decide on the exact boundaries of these zones and their subsequent management. Such techniques have been applied in a study for the Cairngorm National Park Authority in Scotland. Maps of remoteness (derived using the methods described above) were combined with maps of the perceived naturalness of the land cover, the absence of modern human artefacts (derived from viewshed analyses – see Box 6.10), and the rugged and challenging nature of the terrain, to describe the wildness of the Cairngorm landscape. This is being used as a means of describing landscape quality within the park and as a basis for landscape protection. The results of the mapping work are shown in Figure 6.13 and have been placed on the Internet using MapTube for the public to view (see **Chapter 8**).

### FURTHER INFORMATION

Accessibility and remoteness from health care in Australia: www6.health.gov.au/internet/main/publishing.nsf/Content/health-historicpubs-hfsocc-ocpanew14a.htm
Using remoteness indices in mapping wilderness: www.wilderness.net/library/documents/Fritz_2-15.pdf
Wildness in the Cairngorm national park: www.maptube.org

*Source*: Steve Carver and Steffen Fritz, University of Leeds

## ■ INTEGRATING DATA – MAP OVERLAY

The ability to integrate data from two sources using map overlay is perhaps the key GIS analysis function. Using GIS, it is possible to take two different thematic map layers of the same area and overlay them one on top of the other to form a new layer. The techniques of GIS map overlay may be likened to sieve mapping, the overlaying of tracings of paper maps on a light table. Map overlay has its origins in the work of McHarg (1969), who used large and complex manual map overlays in landscape architecture (see **Chapter 9**).

Map overlay has many applications. At one level it can be used for the visual comparison of data layers. For example, the results of the hotel query 'Where are all the "luxury" hotels?' may be plotted on top of, or overlaid, on a map of the road network to give some spatial context to the results. In this case no new data are produced. In fact, the location of ski pistes or land use layers could just as easily be used.

This technique is used for the overlay of vector data (for example, pipelines) on a raster background image (often a scanned topographic map).

Overlays where new spatial data sets are created involve the merging of data from two or more input data layers to create a new output data layer. This type of overlay may be used in a variety of ways. For example, obtaining an answer to the question 'Which hotels are within 200 m of a main road?' requires the use of several operations. First, a buffering operation must be applied to find all the areas of land within 200 m of a main road, then an overlay function used to combine this buffer zone with the hotel data layer. This will allow the identification of hotels within the buffer zone. Alternatively, the selection of a site for a new ski piste may require the overlay of several data sets to investigate criteria of land use, hotel location, slope and aspect.

As with many other operations and analyses in GIS there are differences in the way map overlays are performed between the raster and vector worlds.

In vector-based systems map overlay is time-consuming, complex and computationally expensive. In raster-based systems it is just the opposite – quick, straightforward and efficient.

## Vector overlay

Vector map overlay relies heavily on the two associated disciplines of geometry and topology. The data layers being overlaid need to be topologically correct

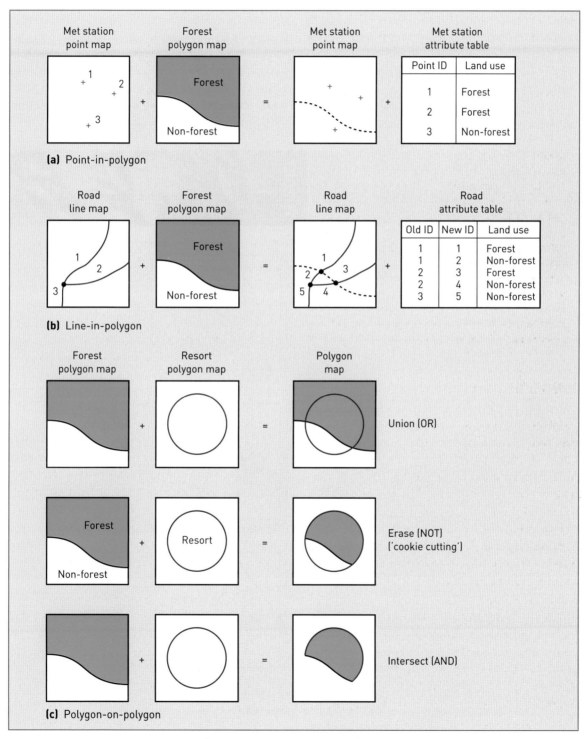

**Figure 6.14** Vector overlays

(see **Chapter 3**) so that lines meet at nodes and all polygon boundaries are closed. To create topology for a new data layer produced as a result of the overlay process, the intersections of lines and polygons from the input layers need to be calculated using geometry. For complex data this is no small task and requires considerable computational power. Figure 6.14 on p.189 shows the three main types of vector overlay; point-in-polygon, line-in-polygon and polygon-on-polygon. This figure also illustrates the complexity of the overlay operations. The overlay of two or more data layers representing simple spatial features results in a more complex output layer. This will contain more polygons, more intersections and more line segments than either of the input layers.

Point-in-polygon overlay is used to find out the polygon in which a point falls. In the Happy Valley GIS meteorological stations are represented as points and land use as polygons. Using point-in-polygon overlay on these vector data layers it is possible to find out in which land use polygon each meteorological station is located. Figure 6.14 illustrates this overlay process. On the output map a new set of rain gauge points is created with additional attributes describing land use. The algorithm for point-in-polygon is explained in Box 6.4.

## BOX 6.4 Point-in-polygon analysis

THEORY

There are several methods of solving point-in-polygon problems but the most common technique used in GIS is the 'half line' or Jordan method (Laurini and Thompson, 1992; Martin, 1995). Consider a circular polygon with a point at its centre, and another point outside the circular polygon (Figure 6.15a). If a line (the 'half line') is extended from the central point to the edge of the data layer it will cross the polygon boundary once. A line extended from the point outside the circle will not cross the polygon boundary. Now consider a more complex polygon with crenellated edges (Figure 6.15b). A line extended from the central point will always cross the polygon boundary an odd number of times, whilst a line extended from the point outside the polygon will cross the boundary an even number of times. Thus, the Jordan algorithm is based on the rule that if the number of crossings is an odd number a point is inside the polygon, and if the number of crossings is an even number, a point is outside the polygon. In practice, the 'half line' is drawn parallel to the *x* axis and the method works for even the most complicated polygons containing holes and islands (Laurini and Thompson, 1992).

A difficulty with this method arises when points lie exactly on the polygon boundary (Figure 6.15c). Are these considered in or out? The answer may depend on the decision rule implemented in the software you are using, and some systems will say in and others out. Laurini and Thompson (1992) suggest that, where the point has been produced using a mouse, it can be moved slightly to facilitate the in or out decision. Another problem situation is where the 'half-line' is coincident with the polygon boundary, making the counting of crossings impossible (Figure 6.15c).

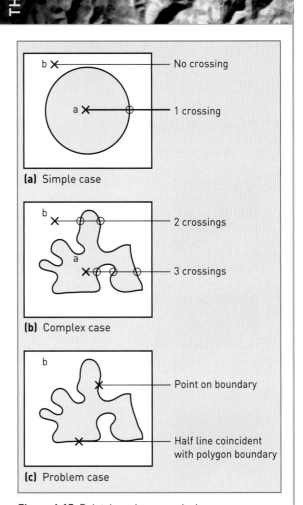

(a) Simple case

b ✕ — No crossing

a ✕ ⊕ — 1 crossing

(b) Complex case

b ✕ ⊕ ⊕ — 2 crossings

a ✕ ⊕ ⊕ ⊕ — 3 crossings

(c) Problem case

b — Point on boundary

— Half line coincident with polygon boundary

**Figure 6.15** Point-in-polygon analysis

Line-in-polygon overlay is more complicated. Imagine that we need to know where roads pass through forest areas to plan a scenic forest drive. To do this we need to overlay the road data on a data layer containing forest polygons. The output map will contain roads split into smaller segments representing 'roads in forest areas' and 'roads outside forest areas'. Topological information must be retained in the output map (Figure 6.14b), therefore this is more complex than either of the two input maps. The output map will contain a database record for each new road segment.

Polygon-on-polygon overlay could be used to examine the areas of forestry in the Happy Valley resort. Two input data layers – a forest data layer containing forest polygons, and the resort boundary layer – are required. Three different outputs could be obtained (Figure 6.14c):

1  The output data layer could contain all the polygons from both of the input maps. In this case the question posed is 'Where are areas of forestry OR areas which are within Happy Valley?' This corresponds to the Boolean OR operation, or in mathematical set terms, UNION. This may be useful if the Happy Valley management committee was interested in buying new forest areas to extend the scenic forest drive.

2  The output data layer could contain the whole of the resort area, and forest within this. The boundary of the resort would be used as the edge of the output map, and forest areas would be

cut away if they fall outside it. This operation is referred to as 'cookie cutting'. It is equivalent to the mathematical IDENTITY operation and the identity of the resort boundary is retained in the output. The questions being answered are 'Where is the resort boundary, and where are areas of forest within this?' This overlay might be used in preparation for calculation of the percentage of the area of the resort covered by forest.

3  The output data layer could contain areas that meet both criteria: that is, areas that are both forest and within Happy Valley. An output map would be produced showing the whole of any forest polygons that are entirely within the resort boundary, and 'cut' away forest polygons that cross the resort boundary. This is the mathematical INTERSECT operation, and the output map shows where the two input layers intersect. 'Where are forest areas within Happy Valley?' is the question being answered. As a thematic data layer showing forestry in Happy Valley this may be useful for further analysis of the condition of the resort's forestry resources.

Overlay operations are seldom used in isolation. In practice, it is common to query a data layer first, then perform an overlay. To obtain areas of forestry used in the examples above, it would be necessary to extract these areas from the land use data layer first using a query. Box 6.5 discusses the use of polygon-on-polygon overlay in the nuclear waste case study.

## BOX 6.5  Siting a nuclear waste repository
CASE STUDY

Polygon-on-polygon overlay is often used for site selection or suitability analysis. In the nuclear waste case study the aim was to find those areas of Britain that simultaneously satisfied four siting constraints. The constraints are summarized in Table 6.4. Using a vector GIS to solve this problem the approach can

be expressed in terms of GIS operations and Boolean overlays. This is illustrated in the flow chart shown in Figure 6.16. Examples of buffering and polygon intersect overlays in identifying areas of suitable geology that lie within transport route buffer zones are given in Figure 6.17.

**BOX 6.5**

**TABLE 6.4 Siting criteria for a nuclear waste repository**

| Theme | Criteria |
|---|---|
| Geology | Chosen site must be in an area of suitable geology |
| Accessibility | Chosen site must be easily accessible |
| Population | Chosen site must be away from areas of high population density |
| Conservation | Chosen site must be outside any areas designated for conservation value (for example National Park, Site of Special Scientific Interest) |

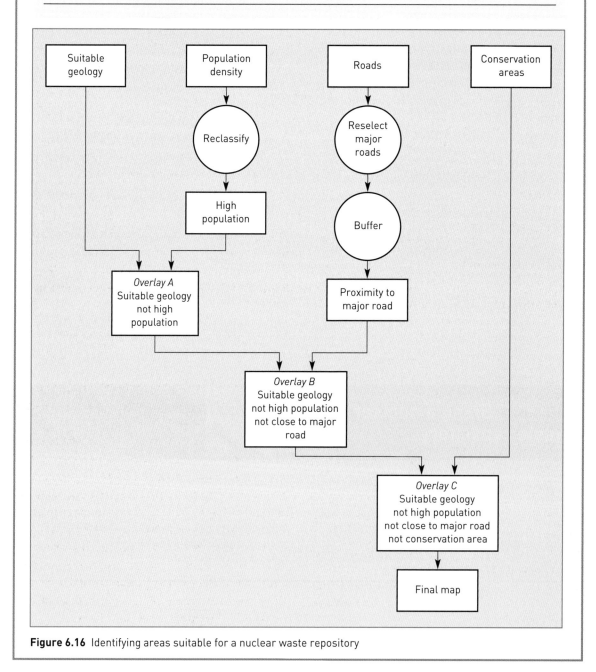

**Figure 6.16** Identifying areas suitable for a nuclear waste repository

BOX 6.5

Legend

□ 3 km railway buffer

▨ Surface clay geology

▨ Intersect of rail buffer and clay geology

0  5  10    20    30    40
Kilometres

N

**Figure 6.17** Radioactive waste case study: results of vector overlay showing the intersect of the rail buffer and clay geology

One problem with vector overlay is the possible generation of sliver (or 'weird') polygons (Chrisman, 2002). These appear after the overlay of two data sets that contain the same spatial entities. If the Happy Valley resort boundary were digitized by two different people, two separate representations of the area would be created. If these were overlaid, long thin polygons would be seen along the boundary instead of a single line. These 'sliver' polygons arise from inconsistencies and inaccuracies in the digitized data. Frequently such errors go undetected, but they can become apparent during vector overlay operations. Sliver polygons are considered further in **Chapter 10**. Other overlay problems, which relate to both vector and raster methods, are considered in Box 6.7.

## Raster overlay

In the raster data structure everything is represented by cells – a point is represented by a single cell, a line by a string of cells and an area by a group of cells. Therefore, the methods of performing overlays are different from those in vector GIS. Raster map overlay introduces the idea of map algebra or 'mapematics' (Berry, 1993). Using map algebra input data layers may be added, subtracted, multiplied or divided to produce output data. Mathematical operations are performed on individual cell values from two or more input layers to produce an output value. Thus, the most important consideration in raster overlay is the appropriate coding of point, line and area features in the input data layers.

Imagine that four of the Happy Valley data layers have been rasterized: the location of meteorological stations, the road network, the land use layer and the resort boundary. The meteorological stations are represented in a raster data layer where the value of '1' has been given to cells containing the stations. Roads are coded '2' in the roads raster layer and cells have been linked to create chains. Each cell in the land use layer has a value reflecting the type of land use present. Settlement is coded '1', water '2', agricultural land '4' and forestry '5'. The resort area has been given a value '10'. On all data layers '0' is the value given to cells that do not contain features of interest.

To find out which meteorological stations are contained within the Happy Valley resort an equivalent to the vector point-in-polygon overlay is required. To do this, one approach would be to add the two data layers (Figure 6.18a). The output map would contain cells with the following values:

- 0 for cells outside the resort boundary and without a meteorological station;
- 1 for cells containing meteorological stations but outside the resort boundary;
- 10 for cells inside the resort boundary but without a meteorological station;
- 11 for cells inside the resort boundary and containing a meteorological station.

In an operation equivalent to the vector line-in-polygon method (Figure 6.18b), the sections of roads that pass through forest areas could be obtained. This would require the roads data layer, and a reclassified version of the land use map that contained only forest areas. Again the two maps could be added. The output map would contain cells with the following values:

- 0 for cells with neither roads nor forest present;
- 2 for cells with roads, but outside forest areas;
- 5 for cells with forest present, but roads absent;
- 7 for cells with both forest and roads present.

If the value '2' for a road were added to land use codes, the new value for a cell could be the same as that for another land use type (for example, a road value 2 + land use value 2 (water) = 4 (which is the same as the value here for agricultural land)). Thus, the coding of raster images used in overlay is very important, and frequently users employ Boolean images (using only codes 1 and 0) so that algebraic equations will produce meaningful answers.

Polygon-on-polygon analysis is conducted in just the same way (Figure 6.18c). Again, the coding of input layers is the key to understanding the output produced by raster overlay. For example, adding the forest layer and the resort boundary would produce an output layer with the following codes:

- 0 for cells outside the resort boundary and with forest absent;
- 5 for cells outside the resort boundary and with forest present;
- 10 for cells inside the resort boundary and with forest absent;
- 15 for cells inside the resort boundary and with forest present.

The output map is equivalent to a union polygon-on-polygon overlay in vector GIS. Reclassification will produce variants on this, and other overlay operations are available, by multiplying, subtracting or dividing data layers, or by the implementation of a 'cover' or 'stamp' operation (which is equivalent to a 'cookie cutting' operation in vector GIS).

Box 6.6 on p. 196 illustrates some of the options available in raster overlay with reference to the nuclear waste siting case study.

The algebraic manipulation of images in raster GIS is a powerful and flexible way of combining data and organizing analysis. Equations can be written with maps as variables to allow the development of spatial models. Berry (1993) and Chrisman (2002) provide examples, and **Chapter 12** looks at this approach in more detail in the context of designing GIS analysis.

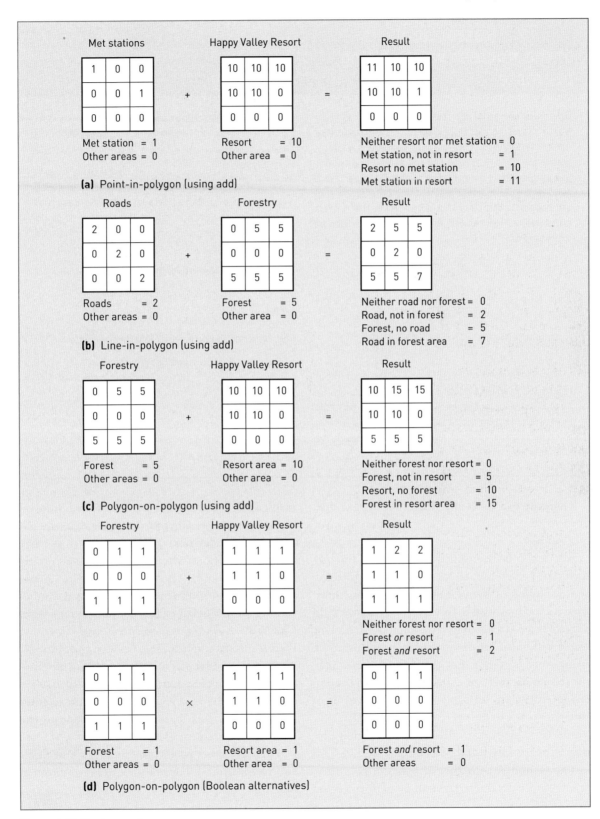

**Figure 6.18** Raster overlays

## BOX 6.6 Raster analysis of the nuclear waste siting example

CASE STUDY

One solution to the process of selecting a nuclear waste site using raster data would be to create four binary data layers. These would represent geology, population density, a road and rail accessibility buffer and areas without conservation status. On these layers cells where the nuclear waste siting criteria are met are coded '1' and cells that do not meet the criteria are coded '0'. The solution to the problem can then be expressed as:

Output cell value = geology × population ×
                    accessibility × conversation

Thus, for a cell in an area meeting all four criteria the value in the output layer will be:

Output cell value = 1 × 1 × 1 × 1
                  = 1

Any cell coded as zero in any of the four input layers will cause the equation above to return a value of zero in the output layer. For example, if a cell falls in an area meeting the first three criteria, but failing the conservation criteria, the output value would be calculated by:

Output cell value = geology × population ×
                    accessibility × conversation
                  = 1 × 1 × 1 × 0
                  = 0

An alternative approach to the overlay process would be to add the data layers. In this case the output cell values would reflect the number of siting criteria met by each cell. The method is summarized below for an output pixel meeting all four criteria:

Output cell value = geology × population ×
                    accessibility × conversation
                  = 1 + 1 + 1 + 1
                  = 4

And, for a cell in an area meeting three criteria:

Output cell value = geology × population ×
                    accessibility × conversation
                  = 1 + 1 + 1 + 0
                  = 3

---

There are two issues affecting raster overlay that need to be considered by users: *resolution and scales of measurement.*

Resolution is determined by the size of the cell used. SPOT satellite data, for example, are collected at a resolution of 10 m (see **Chapter 2**). For some analyses you may wish to overlay a SPOT image with data collected at a different resolution, say 40 m. The result will be an output grid with a resolution of 10 m, which is greater than the resolution at which the second data set was collected. Since you cannot disaggregate data with any degree of certainty, a better approach to the overlay of these two data sets would be to aggregate cells in the SPOT image to match the resolution of the second data set.

The second issue is that of scales of measurement. Nonsensical overlays can be performed on map layers coded using nominal, ordinal, interval and ratio scales (see **Chapter 2**) if these scales are not sufficiently understood. For example, it is possible to add, subtract or multiply two maps, one showing land use coded using a nominal scale (where 1 represents settlement and 2 represents water) and another showing mean annual rainfall coded on a ratio scale. The result, however, is complete nonsense because there is no logical relationship between the numbers. A mean annual rainfall of 1000 mm minus land use type 1 is meaningless! Care is therefore needed when overlaying raster data layers to determine whether the operation makes real sense according to the scales of measurement used.

There are other problems that affect both raster and vector overlay. Four of these are introduced in Box 6.7.

# BOX 6.7 Problems affecting vector and raster overlay

**PRACTICE**

**1** *The modifiable areal unit problem.* The modifiable areal unit problem (MAUP) occurs when arbitrarily defined boundaries are used for the measurement and reporting of spatial phenomena (Openshaw, 1984). Data used in GIS are frequently continuous in nature. This means that the value of the entity recorded varies continuously over space. Population data are an example. Although individual people are measured in population censuses, data are normally reported for areas (for example, in the UK enumeration districts, wards, districts and counties are used). The boundaries of these units are arbitrary (from a statistical point of view) and subject to change. They do not necessarily coincide with breaks in the data. Thus, changing the boundaries of units and disaggregation or aggregation of units can affect the appearance of the data set.

The overlay of two data sets, which both have modifiable units, should be undertaken with care, since overlay will almost inevitably lead to change in boundaries. The Department of the Environment (1987) suggests that results should be confirmed by repeating the analysis with different areal units.

**2** *Ecological fallacy.* A problem associated with the MAUP is the ecological fallacy (Openshaw, 1984). This occurs when it is inferred that data for areas under study can be applied to the individuals within those areas. For example, if analysis of geodemographic data identifies the average income level in a postal zone, it is an ecological fallacy to assume that all the individuals within that postal zone have that level of income since the average income, level will be calculated from a range of incomes.

Only when totally homogeneous data are aggregated into zonal data can the MAUP and the ecological fallacy be avoided. As geographical data are very rarely homogeneous, it is important to recognize these problems when performing overlay analyses on polygon data. How aggregation systems are defined, zones constructed and boundaries drawn will affect the results of analysis (see **Chapter 10**). This problem has long been recognized, and there are many historical examples of boundaries being redrawn or small units aggregated in different ways to produce desired results (Monmonier, 1996).

**3** *Selecting threshold criteria.* In the radioactive waste disposal case study four different map layers are combined in an overlay procedure to identify areas of the UK that simultaneously satisfy certain criteria. The problem is how to define the threshold criteria used for each constraint. For example, the choice of a population density limit of 490 persons per square kilometre is based on an adaptation of the Nuclear Installations Inspectorate (NII) nuclear power station siting guidelines (Openshaw *et al.*, 1989). Who is to say that 490 persons per square kilometre is a better threshold than 500 or even 491? Differences in the thresholds applied will produce differences in the output map. Fortunately, it is possible to investigate how much difference small changes in thresholds make to output maps. Thresholds can be re-applied to one or more of the input layers and overlay analysis repeated to view changes in the output. This form of investigatory analysis is termed 'what if?' modelling. Figure 1.8 shows results obtained from this type of 'what if?' modelling.

**4** *Visual complexity.* Another map overlay problem is visual complexity. When two complex maps are overlain, the output is likely to be more complex than either of the two input maps. The sheer complexity of some GIS outputs makes them very difficult to interpret. Generalization may be necessary, though with generalization comes a loss of detail.

## SPATIAL INTERPOLATION

Spatial interpolation is the procedure of estimating the values of properties at unsampled sites within an area covered by existing observations (Waters, 1989). In an ideal situation a spatial data set would provide an observed value at every spatial location. Satellite or aerial photography goes some way to providing such data; however, more often data are stratified (consisting of regularly spaced observations but not covering every spatial location), patchy (clusters of observations at specific locations) or even random (randomly spaced observation across the study area). The role for interpolation in GIS is to fill in the gaps between observed data points.

A common application of interpolation is for the construction of height contours. Contours on a topographic map are drawn from a finite number of height observations taken from surveys and aerial photographs. The height of the land surface between these points is estimated using an interpolation method and represented on a map using contours. Traditionally, contour maps were produced by hand, but today they are most often drawn by computer. In the old hand–eye method, often referred to as *line threading* or *eye-balling*, contour lines were drawn between adjacent spot heights and divided into the chosen contour interval by assuming that the slope between adjacent spot heights remained constant. This technique suffered from a number of problems:

the inaccurate assumption that slope is constant, human error, subjectivity and the amount of time needed if a large number of data points required interpolation. In the Happy Valley example interpolation could be used to estimate snow depth from observations obtained throughout the valley.

Most GIS packages offer a number of interpolation methods for use with point, line and area data. Box 6.8 presents a widely used classification of interpolation methods that is useful when deciding which method is appropriate for a particular data set (after Lam, 1983; Burrough, 1986).

Whichever interpolation technique is used, the data derived are only an estimate of what the real values should be at a particular location. The quality of any analysis that relies on interpolated data is, therefore, subject to a degree of uncertainty. It is essential for the user to appreciate the limitations of interpolated data when using the results in further GIS analysis. Interpolation techniques are reviewed in full in other books and papers (Lam, 1983; Burrough, 1986; Davis, 1986; Waters, 1989) so brief details of just four of the most frequently used methods – Thiessen polygons, TIN, spatial moving average and trend surfaces – will be presented here. Figure 6.19 shows the results of applying these four methods to a random sample of 100 data points extracted from a digital elevation model (Figure 6.19a).

Thiessen polygons (or Voronoi polygons) are an exact method of interpolation that assumes that the

## BOX 6.8 A classification of interpolation methods

**1** *Local or global.* Global interpolation methods apply a single mathematical function to all observed data points and generally produce smooth surfaces. Local methods apply a single mathematical function repeatedly to small subsets of the total set of observed data points and then link these regional surfaces to create a composite surface covering the whole study area.

**2** *Exact or approximate.* Exact methods of point interpolation honour all the observed data points for which data are available. This means that the surface produced passes exactly through the observed data points, and must not smooth or alter their values. Exact methods are most appropriate when there is a high degree of certainty attached to the measurements made at the observed data points. Approximate methods of interpolation do not have to honour the observed data points, but can smooth or alter them to fit a general trend. Approximate methods are more appropriate where there is a degree of uncertainty surrounding the measurements made at the sample points.

**3** *Gradual or abrupt.* Gradual and abrupt interpolation methods are distinguished by the continuity of the surface produced. Gradual methods produce a smooth surface between sample points, whereas abrupt methods produce surfaces with a stepped appearance. Terrain models may require both methods as it may be necessary to reproduce both gradual changes between observed points – for example, in rolling terrain, and abrupt changes where cliffs, ridges and valleys occur.

**4** *Deterministic or stochastic.* Deterministic methods of interpolation can be used when there is sufficient knowledge about the geographical surface being modelled to allow its character to be described as a mathematical function. Unfortunately, this is rarely the case for surfaces used to represent real-world features. To handle this uncertainty stochastic (random) models are used to incorporate random variation in the interpolated surface.

---

values of unsampled locations are equal to the value of the nearest sampled point. This is a local interpolator, because the global characteristics of the data set exert no influence over the interpolation process. It is also an abrupt method of interpolation as sharp boundaries are present between the interpolated polygons.

Thiessen polygons are created by subdividing lines joining nearest neighbour points, drawing perpendicular bisectors through these lines, and then using these bisectors to assemble polygon edges as shown in Figure 6.20 on p.202 (Laurini and Thompson, 1992). If observed data points are regularly spaced, a regular lattice of square polygons will result. If the observed data points are irregularly spaced, a surface of irregular polygons will be produced (Figure 6.19b).

The most common use of Thiessen polygons is to establish area territories for a set of points. Examples of applications include the transformation of point climate stations to watersheds and the construction of areas of influence around population centres (Laurini and Thompson, 1992; Chrisman, 2002). Although Thiessen polygons can be drawn around

elevation observations, this is not the most appropriate method to use, because elevation data exhibit gradual rather than abrupt properties. This can be seen if you compare the Thiessen polygon interpolated altitude data in Figure 6.19b with the original terrain surface in Figure 6.19a.

A triangulated irregular network or TIN is an elegant way of constructing a surface from a set of irregularly spaced data points. This method of spatial interpolation is often used to generate digital terrain models (**Chapter 3**). An example 3D perspective TIN, created using the same data as in these interpolation examples, is shown in Figure 3.23. The TIN model is an exact interpolation method based on local data points. In this method adjacent data points are connected by lines to form a network of irregular triangles (Figure 6.19c), hence the name. Because the value at each of the data points (forming the corners of the triangles) is known and the distance between these points can be calculated, a linear equation and trigonometry can be used to work out an interpolated value for any other point within the boundary of the TIN. This produces a much better interpolated surface for the re-sampled altitude data

**(a)** Original terrain surface with sample points

**(b)** Thiessen polygons

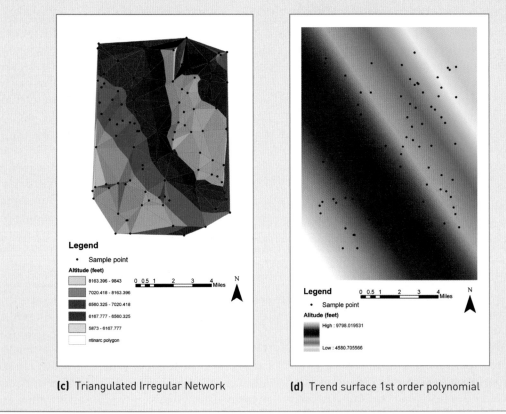

**(c)** Triangulated Irregular Network

**(d)** Trend surface 1st order polynomial

**Figure 6.19** Four different interpolation methods applied to sampled terrain data

**(e)** Trend surface 2nd order polynomial

**(f)** Trend surface 3rd order polynomial

**(g)** Trend surface 4th order polynomial    **(h)** Inverse distance weighted

**Figure 6.19** *Continued*

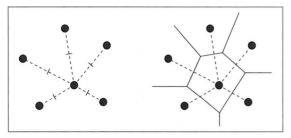

**Figure 6.20** Constructing a Thiessen polygon net

in Figure 6.19a. However, because the TIN method relies on interpolating values between neighbouring data points it is impossible to extrapolate outside the data points making up the 'convex hull' or boundary of the sample points shown in Figure 6.19c.

The spatial moving average is perhaps the commonest interpolation method used by GIS. It involves calculating a value for a location based on the range of values attached to neighbouring points that fall within a user-defined range. This neighbourhood filter is passed systematically across the region of interest, calculating new interpolated values as it goes. This is shown in diagrammatic form in Figure

6.21 for both vector and raster GIS. The spatial average method produces a surface that suppresses the values of known data points to reveal global patterns in the data. This is an example of an approximate point interpolation method since it does not honour the values of known data points and recalculates these values as the filter passes over them.

A circular filter is the most obvious choice, because known data points in all directions have an equal chance of falling within the radius of the interpolated location (Figure 6.21a). In raster GIS a square or rectangular filter is often employed since this is easier to calculate (Figure 6.21b). The size of the filter is based on assumptions about the distance over which local variability is deemed important. A distance-weighting function may be applied to enhance the influence of closer data points. For example, a point 1 km away from the interpolated location will be considered more important than one 2 km away. Burrough (1986) provides details of a range of variations on this method. An interpolated surface, derived using a distance-weighted spatial moving average approach, is shown in Figure 6.19h.

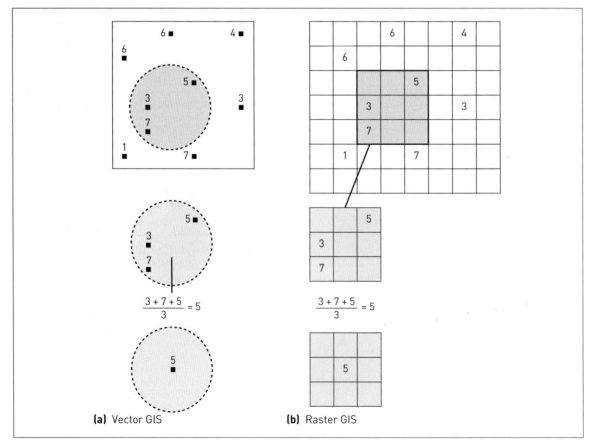

**(a)** Vector GIS   **(b)** Raster GIS

**Figure 6.21** Spatial moving average in (a) vector and (b) raster GIS

The spatial moving average method is best suited to those instances where the values of the known data points are not exact and may be subject to measurement error, but where they nonetheless represent variation in a global pattern. Examples of suitable applications include the interpolation of census data, questionnaire responses and field survey measurements (such as soil pH or rainfall infiltration rates).

The trend surface method attempts to fit a mathematically defined surface through all the data points so that the difference between the interpolated value at a data point and its original value is minimized. It is an example of an approximate and global interpolator. The method does not necessarily honour all of the data points but does use all the data points in its calculation. It is a type of regression model that fits a surface through a three-dimensional set of data points $(x,y,z)$ rather than fitting a line or curve through a two-dimensional set of data points $(x,y)$. This is demonstrated for a two-dimensional example in Figure 6.22. A simple single-order polynomial trend surface fits a flat inclined plane through the data points as shown in Figure 6.19d. By increasing the number of polynomials increasingly complex surfaces can be fitted to the data and the goodness of fit increased. A second-order polynomial fits a surface that is curved in one dimension only to fit the greatest trend of variation in the data (Figure 6.19e). A third-order polynomial fits a surface that is curved in two dimensions (Figure 6.19f), a fourth-order in three (Figure 6.19g), and so on. Increasing the order of the polynomial is limited

by the number of data points and can produce overly complex and unrealistic looking surfaces. Splines are a related method of interpolation that operate in a similar way, except they work on local data points.

Comparing the surfaces generated by these four different methods with the original DEM gives an appreciation of the difficulties involved in the interpolation of point data. Error maps of the absolute difference between estimated elevation and the original DEM can be produced to show where the errors are greatest. Error maps for each of the interpolations shown in Figure 6.19 are shown in Figure 6.23. The TIN method shows the least error overall, but does not allow extrapolation outside the convex hull of the data points. All three methods show low levels of inaccuracy in the areas of low relief, but they perform worse in areas of high relief in the south-west of the DEM. The biggest errors occur around the edge of the image in Figure 6.23b where the Thiessen polygon, trend surface and distance weighted average methods attempt to interpolate beyond the data points. In this example, the random sample of 100 points was taken from the central section of the DEM (Figure 6.23a) to demonstrate the difficulty of interpolation outside the sample region. Even within the sample an 'edge effect' can still be seen. For this reason it is recommended that sample points from outside the region of interest are collected with data that will later be interpolated to produce a surface. This is especially important for the TIN method because it does not allow interpolation outside the convex hull (Figure 6.19c).

# ■ ANALYSIS OF SURFACES

Consideration of techniques available in GIS for surface analysis follows on logically from the discussion of interpolation, since interpolation techniques will invariably have been used to create a surface for analysis. Since the Earth is three-dimensional, it would seem that all GIS applications should include some element of three-dimensional analysis. However, software packages able to handle and analyse true three-dimensional data are limited, and, as **Chapter 3** outlined, analysis in GIS is more likely to be 2.5D, since the surfaces that are produced are simply that – surfaces. There is no underlying or overlying information. This prevents the analysis of, for example, geological or atmospheric data, and

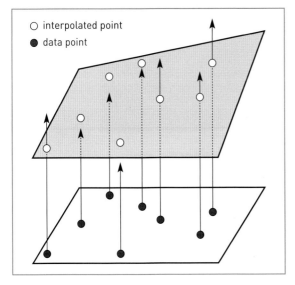

**Figure 6.22** Fitting a trend surface

**(a)** Original terrain surface with sample points

**(b)** Thiessen polygons

**(c)** Triangulated Irregular Network

**(d)** Trend surface 1st order polynomial

**Figure 6.23** Error maps showing interpolation errors for the four methods applied to sampled terrain data in Figure 6.19

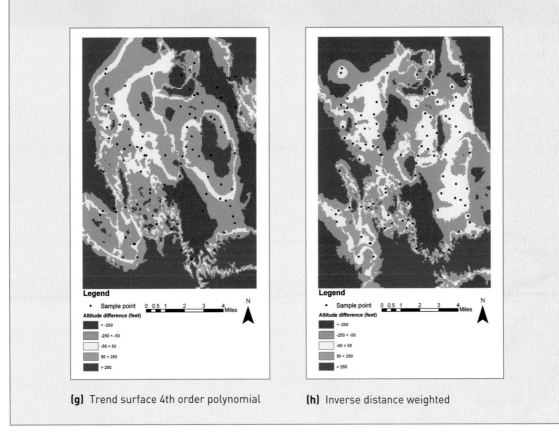

**(e)** Trend surface 2nd order polynomial

**(f)** Trend surface 3rd order polynomial

**(g)** Trend surface 4th order polynomial

**(h)** Inverse distance weighted

**Figure 6.23** *Continued*

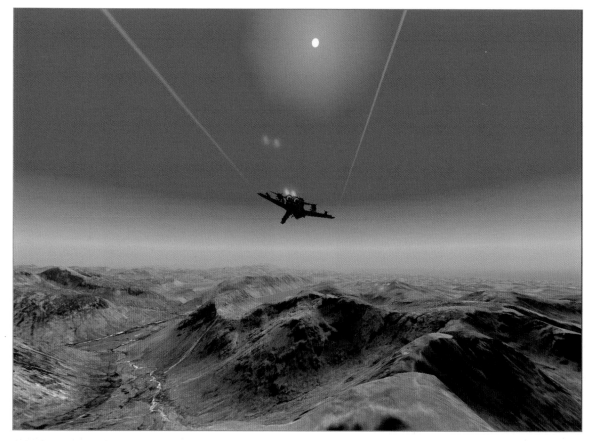

**Figure 6.24** Example of 3D terrain rendering
*Source*: Horizon Simulation Ltd, with permission

even to add realistic-looking trees with height to a GIS terrain model, a CAD or other design package may be necessary (see Figure 6.24).

Despite these limitations, there are still many analysis functions in GIS that operate on surfaces. These range from draping functions that allow the

**Figure 6.25** Visualizing a new ski piste in Happy Valley

results of other analyses to be overlain on a surface, to slope and aspect calculation, and visibility analysis. The selection of a new ski piste site gives some ideas of how these might be applied. Consider first the value of draping the results of the site selection analysis over a terrain surface compared with presentation of the results on a 'flat' map. The visualization of the piste will be much enhanced when it can be seen in context with the terrain it descends. Slope calculations may help with the designation of the piste as an easy or difficult ski run. Aspect calculations will help to analyse the snow retention characteristics. Visibility analysis could be applied to examine what a skier would see from the piste, and to determine from which other locations the ski piste would be visible. Figure 6.25 demonstrates the possibilities with an example from the Happy Valley GIS.

Slope, aspect and visibility are considered here as these are the most commonly used applications of terrain models in GIS. Other applications of DTMs – for example, hydrological modelling – are considered further in **Chapter 7**.

## Calculating slope and aspect

Slope is the steepness or gradient of a unit of terrain, usually measured as an angle in degrees or as a percentage. Aspect is the direction in which a unit of terrain faces, usually expressed in degrees from north. These two variables are important for many GIS applications. As suggested above, both may be required in the planning of a new ski piste in Happy Valley. Slope values may be important for the classification of the slope for skiing. Aspect is important to ensure that the piste selected will retain snow cover throughout the ski season (a completely south-facing slope may not be suitable, since snow melt would be more pronounced than on a north-facing slope in the northern hemisphere).

Slope and aspect are calculated in two ways according to the type of DTM being used. In raster DTMs slope and aspect are calculated using a 3 × 3 window that is passed over a database to determine a 'best fit tilted plane' for the cell at the centre of the window. This allows the calculation of constants for the equation

$$z = a + bx + cy$$

where $z$ = height at the point of interest (the centre of the window), $(x,y)$ = co-ordinates of the point at the centre of the window, and $a, b, c$ = constants.

Slope and aspect for the centre cell can then be calculated ($S$ = slope, $A$ = aspect) using the formulae

$$S = b^2 + c^2$$
$$A = \tan^{-}1\left(\frac{c}{b}\right)$$

There are many other methods of calculating slope and aspect from a DEM including the neighbourhood method, quadratic surface method, best fit plane method and maximum slope method (Wilson and Gallant, 2000).

In the vector TIN model slope and aspect variables are usually calculated using a series of linear equations when the TIN is generated. The equations calculate the slope and aspect of the individual triangles formed by the TIN. Figure 6.26 shows an example of a shaded slope and aspect image.

Slope and aspect may also be calculated as a first step to more complex terrain analysis. Parameters such as analytical hillshading, the rate of change of slope, convexity or curvature (Weibel and Heller, 1991) may be necessary for terrain visualization, landform analysis or classification.

Analytical hillshading is a well-established technique used to depict the general pattern of terrain in maps. It works by calculating the location of shadow and the amount of sun incident on the terrain surface when the sun is in a particular position in the sky. Figure 6.27 shows examples of maps using this method. In the past, these terrain 'highlights' were drawn on a map by a skilled cartographer (see Figure 6.27a and b), but they can now be calculated using a GIS (see Figure 6.27c). This requires a DTM together with the sun's position in the sky. The sun's position is represented by altitude (angle above the horizon) and azimuth (angle relative to north). Analytical hillshading can also be used to derive relative estimates of the amount of sunlight (incoming solar radiation) received by a particular slope or cell in a DEM at a particular time or across a specified time period. This can be useful in modelling snow melt, freeze-thaw cycles, rates of photosynthesis, suitability for certain crop types and yields from solar energy installations. Figure 6.28 gives examples of analytical hillshading models for Happy Valley during the course of a day in the ski season.

Curvature describes the rate of change of slope and/or aspect. Slope curvature can be calculated in plan or profile or a combination of both. Profile curvature is the rate of change of slope relative to a linear or flat profile. Negative curvature indicates a concave slope profile and positive curvature indicates a convex profile. Plan curvature describes the rate of change of aspect relative to a given datum. Negative curvature indicates concave plan form (such as troughs and valleys) and positive curvature indicates convex plan form (such as spurs and ridges). A knowledge of slope curvature is important for studies of slope-forming processes and hydrology as it controls (and is, in turn, determined by) mass movement, soil thickness and runoff and flow rates. Combinations of profile and plan curvature can be used to identify and extract features from a DTM such as peaks, passes, ridges and troughs. This can be an important tool in helping geomorphologists understand landscape form and process (Arrell, 2001; Fisher *et al.*, 2004). Examples for a sample DEM are shown here in Figure 6.29 and Box 6.9 on p. 211 outlines the use of GIS for the identification and measurement of mountains. In the Happy Valley example, the management team might well be interested in slope curvature in order to help model snow accumulation and avalanche risk as well as helping with ski piste design by identifying potential jumps (convex profiles) and compressions (concave profiles).

**(a)** Altitude

**(b)** Slope

**(c)** Aspect

**Figure 6.26** Three-dimensional analysis: shaded slope and aspect image

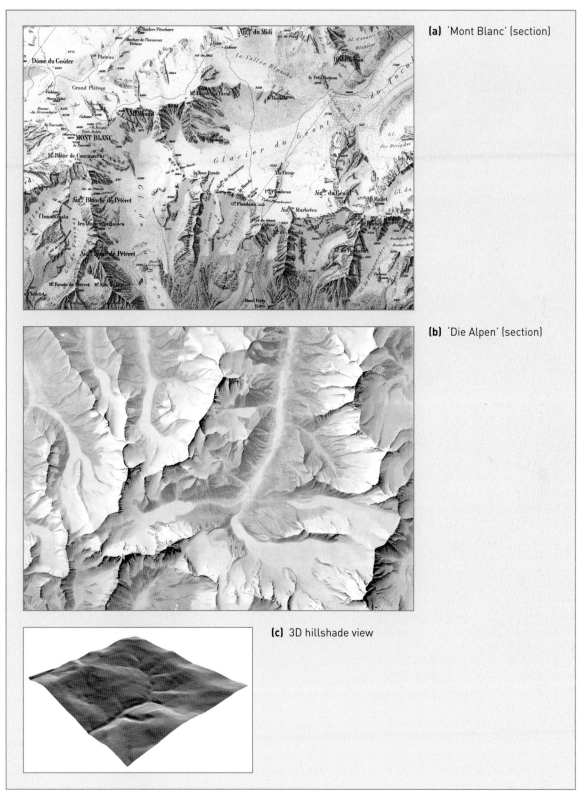

**(a)** 'Mont Blanc' (section)

**(b)** 'Die Alpen' (section)

**(c)** 3D hillshade view

**Figure 6.27** Example of cartographic hillshading techniques
*Sources*: (a, b) Institute of Cartography, ETH Zurich, Switzerland, used by permission

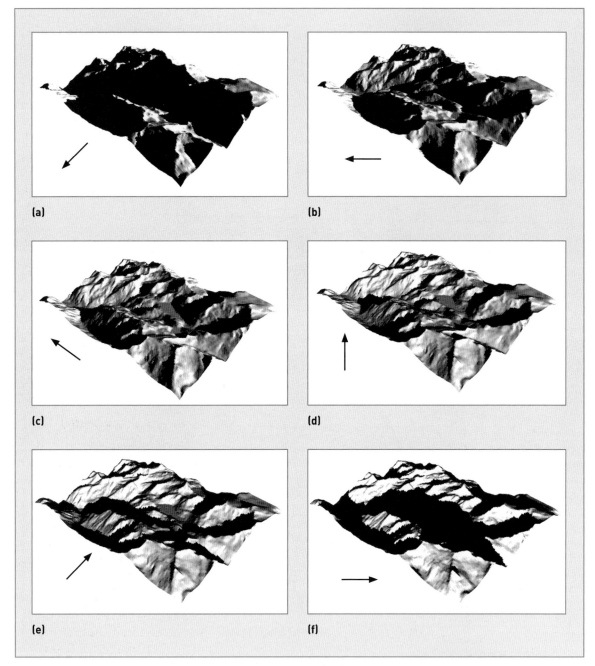

(a)

(b)

(c)

(d)

(e)

(f)

**Figure 6.28** Modelling incoming solar radiation in Happy Valley (a–f) representing morning to evening during a winter's day, arrows show sun direction.

(a) Profile curvature    (b) Plan curvature    (c) Feature extraction

**Figure 6.29** Slope curvature and feature extraction

# BOX 6.9 Identifying mountains with GIS

## Jo Wood

Digital Terrain Models contain information about the height of parts of the Earth's surface by storing elevation at known intervals on the ground. GIS can process this information to identify and measure features in the landscape that have some meaning to us. This case study considers how we can use GIS to identify and measure mountains in a landscape. There are a number of reasons why we might need to identify mountains, ranging from tourism applications that chose footpath routes through a landscape, to the study of the human relationship with the landscape (known as *ethnophysiography*), to the allocation of government subsidies to mountainous regions.

Identifying mountains is not an easy task, because the concept of what we mean by a mountain is not well defined. We might all have an idea of a typical mountain (rocky? snow covered? high? pointed?), but that idea is likely to vary between different people or within different contexts. So, the problem we face when using a GIS is what rules do we use to partition a landscape into areas containing mountain features. Two approaches are considered here: using the values in a DTM to measure shape, and comparing relative height of different points on a DTM.

## USING THE VALUES IN A DTM TO MEASURE SHAPE

We could consider mountains to be parts of a landscape surface that are 'convex-up' in all directions. Measuring such convexity is not dissimilar to measuring slope or aspect: a mathematical function is fitted through the set of elevation values contained within a local 'window' of a DTM. The second derivative of this function (the rate of change of local slope) can be used to measure curvature at the central value. This measure can also be used to identify other features on a surface such as ridges, channels, pits and passes. However, before we can turn convexity measures into classification of mountain regions we have to consider a further problem. We know that the concept of a mountain is not sufficiently modelled by shape alone as we also need to consider something of its size. A landscape is likely to consist of many features that are convex-up, but most of those will be small in extent and height. We can therefore improve our classification process by specifying a scale at which we are willing to consider convex-up features to be possible candidates for mountain features. Figure 6.30 shows 3 scales of feature classification applied to a DTM of the English Lake District.

## BOX 6.9

The finest scale of analysis (Figure 6.30a) that measures features over a 150 m distance identifies large numbers of small features in the landscape. Some of these features may well be considered mountain peaks by some people, but many more are simply too small to provide a useful indication of 'mountainness'. Broader scale analysis (Figure 6.30b) provides a classification that is perhaps closer to a workable definition of a mountain where peaks are associated with named features in the landscape. The broadest scale of analysis (e.g. Figure 6.30c) begins to generalize individual mountain features into larger mountain systems. Such analysis allows us to consider the scale at which we might be most interested. We can extend this idea by comparing feature classifications over a range of scales to generate maps of fuzzy feature classification. We may be more certain that a point on the landscape that is classified as a peak at all scales is truly a peak, than one that only appears to be a peak at one particular scale. This allows the GIS to model the inherent uncertainty present in our vague definition of what makes a mountain.

### COMPARING RELATIVE HEIGHT OF DIFFERENT POINTS ON A DTM

Imagine standing on some point in a landscape that may or may not be the summit of a mountain. Now imagine walking from that point to somewhere else in the landscape that is just higher than your start point. If the path you take is such that it minimizes the amount of height you lose before re-ascending to the higher point, we can record that loss in height ('relative drop') and associate it with the start point.

The lowest point on that path will be located at the highest pass surrounding the start point. A GIS can apply this algorithm to every cell in a DTM with relative ease. Most locations on a surface will have a relative drop of 0 since they will be adjacent to at least one point of higher elevation that does not involve any descent at all. However, local peaks will have non-zero drop heights since, to move from them, you would have to descend some distance to the surrounding pass. The larger the relative drop, the more prominent the peak feature in the landscape. Figure 6.31 shows the relative drops of 5 m or more calculated for all cells in a DTM of the northern Lake District.

The measurement of relative drop allows us to identify peak summits, the significance of which can be controlled by selecting only those above a certain relative drop. The process can be extended by identifying contributing areas surrounding each summit as the area contained within the contour that passes through the highest surrounding pass. These in turn can be calculated for a range of relative drop thresholds to build up a fuzzy peak classification in much the same way as was applied to convexity measures. This gives us an alternative procedure for identifying potential mountains in a region that can accommodate the vagueness in their definition.

GIS will never be able to provide the 'true' objective definition of vague features like mountains. However, appropriate use of GIS allows us to encode a set of rules that can accommodate such vagueness. These rules allow us to consider what it is that contributes to our understanding of the landscape (for example, the role played by shape, the impact of scale, the influence of relative position and height of features).

**(a)** Features at a 3 × 3 pixel scale (150 m)

**(b)** Features at a 31 × 31 pixel scale (1.75 km)

**(c)** Features at a 121 × 121 pixel scale (6 km)

Peak
Ridge
Pass
Channel
Pit

**Figure 6.30** Three scales of DTM feature classification of the northern Lake District using convexity measures to characterize shape. Areas coloured yellow and red can be considered candidates for mountain features.
*Source*: Ordnance Survey

**BOX 6.9**

**Figure 6.31** Summits identified by relative drop from a DTM of the northern Lake District. Numbers indicate relative drop in metres of all summits greater than 5 m above their surroundings. Names indicate peaks identified in Wainwright (2003)
*Source*: Ordnance Survey

**FURTHER INFORMATION**

The analysis described in this case study was carried out with the GIS LandSerf written by Jo Wood. It is freely available from www.landserf.org

Further details on the algorithms used can be found in:

Fisher P, Wood J, Cheng T (2004) Where is Helvellyn? Fuzziness of multi-scale landscape morphology. *Transactions of the Institute of British Geographers* 29 (1): 106–28

Wood J (1998) Modelling the continuity of surface form using digital elevation models. In: Poiker T, Chrisman N (eds) *Proceedings, 8th International Symposium on Spatial Data Handling*, IGU, Vancouver, pp. 725–36

Some discussion on what defines a mountain can be found in the following:

Fisher P, Wood J (1998) What is a mountain? Or the Englishman who went up a Boolean geographical concept but realised it was fuzzy. *Geography* 83 (3): 247–56

Smith B, Mark D (2003) Do mountains exist? Towards an ontology of landforms. *Environment and Planning B: Planning and Design* 30: 411–27

*Source*: Jo Wood, City University, London

## Visibility analysis

One of the common uses of terrain models is in visibility analysis, the identification of areas of terrain that can be seen from a particular point on a terrain surface. This is used in a variety of applications ranging from locating radio transmitters and cellular communications (mobile phone) masts for maximum coverage to minimizing the impact of commercial forestry in protected areas (Fishwick and Clayson, 1995; DeMers, 1997). In Happy Valley, visibility analysis could be used to determine the areas of the resort that would be visible from the top of the proposed new ski piste and so determine its scenic qualities. Conversely, it could be used to determine other locations in the valley that could see the top of

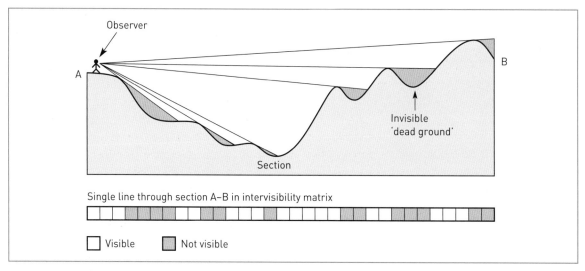

**Figure 6.32** Ray tracing for visibility analysis

the new ski piste, or any other point along the piste and so determine the visual impact of this new development and its associated infrastructure.

DeMers (2005) explains how these analyses work. The location of the observer is connected to each possible target location in the terrain (i.e. a line is drawn from the observer at the top of the ski piste to all other locations in the area). The line, or ray, is followed from each target back to the observer, looking for locations that are higher (Figure 6.32).

Higher points will obscure what is behind them. Thus, through repeated 'ray tracing' a viewshed map can be built (Clarke, 1990). Figure 6.33 shows the viewsheds calculated for both point and line features placed within the DEM data used in Figure 6.19a.

The method for calculating a visibility surface is similar in both raster and vector GIS and in each case the results of analysis would be presented as a viewshed map or Zone of Visual Influence (ZVI). This is a Boolean polygon map that indicates which

**Figure 6.33** Results of viewshed analyses for single point and linear feature

areas of terrain are visible from the location of interest. This map may be more easily interpreted and understood by draping it over the top of a terrain surface for visualization.

Some GIS packages offer more sophisticated visibility analyses that allow the user to specify the height of the viewing position as an offset above the terrain surface height at that location. In this manner it is easy to calculate from where the tops of tall structures such as ski lift pylons, or tall chimneys, can be seen. It also allows the eye level of the viewer to be incorporated in the analysis which, while only small, can have a significant impact on the results. This impact may be particularly noticeable near the edges of a viewshed where viewing angles are particularly acute, and even small height changes in the intervening terrain can make the difference between an object being visible or not.

Some GIS packages allow the incorporation of barrier features, such as buildings and trees, into the visibility analysis to produce a more realistic result based on the computation of 'relief shadows' (areas which lie in the 'shadow' of solid barrier objects). Some GIS even allow you to 'walk' or 'fly' through a terrain model to visualize what the view would be like at various points on or above the DTM. This is enhanced by the ability to drape other data onto the surface of a DTM (such as a land-use map or aerial photograph) and adds texture effects. Thus, the realism of views is increased (see Figure 3.31). The computing overheads for traditional viewshed analyses in GIS tend to be large such that some animations and visualizations require many hours of processing time to develop. Some work has focused on developing faster viewshed tools using parallel processing and grid computing approaches (Kidner *et al.*, 1997) or utilizing more efficient algorithms borrowed from the computer games industry. The latter approach allows real-time viewshed exploration with large, high resolution DEMs on standard desktop workstations.

A short case study detailing the use of GIS and visibility analysis in evaluating the impact of a proposed wind farm development in Australia is presented in Box 6.10.

---

## BOX 6.10 Wind farm visibility in Cape Bridgewater, Australia

CASE STUDY

### Ian Bishop

In 2001 Pacific-Hydro Ltd sought approval to construct what would have then been Australia's largest wind farm: 100 turbines with maximum height of 100 m (67 m to hub, 33 m blades) near Portland in south-western Victoria. The area (Figure 6.34) is blessed with magnificent coastal scenery around the three capes: Cape Bridgewater, Cape Nelson and Cape Sir William Grant.

The development was supported by the local council but opposed by many local residents including the Friends of the Cape who commissioned a number of studies into the visual impact of the proposal. Whilst noise and bird kill were minor issues in the campaign, the main focus of the objections was the impact on the visual amenity of the coastline. The 7 turbines planned for Cape Bridgewater itself were perhaps the single greatest bone of contention, but the Friends generally felt that any part of this development was inappropriate.

A large number of photomontage-based representations of the proposed development were commissioned by Pacific Hydro to illustrate the scale of the installations. However, no visibility mapping was included in the Environmental Effects Statement. The Department of Geomatics at the University of Melbourne was therefore commissioned by Friends of the Capes to provide visibility mapping. Although calibration studies on visual impact existed (for instance, Bishop, 2002) we were asked only to map seen areas and the level of exposure without making any assessment based on effect drop-off with distance and/or visual absorption. This case study describes the visibility analysis undertaken for the Cape Bridgewater proposal.

A GIS-based visibility analysis can be undertaken in a number of ways. In relation to the degree of visibility both the number of visible towers and the extent of their visibility (for example, the top of the blades, whole rotor or whole turbine) can be important. The process itself is simple

→

**BOX 6.10**

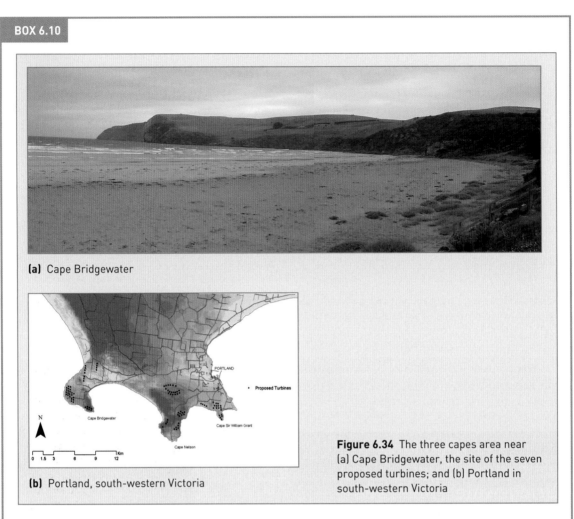

**(a)** Cape Bridgewater

**(b)** Portland, south-western Victoria

**Figure 6.34** The three capes area near (a) Cape Bridgewater, the site of the seven proposed turbines; and (b) Portland in south-western Victoria

enough and the usual algorithms well described in De Floriani and Magillo (1999). The maps here were generated using the 3D Analyst extension to ArcView 3.2 (Ormsby and Alvi, 1999) but other GIS packages can produce a similar result. The simplest analysis involves a single tower (stored in the GIS as a point) with a given height equal to the maximum blade height (100 m). This is assessed in terms of visibility from points 1.7 m (approximate eye height) above the terrain. This shows simply a binary result: areas which can see the top of the turbine and those that cannot. The terrain model (stored in the GIS as a Grid or Raster) used for the analysis can be a simple digital terrain model (DTM) based solely on the land surface, or it can be a digital surface model (DSM) in which the heights (known or assumed) of buildings and vegetation have been added to the terrain in their mapped (hopefully true) locations.

The same mapping can be done at different tower heights. If, for example, the height of the rotor hub is used, the positive visibility means that a viewer can see at least the upper half of the rotor, perhaps more. Non-visibility means that a viewer at the mapped location can see either less of the turbine than this, or none at all. Doing this analysis at several different heights and adding them together gives a mapping of the degree of exposure of the turbine being analysed (Figure 6.35).

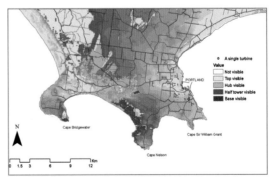

**Figure 6.35** Degree of visibility for the single turbine on Cape Bridgewater

BOX 6.10

(a)                                                  (b)

**Figure 6.36** A comparison of the visibility of the seven Cape Bridgewater turbines based on line of sight alone (a) and when moderated by distance (b). The numbers have been set to a maximum of 70 in each case (up to 10 per turbine) for comparability.

A map showing the number of visible turbines is similar in appearance but easier to construct since the software keeps a record of how many lines of visibility there are between the tower points and the DTM cells. This mapping could also be based on a particular height value (i.e. exposure level) for all the towers. The two types of mapping could be combined, but it would be necessary to make an assumption about visual equivalence. For example, it could be argued that two turbines seen from 80 m upward is equivalent to one turbine seen from the ground upward. Maps of this kind, along with much other evidence, were taken to the administrative tribunal hearing the case. Whether they had any influence on the final decision – to allow the development of all the proposed turbines except the seven on Cape Bridgewater – there is no way to know. It is arguable that such analysis is a good starting point, but it does not tell much about the actual visual impact of the development. The areas of high visibility could, for example, be highly inaccessible, whereas the key scenic zones might have low visibility levels. Such comparisons can be made subjectively from the maps, especially by people familiar with the local environment and social context. However, such factors can also be included in the GIS-based analysis (as done in the case of transmission lines by Hadrian *et al.*, 1988).

The single most meaningful additional variable is distance. Few people would dispute that as a turbine recedes into the distance its visual impact also diminishes. The GIS can deal with this easily since the distance to each turbine is quickly mapped. The more complex part is choosing the appropriate rate of decline of impact. A common option is to assume no decline in impact out to a certain distance and then an immediate drop to zero impact. This is the implicit approach in definition of zones of visual influence (ZVI). Somewhat closer to reality might be a linear decline. Physics suggests the amount of retinal space taken by an object declines as the inverse of the distance-squared. Perceptual studies are inconclusive except to make clear that the rate of decline is less than $1/d^2$ (for example, Hull and Bishop, 1988; Bishop and Miller, 2007). To illustrate how the process can be applied, Figure 6.36 compares the result under no decline with the assumption of decline with distance approximated by $1/d$ and reaching 50 per cent impact at 4 km as suggested by Bishop (2002). Further steps towards a full impact model would require modulation via maps of population density, main travel routes, local and tourist view significance, visual absorption and so on. Perhaps, after all, it is not surprising that many times the process stops with the simple visibility mapping or the ZVI.

*Source*: Ian Bishop, The University of Melbourne, Australia

# ■ NETWORK ANALYSIS

A network is a set of interconnected lines making up a set of features through which resources can flow. Rivers are one example, but roads, pipelines and cables also form networks that can be modelled in GIS. **Chapter 3** considered the structuring of network data in vector GIS. Here we consider the applications and analysis of network data and how answers to networking problems could be obtained using a raster GIS. There are several classic network-type problems, including identifying shortest paths, the travelling salesperson problem, allocation modelling and route tracing. In the Happy Valley context, network analysis applications could be used to find the shortest route from one hotel to another, to plan a circular forest drive, to route waste collection vehicles or to allocate hotel guests to their nearest ski school.

## The shortest path problem

The shortest path between one point and another on a network may be based on shortest distance, in which case either raster or vector GIS could attempt a solution. A raster GIS could provide an answer from a proximity analysis. Impediments to travel can be added to a raster grid by increasing the value of cells that are barriers to travel, then finding a 'least cost' route through a grid. Networks structured in vector GIS offer more flexibility and a more thorough analysis of impediments such as traffic restrictions and congestion. However, the shortest path may not be defined simply in terms of distance. For example, for an emergency vehicle to reach an accident the quickest route may be needed and this may require the traverse of less congested minor roads.

Shortest path methods work by evaluating the links and turns required to traverse a network between required stops. Several potential paths may be considered before the route with the least cumulative impedance is constructed from the intervening network. This process is repeated for all required stops until the whole journey path has been defined. Such a method might be used to identify the quickest route from the centre of a resort to a new ski piste.

## The travelling salesperson problem

The travelling salesperson problem is a common application of network analysis. The name arises from one application area where a salesperson needs to visit a specific set of clients in a day, and to do so by the best route (usually the quickest). The waste collection vehicle has the same problem – it needs to visit all the hotels in Happy Valley, then return to base. In each case the question is 'In which order should the stops be visited, and which path should be taken between them?' This is a complex computing task. Imagine a situation where the delivery van has to visit just 10 customers. The possible number of combinations in which the 10 customers could be visited is 9 factorial (9! or $9 \times 8 \times 7 \times 6 \times 5 \times 4 \times 3 \times 2 \times 1 = 362,880$). Add to this the difficulties of determining the route with the least cumulative impedance and this turns into a hefty computing problem. In GIS network analysis the ordering of the stops can be determined by calculating the minimum path between each stop and every other stop in the list based on impedance met in the network. A heuristic (trial and error) method can then be used to order the visits so that the total impedance from the first stop to the last is minimized. In the same way, the calculation of a route for a scenic drive could be aided by the inclusion of specified stops that need to be visited in a particular order.

## Location-allocation modelling

Network analysis may also be used for the allocation of resources by the modelling of supply and demand through a network. To match supply with demand requires the movement of goods, people, information or services through the network. In other words, supply must be moved to the point of demand or vice versa. Allocation methods usually work by allocating links in the network to the nearest supply centre, taking impedance values into account. Supply and demand values can also be used to determine the maximum catchment area of a particular supply centre based on the demand located along adjacent links in the network. Without regard for supply and demand limitations, a given set of supply centres would service a whole network. If limits to supply and demand levels are indicated, then situations can arise where parts of the network are not serviced despite a demand being present. Consider an example in which five ski schools service the network shown in Figure 6.37. If we assume that clients always enrol at the nearest school (based on link and turn impedance values) and there are no limits to the number of places at each school (i.e. unlimited supply), then each link in the network is assigned to the nearest school, and clients staying at hotels along those links will be allocated with the link. If the supply of ski school places is limited at one or all of the schools, the pattern of allocated network links changes according to the nearest ski school with free places. It is possible to have a situation where all the places are filled, leaving some clients without a ski school place. In this case a new ski school needs to be built in the unserviced area, or the number of places increased at one or more of the existing ski schools.

Allocation modelling in network data is the basis for more advanced analyses such as location-allocation modelling. This can be used to identify the optimum location for a new centre to service a shortfall in supply relative to demand. In the above example, demand for ski school places outstrips supply such that parts of the network are left unserviced. If it is not possible to increase supply at existing ski schools, a new school must be built. The obvious location is within the unserviced area identified above, but identifying the optimum location becomes another network problem. This is because, once built, the new school will radically alter the

pattern of ski school catchment areas throughout the whole network. This is a more complex modelling problem that is revisited in **Chapter 7**.

## Route tracing

The ability to trace flows of goods, people, services and information through a network is another useful function of network analysis. Route tracing is particularly useful for networks where flows are unidirectional, such as stream networks, sewerage systems and cable TV networks. In hydrological applications route tracing can be used to determine the streams contributing to a reservoir or to trace pollutants downstream from the site of a spillage. Route tracing can be used to find all the customers serviced by a particular sewer main or find those affected by a broken cable.

Connectivity, the way network links join at network nodes, is the key concept in route tracing. Without the correct connectivity in a network, route tracing and most other forms of network analysis would not work. Directionality is also important for route tracing as this indicates the direction in which the materials are moving along the network. Knowledge of the flow direction is critical to establishing upstream and downstream links in the network. This gives rise to the concept of a directed network in which each link in the system has an associated direction of flow. This is usually achieved during the digitizing process by keeping the direction of digitizing the same as the flow direction in the network.

Once a directional network has been established, tracing the links downstream or upstream of a point on the network is simply a question of following or going against the flow, respectively.

## Quantitative spatial analysis

Quantitative spatial analysis allows ideas about spatial processes and patterns to be tested and is used to help find meaning in spatial data. Quantitative analysis methods can (after Fotheringham *et al.*, 2000):

- reduce large data sets to smaller amounts of more meaningful information;

- explore data to suggest hypotheses or examine the distribution of data. Exploratory data analysis (EDA) techniques are used for this; and

**Figure 6.37** Network data analysis: allocation of ski school to clients on a network

- explore spatial patterns, test hypotheses about these patterns and examine the role of randomness in their generation.

Quantitative methods for spatial data analysis developed distinct from other statistical methods used for non-spatial data because of the distinguishing features of spatial data. Observations of spatial data are not independent, but it is assumed that features that are close together in geographical space are in some way related (this is Tobler's 'First Law of Geography'). Different results can also be obtained when applying the same technique to the same distribution of data, simply by varying the spatial units used (this problem is known as the Modifiable Areal Unit Problem – see Box 6.7).

In some ways GIS has helped to revive the quantitative tradition in the spatial sciences, but it is interesting to note just how few statistical analysis tools there are in GIS software packages. As a result, many users turn to external statistical packages such as R, Minitab and SPSS to perform even relatively simple quantitative analyses (see Figure 6.38). However, all GIS can provide useful data for statistical

analyses simply by outputting spatially referenced data in a form that can be read by another package.

There are three main types of quantitative spatial methods:

1 *Exploratory and descriptive statistics*. These can be used to describe the distribution of spatial phenomena. For example, a histogram may be created to show the distribution of land parcels of different sizes. For exploratory and descriptive statistical models a GIS can provide: attribute data including line length, polygon area and perimeter; distance from one point observation to another point, line or area feature; distances to nearest observations for input into nearest neighbour statistics or cluster analysis; and location of point observations relative to area data for chi-squared analyses.

2 *Predictive statistics*. These are used to look at relationships between spatial phenomena. For example, regression analysis might be used to look at the relationship between altitude and vegetation type. In this instance, geographical data sets of altitude and vegetation type, biomass and diversity can be overlaid in the GIS to provide a scatterplot

**Figure 6.38** R interface
*Source*: The R Project for statistical computing, www.r-project.org

of paired observations that show the relationship between the dependent variable (vegetation type, biomass and diversity) and the independent variable (altitude). The scatterplot can be used for regression analysis and the subsequent model used to predict the vegetation type for areas where there are no data. Multiple input layers of dependent and independent variables can be used as the basis for multiple regression models.

3 *Prescriptive statistics.* These are used to ask 'what if?' questions. They help with the prediction of what might happen in a particular set of circumstances. For example, optimization methods could be used to look at possible market analysis scenarios to inform retail decision-making.

Each analysis technique may be applied at a local or global level. Global application to a whole data set (for example, calculating average rainfall for a whole country) will mask regional and local variations. Local application to parts of a data set (for example, calculating average rainfall for each state or county) provides statistics that can easily be displayed with GIS and that can be used to search for exceptions and 'hotspots'. It is also possible to geographically weight a regression analysis so that greater emphasis is placed on the relationship between local observations (Fotheringham *et al.*, 2000; Fotheringham *et al.*, 2002).

## REFLECTION BOX

- Eight interpolation methods are introduced in Box 6.8 (A classification of interpolation methods). Describe briefly a real or possible application for each of these methods.
- Surfaces can be created from population sampling points to represent population distributions. Could the surface analysis operations introduced in this chapter be used on population surfaces?

Review the section on the analysis of surfaces and make notes on how the operations covered could be applied to population surfaces.
- Imagine you are helping to design a GIS for an emergency service (police, fire or ambulance service). What requirements for analysis might the organization have? How can network operations be used in such an application?

## CONCLUSIONS

This chapter has covered the most common methods of data analysis available in current GIS packages. The next chapter will delve further into some of the more sophisticated modelling capabilities of GIS. However, it is now possible to see how analysis functions might be used singly or together to produce information that can be used to assist decision-makers. The chapter started with some suggested questions for the Happy Valley GIS team. It is useful to revisit these questions, and consider the GIS functions needed to address them. Table 6.5 indicates that there is a choice of methods in many cases, giving more than one way of obtaining an answer to a question. This is true for many GIS problems, and users need to be flexible in their approach to analysis and able to mix and match the functions available to obtain a suitable solution

to their problem. In addition, users need to be aware of the strengths and weaknesses of different types of GIS (raster and vector) for data analysis, and the limitations of some of the functions they routinely apply (for example, length measurements). However, an informed GIS analyst should be able to produce useful answers by combining data analysis functions into an analysis schema. How to do this will be covered further in **Chapter 12**.

Data analysis is an area of continuing development in GIS; as users demand more sophisticated analysis, the vendors attempt to satisfy them. Specialist software packages have emerged which concentrate on only a few of the functions described in this chapter (there are, for example, several examples of packages geared towards network analysis

## CONCLUSIONS

or terrain analysis). New analysis tools frequently start out in the academic domain. For example, the book by Fotheringham and Rogerson (1994), which laments the lack of spatial analysis functions (even exploratory data analysis functions like the calculation of means and production of charts and graphs), suggests some areas for development. There are continued calls for improved GIS functionality, particularly from those interested in spatial analysis. National research agencies have placed spatial analysis on their research agendas (National Research Council, 1997; Harding and Wilkinson, 1997). However, the range of functions currently available is sufficient to meet the needs of the majority of GIS users and if it is not, it is relatively easy to link to specialist packages to find a solution.

TABLE 6.5 **Happy Valley GIS questions and possible solutions**

| Happy Valley GIS question | Suggested GIS functions required to address this question |
| --- | --- |
| Which is the longest ski piste in Happy Valley? | Measurement of length or Query of attribute database (if length is included as an attribute) |
| What is the total area of forestry in the valley? | Area measurement or Reclassification of forestry map, then area calculation |
| How many luxury hotels are there in the valley? | Attribute (aspatial) data query |
| Where are the luxury hotels with less than 20 bedrooms? | Attribute query followed by spatial query or Combinatory AND query |
| Which hotels are within 200 m of a main road? | Buffering, then point-in-polygon overlay, then query or Proximity analysis, then reclassification, then point-in-polygon overlay and query |
| In which land use types are the meteorological stations located? | Point-in-polygon overlay |
| Which roads should I use for a scenic forest drive? | Line-in-polygon overlay |
| What is the predominant land use type in the Happy Valley resort? | Polygon-on-polygon overlay (identity), then query |
| Where could a new ski piste be located? | A map algebra overlay, depending on siting criteria selected |
| What is the predicted snowfall for the new ski piste? | Interpolation, then overlay |
| What are the slope and aspect of the new ski piste? | Three-dimensional analysis – slope and aspect calculation |
| From where will the new ski piste be visible? | Three-dimensional analysis – visibility analysis |
| What is the relationship between hotel room prices and proximity to ski lift stations in the Happy Valley resort? | Quantitative spatial statistics |
| What is the quickest route for the ski bus to pick up 5 sets of customers scattered across the Happy Valley resort? | Network analysis |

## REVISION QUESTIONS

- Using examples, discuss the different types of queries used in GIS.

- Suggest 10 applications for buffering and justify your choice.

- Outline the ways in which overlay is used for the integration of data.

- Discuss the advantages and disadvantages of adopting either a vector or a raster approach to data analysis.

- What is interpolation? Why are interpolation techniques included in GIS?

- How can surface data be analysed in GIS?

- Propose a methodology, using data analysis techniques, for the selection of a site for a new ski resort. You have the following data available: land use (classified into rural and urban), roads (motorways, highways and minor roads), meteorological data (point met stations with average snowfall) and terrain (a 2.5D model). Your proposed site should be in a rural area, with good accessibility by car and bus, in an area of high snowfall and with good scenic potential.

## FURTHER STUDY – ACTIVITIES

- Sketch out some very simple lines and areas on a piece of graph paper and use both raster and vector GIS principles to calculate their length, area and perimeter, respectively. Are there any differences between the different methods?

- Draw a simple pattern of points, a line network and a polygon on a piece of A4 or letter-sized paper using coloured pens. Draw a second complex polygon on a sheet of tracing paper or an overhead transparency and overlay this onto your first sheet of paper. Use four tick points on the paper and overlay to keep them aligned (alternatively you could tape them together with sticky tape). Using coloured pens manually perform a point-in-polygon, line-in-polygon and polygon-on-polygon overlays.

- Imagine you are designing a site search for a new ski resort to compete with the existing Happy Valley resort. Draw a flow chart describing the steps and operations you would take. Assume the key factors are proximity to urban areas (market demand), altitude (reliable snow cover), proximity to road network (accessibility) and conservation area status (planning restrictions).

- Draw a small number of x,y,z data points (approximately 20) on graph paper and create (a) a Thiessen polygon network and (b) a contour map using eye-balling (line threading) techniques. You could use some real data for this such as meteorological station records or spot heights.

- Download and install one of the readily available free terrain analysis software packages such as Landserf, TAS or SAGA (see web links below). Experiment with the different variables you can derive from sample DEM data (such as slope, aspect, curvature) or downloaded data for your area.

- Calculate an intervisibility matrix for the line connecting two points (a,b) selected at random from a topographic contour map using the ray-tracing technique (hint: draw the terrain profile along this line on graph paper to help you and use a ruler to identify visible and invisible slopes). Is point b visible from point a? Which parts of the intervening terrain surface are visible from point a? What would be the effect of adding an offset of 100 m to point b?

## FURTHER STUDY – READING

There are many opportunities to follow up material covered in this chapter. Many GIS textbooks devote useful space to data analysis (for example Laurini and Thompson, 1992; Worboys, 1995; Chrisman, 2002; Burrough and McDonnell, 1998; DeMers, 2005) and there are a few books devoted completely to spatial analysis and GIS (Fotheringham and Rogerson, 1994; Chou, 1996; Mitchell, 1999; Fotheringham *et al.*, 2000).

Map overlay is a technique that has enthralled cartographers and geographers for some time. There are several texts that will help you understand the origins and basics of the method. These include Ian McHarg's now famous book *Design with Nature* (1969) and the writings of geographers such as Waldo Tobler and Mark Monmonier (Tobler, 1976; Monmonier, 1996). The difficulties of automating map overlay operations occupied some of the finest geographical and computer minds in the 1970s, including Nick Chrisman and Mike Goodchild. More recent texts covering the subject include Berry (1993) and Chrisman (2002). The modifiable areal unit problem (MAUP) is described in detail by Openshaw (1984) and a good example of applied GIS overlay technique is given by Openshaw *et al.* (1989) in their pursuit of a solution to the nuclear waste disposal example that we have been using in this book.

For an excellent review of spatial interpolation methods see Lam (1983). Other useful texts include Burrough and McDonnell (1998) and Davis (1986).

A readable review of GIS applications in network analysis relevant to transportation problems is provided by Vincent and Daly (1990). Short sections on network analysis can also be found in Laurini and Thompson (1992) and Chrisman (2002). Spatial interaction modelling is covered by Wilson (1975) and Fotheringham and O'Kelly (1989).

Surface analysis in GIS is a popular subject on which there are many papers and texts. A good starting point for further reading is the book *Mountain Environments and GIS* edited by Price and Heywood (1994), which contains many examples of DTMs at work. This book also contains a concise overview of DTMs by Stocks and Heywood (1994). Alternatively, Maguire *et al.* (1991) contains two relevant chapters on DTMs and three-dimensional modelling (Weibel and Heller, 1991; Raper and Kelk, 1991). A comprehensive review of terrain analysis and its applications can be found in Wilson and Gallant (2000).

Longley *et al.* (2010) offer an excellent chapter on advanced spatial analysis, which covers descriptive statistics, optimization methods and hypothesis testing. Fotheringham *et al.* (2000) explain the role of GIS in spatial analysis and explore current issues in spatial analysis.

Berry J K (1993) *Beyond Mapping: Concepts, Algorithms and Issues in GIS*. GIS World Inc., Colorado

Burrough P A, McDonnell, R (1998) *Principles of Geographical Information Systems*. Oxford University Press, Oxford

Chou Y-H (1996) *Exploring Spatial Analysis in Geographic Information Systems*. OnWord Press, USA

Chrisman N R (2002) *Exploring Geographic Information Systems*. 2nd edn. Wiley, New York

Davis J C (1986) *Statistics and Data Analysis in Geology*. Wiley, New York

DeMers M N (2005) *Fundamentals of Geographic Information Systems*. 3rd edn. Wiley, New York

Fotheringham A S, O'Kelly M E (1989) *Spatial Interaction Models: Formulations and Applications*. Kluwer Academic, Dordrecht

Fotheringham A S, Rogerson P (eds) (1994) *Spatial Analysis and GIS*. Taylor and Francis, London

Fotheringham A S, Brunsdon C, Charlton M (2000) *Quantitative Geography: Perspectives on Spatial Data Analysis*. Sage, London

Lam N S (1983) Spatial interpolation methods: a review. *American Cartographer* 10: 129–49

Laurini R, Thompson D (1992) *Fundamentals of Spatial Information Systems*. Academic Press, London

Longley P A, Goodchild M F, Maguire D J, Rhind D W (2010) *Geographical Information Systems and Science*. 3rd edn. Wiley, Chichester

Maguire D J, Goodchild M F, Rhind D W (eds) (1991) *Geographical Information Systems: Principles and Applications*. Longman, London

McHarg I L (1969) *Design with Nature*. Doubleday, New York

Mitchell A (1999) *The Esri Guide to GIS Analysis: vol. 1: Geographic Patterns and Relationships*. Esri Press, Redlands

Monmonier M (1996) *How to Lie with Maps*. University of Chicago Press, Chicago

Openshaw S (1984) *The Modifiable Areal Unit Problem: Concepts and Techniques in Modern Geography*, vol. 38. GeoBooks, Norwich, UK

Openshaw S, Carver S, Fernie J (1989) *Britain's Nuclear Waste: Safety and Siting*. Belhaven, London

Price M F, Heywood D I (eds) (1994) *Mountain Environments and Geographical Information Systems*. Taylor and Francis, London

Raper J F, Kelk B (1991) Three-dimensional GIS. In: Maguire D J, Goodchild M F, Rhind D W (eds) *Geographical Information Systems: Principles and Applications*. Longman, London, vol. 1, pp. 299–317

Stocks A M, Heywood D I (1994) Terrain modelling for mountains. In: Price M F, Heywood D I (eds) *Mountain Environments and Geographical Information Systems*. Taylor and Francis, London, pp. 25–40

Tobler W R (1976) Spatial interaction patterns. *Journal of Environmental Systems* 6: 271–301

Vincent P, Daly R (1990) GIS and large travelling salesman problems. *Mapping Awareness* 4(1): 19–21

Weibel R, Heller M (1991) Digital terrain modelling. In: Maguire D J, Goodchild M F, Rhind D W (eds) *Geographical Information Systems: Principles and Applications.* Longman, London, vol. 1, pp. 269–97

Wilson A G (1975) Some new forms of spatial interaction model: a review. *Transportation Research* 9: 167–79

Wilson J P, Gallant J (2000) *Terrain Analysis: Principles and Applications.* John Wiley, London

Worboys M F (1995) *GIS: A Computing Perspective.* Taylor and Francis, London

## WEB LINKS

### GIS and spatial analysis:

- WterontheWeb – Interactive online map service: waterontheweb.org/under/gis/IMSindex.html

- Query function explained at: waterontheweb.org/under/gis/query.html

- Ordnance Survey GIS Files: www.ordnancesurvey.co.uk/oswebsite/gisfiles/section4/

- Geospatial analysis website and book: www.desmith.eu/MJdS/output/GeospatialAnalysis.htm

### Software for surface analysis:

- LANDSERF: www.soi.city.ac.uk/~jwo/landserf/

- Whitebox geospatial analysis tools: www.uoguelph.ca/~hydrogeo/Whitebox/index.html

- SAGA: www.saga-gis.org/en/index.html

### Software for statistical analysis:

- R: www.r-project.org/
- GWR: ncg.nuim.ie/ncg/GWR/index.htm

Visit our website **www.pearsoned.co.uk/heywood** for further web links, together with self-assessment questions, activities, data sets for practice opportunities and other interactive resources.

# Analytical modelling in GIS

## Learning outcomes

By the end of this chapter you should be able to

- Explain what a 'model of spatial process' is

- Identify the most common types of process model

- Describe how process models are implemented in GIS

- Give examples of how GIS has been used in the modelling of physical and human processes

- Define diffusion modelling and explain where it can be used

- Describe multi-criteria evaluation and how it is implemented in GIS

- Outline the main problems associated with the use of GIS to build process models

- Describe how public input can be incorporated in GIS analysis

## ■ INTRODUCTION

**Chapters 3** and **6** explored how models of spatial form are represented and analysed using GIS. These models can be used in many ways in data analysis operations; however, they tell us nothing about the processes responsible for creating or changing spatial form. Population change, consumer spending patterns, soil erosion and climate change are examples of processes that have an impact on spatial form. For example, population change affects the growth or decline of a city; changes in spending patterns influence the patterns of out-of-town retailing; soil erosion affects agricultural yields and farming practices; and climate change affects the frequency of flooding experienced in low lying areas. These processes are difficult to represent, partly because they are usually complex. However, if we wish to include them in our decision-making processes, we need to develop models to help us. The models we use are called 'process models'. If we return to the car analogy we now realize that there is more to driving than just switching on the ignition and turning the steering wheel. We need to be able to estimate how far we can go on a tank of fuel or predict what will happen if we hit the brakes too hard on an icy road. In the same way, using a GIS in conjunction with a process model of population change may help us to predict where growth will occur so that we can make plans to meet future needs and take decisions about appropriate action. In this 'What if?' modelling a GIS and a linked model are used together to ask questions about the likely outcomes of particular plans, actions or decisions. For example, by coupling a hydrological model with a GIS we can ask 'If we build a dam here, what will be the impact on river flows, water temperature and sediment yields?' or if we link a transport model with land use data 'If we build a retail superstore here what will happen to local traffic flows during peak periods?'.

This chapter provides a summary of process models before considering how they can be implemented in GIS. These models are then approached from an applications perspective, and three examples are examined: physical process models, human process models and decision-making models. Case studies are used to illustrate the real-world applications of these models. To conclude, the chapter considers some of the advantages and disadvantages of using GIS to construct spatial process models.

## ■ PROCESS MODELS

A process model simulates real-world processes. There are two reasons for constructing such a model. From a pragmatic point of view' decisions need to be made and actions taken about spatial phenomena. Models help this process. From a philosophical point of view a process model may be the only way of evaluating our understanding of the complex behaviour of spatial systems (Beck *et al.*, 1995). In GIS process models may be used in either role. For example, in the nuclear waste and house-hunting case studies GIS has been used to help the decision-making process. In the Zdarske Vrchy case study GIS was used to aid understanding of complex ecological and environmental processes such as the effects of acid rain on forest ecosystems, and the effects of habitat change on the endangered black grouse.

There are many different approaches to process modelling. To decide which approach is appropriate in a particular situation, an understanding of the range of models available and their strengths and weaknesses is required. Thus, a classification of process models is a good first step.

It is a relatively simple task to classify process models into one of two types: *a priori* or *a posteriori* (Hardisty *et al.*, 1993). *A priori* models are used to model processes for which a body of theory has yet to be established. In these situations the model is used to help in the search for theory. Scientists involved in research to establish whether global warming is taking place would perhaps use *a priori* models, as the phenomenon of global warming is still under investigation.

*A posteriori* models, on the other hand, are designed to explore an established theory. These models are usually constructed when attempting to apply theory to new areas. An *a posteriori* model might be developed to help predict avalanches in a mountain area where a new ski piste has been proposed. Avalanche formation theory is reasonably well established, and several models already exist which could be applied to explore the problem.

It is possible for a model to change from *a posteriori* to *a priori* and vice versa. In the avalanche prediction example, the existing model may fail to explain the pattern of avalanche occurrences. If so, it would be necessary to adapt the model to explore the new situation and thus contribute to theory on the formation of avalanches.

Beyond the *a priori/a posteriori* division, developing a further classification of process models becomes quite complex. However, two useful classifications (Hardisty *et al.*, 1993; Steyaert, 1993) have been integrated here to provide a starting point for examining the different types of models. The classification includes natural and scale analogue models, conceptual models and mathematical models.

## Natural and scale analogue models

A natural analogue model uses actual events or real-world objects as a basis for model construction (Hardisty *et al.*, 1993). These events or objects occur either in different places or at different times. A natural analogue model to predict the formation of avalanches in the previously unstudied area of a new ski piste might be constructed by observing how avalanches form in an area of similar character. The impact that avalanches would have on the proposed ski piste could also be examined by looking at experiences of ski piste construction in other areas (see Figure 7.1).

There are also scale analogue models (Steyaert, 1993) such as topographic maps and aerial photographs, which are scaled-down and generalized replicas of reality. These are exactly the kind of analogue models that GIS might use to model the avalanche prediction problem.

**Figure 7.1** Natural analogue model for predicting avalanche hazard

## Conceptual models

Conceptual process models are usually expressed in verbal or graphical form, and attempt to describe in words or pictures quantitative and qualitative interactions between real-world features. The most common conceptual model is a systems diagram, which uses symbols to describe the main components and linkages of the model. Figure 7.2 shows a conceptual model of the avalanche prediction problem as a systems diagram. Conceptual modelling is returned to in **Chapter 12** as a framework for project management in GIS.

## Mathematical models

Mathematical process models use a range of techniques including deterministic, stochastic and optimization methods. In deterministic models, there is only one possible answer for a given set of inputs. For example, a deterministic avalanche prediction model might show a linear relationship between slope angle and size of avalanche. The steeper the slope, the smaller the avalanche which results, since snow build-up on the slope will be less. This model might be created by developing a least-squares regression equation for slope angle against avalanche size (see Figure 7.3 on p. 230). Such deterministic models work well for clearly defined, structured problems in which a limited number of variables interact to cause a predictable outcome. However, few simple linear relationships exist in geographical phenomena. In most situations there is a degree of randomness and/or uncertainty associated with the outcome. This is true in the avalanche example, as slope angle is only one factor amongst many that must be taken into account when trying to predict the potential size of an avalanche. Where there is uncertainty about the nature of the process involved, a mathematical model, known as a *stochastic model*, is needed.

Stochastic models recognize that there could be a range of possible outcomes for a given set of inputs, and express the likelihood of each one happening as a probability. We know that slope angle and size of avalanche are related, but also that the problem is much more complex than suggested by our deterministic model. In the deterministic model the assumption was made that as slopes

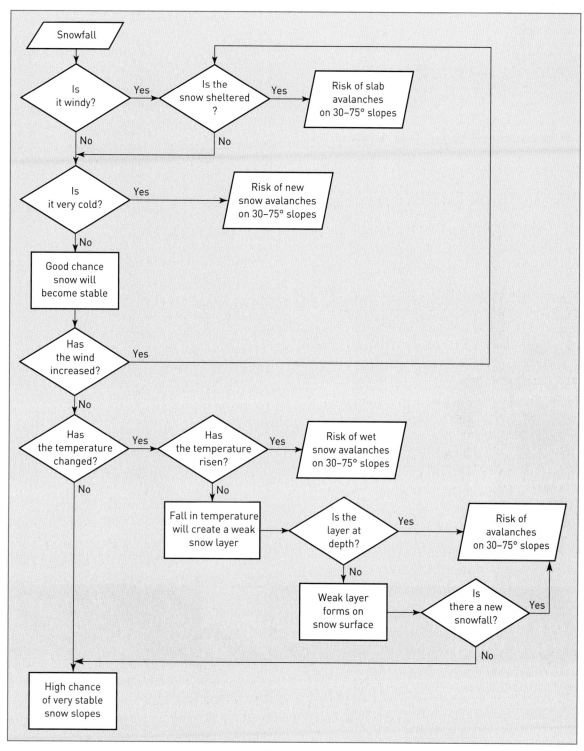

**Figure 7.2** Simplified conceptual model of avalanche prediction

get steeper there will be less build-up of snow and smaller avalanches. However, in reality other variables will be involved: for example, direction of slope, exposure to wind, changes in temperature and underlying topography. The predicted size of an avalanche is based on the probability of a number of these factors interacting in a particular place at a particular time.

**Figure 7.3** Regression model of slope angle against avalanche size

Steyaert (1993) notes that both deterministic and statistical models can be subdivided into steady-state and dynamic models. Steady-state models are fixed for a particular time and dynamic models have at least one element that changes over time. Our avalanche prediction model is clearly a dynamic model, since time plays an important role in the changing nature of the snow cover and hence when an avalanche will occur (see Figure 7.2). For example, changes in temperature over time will affect the stability of the snow and risk of avalanche.

The final type of mathematical model is an optimization model. These models are constructed to maximize or minimize some aspect of the model's output. Assuming a model has been developed which shows where and when avalanches are likely to occur, an optimization model could be used to help identify the area of minimum avalanche risk at a given time. The output from this model could then be used to identify ski routes that minimize exposure to avalanche risk.

## Process modelling and GIS

In GIS all three approaches – natural and scale analogue, conceptual and mathematical modelling – are used to model spatial processes. They may be used in isolation, substituted for each other in an iterative development process or combined in a larger, more complex model. Box 7.1 presents a possible scenario for the development of a snow cover prediction model in Happy Valley.

The example in Box 7.1 shows how different modelling techniques can be used together to build up complex models of spatial processes. Unfortunately, proprietary GIS software provides few, if any, process models as part of the standard set of functions. In many ways this is understandable since process models are heavily adapted to meet the requirements of particular applications. Thus, generic models, which could be made available in GIS, would be far too inflexible for widespread use. However, there are some generic modelling functions in GIS. In addition, many of the analytical functions found in GIS (**Chapter 6**), when coupled with functions provided by other modelling software, provide an environment for constructing application-specific models (Box 7.2).

## BOX 7.1 The development of a snow cover model for Happy Valley

CASE STUDY

The Happy Valley GIS team has attempted to construct a model of snow cover in the valley to assist the siting of a new ski piste. The ski piste must be in an area where snow cover is reliable throughout the winter. To assist the decision where to site the new piste the following models were developed:

**1** The first model of seasonal snow cover variation was developed using GIS. A series of digitized aerial photographs was used to establish how the snow cover in the valley had changed over time. This sequence

of photographs was used as a scale analogue model to predict future snow cover patterns. However, the results were unsatisfactory, as other factors clearly needed to be considered. The valley's meteorological expert suggested that altitude and slope angle were important factors to include in any model.

**2** The next model developed was also a scale analogue model. Extra layers of data were added to the GIS. A DTM provided surface data, and analysis undertaken on this allowed the prediction of snow

---

**BOX 7.1**

melt at different altitudes and on slopes with different aspects.

**3** A stochastic mathematical model was developed next to compare data on snow melt patterns over a number of years. This predicted the probability of a particular slope having full snow cover at any time during the ski season.

**4** Finally, a conceptual model of the whole modelling process was developed by summarizing the stages required in a systems diagram. This conceptual model was used to facilitate revisions to the modelling process following validation of results by field checking.

Datasets and activities relating to the Happy Valley Case Study can be found online at **www.pearsoned. co.uk/heywood**.

---

# BOX 7.2 Linking GIS with other spatial analysis and modelling software

PRACTICE

The GIS toolbox can provide only a limited number of spatial analysis and modelling tools. Therefore, as Maguire (1995) suggests, there is considerable value in integrating GIS with other specialist systems to provide for the needs of the process modeller. There are two main ways in which GIS software can be linked with other spatial analysis and modelling tools, and the approach adopted will depend on the nature of the application:

**1** The spatial analysis or modelling software can be integrated within a GIS environment, or the GIS can be integrated within the modelling or analysis

environment. In this case one software environment will dominate and therefore dictate the structure of the model.

**2** The second approach involves either tightly or loosely coupling the GIS to the other software environment. In tightly coupled models the linkage between the two products will be hidden from the user by an application interface. In a loosely coupled model the user may well have to perform transformations on the data output from the GIS, spatial analysis or modelling software before they can be used in another environment.

---

The remainder of this chapter explores how GIS can be used to help construct spatial models for physical, human and decision-making processes. These will be examined by reference to specific examples: forecasting and diffusion; location-allocation and spatial interaction models; and multi-criteria evaluation techniques. The range of applications considered is not comprehensive; however, the examples illustrate key issues in the model-building process.

## ■ MODELLING PHYSICAL AND ENVIRONMENTAL PROCESSES

All the different types of models introduced in the earlier part of this chapter have been applied to the modelling of physical processes. Jakeman *et al.* (1995) provide many examples of the application of these techniques for modelling change in environmental

systems. Goodchild *et al.* (1993) take a closer look at how environmental and physical models can be constructed in or coupled with GIS. Table 7.1 provides a summary of some of the application areas being dealt with in GIS.

Box 7.3 presents a case study in which the reliance on spatial data and need for spatial analysis functions makes GIS an appropriate tool for the modelling of air quality.

In a spatial context, perhaps one of the most challenging tasks for physical and environmental modellers is *forecasting* what may happen in the future under a given set of conditions. For example, predicting when and where an avalanche may take place and estimating its likely impact is a difficult task. However, if the nature of the system under investigation is reasonably well understood then it can be modelled and future trends, conditions, changes and outcomes forecast.

**TABLE 7.1 Applications of spatial models in GIS**

| Application | Source |
| --- | --- |
| **Physical and environmental applications** | |
| GIS analysis of the potential impacts of climate change on mountain ecosystems and protected areas | Halpin, 1994 |
| GIS analysis of forestry and caribou conflicts in the transboundary region of Mount Revelstoke and Glacier National Parks, Canada | Brown et al., 1994 |
| Simulation of fire growth in mountain environments | Vasconcelos et al., 1994 |
| Integration of geoscientific data using GIS | Bonham-Carter, 1991 |
| Spatial interaction models of atmosphere–ecosystem coupling | Schimel and Burke, 1993 |
| Landslide susceptibility mapping using GIS and the weight-of-evidence model | Lee and Choi, 2004 |
| GRASS GIS-embedded Decision Support Framework for Flood Simulation and Forecasting | García, 2004 |
| Geographical innovations in surface water flow analysis | Richards et al., 1993 |
| **Human applications** | |
| Techniques for modelling population-related raster databases | Martin and Bracken, 1991 |
| Use of census data in the appraisal of residential properties within the UK: a neural network approach | Lewis and Ware, 1997 |
| Design, modelling and use of spatial information systems in planning | Nijkamp and Scholten, 1993 |
| Selecting and calibrating urban models using ARC/INFO | Batty and Xie, 1994a, 1994b |
| An introduction to the fuzzy logic modelling of spatial interaction | See Openshaw and Openshaw, 1997 |
| Intra-urban location and clustering of road accidents using GIS: a Belgian example | Steenberghen et al., 2004 |
| A GIS-based approach for delineating market areas for park and ride facilities | Farhan and Murray, 2005 |
| An investigation of leukaemia clusters by use of a geographical analysis machine | Openshaw et al., 1988 |

# BOX 7.3 Using GIS to help manage air quality through estimating emissions of atmospheric pollutants

CASE STUDY

## Sarah Lindley

Air quality scientists and managers use a suite of tools to help them understand and control air quality in a particular area.

Two of the most important of these tools are:

- *Emissions inventories.* These are developed in order to find out about the nature, distribution and relative importance of different pollutant sources, such as industry and road transport.
- *Dispersion models.* These can be used to understand the ambient concentrations of pollutants that might result from emissions once they are mixed into the atmosphere.

One example of an air quality model commonly used by local authorities in the UK is ADMS-Urban which is used as a software add-on to the GIS package ArcView.

This case study considers the emissions inventory. It outlines the processes and outcomes of using GIS to develop an emissions inventory for oxides of nitrogen (NOx) from anthropogenic (human-made) sources. When oxidized in the atmosphere NOx gives rise to nitrogen dioxide ($NO_2$), which affects both human health and the environment, the latter through its role in acidification. NOx emissions come from a number of sources such as road transport, rail transport, airports, shipping and industry. In this case study the sources considered are those

**BOX 7.3**

associated with the Manchester and Liverpool conurbations in north-west England.

### THE ROLE OF GIS

Emissions sources are usually characterized as being point, line or area sources. This relates to whether emissions come from a single stationary location such as an industrial chimney or a power station (point sources), multiple individual sources that are mobile and which tend to operate on a linear network such as roads or railways (linear sources), or sources that are either too numerous to consider individually such as emissions from houses within a town and/or that are naturally diffuse, such as emissions from soil microbes (area sources). This characterization lends itself nicely to representation as points, lines or areas within GIS.

The process of estimating emissions firstly requires the development of a geospatial database of sources. The database also needs to hold attributes that describe the specific characteristics of sources and the amount of polluting activity associated with each. For road transport sources, for example, this involves the consideration of the location, length and types of roads, the amount of traffic travelling on the roads and the composition of that traffic. Once this information is known, it can be used with an emissions factor to estimate emissions. An emissions factor is an estimate of the amount of pollution that can be expected as a result of a certain type and amount of polluting activity. Looking at the road transport example, this could be expressed as the number of grams of NOx that would be expected if a single diesel-fuelled car travelled 1 km along a highway at a particular speed. Once the amount of pollution associated with traffic travelling on each particular road link is known, the data can then be summed and mapped onto an appropriate spatial framework. Since this requires a detailed geospatial database with sources shown as discrete point, line and area sources and since these sources need to have topology for the overlay analysis required, it is most appropriate to carry out the work using a vector spatial data model. In this case the GIS package ArcInfo was used.

The stages of the GIS-based emissions inventory project can be summarized as follows (see Figure 7.4):

**Figure 7.4** Methodology for estimating emissions within a GIS framework

## BOX 7.3

**1** Develop a geospatial database showing the locations and types of pollutant sources in the study area. Categorize sources as points, lines or areas.

**2** Decide on the spatial units to be used to represent the final emission estimates and create an appropriate data layer, in this case a regular grid of 1-km square vector polygons.

**3** Overlay the polygon layer with the point, line and area emission source layers so that each source knows in which square it is located. This process also means that the length attributes of line features relate only to the part of the line or road that actually falls within a particular square.

**4** Develop and apply emissions models (as database calculations) for a particular source to allow an estimate to be made of the amount of pollution to be expected during a particular time period.

**5** Sum up the emissions from each source category within the particular geographical areas of interest, in this case for each individual 1 × 1 km polygon. Map the results.

**6** Sum up the emissions from all source categories to create a grand total of emissions per individual 1 × 1 km polygon. Map these totals so that the overall distribution of emissions from all sources can be gauged.

**Figure 7.5** Estimates of the emissions of NOx from all sources (top) and motorways (bottom) in the Manchester and Liverpool conurbations, north-west England (1997)

BOX 7.3

## THE RESULTS

Figure 7.5 shows the results of applying the methodology for estimating NOx emissions in the study area. Many of the highest emitting 20 per cent of the 1 × 1 km zones, those with an annual contribution of more than 60 tonnes per annum in 1994 were associated with road transport sources (mainly with the region's motorway (highway) network). Just considering road transport sources, the overall contribution made by motorways was 46 per cent. The mean and maximum rates of NOx emissions on motorways were estimated to be 298 and 716 kg km-1 day-1 for average weekdays and reduced to 179 and 430 kg km-1 day-1 for average Sundays. However, there were also other causes of high emissions totals: for example, the polygon containing Manchester Airport and various industrial sites across the study area. This can be used as an input to dispersion models (with extra information) or perhaps more directly to work out where alternative transport schemes, for example, may be most effective to reduce overall emissions.

In this example, total emission estimates were created for individual 1 × 1 km polygons. An examination of the inventory results was found to produce useful information for air quality management applications, including the identification and explanation of broad patterns of emissions and the occurrence of emission 'hotspots' at different scales of investigation.

GIS overlay was also used to compare the results from this project with those from another inventory project. This revealed the importance of the accuracy and completeness of spatial data used to locate emissions sources, estimate emissions and map emission inventory results (Lindley *et al.*, 1999,

2000). Another interesting issue relates to the effect of different mapping techniques used to investigate the spatial distribution of final emission estimates. The use of different means of generating display categories was found to have a considerable bearing on possible interpretations of the emissions data produced (Lindley and Crabbe, 2004).

This type of work is essential for effective air quality management – for example to help with the determination of air quality management areas for the control of pollutant emissions. Emissions inventories are developed by both central and local government (see the suggested sources for further information). As well as facilitating the production of the actual estimates themselves, GIS can also be used to assess their reliability and also to map the results in a variety of ways.

A further role for GIS is to help investigate spatial patterns of air quality derived from measurements made as part of an air quality monitoring network. Here, interpolation may be important to help interpret data about the actual concentrations of particular pollutants at specific points in space and time (Lindley and Walsh, 2005). Interpolation and dispersion modelling are useful for determining whether health-based air quality standards have been or are likely to be breached.

### SOURCES OF FURTHER INFORMATION

Global inventories: www.geiacenter.org/
UK national inventory: www.naei.org.uk/
Greater Manchester air quality management work:
www.greatairmanchester.org.uk
ADMS-Urban model: www.cerc.co.uk/environmental-software/ADMS-urban-model.html
*Source*: Sarah Lindley, University of Manchester

---

Forecasting is inherently time-dependent. Unfortunately GIS data models handle the time dimension poorly (**Chapter 3**); therefore, most applications that involve making predictions or forecasts are based around some non-GIS model. Often these models are not spatial. They may deal with a single site or assume large homogeneous areas. However, when these models are linked with GIS they can be expanded to include a range of spatial inputs and outputs. Linking existing aspatial models with GIS can be difficult because the models do not necessarily include important key spatial phenomena such as spatial interaction and

spatial autocorrelation when making their predictions. Other models are inherently spatial and are particularly well suited for integration with GIS. The linking of existing forecasting models with GIS provides access to a larger database and tools with which to visualize how a spatial phenomenon diffuses over time. Different methods of visualizing output from forecasting models are presented in **Chapter 8**.

Forecasting models tend to be dynamic, with one or all of the input variables changing over time. For example, in the case of a forest fire, as time elapses from the start of the fire the prevailing weather conditions may change, as will the amount of forest still left to burn. Simulating the geographical spread of a dynamic phenomenon such as a forest fire requires a special type of forecasting model known as a *diffusion model*. These are best explained by reference to examples.

Imagine that the Happy Valley GIS team wishes to explore how GIS can be used to predict the rate of spread of summer forest fires in the area. The team has access to a conceptual fire model as shown in Figure 7.6. This shows that the rate of diffusion of the fire from its starting point is based on the spatial characteristics of the neighbourhood. The density of trees, the nature of the topography of an area, the presence of previously burnt areas and dynamic variables such as wind direction all affect the spread of fire. The spatial diffusion problem faced by the fire model is how to use this knowledge to simulate the movement of the fire over time. One way to achieve this is through close coupling with GIS.

Vasconcelos *et al.* (1994) describe one approach to modelling fire behaviour. Their method involved three stages:

1 A raster approach was used to create a spatial database of the criteria influencing fire behaviour. These criteria were determined from an aspatial model and included data on fuel, topography and weather. Combining these with spatial data allowed the creation of data layers for a GIS.

2 The value associated with each cell in the raster images was then fed into an established aspatial fire-modelling program known as BEHAVE (Burgan and Rothermel, 1984). The output was a map showing the status of the fire after a given time.

3 The modelling capabilities of GIS were then used to 'spread' the fire to adjacent cells on a cell-by-cell basis. The extent of the spread depended on the amount of fuel present, the topography of the land and recent weather. This diffusion process was continued until a specified time limit had been reached or all fuel sources for the fire had been consumed.

Using this approach, the GIS was used to produce three summary maps that showed the rate of spread

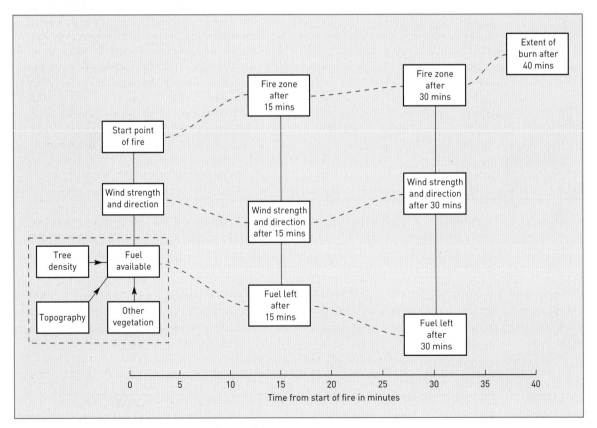

**Figure 7.6** A simplified conceptual forest fire model

of the fire, the spatial intensity of the fire and the direction of maximum spread.

In the forest fire example above, the focus was on how GIS can be used to provide input data, facilitate

spatial diffusion and assist with the visualization of results. Box 7.4 provides another example of how GIS has been used to provide data for input to hydrological forecasting models.

---

# BOX 7.4 Using GIS to provide terrain variables for hydrological modelling

**PRACTICE**

The hydrological cycle is an example of a dynamic natural system that can be modelled with the help of GIS. Not all hydrological models involve a spatial element, even though the inputs and controls of the hydrological cycle are inherently spatial (for example, rainfall is never distributed evenly across a catchment). GIS can provide variables for input to hydrological modelling using spatial data input, management, analysis and display functions that may be lacking in other modelling systems. This is particularly true for terrain variables such as slope, aspect and watershed analysis.

Terrain is fundamental to hydrology, since surface water always flows down a slope. This is the golden rule of hydrology. The shape of the ground surface determines how water flows downhill – at what speed, in what direction, how much infiltrates into the soil and how different flow paths accumulate. DTMs are extremely helpful for determining direction of slope. Analysis of a DTM in GIS provides basic data on height, slope and aspect, from which it is possible to calculate the direction of surface runoff and determine how water accumulates to form streams and rivers. Areas of divergent flow indicate the location of spurs, ridges, peaks and passes which can be linked to delimit catchment boundaries or watersheds. Surface flow directions can be modelled using the Deterministic 8 or D8 algorithm (O'Callaghan and Mark, 1984). This adds the accumulated upslope area of every cell to the neighbouring cell with the steepest gradient on the basis of the eight possible directions from the centre cell in a $9 \times 9$ grid as shown in Figure 7.7a. This results in a flow accumulation matrix that can be used to derive an approximate stream network based on the assumption that a stream will begin to form once water from a certain number of upslope cells has accumulated (Figure 7.7b). Cells with an upslope contributing area which is greater than a given threshold are labelled 'stream cells', and used to define the stream network. The raster grid can then be vectorized to allow the network to be used in other

analysis or for the production of cartographic output (Figure 7.7c).

When defining a realistic stream network one problem is choosing an appropriate threshold so that the density of the derived network matches, as near as possible, the density of the real stream network. It is also important to recognize that the density of the drainage network from place to place and from time to time will be affected by differences in hydrological process rates and temporal changes in variables such as soil moisture.

Problems with the D8 algorithm may occur in low-relief areas where there might be more than one lowest cell or large flat areas. Errors in runoff models like this are dealt with further in **Chapter 10**. A number of alternative models have been developed that incorporate random and multiple flow directions (for example, Rho8 and Deterministic Infinity models) together with route tracing models (such as the Kinematic Routing Algorithm). These and other models are reviewed by Wilson and Gallant (2000) and Hengl and Reuter (2009). Once a flow direction and accumulation matrix have been created using one of these models, it is possible to derive other useful information on topographic indices such as the wetness index or stream power index, or perform query-based analyses.

Queries at single or multiple points on the DEM using the flow direction and flow accumulation matrices can be performed to identify contributing or dispersal areas. Contributing areas are those areas upslope of a point, line or area. Subcatchments contributing to a point on a stream network can be defined using contributing areas. Dispersal areas are those areas into which water from a point, line or area flows. The route that a unit of precipitation follows across a DEM and through a stream network to its outflow can be traced in this manner. Both contributing areas and dispersal areas can be useful in identifying point and non-point pollution sources and then tracing the pathways of water-borne pollutants such as industrial effluent (for instance, point discharge

**BOX 7.4**

**(a)** D8 algorithm

**(b)** Simple example of flow accumulation

**(c)** Stream network delineation

**(i)** Altitude

**(ii)** Slope

**(iii)** Aspect

**(iv)** Convergence

**(v)** Analytical hillshading

**(vi)** Surface specific points

**(vii)** Flow accumulation (upslope area)

**(viii)** Wetness index

**(ix)** Stream power index

**(x)** Slope length factor

**(xi)** Upstream contributing area

**(xii)** Downstream dispersal area

**(d)** Matrix of catchment variables determined from DEM (i–xiii)

**(xiii)** Solar radiation

**Figure 7.7** Derivation of catchment variables using a DEM

**BOX 7.4**

from a pipe or outfall) or agricultural pesticides (non-point or area applications across whole fields) through the catchment to delimit the downstream area affected and identify possible management strategies. Examples of topographic indices and query-based analyses are also given in Figure 7.7d.

Other digital information may be added to a DTM (such as soil type and land cover) to determine likely infiltration rates and the hydrologic response to given precipitation inputs. This is more realistic than assuming that all the water falling as rain onto a DTM surface will run off as surface flow as in the

models described above. Data on infiltration rates, evapotranspiration, interception, groundwater percolation and subsurface storage can be modelled using DTM data in combination with data on soils, land use, vegetation, climate and human action.

Thus, GIS can contribute to hydrological forecasting models by providing a suite of tools for the manipulation and generation of terrain-related variables. Wilson and Gallant (2000) provide an excellent review of terrain analysis and derivation of variables of interest to hydrologists.

## REFLECTION BOX

- Think about the relevance of natural and scale analogue, conceptual and mathematical models in your own discipline or field of interest. Can you identify models which fit into these categories?
- Write your own explanations of the following terms:
  - Deterministic model
  - Stochastic model
  - Optimization model.

- For GIS software you have access to, or information about, look for details of modelling functions. Are these provided as part of the basic software package, available as optional extra functions, or not provided at all?
- What forecasts are necessary in your discipline? Could GIS help with the development of these forecasts? What problems might there be in developing suitable forecasts?

A recent development in forecasting models has been to couple GIS with artificial intelligence (AI) techniques such as neural networks (Hewitson and Crane, 1994). Neural networks are essentially classification systems that look for pattern and order in complex multivariate data sets. If the data are longitudinal records of inputs and outputs to a system they can be used to 'train' a neural network to recognize complex relationships between variables. The resulting neural network can then be used to predict system outputs from a given set of inputs and knowledge of preceding conditions (Openshaw and Openshaw, 1997).

Several problems make accurate forecasting very difficult. It is generally accepted that the accuracy and precision of forecasts decreases with the length of the prediction. A classic example is the weather forecast. With improved understanding of weather systems and better data, weather forecasters can predict what the weather is going to do in the next

24 to 48 hours fairly accurately. Beyond 48 hours the accuracy of their predictions begins to fall to such an extent that long-term, five-day forecasts are only about 40 per cent correct (Tyler, 1989). The introduction of random or stochastic processes, as advocated by proponents of chaos theory (Stewart, 1989), may play an increasingly important role in improving the accuracy of longer-term predictions, especially models of natural systems like weather and climate.

## ■ MODELLING HUMAN PROCESSES

The complete range of modelling techniques introduced earlier in this chapter has also been applied to the modelling of human processes (Table 7.1). However, perhaps the most difficult aspect of modelling humans is simulating their spatial behaviour. For example, how do people move through space when influenced and constrained by roads and traffic, buildings and fences, or other people, and how do people make decisions about where to live, which shops to buy food from, or where to go on holiday. *Spatial interaction models* offer one method of modelling spatial behaviour, and the links between these and GIS are the focus of this section. Spatial interaction models are used to help understand and predict the location of activities and the movement of materials, people and information (Birkin *et al.*, 1996). They can be used, for example, to predict the flow of people from their home to shops. In this case, the assumption is made that, as long as the number and characteristics of shops and people remain the same, so does the predicted behaviour of people. It is also assumed that people will use the shops closest to their home. If one parameter in the model is changed (for example, a new shopping centre may be added to increase the choice of shops available) the predicted behaviour of people will change. Spatial interaction models rely on the fact that the nature of this change can be predicted since it can be compared with the behaviour of people elsewhere. In this respect, spatial interaction models are a form of natural analogue model. Time does not have a role in the simplest spatial interaction models, therefore these models can be implemented in GIS. To explore how spatial interaction models work and identify the benefits of coupling them with GIS it is useful to consider an example.

Siting a new supermarket in Happy Valley will influence the shopping patterns of residents of Happy Valley and visitors to the area. Skiers staying in Happy Valley will not normally travel many miles to buy groceries but will tend to choose the nearest store that satisfies their requirements for value for money, choice, quality and convenience. Given that their primary reason for visiting Happy Valley is a skiing holiday, convenience may be the most important factor. In this case the most convenient supermarket may be one closest to their apartment. If the new supermarket is located close to the main apartment complex there is a good chance that a large number of people will make use of the facility. The amount of trade done by other stores in the area may be reduced as a result. Therefore, the distance between the origin of a journey (the apartment) and the destination (the supermarket) will affect the amount of 'interaction'

that takes place between the two locations. If the new supermarket is in a poor location, for example if it is too far away from the main ski village, the skiers will continue to use existing stores that are closer. The only way a new supermarket situated away from the apartment blocks may be able to attract customers is to offer additional services and facilities to induce customers to travel further. The 'attractiveness' of a store can be represented in a spatial interaction model as an attractiveness index. This can be estimated from surveys of consumer opinion and by counting the number of people visiting the destination and determining how far they have travelled.

The concepts of supply and demand are also important for spatial interaction modelling. Demand occurs at an origin (for example, the apartment block), while supply is located at destinations (the supermarket). Production, the capacity of the origin to produce a trip for a particular activity, is also important, and is generally some function of the population of the origin. If the population of the Happy Valley ski village is over 2000 people, its capacity to produce trips to the supermarket will be greater than a nearby farming hamlet that is home to three families.

Distance between an origin and a destination does not have a fixed effect on spatial interaction. Distance operates as a 'friction' value: the greater the distance, the longer the journey and the more time and effort (i.e. cost) required to make it. The effect of distance also depends on the type of activity and customers. For example, skiers would use the nearest supermarket if all they needed was bread and milk, but for a good bottle of wine they may be prepared to travel further.

The effects of distance and attractiveness can be studied using *gravity models* (Birkin *et al.*, 1996). These use a distance-decay function (derived from Newton's law of gravitation) to compute interactions given the relative attractiveness of different destinations. The distance-decay function is used to exaggerate the distance between origins and destinations. In the Happy Valley supermarket example the distance-decay function applied should be large as skiers are prepared to travel only short distances to use a particular store. Other activities such as shopping for skis, clothes or presents demand a smaller distance-decay function because skiers will be more willing to travel greater distances for these items. Choosing the correct distance-decay function to use for a particular activity is essential in developing a sensible model. Gravity models can be calibrated using data on trip lengths derived from customer surveys since the function is related to the average trip length for the activity.

The basic data required to develop spatial interaction or gravity models are a set of origins and destinations. These can be provided by a GIS as point entities. In the case of our supermarket example, one data layer would represent the proposed site for the new supermarket and a second data layer the distribution of the apartments in Happy Valley. In addition, the GIS could be used to provide a further data layer representing the transport networks (including streets and footpaths) along which goods and people move. GIS can be used to calculate real distances between origins and destinations. In addition, the GIS database can be used to store attribute information about individual locations. In the case

of the new supermarket these attributes could include the type of goods and services the store offers. Attributes for the apartments could reflect the socio-economic characteristics of skiers who rent them. This additional information could be used to determine the demand, supply and attractiveness characteristics for the model. GIS can also be used to display the results of the modelling process as a series of accessibility contours for both origin and destination locations, or as interaction probability surfaces (showing the probability of skiers in a given location making trips to the proposed supermarket).

Birkin *et al.* (1996) provide numerous examples of how spatial interaction and gravity models have been coupled with GIS. Box 7.5 presents a case study of the use of GIS for retail sales forecasting which illustrates the coupling of GIS with other software (Box 7.2). Harder (1997) and Mitchell (1997) offer other examples of the application of GIS to the modelling of human processes in areas such as health care, schools and recycling.

---

## BOX 7.5 GIS-enabled retail sales forecasting

**CASE STUDY**

### Tony Hernandez

A major challenge for retailers is estimating and forecasting the likely sales revenue that they may generate from an existing network of stores and from proposed new stores. Sales forecasts are typically used to set sales targets and are often critical in strategic decisions relating to store openings and expansions. The aim of retail sales forecasting is to identify the factors that determine retail sales for a given set of stores and to produce models of their relationship with retail sales. Regression analysis is one of the most popular methods used, with sales (or some other measure of productivity such as sales per square foot) as the dependent variable and a mix of internal and external variables as independent variables. Internal variables may include the size of the stores, the number of car parking spaces, the number of staff working in the store; and external variables typically include resident population, income and number of competitors within the store's trade area. Sales forecasting models allow analysts to assess the performance of existing stores by comparing the actual sales generated by a given

store with the predicted value from the model. The difference between actual and predicted (called the *residual*) provides an indication if a store is under-performing (i.e. generating less actual sales than the model predicts) or over-performing (generating more actual sales than predicted) – see Figure 7.8. The challenge for retailers is in building robust sales forecasting models that can be applied easily across their entire store portfolio. This case study outlines how 'The Speciality Depot' (name of actual retailer is disguised) used a GIS-based approach to operationalize sales forecasting applications. The Speciality Depot is an established international retail organization operating in a number of countries across North America and Europe. The company operates over 500 stores across Canada.

The research approach can be divided into four key stages:

### 1 CREATING A DATABASE OF STORE OPERATIONS AND MARKET CHARACTERISTICS

The first task was to create a comprehensive database of information on the *internal operation* of the store and the *external trading environment*.    →

BOX 7.5

**Figure 7.8** Residuals (percentage difference between actual and predicted sales) for The Specialty Depot's store network in Toronto, Ontario, Canada

### Internal operation

Many of the data sets that relate to the internal operation of the store were already available within the organization. These included data on the existing level of sales, the number of staff, car parking spaces, the number of checkouts, the mix of products and services available within every store.

### External trading environment

GIS was used to generate many of the external trading characteristics:

1 *Trade areas.* 'Realistic' trade areas were required for each store. A very large data set of customer records for each store was made available and every customer was geocoded at postcode level. GIS was used to select the closest 60 per cent and 80 per cent of customer records to each store, and a simple convex-hull created to define the primary and secondary trade areas (polygons within which the stores draw

most of their custom). Every trade area was validated manually and anomalies in the convex-hull method were minimized by user-intervention. Whilst the GIS automated the lengthy process of selecting customer records around each store, the resulting trade areas were not 'finalized' until each had been checked and signed-off by the analyst and senior management.

2 *Competitors.* A comprehensive database of competitor stores was used to generate a count and the amount of retail square footage operated by competitors surrounding every store. These were categorized into major and minor competitor categories.

3 *Consumers.* Data from the Canadian Census of Population were used to create profiles of potential consumers for each of the stores, including variables: for example, relating to population size and age structure, income, family status and dwelling type. Additional customized data sets were used to estimate daytime population and the amount of disposable income within each trade area.

BOX 7.5

## 2 UNDERTAKING STATISTICAL ANALYSIS AND BUILDING REGRESSION MODELS

The data set created in the first stage was exported from the GIS and imported into a standard statistical software package (such as SPSS, SPLUS, SAS). A series of exploratory analyses were undertaken to examine the data set, correlate variables and transform variables where necessary (for example, to normalize skewed distributions). The model building was undertaken outside the GIS environment, which is typical for retail store location research projects. Due to significant differences in the operating environment of the stores, the large data set was subdivided into three groups – a group of stores in the downtown, stores in the suburbs, and stores in small town markets. This resulted in the creation of three separate sales forecasting models, with each model built using the most significant set of variables. In the downtown store model, for example, the size of the daytime population was included, but this was not used in the suburban store model. During the model-building process a few stores were excluded from the analysis as they were clearly 'unique' stores (statistical outliers). The models were tested to ensure that they did not violate 'classic' statistical rules. At the end of this stage, each of the three sales forecasting models consisted of a small set of key variables (in equation form as illustrated below) that could be used to predict sales.

Generic Sales Forecasting Regression Model:
$$P_i = f(IS_i, ES_i)$$

where: $Pi$ is a measure of performance at store $i$; $IS_i$ measure/s of internal characteristics at store $i$; $ES_i$ measure/s of external characteristics at store $i$.

## 3 PROGRAMMING THE GIS USER-INTERFACE TO OPERATIONALIZE THE MODELS

The third stage involved programming these three equations into a GIS guided user-interface so that users could quickly create sales forecasts for existing stores if the store trade area changed (for instance, if a new competitor opened) or a new planned store at a given location (to predict sales for this new location).

The Sales Forecasting GIS-Interface (SFGI) was built using a macro-GIS language to create a customized interface available to the user as part of a standard GIS package. When the user runs the macro, the interface and all necessary data are automatically loaded within the GIS. The interface was designed to allow a user to work through the sales forecast in a number of steps:

1 Navigating the map window and selecting new or existing store locations.

2 Determining the type of model to run (downtown, suburb or small town).

3 Changing model parameters (optional).

4 Running the model (which would automatically collect all necessary information from the underlying data layers, and then run the model equation).

5 Reporting the model results on-screen.

The interface allows the user to quickly re-run models using different parameters or different locations – and to compare the results of these models.

## 4 USER-TESTING AND DEPLOYMENT

The GIS interface was tested with a number of retail analysts who provided feedback on how the interface functioned. The testing stage highlighted a number of shortfalls in terms of the potential for users to alter the models by changing parameters without being warned. As a result a number of user warning pop-up boxes were added to ensure that the user had to confirm that they wanted to change the base model settings. The final deployment involved generating user documentation and providing training.

This case study highlights how GIS can be used within store location research to manage the underlying geographical data and to operationalize a set of sales forecasting models. Whilst the models were not built within the GIS itself, the user-interface allows users to alter model parameters and run customized models. A challenge with such customized decision support is the need to re-run the modelling exercise periodical (typically within a two-year timeframe) to ensure that the models are up to date and relevant. The GIS approach outlined has been used successfully by The Speciality Depot, allowing a number of business managers and company executives to undertake sales forecasting 'on-the-fly': for example, using their laptop computers to run models when visiting potential 'new store' sites. The project also raised the profile of GIS, modelling and geographical data within the company.

*Source*: Tony Hernandez, Ryerson University

## ■ MODELLING THE DECISION-MAKING PROCESS

Outputs from process models of both human and physical systems are the raw information that will be required to assist various types of decision-making. Output from a location-allocation model might show potential locations for a new supermarket within Happy Valley. This would need to be integrated with other regional planning data to ensure that the resort is being developed in line with regional strategic planning objectives. Moreover, it may be that a choice has to be made between the development of a supermarket or the construction of a new ice rink. As each development will have different spatial impacts on the environment, it may be necessary to evaluate what these will be to help decide which of the two development options should be pursued. In the same way, areas of high fire risk output from the Happy Valley forest fire model may need to be integrated with other environmental data to determine priority areas for nature conservation.

Map overlay is the traditional technique for integrating data for use in spatial decision making. For example, in siting nuclear waste facilities, the criteria defining geological suitability and conservation area status can be combined using the overlay procedures outlined in **Chapter 6**. However, overlay analyses suffer from certain limitations when dealing with decision-making problems of a less well-defined nature. These are summarized by Janssen and Rievelt (1990):

- Digital map overlays are difficult to comprehend when more than four or five factors are involved.

- Most overlay procedures in GIS do not allow for the fact that variables may not be equally important.

- When mapping variables for overlay analysis, decisions about threshold values (see Box 6.7) are important (the outcome of polygon overlay depends strongly on the choice of threshold values).

One way to address these problems is to use *multi-criteria evaluation* (MCE) techniques to either supplement or replace standard map overlay in GIS. MCE techniques allow map layers to be weighted to reflect their relative importance and, unlike polygon overlay in vector GIS, they do not rely on threshold values. Therefore, MCE provides a framework for exploring solutions to decision-making problems, which may be poorly defined.

MCE is a method for combining data according to their importance in making a given decision (Heywood *et al.*, 1995). At a conceptual level, MCE methods involve qualitative or quantitative weighting, scoring or ranking of criteria to reflect their importance to either a single or a multiple set of objectives (Eastman *et al.*, 1993). It is not the intention here to introduce the full range of MCE techniques, since these are documented elsewhere (Voogd, 1983; Janssen, 1992; Eastman *et al.*, 1993). However, an understanding of the key concepts is important to appreciate how MCE can be applied in a GIS context.

In essence, MCE techniques are numerical algorithms that define the 'suitability' of a particular solution on the basis of the input criteria and weights together with some mathematical or logical means of determining trade-offs when conflicts arise. These techniques have been available since the early 1970s and a whole literature has developed that is devoted to MCE and related fields (Nijkamp, 1980; Voogd, 1983). It is the overlap with GIS that is of interest, since the application of MCE techniques within a GIS framework permits the limitations of standard map overlay (defined above) to be addressed. An example of how MCE is implemented in GIS is given in Box 7.6. MCE, in common with other scientific disciplines, has developed its own specialized terminology. The examples below inevitably contain some of this 'jargon', but this is kept to a minimum and defined as necessary.

The first step in MCE is to define a problem. In the house-hunting case study this is where to look for a new home. A range of criteria that will influence the decision must then be defined. The criteria can be thought of as data layers for a GIS, and in the house-hunting example these included proximity to schools and roads, and quality of neighbourhood. The criteria selected should reflect the characteristics of the neighbourhood in which the decision maker wishes to live. For example, a decision maker with a young family may feel it is important to be near a school, though in a rural area. A method of weighting or scoring the criteria to assess their

# BOX 7.6 Implementing MCE in GIS

Perhaps the simplest MCE algorithm is the weighted linear summation technique. The steps involved in applying this model within a raster GIS are illustrated in Figure 7.9 and can be described as follows:

1 *Selection of criteria.* Most MCE analyses in GIS start with the identification of the data layers that are important to the problem. These criteria are represented as separate raster data layers in the GIS.

2 *Standardization of criterion scores.* Most MCE analyses, especially those using quantitative and mixed data sources, require some form of standardization of the scales of measurement used by the data layers. This is necessary to facilitate the comparison of factors measured using different units and scales of measurement (it would be nonsense to directly compare population density with distance from the

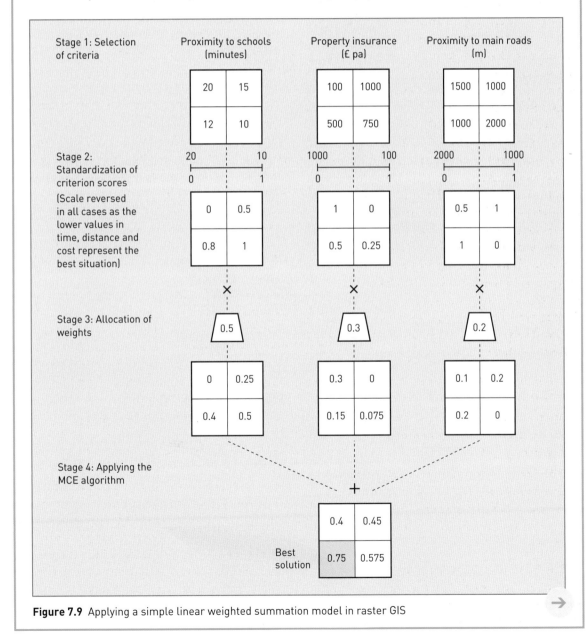

**Figure 7.9** Applying a simple linear weighted summation model in raster GIS

BOX 7.6

nearest railway line). Therefore, standardization of the scales will enable meaningful comparisons to be made between the data layers. Standardization can be done in a number of ways, but it is normal to apply a linear stretch routine to re-scale the values in the raster map between the maximum and minimum values present. Care needs to be taken to observe 'polarity' such that beneficial factors are represented on a scale that gives a high value to high benefit and a low value to low benefit, whilst cost factors are represented on a scale that gives a low value to high cost and a high value to low cost. Another common method is to bring all the values on a data layer to an interval value between 0 and 1. For example, if the values on the proximity to an urban map range between 0 and 10 miles, using a standardized scale a value of 0 miles will equate to 0 and a value of 10

miles to 1. A value of 5 miles would score 0.5.

**3** *Allocation of weights.* Weights are allocated which reflect the relative importance of data layers. A weighting of 80 per cent may be expressed as 0.8.

**4** *Applying the MCE algorithm.* An MCE algorithm may then multiply these standardized scores by the weights for each of the data layers in stage 1 and sum these to allocate a score to each pixel on the output map. The map produced will be a surface with values ranging from 0 to 1. In the case of the house-hunting example, areas with a value close to 1 represent those areas where the home buyer should start looking for a house. Further evaluation of the results may be carried out by ranking the values in the results map and reclassifying the ranked map to show the top-ranked sites.

importance (a decision rule) must then be constructed. This is done by adding weights (expressed as percentages or proportions) to reflect the importance of each criteria (Figure 7.10). A high level of importance (50 per cent) may be attached for

proximity to a school and a low level of importance (20 per cent) for proximity to the road network. Application of an MCE method will result in a list of neighbourhoods in which the home buyer could choose to live.

| Criteria | Weights |
|---|---|
| 1 Proximity to schools | 0.5    (50%) |
| 2 Property insurance | 0.3    (30%) |
| 3 Proximity to main roads | 0.2    (20%) |

**Figure 7.10** Weighting data layers in the house hunting case study

MCE could also have been applied in the nuclear waste case study. The problem was defined and a solution developed in **Chapter 6** using polygon overlay. The results of this overlay procedure gave a set of areas that satisfy certain criteria relating to geology, population, accessibility and conservation. The number of potentially suitable sites within the suitable area identified may total several thousand. Once the suitable area is defined, the question then becomes how to identify a limited number of best sites from the thousands available.

In the case of nuclear waste disposal it is possible to identify a whole range of factors that may be important in determining a site's suitability, including population density within a 50-kilometre radius, quality of on-site access, distance from conservation area(s), distance from nearest railway line, accessibility to waste producers and even marginality of the local parliamentary constituency! Satisfying certain deterministic criteria applied to these factors is not essential at this stage, but we may wish to optimize them such that the best sites are those which possess all the best qualities. This is not the simple ranking and sorting procedure that it may at first sound, since many of the factors specified may be of differing importance and may be conflicting. For example, consider the factors regarding population density within a 50-kilometre radius and distance to the nearest railway line. We may wish to optimize the siting solution by identifying those sites in areas of low population density within the 50-kilometre radius, and at the same time minimize the distance from the nearest railway line. In many areas these are conflicting criteria since railways usually run through and link highly populated areas. It is likely that the sites with very low population densities will be a long way from the nearest railway line, whilst the sites near to a railway line will have higher population densities. The solution to this problem is therefore to use MCE techniques, which allow the factors to be weighted according to their relative importance. Thus, the trade-offs necessary to determine which sites are actually the best available are simplified.

GIS is an ideal framework in which to use MCE to model spatial decision-making problems as it provides the data management and display facilities lacking in MCE software. In return, MCE provides GIS with the means of evaluating complex multiple criteria decision problems where conflicting criteria and objectives are present. Together, GIS and MCE-based systems can provide the decision maker with a more rational, objective and unbiased approach to spatial decision-making and support than hitherto (Carver, 1991b; Eastman *et al.*, 1993).

MCE, like any other technique, is not without its problems. The main problems are the choice of MCE algorithm and the specification of weights. Different MCE algorithms may produce slightly different results when used with the same data and the same weights. This is due to the differences in the way the trade-offs are performed. The most intractable problem, however, is in the specification of (factor) weights since these strongly determine the outcome of most analyses. Different people will view the problem differently, and so specify different weighting schemes. In the context of the nuclear waste example, it is possible to imagine that the weighting schemes specified by the nuclear industry and the environmental lobby will be entirely different. The nuclear industry is more likely to place greater emphasis on economic and engineering factors, whereas the environmental lobby is more likely to stress social and environmental factors. In this case, the results will be very different and further analysis would be required to identify least-conflict solutions (Carver, 1991b). This has led to developments in public participation GIS in which public inputs are the key to responsible GIS applications. An introduction to the theory and concepts surrounding Public Participation GIS (PPGIS) are given in Box 7.7 and Box 7.8 provides a case study of PPGIS in use, focusing on the difficulty of translating 'vernacular' geographies into information for decision making.

## ■ PROBLEMS WITH USING GIS TO MODEL SPATIAL PROCESSES

The use of GIS to facilitate the development of spatial process models is not without problems. We have already alluded to some of these, such as the availability of appropriate modelling tools with the GIS toolbox. However, whilst technical problems can be solved (by integrating or coupling GIS with appropriate software tools), there are other, more conceptual problems that the process modeller wishing to use GIS should be aware of.

# BOX 7.7  Public participation GIS: information is power

**THEORY**

One of the many criticisms levelled at GIS is that it is an elitist technology accessible only to those organizations with sufficient resources to pay for systems, data and technical staff (Pickles, 1995). GIS has also been criticized for its lack of knowledge-based input (Taylor, 1990). As GIS moves into the public sphere via the Internet, it is now possible to counter these two claims by giving the public much greater access to GIS and spatial datasets.

The public is an enormous data resource that can be of tremendous benefit to decision makers. Participatory approaches use this resource to help populate GIS databases with information on local knowledge and grassroots opinion. In return, decision makers are able to make decisions that are more in tune with local feelings and needs.

Early public participation GIS (PPGIS) involved groups of decision makers sitting around a single GIS workstation and asking them to explore a problem with the aid of relevant GIS datasets. An example of this was the Zdarske Vrchy case study where local resource managers used GIS to explore decisions about recreational management (Petch *et al.*, 1995). Shiffer (1995a, 1995b) extended the PPGIS model to include multimedia to draw the public further into the decision-making process. Shiffer used video and audio representations of aircraft during take-off and landing linked to a GIS database of residential areas around a proposed airport site.

Recent PPGIS have used Internet-based approaches because of the advantages this offers in terms of access for the public 24 hours a day, seven days a week. Open access systems via the Internet may also help break down some of the psychological barriers to participation present in face-to-face meetings that lead to 'us and them' type stand-offs between the decision-making authorities and the public.

A model of public participation in spatial decision-making facilitated by Internet-based PPGIS is that of 'explore, experiment and formulate' (Kingston *et al.*, 2000). In this model the participating public can *explore* the decision problem using the GIS database and *experiment* with possible alternative solutions before *formulating* their own informed decision as to the best course of action. The information accessed and the choices made in reaching this decision are fed back to the decision makers via the Internet along with the decision made. This allows much greater insight into public opinion than is possible from more traditional methods such as local meetings and consultative documents.

The nuclear waste case study is an example of a spatial decision problem that would benefit significantly from greater public input, perhaps via Internet-based PPGIS. It is clear that nobody wants to live next door to a nuclear waste disposal site because of the possible health risk (real or imagined) and the likely effects on property prices. We have seen in previous chapters how GIS can be used to identify possible sites for such a facility based on geographical criteria in a map overlay model. We have also seen how MCE methods may be integrated with GIS to model the suitability of individual sites based on the trade-offs between weighted factor maps. By incorporating MCE approaches into Internet-based PPGIS it should be possible to survey the opinions of a large number of people across the country concerning the general geographical factors (and their weights) relevant to the problem of identifying a suitable site for a nuclear waste repository.

PPGIS using the Internet does have problems. These include difficulties regarding the representativeness and validity of responses gained. At present not everyone has equal access to the Internet nor do they have the necessary training in order to use it. Distinguishing valid responses from those made by people just 'playing around' or wishing to bias the results one way or another is difficult. Perhaps the biggest problem is the willingness of those people already in positions of power (politicians, government ministers, planners and local officials) to place greater decision-making powers in the hands of the public through such systems. Governments around the world are placing increased emphasis on enhancing services and public accountability by electronic means. It will be interesting to see just how far down this route to 'cyber-democracy' they are prepared to go.

## FURTHER INFORMATION

Locating a nuclear waste disposal facility online decision support system: www.ccg.leeds.ac.uk/teaching/nuclearwaste/

# BOX 7.8 Fuzzy GIS and the fear of crime

## Andrew Evans

Standard GIS were designed by scientists and computer programmers who see the world in terms of numbers and clear lines. Real people, however, tend to exist in a very different 'vernacular' geography. Millions of us go 'downtown for the night' or 'to the shops on Saturday', avoid 'the rough end of the city' and park away from 'high crime areas'. But we are unable to clearly define what these terms mean geographically, despite the fact that they are intimately tied into our behaviour. These vernacular geographical terms are a good thing: the use of terms like 'Uptown' or 'the rough end of town' allows us to communicate locational information that includes summaries of associated environmental, architectural and socio-economic contexts. They also place us in a socio-linguistic community with shared understandings and, unfortunately, prejudices. Vernacular geographical terms are not simply locational labels, they often represent zones in which we constrain our activities, and they tell members of our community that this constraint should be added to their shared knowledge and acted upon. This private and shared vernacular geography affects millions of people daily, yet, because it has a difficult and subjective nature, it is hard to tie directly to the kinds of data we use to make scientific decisions or policies. Because of this, we need a new kind of GIS – one that talks about the world in the same way as normal people; one which allows us to capture how normal people understand the world.

Let us take, as an example, areas in which people are afraid of crime. Crime surveys frequently show that respondents have a higher fear of crime than is justified by actual crime figures (for a review of the extensive literature, see Hale, 1996). The fear of crime can have a significant impact on people's lives. Twenty-four per cent of respondents in the 2004/05 British Crime Survey claimed they didn't go out alone at night because of the fear of crime. Eleven per cent of respondents claimed they either didn't walk alone during the day for fear of crime, or went out once a month or less because of it (Allen, 2006). Concern about crime has significant influence on many people's lives, and influences which areas people travel to at different times.

However, one of the biggest difficulties in tackling the fear of crime is that it is hard to measure. If we ask people to delimit and explain areas where they think crime levels are high, most will draw on a range of continuous and discrete properties of the areas, often at different scales, as well as various experiences, architectural details, and media coverage. The areas defined may have boundaries at prominent landscape features but are more likely to be vague or ill-defined geographically. The intensity with which an area is felt to be in a category like 'high crime' often declines over some distance, and the zones themselves often have more or less 'high crime' areas within them. In short, such areas are 'Fuzzy' or 'Vague' – they have indistinct boundaries and a varying degree of importance: what we need therefore is a 'Fuzzy' or 'Vague' GIS to cope with them, and this is generally true of most 'vernacular' areas.

### IDENTIFYING 'HIGH CRIME' AREAS IN LEEDS, UK

Such a GIS was developed for a pilot study to look at where people felt 'high crime' areas were in the city of Leeds. Leeds is a UK city with a population of over 700,000, and had 118,559 reported crimes in 2001/2002 when the study was completed. Most of these crimes were in the city centre; however, this doesn't tend to be in the group of areas that people think of as having 'high crime'. To capture these areas a new type of web-based GIS (called 'Tagger') was developed. The GIS was based around a 'spraycan' interface (Figure 7.11a) which allowed people to spray areas of varying intensity on a map and attach some attribute data to them – in this case what they felt about the areas. The sprayed areas were then sent to a server across the web, collated with other people's areas, and these could be queried by users via a query interface (Figure 7.11b). This allowed them to see an aggregate map of everyone's sprayed areas. When they clicked on the map, the comments associated with the most intense spraying for that point floated to the top of a comments list.

The real crime intensities for the city can be seen in Figure 7.12a. The aggregate map of user-sprayed areas can be seen in Figure 7.12b. At each pixel the user-sprayed map shows the intensity levels for all users summed and divided by the total number of users – the average areas regarded as being 'high crime' zones. The colour scale has been

→

**BOX 7.8**

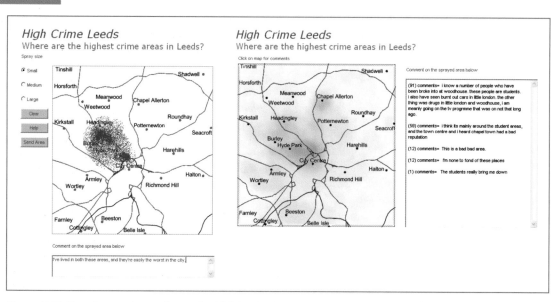

**Figure 7.11** User inputted area of perceived 'high crime' together with attribute data. (left) Simple area and comment about why the specific areas were chosen; (right) Output showing all user areas averaged, together with ranked comments for one area

stretched so that white areas are those that were not sprayed and the areas that were sprayed the most are black. The crime data in Figure 7.12a is rescaled between its minimum and maximum in the same way for comparison. The sprayed data undergoes smoothing to change it from sprayed dots to a density surface, and some compression before it is

transmitted across the Web, so the same treatment has been applied to the crime data.

Analysing such data is not simple, and little work has been done on it. However, as a very simple example to show its potential in decision making, we can compare the absolute crime levels and the perceived areas directly. Figure 7.12c shows

(a)                    (b)                    (c)

**Figure 7.12** Total crime densities for Leeds for all crimes committed in 2002. (a) Blacker areas are higher in crimes. The circular high is genuine and mainly reflects the location of the inner ring road around the city. (b) Areas sprayed as 'high crime' areas by users in a pilot running from August to September 2002. Blacker areas are those felt to be higher in crime. (c) The difference between (a) and (b), generated after stretching the highest perceived crime area levels to the highest real crime levels and the lowest perceived crime levels to the lowest crime levels. Red areas will tend to have higher crime levels than expected, blue areas lower. UK Census Wards are shown for reference.

a simple subtraction of the two data sets after they have been rescaled to the same range. Such a map could be used to show the broad, quasi-quantitative differences between the perceived and actual crime levels for use in decision making.

In addition to this simple example analysis, the users of the system gained directly from it. They were allowed to see the aggregated map and query it for other users' comments. This allowed users to answer questions like: 'how scared of crime are my neighbours?' and 'does anyone else feel the same way as me?'. Whilst there are obvious difficulties in presenting such data back to users from ethical, socio-economic and (potentially) libel and justice-related points of view, in this pilot it was felt appropriate to understand how users responded to the system. Feedback indicates that users liked and understood all aspects of the GIS (both input and querying) and found it easy to use.

Systems like that presented here aim to draw together professional and popular ways of seeing the world, to their mutual advantage. For the GIS professional, capturing popular understanding of the world should reveal the motivating forces behind people's spatial behaviour, and allow for better policy development. For the general population it should let them use their own language rather than the vocabulary of GIS professionals to highlight important things. Given this, capturing vernacular geography promises to elevate both policy analysis and democratic decision making, and to give a louder voice to a geography that is rich and important.

### FURTHER INFORMATION

Source code and further reading can be found via: www.ccg.leeds.ac.uk/groups/democracy/
Evans, A. J. and Waters, T. (2007) Mapping vernacular geography: web-based GIS tools for capturing 'fuzzy' or 'vague' entities. *International Journal of Technology, Policy and Management* 7 (2): 134–50. http://dx.doi.org/10.1504/IJTPM.2007.014547

*Source*: Andrew Evans, University of Leeds

These include:

- the quality of source data for model calibration;
- the availability of real world data for model validation;
- the implementation within a GIS;
- the complexity of modelling reality; and
- the conceptual and technical problems of building multi-dimensional models.

The data requirements of process models can be large and problematic. For example, many forecast models rely on the availability of longitudinal records to calculate past trends and relationships. These records are used to predict future conditions. Suitably long records in many cases may not exist. Similarly, models that have been developed in a non-spatial context may require detailed data inputs which, again, do not exist as spatial data sets. An example of this can be seen in hydrological modelling – where do you obtain detailed spatial data on evapotranspiration rates or soil moisture conditions? Another example is the Universal Soil Loss Equation, which has often been applied within a GIS context. The equation is written as:

$$A = R \times K \times LS \times C \times P$$

where: $A$ = average annual soil loss; $R$ = rainfall factor; $K$ = soil erodibility factor; $LS$ = slope length–steepness factor; $C$ = cropping and management factor; and $P$ = conservation practices factor.

Finding the relationship between the factors in the soil loss equation and making the necessary measurements is simple enough for small experimental soil plots. However, extending the measurements to a large study area, where the controlling factors can vary by orders of magnitude, is difficult (see Figure 7.13). Often surrogate measures and interpolated data are used. For example, spatial variations in evapotranspiration may be estimated by combining previous field studies on different vegetation types and weather with land cover information derived from satellite imagery.

All process models need to be validated or checked to make sure that they work and that the results they produce make sense. The easiest way to validate a model is to compare its predictions with a real event (for example, the predictions of a forest fire model could be compared with a real fire situation such as in Figure 7.14). Starting a fire for model validation would not, however, be a feasible validation method. Equally, if we were trying to model the potential effects of global warming we would have a long time to wait before we could compare the results of our model with reality.

**(a)** Severe soil erosion on agricultural land (loose dry soil from the bare field in the background has been removed by wind action and deposited in the foreground burying the five bar gate)

**(b)** Example of mapped USLE output

**Figure 7.13** Soil erosion models
*Source*: (b) SciLands GmbH, www.scilands.de

**Figure 7.14** Forest fire in the Canadian Rockies
*Source*: Courtesy of Sheelagh M Coles

In the case of some processes, such as the forest fire example, we can use historical events to validate our model, if the historical records of the event provide us with enough detail. Therefore, we could use our fire model to simulate a fire that happened in 1950, and compare our results with a survey done after the fire had burnt out. However, this approach is unable to solve all problems. In some cases the best option is to produce a range of scenarios showing what the likely outcome may be and hope that reality is somewhere between the worst and best cases. The main GIS problem associated with validating spatial process models is therefore lack of appropriate spatial data.

Implementation of process models within GIS enforces a spatial context. When models were originally developed for non-spatial applications this can create conceptual and operational problems. The original model may not consider geographical processes such as spatial interaction, diffusion or spatial autocorrelation because it was designed for single, non-spatial applications. As a result, the extension of models into a spatial context must be undertaken with care.

Some GIS include visual modelling tools that allow users to construct spatial models within a modelling window. An example of such a model building interface is shown in Figure 7.15. Data and process icons can be combined using on-screen 'drag and drop' techniques to build up a model flow

**Figure 7.15** Example of a model builder interface (Idrisi32 Macro Modeler) showing a dynamic urban growth model based on land use and suitability map inputs used to produce a map of urban growth areas
*Source*: Screenshot shows IDRISI interface. Reproduced by permission of Clark Labs, Clark University

chart. Once a model has been constructed in this manner, it is interpreted by the GIS and processed to generate the required output. These visual model builders have made spatial modelling accessible to a wider number of GIS users, but at the same time raise important questions about the users' ability to understand exactly what they require from the GIS and how to achieve this.

The human and physical worlds are extremely complex and interlinked. Theories of environmental determinism suggest that human development is strongly influenced (if not determined) by the physical environment. More anthropocentric theories stress the increasing level of control that the human race has over the natural environment and how the latter has been irrevocably changed by human activity. Whichever theory you subscribe to, there is no escaping that the two systems, human and natural, cannot really be considered in isolation from each other. Agriculture is a good example. A crop may not grow naturally in an area that receives a low rainfall, so a more drought-tolerant crop

may be grown instead. Thus, the environment determines human activity. Alternatively, it may be possible to irrigate the land to allow less drought-tolerant crops to be grown. Thus, human action overcomes environmental factors. Therefore, complex models are necessary to simulate the complexity of human and environment interactions. As the model becomes more comprehensive, there will be greater data requirements, increased complexity of intra-model interactions, greater opportunity for errors in both model data and model processes, and an increased likelihood that the model outputs will have little relevance to reality. Thus, there is much sense in the saying 'keep it simple'.

For many problems that require a modelling solution the models commonly used by GIS software may be inappropriate. For example, **Chapter 3** discussed the limitations of standard GIS data models regarding their handling of time and 3D data. There are certain advanced applications where further problems arise. Global modelling applications are a case in point. Current GIS data models employ a 'flat Earth' approach. Although most GIS packages allow surface modelling of some description (2.5D) and some even permit true solid 3D modelling, rarely do commercially available systems provide facilities for adequately modelling global phenomena. To do this with any degree of realism requires a spherical data model with no edges. Experimental global data models exist. These use a sphere constructed of triangular or hexagonal cells (in a similar fashion to a radar dome or leather football) to approximate the Earth's surface (Goodchild and Yang, 1989). These have found applications in global climate modelling and atmospheric and oceanic circulation models.

## REFLECTION BOX

- What is a spatial interaction model? Why are spatial interaction models important in the modelling of human processes?
- What problems can be faced when using map overlay techniques to support decision making in complex planning situations? How can MCE help overcome some of these problems?
- 'Information is power'. Reflect on the relevance of this statement to disciplines (e.g. retailing, homeland security, environmental management). Does sharing of data and information take place? What are the implications of this for GIS analysis and decision making?

- Return to the house hunting case study (Figure 1.13). The criteria established for decision making were:
  - proximity to a school;
  - proximity to a main road;
  - proximity to an urban area; and
  - location in low crime area.
- Talk to your family or flatmates and ask them for their views on these criteria. How close to a school/main road/urban area would they like to be? Which of the criteria do they think is most important? How could you incorporate their views into GIS analysis to find a suitable home?

## CONCLUSIONS

Spatial process models are constructed for a variety of different purposes: to help structure ideas, to improve our understanding of a problem or to communicate our ideas to others. Models do not provide answers, but offer a continuous and iterative process from which we gain knowledge about the real world. We can only claim that a model has worked if it has served its purpose. The application of a model allows identification of its weaknesses and allows adjustments to be made. This cyclical nature of model building is often described as a 'feedback process'.

GIS provide a good framework for model building. GIS may need to be coupled with other modelling applications or may be able to provide input data and output capabilities to supplement other modelling software. Since most models have an applications focus, there is a clear need for greater availability of modelling and coupling software within GIS.

The main power of GIS in analytical modelling is that it can provide a spatial database of model variables, and can be used to diffuse model output over geographical space. GIS also offers transformation tools that can be used to integrate disparate data sets required for some models. Data for model calibration can also be handled by GIS.

The problems of using GIS in analytical modelling include the populating of spatial databases (data encoding) at a scale or resolution appropriate for the model. In addition, many models are aspatial in character and do not translate easily to a spatial context. The other major limitation of current GIS is its inability to handle temporal data.

If one single characteristic of GIS has to be selected as a key advantage for linking GIS with analytical models, it would probably be the options for the visualization of data inputs and information outputs. The range of visualization options available in GIS are discussed in more detail in the next chapter.

## REVISION QUESTIONS

- Outline the differences between models of spatial form and models of spatial process.

- Present, with examples, a classification of process models. Why is it important to develop such a classification scheme?

- Outline the ways in which process models can be linked with GIS.

- Describe how spatial process models can be used to forecast the behaviour of physical systems.

- How can GIS be used in the modelling of human processes?

- Describe the MCE approach to modelling the decision-making process.

- Discuss three problems with the implementation of process models in GIS.

- What is PPGIS? When might PPGIS be used?

## FURTHER STUDY – ACTIVITIES

- Use the Internet or a library to find examples of papers describing the use of GIS in your discipline or area of interest. Produce your own version of Table 7.1 to summarize your findings.

- Draw and annotate a systems diagram for (a) a simple overlay model, and (b) a rule-based model.

- Download one of the free terrain analysis packages from the list of web links provided in **Chapter 6** and install it on your PC. Use the software to explore and experiment with the different terrain variables used in hydrological modelling.

- Try loading a DEM downloaded from the Internet into the terrain analysis software and calculate catchment variables.

- Use the Internet to access some of the PPGIS websites listed below.

## FURTHER STUDY – READING

There are several good sources of information that can be used to follow up material in this chapter. Goodchild *et al.* (1993), Ripple (1994) and Heit and Shortreid (1991) include details of many environmental GIS applications involving modelling. Specific applications of forecasting and diffusion models are presented in Haines-Young *et al.* (1994) and Price and Heywood (1994). Wilson and Gallant (2000) and Hengl and Reuter (2009) both provide excellent and detailed reviews of terrain analysis applications. For a review of modelling in human applications Birkin *et al.* (1996), Longley and Clarke (1995) and Longley and Batty (2003) provide useful chapters. Batty and Xie (1994a and 1994b) present two interesting papers on embedding urban models within GIS. Fotheringham and O'Kelly (1989) provide a comprehensive review of spatial interaction models.

Several research papers on the integration of GIS and multi-criteria evaluation are available. The papers by Janssen and Rievelt (1990) and Carver (1991b) are essential reading, whilst Thill (1999) and Malczewski (1999) provide comprehensive overviews of the use and application of MCE within GIS.

Batty M, Xie Y (1994a) Modelling inside GIS: Part 1. Model structures, exploratory spatial data analysis and aggregation. *International Journal of Geographical Information Systems* 8 (4): 291–307

Batty M, Xie Y (1994b) Modelling inside GIS: Part 2. Selecting and calibrating urban models using ARC/INFO. *International Journal of Geographical Information Systems* 8 (5): 451–70

Birkin M, Clarke G, Clarke M, Wilson A (1996) *Intelligent GIS: Location Decisions and Strategic Planning.* GeoInformation International, Cambridge, UK

Carver S J (1991b) Integrating multi-criteria evaluation with geographical information systems. *International Journal of Geographical Information Systems* 5 (3): 321–39

Fotheringham A S, O'Kelly M E (1989) *Spatial Interaction Models: Formulations and Applications.* Kluwer Academic, Dordrecht

Goodchild M F, Parks B O, Steyaert L T (eds) (1993) *Environmental Modelling with GIS.* Oxford University Press, Oxford and New York

Haines-Young R, Green R D, Cousins S H (eds) (1994) *Landscape Ecology and GIS.* Taylor and Francis, London

Heit M, Shortreid A (1991) *GIS Applications in Natural Resources.* GIS World Inc., Colorado

Hengl T, Reuter H I (2009) *Geomorphometry: Concepts, Software, Applications.* Developments in Soil Science Volume 33. Elsevier, Amsterdam

Janssen R, Rievelt P (1990) Multicriteria analysis and GIS: an application to agricultural land use in the Netherlands. In: Scholten H J, Stillwell J C H (eds) *Geographical Information Systems for Urban and Regional Planning.* Kluwer Academic, Dordrecht

Longley P A, Batty M (2003) *Advanced Spatial Analysis: The CASA book of Spatial Analysis.* Esri Press, Redlands

Longley P, Clarke G (eds) (1995) *GIS for Business and Service Planning*. GeoInformation International, Cambridge, UK

Malczewski J (1999) *GIS and Multi-criteria Decision Analysis*. Wiley, London.

Price M F, Heywood D I (eds) (1994) *Mountain Environments and Geographical Information Systems*. Taylor and Francis, London

Ripple W J (ed) (1994) *The GIS Applications Book: Examples in Natural Resources: a Compendium*. American Society for Photogrammetry and Remote Sensing, Maryland

Thill J-C (1999) *Spatial Multicriteria Decision Making and Analysis: A Geographic Information Systems Approach*. Ashgate, Aldershot.

Wilson J P, Gallant J (2000) *Terrain Analysis: Principles and Applications*. John Wiley, London

## WEB LINKS

### Modelling environmental processes:

- University of Texas Center for Research in Water Resources: www.ce.utexas.edu/prof/maidment/gishydro/home.html

- GIS Lounge, hydrology links: gislounge.com/hydrology-and-gis/

- PCRaster: pcraster.geog.uu.nl/

### Modelling human processes:

- Research projects at the Centre for Advanced Spatial Analysis: www.casa.ucl.ac.uk/projects/index.asp

### PPGIS:

- The PPGIS web ring: t.webring.com/hub?ring=ppgis

- Integrated approaches to participatory development (IAPAD): www.iapad.org/

Visit our website **www.pearsoned.co.uk/heywood** for further web links, together with self-assessment questions, activities, data sets for practice opportunities and other interactive resources.

# CHAPTER 8

# Output: from new maps to enhanced decisions

## Learning outcomes

By the end of this chapter you should be able to:

- List the main forms of GIS output

- Describe the basic elements of a map

- Outline how three-dimensional output is handled by GIS

- Describe the output media used by GIS

- Explain why maps are important decision aids

- Explain what is meant by Spatial Decision Support Systems (SDSS)

- Describe how the Internet has affected GIS

- Explain what a VRGIS is

# ■ INTRODUCTION

Informed decision making and problem solving rely on the effective communication and exchange of ideas and information. A decision about the siting of a nuclear waste repository requires high-quality representation of the findings of site analysis, it requires presentation of these results to experts and the public and it will require an exchange of ideas and opinions at public meetings. Planning the development of an out-of-town retail park similarly requires presentation of the results of analysis in a format that can be understood by all parties involved in the decision-making process. There are many examples of GIS output being used to support decision making – from the environmental sciences, transport planning, archaeology, business studies and other disciplines. And for smaller-scale projects, perhaps individual research activities, effective presentation of results will help others to understand the results and the implications of any findings.

If GIS is to assist these activities, the information it produces must be meaningful to users. Users must be aware of the different forms information from GIS can take (including maps, statistics and tables of numbers). The term 'output' is often used to describe the ways in which information from a GIS can be presented. However, unflattering though the term may appear, it should be remembered that there is a continuity between data, information, decision making and problem solving, and that output – good or bad – will influence the decision-making process.

In the car analogy this is where, having used your car to reach your destination, you need to decide what to do next. If your destination was Happy Valley, you could go snowboarding or downhill skiing, ice-skating or tobogganing. Alternatively, you could relax in one of the cafés and do some shopping. In this context the car journey is not the end in itself, rather the beginning of a new adventure. In a similar way, users should see GIS output not as the final goal of a GIS project but rather as the starting point for informed decision making or problem solving.

The most common form of output from GIS is a map. In many cases the map will be thematic and will illustrate the spatial variation or pattern in a particular variable. For example, a map produced by the Happy Valley fire model will show areas of forest likely to be affected by fire. In other cases, GIS may be used to produce topographic maps such as those used by walkers and skiers which is to be used as a navigation aid. GIS output may also be a single number or list of numbers – for example, a nearest neighbour statistic used to help determine the location of the new Happy Valley supermarket or the correlation between a set of mapped variables such as avalanche occurrences and slope aspect. Alternatively, GIS output may be a three-dimensional model of a landscape, produced by draping land use data over height data provided by a digital terrain surface. Figure 8.1 shows a range of GIS outputs produced using the Happy Valley GIS.

As the most common method of visualizing information generated by GIS is the map, this chapter starts by considering the basic principles of map design. An understanding of these principles is essential for the effective communication of information and ideas. In addition, an understanding of the complexity of the map design process also helps appreciation of the power of maps as a visualization tool. The advantages and disadvantages of alternative forms of cartographic output are then reviewed together with other non-cartographic techniques. The chapter concludes with a brief discussion of the role of GIS output in supporting decision making.

# ■ MAPS AS OUTPUT

Despite recent advances in computer visualization, the map is still the most elegant and compact method of displaying spatial data. The role of the map is to communicate spatial information to the user. This information may include location, size, shape, pattern, distribution and trends in spatial objects. In designing a map so that it best achieves its objectives, it is necessary to consider a number of key map design elements (Robinson *et al.*, 1995). These include:

- the frame of reference;
- the projection used;
- the features to be mapped;
- the level of generalization;
- annotation used; and
- symbolism employed.

**(a)** Forest map

**(b)** Hotel distance surface

**(c)** Terrain surface

**(d)** Snow depth model

**(e)** 3D visualization

**Figure 8.1** Examples of GIS output from the Happy Valley GIS

When designing GIS maps we have the element of choice firmly on our side. The choices regarding frame of reference, projection, scale, generalization, content and symbolism are ours. We are in control and can manipulate the individual components of the map to suit our purpose. However, with choice comes responsibility, and the choices made ultimately determine the effectiveness of the map as a communication tool. Despite wide recognition by many that one of the strengths of GIS is its power as a visualization tool (Buttenfield and Mackaness, 1991; Hearnshaw and Unwin, 1994) there is comparatively little literature on producing quality output maps with GIS. This is perhaps understandable as the topic is covered in detail within the

cartography literature (see for example Monmonier, 1993; MacEachren and Taylor, 1994; Robinson *et al.*, 1995; Wood and Keller, 1996). We do not go into great detail here on the cartographic principles of map production as that would constitute a book in itself, but summarize the main points and direct you towards appropriate reading for further study.

Since a map can be regarded as a scale model of the real world, it needs some form of spatial referencing so that the user can fix its location in 'real-world space'. A number of graphic and non-graphic devices are used for this purpose. A grid may be used to give a spatial frame of reference with lines representing latitude and longitude or planar (x,y) co-ordinates. The spacing and labelling of the grid lines provide information on scale, and the orientation of the grid lines indicates which way the map is facing. A north arrow may be used to show orientation without a grid, whilst a numeric scale or a graphical scale bar is often used to provide information regarding the size of the map relative to the real world. An inset map showing the location of the main map area within its wider geographical setting is a very useful device in indicating approximate geographic location. Figure 8.2 shows how all of these devices may be used and Figure 8.3 how these may look in a GIS-produced map. A quick look at any published map should reveal the use of one or more examples.

Decisions also need to made about the projection used. **Chapter 2** explained how map projections can radically affect basic measurements such as area and distance and how easily GIS can convert from one projection to another. When drawing a map an appropriate map projection must be chosen.

Whilst this may not be much of an issue for large-scale maps of small areas, it does become an issue for small-scale maps of continents or the whole world. For large-scale maps of small areas, planar co-ordinates derived from a local projection system are the obvious choice. For maps of much larger areas, hard decisions often need to be made about which aspect of the map is most important: area, distances, directions or lines of true scale.

Returning to the idea that maps are communication devices for spatial information, it is important that the map shows only what is necessary to get across the intended message. A map should not contain anything that may detract from its intended purpose. In other words, 'there should be none of the extraneous objects or fancy flourishes typical of maps showing the location of hidden pirate treasure' (DeMers, 1997: 390). Needless and confusing embellishment is often referred to as 'chart junk' and should be avoided (Tufte, 1983).

**Figure 8.2** Map components

**Figure 8.3** Example map

The level of detail shown in a map can also determine how well a map communicates information. Too much detail and the map can become difficult to read, too little and essential information is lost. Generalization is important in this context. This is covered in detail in **Chapter 2**. In creating quality cartographic output, it is necessary to consider both the level of detail used in drawing geographical features (for instance, the number of co-ordinates used to represent a line) and the relative positions in which they are drawn on the map. The high level of precision at which co-ordinates are stored in the GIS data model does not help. Simply plotting these co-ordinates directly onto the map can produce too much detail. For example, in a map showing the road network of the UK that is to be published in a report on nuclear waste disposal, it may be necessary to remove some of the less important minor roads to avoid confusing the overall pattern. In addition, it may be necessary to generalize the detail of main roads by thinning out selected vertices (see Figure 5.18). If the railway network was also to be portrayed on the same map, there is conflict for space where roads and railways run parallel to each other. Plotted at a small scale, the roads and railway lines that follow close and parallel courses will plot directly on top of each other. To avoid confusion between the two features we need to add a slight offset to the different line segments so that they appear to run close to each other but not directly one on top of the other.

As Wood (1993) points out, what you choose to 'leave out' is as important as what you 'leave in'. After all, maps are communication tools and, if you choose not to tell someone something, there is usually a good reason. In many cases it may be that in omitting detail you can claim to be on the side of the user by improving the clarity of your message. For example, skiers in Happy Valley do not need a trail map that shows the location of every tree. However, the map generalization process is not always as honest as one might expect. Remember the Happy Valley ski area map introduced in **Chapter 2** that showed the location of only those restaurants that were owned by the Happy Valley ski company – a generalization that is not really in the best interests of skiers. Monmonier (1995) and MacEachren (1994) provide further insights into how generalizing map information may be done with less than honest intent.

The chosen method of map symbolism can also be important. Symbols on a map are either points, lines or areas. Each symbol can differ in size, shape, density, texture, orientation and colour (Figure 8.4). A poor match between the real world and the symbol used to depict it may confuse the user as to its true nature. It is for this reason that rivers are coloured blue and forests green. If they were coloured red and grey respectively, the user could be forgiven for mistaking the rivers for roads and the forests for urban areas. In our Happy Valley example, it would be a grave mistake to mix up the colours used to denote the standards of individual ski runs on the resort map. The standard

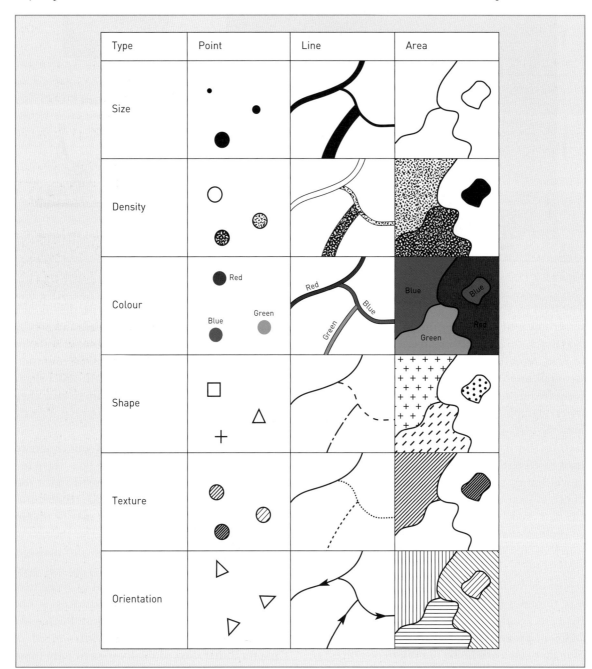

**Figure 8.4** Cartographic symbolism
*Source:* After Bernhardsen (1992)

international colour scheme is green (easy), blue (intermediate), red (advanced), black (expert). Reversing this colour scheme could easily result in problems and even legal action from skiers injured on pistes that they thought they would be able to ski safely. Thus, colour is important in influencing the user's understanding of the map. The overall impact of the map can also be affected by colour and symbolism. Black and white maps that use differently shaped symbols and shading patterns can be effective for simple patterns and are certainly easier and cheaper to reproduce than colour maps. However, colour may be necessary to effectively represent complex spatial patterns where detail would be lost in black and white maps.

The shape and pattern of the symbolism used also need to bear some relation to the feature being represented. For example, it would be bad cartographic practice to use an anchor symbol to show the location of an airport and a lighthouse symbol to show the location of a public telephone.

Returning to our road and rail example, if the railway lines were plotted as dashed black lines (– – – – – –) and the main roads as black lines with short perpendicular hatching (++++++) instead of the other way around, the map user may get the wrong impression since the crosshatched line symbol bears a closer resemblance to real railway lines. Careful choice of shape and pattern of symbolism used can influence the user's impression of the map.

Finally, density and texture of shading can affect the impression given by a map. In a monochrome choropleth map, high values may be represented by dense shading patterns (high density and coarse texture), whereas low values may be represented by lighter shading patterns (low density and fine texture). The case of choropleth mapping is discussed further in Box 8.1.

To ensure correct interpretation, whatever the symbolism used, it is essential that the map user is given the necessary information by a key or legend.

## BOX 8.1  Choices in choropleth mapping

**THEORY**

A thematic map that displays a quantitative attribute using ordinal classes is called a choropleth map. Each class is assigned a symbolism, and this is applied over an area feature (Chrisman, 2002). Choropleth maps are frequently used to present classified data on socio-economic variables such as population or unemployment where these are mapped by regions, but can be adapted to a wide range of uses. Conventions in choropleth mapping have long been a source of discussion amongst cartographers and geographers alike (Monmonier, 1972; Evans, 1977; Kostblade, 1981). Three basic problems present themselves in drawing this kind of map. These are:

1 *Choice of shading patterns*. Shading pattern can influence the user's reading of the map. It is normal to use 'the darker the greater' convention when specifying shading patterns. However, confusion may arise between maps where a high value (for example, in unemployment or pollution values) is seen as negative, and those where a high value is seen as positive, such as per capita income or crop yields. The use of colour has similar connotations. Bright colours are often used for and interpreted as high values. There

is also a high artistic element in the choice of shading patterns. Mixed use of different shading patterns and textures in the same map can be extremely messy, whilst unfortunate interactions between adjacent shading patterns can lead to a visual effect known as *moiré vibration*. Here, shading patterns made up of closely spaced parallel lines give the illusion of movement in the shading pattern. Another convention is to avoid the use of solid white and solid black shading patterns in monochrome maps.

2 *Choice of classification system*. In specifying the classification system of the shading scheme the cartographer needs to specify both the number of classes used and the class intervals allocated. As a rule, the greater the number of classes used, the more confusing the map. In most cases a maximum of five classes is deemed sufficient to display the variation in the data without creating an unnecessarily complex map. There are many different methods of specifying class intervals (Table 8.1). Some are based on arithmetic rules, whilst others try to match the statistical patterns in the data being presented.

→

**BOX 8.1**

**TABLE 8.1 Examples of class interval systems**

| Interval system | Description |
|---|---|
| Equal interval | Splits data into user-specified number of classes of equal width. Can create classes with unequal distribution of observations and/or classes containing no data |
| Percentile (quartile) | Data are divided so that an equal number of observations fall into each class. When four classes are used this method is referred to as *quartiles*. Efficient use is made of each class to produce a visually attractive map. The range of the intervals can be irregular and no indication is given of the frequency distribution of the data |
| Nested means | Divides and subdivides data on the basis of mean values to give 2, 4, 8, 16, etc. classes. Provides compromise between equal number of observations per class and equal class widths |
| Natural breaks | Splits data into classes based on natural breaks represented in the data histogram. Provides commonsense method when natural breaks occur but often data are too scattered to provide clear breaks |
| Box-and-whisker | Splits data using mean, upper and lower quartiles, outlier and extreme values derived from box-and-whisker plot. Accurately describes distribution of the data |

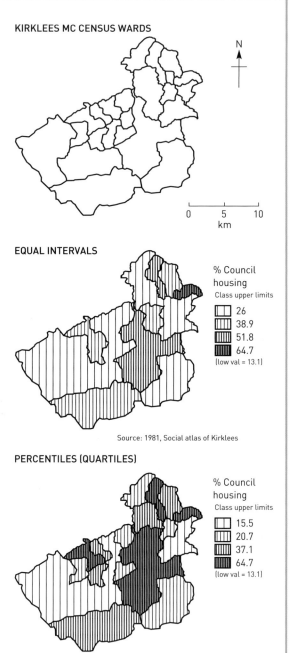

KIRKLEES MC CENSUS WARDS

EQUAL INTERVALS

% Council housing
Class upper limits
26
38.9
51.8
64.7
(low val = 13.1)

Source: 1981, Social atlas of Kirklees

PERCENTILES (QUARTILES)

% Council housing
Class upper limits
15.5
20.7
37.1
64.7
(low val = 13.1)

Source: 1981, Social atlas of Kirklees

**Figure 8.5** Percentage of council housing by census ward

**3** *Choice of spatial unit.* The choice of spatial units used to map the data may be outside the direct control of the cartographer, as it is very difficult to disaggregate data beyond its minimum resolution or translate the data into another set of non-matching units. However, some freedom of choice does exist in decisions about aggregation of the data into larger spatial units. Amalgamation of data in this manner can lead to a loss of information by masking internal pattern and variation that may be important. The modifiable areal unit problem (MAUP) can also mean that the boundaries of the spatial units themselves act to hide underlying patterns in the data (Box

6.7). If aggregating data into larger spatial units, it is normal to do this on the basis of formal relationships: for example, where administrative districts nest into counties which nest into countries or states in a fixed and formal manner. However, there is nothing to say that the cartographer has to maintain these more formal nested relationships. Instead data can be aggregated on a more logical geographical basis.

BOX 8.1

For example, adjacent districts sharing similar characteristics could be aggregated to avoid some of the problems of the MAUP and produce choropleth maps that better reflect the underlying geographical pattern of the data. An automated system for performing this aggregation or Zone DESign (ZDES) has been developed by Openshaw and Rao (1995).

As a result of these choices, choropleth mapping is a technique for the communication of spatial patterns in attribute data that is open to abuse. Unscrupulous cartographers can draw choropleth maps that have shading patterns, class intervals and spatial units specifically chosen to (mis)represent a particular view of the data. This is common in maps making political capital from dubious statistics (Monmonier, 1996). The maps in Figure 8.5 are based on the same data and spatial units but exhibit different patterns, simply due to the choice of class interval. The user therefore needs to be aware of these issues and study the map and its legend critically before arriving at any conclusions about the data presented.

This tells the user what the shading patterns, colours, line and point symbols mean and gives information about the class interval system used (see Figures 8.1, 8.2 and 8.5). Without a key it would be difficult to interpret the mapped information accurately and confidently.

Just as irrelevant detail and inappropriate symbolism can swamp the message in a map, appropriate and economic use of annotation can help bring a map 'to life' and provide added meaning. Annotation is textual and graphical information that either labels the map and features presented within it, or provides the user with supplementary information. The visual and numeric guides to the frame of reference described above may be considered a form of annotation. More obvious uses of annotation include the map title, legend and labels given to mapped features. Supplementary text may be added to a map to provide information about the data used and the mapping methods. For example, information may be provided about data sources, quality estimates, projection system, reference system, copyright and other issues. 'Parts undiscovered' printed on a map showing the geography of the 'Great Solar Eclipse' of 14 July 1748 leaves the user in no doubt as to the quality of the mapped information in certain geographical regions. This is referred to as 'metadata'. Metadata and quality estimates are discussed further in **Chapter 10**. The inclusion of metadata on digital GIS maps is important because in their digital form they are more readily adapted for other purposes. Therefore, metadata acts in a similar way to the health and safety warnings found on consumer products – it guides users as to the appropriate uses of the product.

Map design is as much an art as it is a science. The primary objective is to ensure effective communication of the map's theme(s). This involves many technical decisions about scale, generalization and reference systems, but at the same time artistic decisions about positioning of map elements (title, legend, scale bar and north arrow) and the colour and symbolism used. The artistic element can largely be a matter of personal preference, but certain conventions apply in certain circumstances. A good starting point for developing an appreciation of cartographic design is the book by Wood and Keller (1996), which looks at map design from an applied and theoretical perspective.

## Alternative forms of cartographic output

Traditional maps normally assume that the observer is positioned directly overhead and is looking vertically down at the surface of the Earth. Maps are drawn as plan views and at a fixed spatial scale. In recent years new forms of cartographic output have become more common. These include cartograms, three-dimensional views and animation. These are considered below.

At first sight cartograms look like normal maps. However, instead of showing the location of objects or variables using standard Euclidean co-ordinates, the objects depicted are mapped relative to another variable such as population density or relative distance. In this manner, distances, directions and other spatial relationships are relative rather than absolute. A common use of cartograms is for drawing maps of urban mass-transport networks, such as the Tyne and Wear Metro Map (Figure 3.27a). This type of cartogram is known as a *routed line cartogram*. Routes are plotted as generalized lines showing the sequence of stops and connections rather than showing the actual route taken. In this manner the user's understanding of the map and the spatial relationships between stops is greatly improved but often at the expense of true scale of distance between stops and the actual route taken. You can appreciate this if you compare the Tyne and Wear Metro Map with the map showing the true geographic location of the underground routes in Figure 3.27b.

A related form of cartogram shows space distorted from a central point on the basis of time taken or cost incurred to travel to a series of destination areas. These are known as *central point linear cartograms*. They have been used to show the relative costs of air travel to different parts of the world from one particular airport and how ticket prices are not uniformly determined by distance travelled. Another use of cartograms is to depict the size of areas relative to their importance according to some non-spatial variable such as population or per capita income. These are referred to as *area cartograms*. Examples of area cartograms can be seen in Dorling (1995) and Figure 8.6. Cartograms provide an excellent means of visualizing patterns and differences, but they radically distort geographical space by shifting the relative positions, orientation, size, shape and sometimes connectivity between areas.

Three-dimensional (3D) views of spatial information are becoming a popular method of visualization and have been driven by developments in computer graphics and gaming. These display (x,y,z) co-ordinate information using an orthographic projection based on the angle of view (0–360 degrees around the compass), view azimuth (0–90 degrees from horizontal to vertical) and view distance (distance of viewer from the image space). The resulting image is shown as a fishnet or wire-frame diagram, an example of which is shown in Figure 3.30. This is an excellent means of displaying topographic data. It is also possible to drape other images over the wire frame map. Thus, added realism may be created by draping land cover data derived from satellite imagery or aerial photographs over 3D terrain. In the case of our Happy Valley ski example, the 3D topographic maps issued to skiers to help them navigate their way around the many lifts and pistes within the resort are an example of this kind of GIS output. To produce these, the Happy Valley GIS team has created several 3D wire-frame images of the ski area upon which it draped digitized aerial photographs together with relevant coded vector data for ski lifts, ski pistes and hotels. An artist and graphic designer produced the

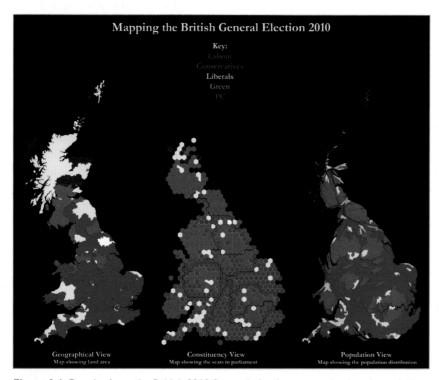

**Figure 8.6** Results from the British 2010 General election mapped using choropleth mapping and area and gridded cartograms (after Dorling, 1995)
*Source*: Benjamin Hennig, Department of Geography, University of Sheffield, www.sheffield.ac.uk/mediacentre/2010/1615.html

final topographic maps complete with legend, annotation and corporate logo.

Animation can be useful when there is a strong temporal element to the data being displayed. Tobler (1970) first used animation as a means of identifying patterns in spatio-temporal data sets. A series of maps is drawn, each showing the spatial pattern of the observed variable at a particular time. When 'played back' in quick succession as a series of frames, the change of the spatial pattern over time can be seen. This technique has proved useful for identifying temporal patterns in point data sets and in visualizing diffusion processes. 'Computer movies' have been used to analyse urban growth (Tobler, 1970) and crime incidents (Dorling and Openshaw, 1992). Animation may be used in a similar way to display all manner of spatio-temporal processes, from the development of forest fires and pollution plumes to changes in hydrological variables and population densities. By combining the two techniques of 3D mapping and animation, modern GIS packages can create realistic 'fly-through' animation using a sequence of three-dimensional images, each viewed from a different perspective. This is a powerful means of terrain visualization, similar to the interactive computer animations seen in flight simulators. Box 8.2 describes other developments in 3D mapping and the use of animation for geovisualization in the retail sector.

---

## BOX 8.2  Retail geovisualization: 3D and animation for retail decision support

CASE STUDY

### Tony Hernandez

Geovisualization, the development of techniques and technologies that facilitate the visual exploration, analysis, synthesis and presentation of geospatial data, is now an important area of GIS research. This case study outlines the application of geovisualization within retail decision support by examining two techniques – 3D mapping and data animation.

### 3D MAPPING

The majority of maps used by retail organizations show an abstract 2D plan view of data, such as a choropleth map. Three-dimensional mapping allows retailers to visualize data values more effectively by scaling each map object (using the z-axis) to highlight intensity and variation in data. In order to convert 2D maps into 3D maps a number of geometric objects need to be created:

- Two-dimensional polygons typically need to be triangulated, with each polygon divided into the minimum set of triangles that define the area (see **Chapter 6** for more on TINs and DEMs).
- Two-dimensional points require the generation of entirely new objects: for example, a 3D sphere or cube around the centroid (often scaled by the data value).
- Two-dimensional surfaces may require points to be interpolated to create a smoothed surface of data values.

Generating 3D objects is computationally intensive, and this, in part, explains why geovisualization has traditionally taken place on high-end computer workstations. For most geovisualization applications polygons need to be generalized before being translated into 3D objects. Numerous algorithms are available for this, and a number of these (including Douglas–Peucker) are available in standard GIS packages. Generalization should minimize geometric complexity, whilst maintaining the integrity of the geographical shape and the topology of the entire data set. Figure 8.7, which was produced using CSCA GeoVisualizer software, shows a comparative 3D view of the retail trade areas for two stores operating in the same market. The height and colour of the 3D bars and cubes represent the size of the retail stores in square feet. These objects are positioned on a probabilistic trade area surface and the major road network is draped as a cosmetic map layer. The probabilistic trade area surfaces were generated using a Huff-type spatial interaction model. Figure 8.7 shows the different shape and size of the trade areas for 'Store A' and 'Store B'. Contour lines highlight variation in the surface. The hollow in the surface next to 'Store B' shows the impact of a nearby competitor (with the likelihood of customers visiting 'Store B' decreasing sharply around the competitor store). With this type of visualization a retail analyst can compare the trading environment of two or more stores (or two or more trade area models for the same store).

BOX 8.2

**Figure 8.7** 3D view of the retail trade areas for two stores operating within the same market

## DATA ANIMATION AND DYNAMIC LEGENDS

Animation helps the user discover trends, patterns and anomalies in complex spatial data. Data displays that change continuously, either with or without user control, are 'dynamic representations'. Dynamic data representation has changed the way users obtain and interact with information across the full range of display technologies. One form of dynamic representation is the animated map. Animated maps display sequences of data (for example, a sequence of maps might show the growth of population in a given area over a set time period). An argument for using animation is that it enables effective display of temporal data, with changes in real-world time reflected by changes in display time. Dynamic representations also permit users to explore geospatial data by interacting with map displays: for example, by rotating and zooming-in and -out of the display during data animation. In order to create smooth animation, it is often necessary to interpolate a number of virtual values between actual values. For example, imagine that a user wants to display a map of retail sales for a network of 30 stores, covering 10 years of annual sales data. If the user created an annual sequence of this data, there would be 10 time snapshots. It is likely that the resulting animated sequence would not appear particularly smooth as the data would change abruptly 10 times. To overcome this an additional sequence of virtual snapshots that

fit between each actual time snapshot can be created. For example, with 9 additional snapshots between each time period the map display would update 100 times, producing a smoother animation.

When dealing with temporal data it is necessary to include tools to allow the user to navigate through time. Video-recorder type buttons are often provided to help the user play, fast-forward and rewind the temporal sequence, along with other controls to set the speed, and start and end time.

The ability to produce animations of data sequences and produce digital movie files is particularly important in terms of the sharing of information between different departments or individuals within a given company. For example, an analyst may produce an animated data movie showing the growth of a store network and the changing level of monthly customer penetration over a three-year time frame. This could be viewed during a capital investment board meeting to highlight the business case for investment in new store locations. A number of other examples of animated 2D and 3D maps can be viewed on the Internet (see the web links below).

When displaying animated data sequences the user needs a legend or some other means of assessing the range and intensity of the values being displayed at any one time. In the case of a static map, a single legend for the single array of data values will suffice.

**BOX 8.2**

When visualizing a series of time periods, multiple arrays of data values are being displayed. As a result, dynamic legends are often used. Since a legend with a minimum and maximum value for time period $t_1$ may not cover the range of data in time period $t_2$ to $t_n$, there are two approaches that can be adopted:

**1** The legend is updated for every new time period (dynamic) – this can be very difficult for the end user to interpret as the objects on the map change shape and colour whilst the legend also changes. This creates difficulties for the user, particularly if the data are only displayed for a short period of time since the user may have little time to make sense of the map.

**2** The second approach is to take the minimum and maximum values across all of the data arrays, and keep the legend static (fixed). In this case, if the data range is large, or there is sporadic or dramatic change in the data values, the fixed legend may not provide useful information across a number of the time periods.

Three-dimensional and animated 2D/3D mapping provides retail analysts with new ways of displaying and interacting with geographical data. Technological barriers that once limited 3D graphical display and animation to high-end computer workstations are being overcome, and the potential to develop retail geovisualization applications is likely to increase significantly. The challenge will be in generating retail applications that provide a set of geovisualization tools that enhance the way in which decisions are made.

**FURTHER INFORMATION**

Business Geomatics: www.businessgeomatics.com

*Source*: Tony Hernandez, Ryerson University

---

**REFLECTION BOX**

- Find three different maps in an atlas or textbook you have to hand, or use maps downloaded from the Internet. Make notes on the key map design features (frame of reference, projection, features represented, generalization, annotation and symbolism) used in each example. Is there anything you might add, or anything you might leave out?
- Imagine that you need to prepare a map to present the proposed location of a new hotel in Happy Valley. What features would you include on your map? Why? What would you leave out? Why?

- Look critically at the maps presented in this chapter. Identify one or two examples which you think make effective use of colour and explain why you have chosen them. Identify one or two examples in which the use of colour could be improved and make suggestions for improvements.
- Is map design an art or a science?

---

## ■ NON-CARTOGRAPHIC OUTPUT

Whilst the map is undoubtedly the main form of GIS output, we must not ignore other non-cartographic forms. Tables and charts containing spatial and non-spatial attribute information derived from the GIS database are important forms of output. Numerical or character information linked to spatial objects (points, lines, areas or surfaces) are often best displayed as tables or in charts. In many cases the output from a GIS analysis may be a single number or character string – for example, a nearest neighbour statistic or the character string returned from a spatial query. This kind of information is best displayed temporarily on the computer screen and then transferred to a document or report as required.

A recent development in the display of non-cartographic information is the *linked display*. This approach makes use of the dynamic linking capabilities provided by window-based operating systems and graphical user interfaces (GUIs) to connect different data display windows. Thus, actions in one window are reflected in displays of linked data in other windows. For example, if a user is simultaneously viewing a choropleth map in one window and a scatterplot of the same data in another, then queries performed on the scatterplot will be echoed in the choropleth map, and vice versa. To illustrate the value of this, we return to avalanche data in the Happy Valley case study. The piste wardens are interested in identifying patterns in the occurrence of avalanches that are not immediately apparent when inspecting the raw data. By creating multiple scatterplots of the variables collected for each avalanche recorded in Happy Valley, and dynamically linking them to a map of avalanche events, the piste wardens are able to investigate statistical and spatial patterns simultaneously. Many GIS packages are now exploiting this functionality to dynamically link tables and charts to map displays (see Figure 8.8). The advantage of this approach to data output is that it provides the user with a highly interactive method of exploring the spatial and aspatial aspects of data simultaneously.

## ■ SPATIAL MULTIMEDIA

Some GIS packages offer facilities for the display and playback of multimedia to supplement cartographic images. Photographic images, text, graphics, sound and video can be linked to map displays to add value to a traditional map. These media provide the user with additional contextual information about the spatial detail in the map and may act as an aid to understanding (Figure 8.9). Certainly, being able to look at photographs and watch video of an area is helpful in developing an improved understanding of what a mapped region actually looks like, especially if you have never visited the area in question. For example, the Happy Valley marketing department has developed a multimedia ski map for the ski area that includes video descents of all the ski pistes, which are started by selecting the piste on the trail map. Outside pure marketing applications such

**Figure 8.8** Example of linked displays in GIS
*Source*: Screenshot shows Esri Graphical User Interface (GUI). ArcMap, ArcView and ArcInfo Graphical User Interfaces are the intellectual property of Esri and is used herein with permission.Copyright © 2011 Esri all rights reserved.

**Figure 8.9** Example of multimedia content in GIS displays
*Source*: Screenshot shows Esri Graphical User Interface (GUI). ArcMap, ArcView and ArcInfo Graphical User Interfaces are the intellectual property of Esri and is used herein with permission.Copyright © 2011 Esri all rights reserved.

spatially referenced multimedia has many benefits. In a utility organization, viewing pipeline condition from photographs or video may be a useful exercise when planning a programme of repairs in a particular area. Heywood *et al.* (1997) consider how multimedia can be used with GIS output in an educational context to inform users about environmental problems. Others have explored how sound can be used to add value to mapped information. Shiffer (1995b) incorporated the sound associated with the take-off and landing of aircraft into a GIS developed to explore the environmental impacts of airports. Whilst multimedia may be exciting, too much can combine to produce a kind of technological overload.

## ■ MECHANISMS OF DELIVERY

In the previous sections of this chapter we have reviewed several forms of GIS output, ranging from maps to animated 'fly-through' visualizations and multimedia presentations. The mechanism, or medium, of delivery of these varies markedly. Two basic types of output medium can be defined: those that are permanent (known as hard copy), and those that are ephemeral (soft copy).

Hard copy may be maps or other GIS output that are printed or plotted on paper or film for the user to take away. A bewildering array of output hardware is available on the market. These devices include dot-matrix printers, bubble-jet and ink-jet printers, laser printers, pen plotters, thermal printers, photographic image setters and electrostatic colour plotters. Each produces a different quality of output using various technologies at varying costs. Table 8.2 summarizes a range of hard-copy output devices and their characteristics.

Ephemeral output is essentially based around the computer screen. It is ephemeral in the sense that the image is digitally reproduced on the computer screen and cannot be taken away, but is viewed by the user for a time and then cleared from the screen. This is typically used by analysts sitting in front of a GIS terminal, or by users accessing images over the Web, using browser software. Increasingly map-based information is being served using wireless technology to users with Personal Digital Assistants (PDAs) and mobile phones. Linked to a GPS, these devices and systems can combine instant local maps and user locations to present almost boundless possibilities for applications from personal navigation systems to real-time monitoring.

There is a wide variety of screen types that can be used for display of output. These include cathode ray tubes (CRT) and liquid crystal displays (LCD). Screens may be monochrome (black-and-white or green-screen) or colour. Most operate by representing the image as a grid of pixels (picture elements) in a similar fashion to the raster data model. Screen technology is developing very rapidly and many new types of screen are under development that will benefit GIS. These include flexible screens, low power consumption for portable devices, and head-up displays for wearable systems.

Slide shows are a commonly used ephemeral method of demonstrating a particular GIS application to interested parties. These are simply an ordered selection of screen shots illustrating a sequence of operations and GIS outputs. More advanced demonstrations may be interactive, allowing the user (who may have little or no GIS training) to choose display and/or analysis options and carry out spatial search and query operations on-screen. Multimedia may be included in such systems as appropriate to add contextual information.

Recent research has focused on the development of *virtual reality* (VR) displays of spatial data. Much of this research and development work is taking place on the Web using the virtual reality modelling language (VRML) (Raper, 1997). An interesting development in this field has been the application of VR and the Web in producing so-called 'virtual field courses' for student use (Fisher *et al.*, 1997). The merit of these websites is that they can either serve as preparatory tools for real field studies or be used to allow students to undertake 'field work' in remote or dangerous locations such as Antarctica, where conventional field courses would be impractical and too costly (see Figure 8.10). Box 8.3 describes developments in Virtual Reality GIS, whilst Box 8.4 describes 'Virtual London'.

**TABLE 8.2 Hard-copy output devices**

| Device | Colour | Quality | Resolution | Cost |
|---|---|---|---|---|
| Dot-matrix | mainly mono | low | low | low |
| Ink-jet | mono/colour | medium | medium | low |
| Pen plotter | mono/colour | high | high | medium |
| Laser | mono | high | high | medium |
| Colour laser | colour | high | high | high |
| Thermal wax | colour | high | high | high |
| Electrostatic | colour | high | high | high |

**(a)** Panoramap          **(b)** Landserf

**Figure 8.10** The Virtual Field Course
*Source*: From the Virtual Field Course website, www.george.ac.uk/vfc.

## BOX 8.3 Virtual reality GIS

PRACTICE

Virtual Reality (VR) produces 'virtual worlds' that are used to imitate reality. Virtual Reality GIS (VRGIS) simulates the real world using a GIS database and computer graphics to produce interactive and navigable 3D visualizations. In practice, VRGIS is a standard GIS database with VR as the user interface.

Recent advances in computing power, the availability of low-cost yet powerful desktop GIS packages and dedicated VR programming languages such as X3D (and it's predecessor VRML – Virtual Reality Modelling Language) have made the combination of GIS and VR possible. VRGIS provides the power and functionality of GIS together with the intuitive and easy-to-use interface of Virtual Reality.

Imagine being able to interrogate a GIS database or retrieve data simply by pointing with your virtual 'finger' or grabbing it with your 'hand'. Or being able to perform a viewshed analysis by 'walking' to the viewpoint and turning through 360 degrees to see what is visible. These are just two examples of what might be possible with VRGIS. Some of the features of VRGIS are (after Faust, 1995):

- realistic representation of real geographic areas;
- free movement of the user within and outside the selected geographic terrain;
- standard GIS capabilities (such as query, select and spatial analysis) in a 3D database; and
- a visibility function that is a natural and integral part of the user interface.

However, many applications are not 'true' VRGIS but are 'loosely coupled' systems in which spatial data are transferred from standard GIS packages using VRML for display using VR software (Berger *et al.*, 1996). In this case the 3D visualization capabilities of the VR software tools are used to augment the 2D cartographic tools of the GIS.

The number and type of applications of VRGIS have increased rapidly over the last few years. Examples include exploration of past and future urban landscapes, environmental impact assessment, visualization of abstract data in the context of real world variables, landscape construction, real-time viewshed exploration and education.

# BOX 8.4 Virtual London

## Michael Batty

Maps offer a convenient and user-friendly view of spatial data, but they are limited in terms of engaging a wider set of users. It is not surprising, therefore, that GI scientists are working to make visual representation ever more realistic. Moving from two to three dimensions, and beyond to four dimensions, extends GIS and GIScience to embrace myriad of developments in computer graphics, computer-aided design (CAD), multimedia, and virtual reality (VR).

To make maps more realistic, one approach is to view the data in three dimensions as surfaces (often regarded as 2.5D). Another is to develop geometric renditions with the third dimension captured as a wire frame. The fourth dimension (time) may be captured using 'temporal slices'. These kinds of representations are being created for many different kinds of user – on the desktop (where the user can be immersed into the media at different levels of intensity), across the Internet or in theatre-like contexts (where many users can engage and manipulate the same environment).

These representations usually require specialist hardware particularly where users wish to immerse themselves in a virtual view of the data. Headsets and other peripheral technologies connected to a desktop or workstation are examples, as are purpose-built theatres (such as CAVES) and cinemascope-style panoramic screens. Much of this technology is now being networked and it is possible to share virtual realities in visual (or even audio-tactile) ways across the Internet through virtual worlds. In addition, with new screen and projector technologies, holographic-type representations that enable users to be quite separate from the environments that they are able to experience in virtual-digital terms, are moving closer (Batty and Hudson-Smith, 2005).

### CONSTRUCTING VIRTUAL ENVIRONMENTS

VR technologies have been embodied in GIS software through extensions to the third dimension. These enable users to extrude the geography and the geometry into forms that mirror what we experience in the real world. Initial developments focused on maps as surfaces. Then map parcels and street lines were extruded into building blocks (adding natural features along the way) and then these objects were rendered in ever-more realistic ways. GIS has already reached other similar software coming from the CAD area, and is complemented by various techniques from image processing and photorealism which seek to represent the environment in the most obvious and realistic ways. Many of the examples of 3D landscape and cityscapes developed in GIS have this form. However, the fact that these are simply one from a multitude of views of spatial data does lead to the notion that many different views of the data in 3D might be assembled. For example, an iconic model of a city based on 3D blocks and their rendition (but containing many other data layers of different abstraction) may be developed and often enabled in different windows on the same screen, or on different but co-ordinated screens in various semi-immersive environments (Batty *et al.*, 2001).

### VIRTUAL LONDON MODEL

A good illustration of these principles is the model built for central London called 'Virtual London'. This model includes 50,000 building blocks which form the morphology of some 20 km$^2$ of the centre of the city. The GIS layer concept is central to the way the model is constructed:

- a digital elevation model provides the 'bare earth';
- an aerial photograph is draped across this to give some local rendering;
- vector-based parcel and streetline map data are overlain; and
- using LiDAR data blocks are extruded up to their median roof heights.

The model provides a wire frame which can be rendered by importing CAD files which have been prepared at greater levels of detail or by developing detailed rendering using photorealistic geomatic techniques. Any of the model's locations can be tagged at any (*x,y,z*) point (or area or line) with various attributes. In this way different attribute surfaces can be visualized in 2D or 3D or in other ingenious representations.

## BOX 8.4

### APPLICATIONS OF VIRTUAL LONDON

To illustrate the visualization of many different kinds of data layers within the model, a variety of air pollution surfaces have been inserted into the geometric view of the model. These have been constructed using ArcGIS by a group of health researchers at Kings College London whose aims are very different from those of the Virtual London project. Their concern is visualizing hot spots of pollution, exploring how different pollution episodes affect different parts of the city, and ultimately tracing the impact of pollution on the health of the resident and working populations. The Virtual London model provides the framework for such studies and Figure 8.11 shows a sample of the pollutant NOx (nitrogen dioxide) layered onto the geometry of the model on the south bank of the river around London's Millennium Eye (the Big Wheel). It shows how intense this kind of pollution is along streets and near significant buildings. Such visualizations have been repeated over and over again with different data sets (see Hudson-Smith *et al.*, 2005).

Other innovative extensions for GIS which are made possible by the use of 3D include:

- showing which buildings might be affected by rises in the level of the River Thames due to global warming;
- testing the impact of high buildings using viewshed analysis;
- analysing deprivation;

- assessing the impact of new public transport lines; and
- going down as well as up in terms of the third dimension to represent the inside as well as the outside of buildings.

The model can also be imported into other media such as virtual worlds. This has been done for educational purposes using the London Grid for Learning. A virtual world is a 3D environment, remotely located on the Internet, which can be accessed in the usual way through a Web browser but which show the user's presence in the 3D scene as an avatar. The scene is configured from the user's own location and perspective and in this way, others logging onto the world can see the user from their perspective (Batty and Hudson-Smith, 2005). Such worlds enable dialogue – usually through text commands but increasingly through audio. Figure 8.12b shows Virtual London imported into such a virtual world. Figure 8.12a shows the model and 8.12b shows a portion of the model inserted in the world which is configured as a virtual exhibition space. Two avatars are logged into this world and it is possible for them to engage in dialogue concerning the design of the space. Key buildings can be moved and the actors/avatars engage in debate about the best position for such buildings. These environments are rapidly becoming the basis for collaborative design. The model can also be printed as hard copy (Figure 8.12c).

The lessons from such applications relate to the way GIS is penetrating many of the traditional areas of computer graphics that focus on digital display but whose pictures and views are empty of real data. This represents a sea change in visualization: spatial data can now be visualized in countless ways of which – through 3D environments – is a particularly important one. At the same time ever-more visual realizations are enabling such developments to generate much greater involvement and participation in promoting the public interest.

### FURTHER INFORMATION

Centre for Advanced Spatial Analysis: www.casa.ucl.ac.uk/
CASA book: www.casabook.com/
Note: The author wishes to thank Andy Hudson-Smith and Steve Evans of CASA at UCL who are the architects of *Virtual London*.

*Source*: Michael Batty, University College London

**Figure 8.11** Layering a Pollution Surface (Nitrogen Dioxide NOx) onto Virtual London

**(a)** GIS Model

**(b)** Multi-User Virtual World/ Exhibition Space

**(c)** Hardcopy model from a CAD/CAM printer

**Figure 8.12** Different Renditions of Virtual London

## Serving maps on the Web

The use of the Internet to disseminate maps and spatial information has been perhaps the biggest development in GIS over the past few years. Initially the Internet was used to give wider access to GIS images and associated reports in their static form as words and pictures. The industry has been quick to develop the possibilities of the Web for more interactive access to GIS software and data. GIS-based websites now allow users to interrogate map data, compose new maps and update and share spatial data. Indeed, one of the most frequent searches on the Internet is for maps and map data. Most GIS packages have Internet mapping additions to complement stand-alone systems. These allow users to serve map-based content to members of their organizations and clients without the need for expensive GIS software and data on local computers. For some organizations this is the ideal way of sharing data between various departments and users, particularly if other networking solutions are precluded by large distances between buildings and offices. The rise of web services, such as Google Maps, Google Earth and Bing, has done much to put maps and aerial imagery within everyone's reach. They are used for everyday activities such as searching for shops and restaurants, confirming addresses, or obtaining driving directions and distances. Where coverage is available, Google Streetview allows users to view ground-level imagery, and global terrain models in Google Earth allow oblique views of maps and imagery to be draped over local terrain (see Figure 8.13). Georeferencing tools are available in Google Earth and Google's Picasa photo management software to allow the 'geo-tagging' of digital photographs with information about where the picture was taken. This permits the display and sharing of images according to geographic location. Other Web-based mapping tools and visualization software, like Google Earth and NASA's World Wind, allow the overlay of third-party map layers and objects onto a map base using XML-based languages such as KML (Keyhole Markup Language). Some of these tools are also available for mobile phones, along with cut-down GIS software (for example, Esri's ArcGIS) and applications from data suppliers (such as the British Geological Survey's iphone app 'iGeology'). Box 8.5 provides examples of how Web-based GIS might be employed by the Happy Valley Ski Company. Web-based GIS is discussed further in **Chapter 13**, and more details are provided in **Chapter 4**.

(a)

(b)

**Figure 8.13** Google maps with Streetview (a) and Google Earth terrain view (b)
*Source*: Courtesy of Google UK

## BOX 8.5 Happy Valley Internet GIS

CASE STUDY

The Happy Valley management team wish to improve the sharing of data within the company; improve resort monitoring and management; improve the marketing of the ski resort to potential customers; and improve the service they provide to their existing customer base. They have employed a Web programmer to work with the GIS and marketing teams to address these broad aims through the features and functions of the website.

### DATA SHARING BETWEEN DEPARTMENTS

Data on the physical and human resources of Happy Valley, together with information on customers and finance, are important resources that need to be continuously available to employees in a range of departments. Copies of GIS and non-spatial data are placed on the company's intranet to allow employees to access essential information

**BOX 8.5**

from their desktops. Changes to the database are password protected to make sure that only genuine changes made by authorized users are permitted. Updates made to the central database are immediately available to all users.

## DYNAMIC PISTE MAPS

GIS-based piste maps are available from the company's Web page. These allow existing and potential customers to check the condition of ski runs before making a visit. The condition of the ski runs is monitored by ski rangers who log the information on PDAs equipped with GPS receivers. Wireless communications transmit information about piste condition and location from the PDAs to the piste database. The GPS location is used to link the new information on piste condition to the relevant stretch of piste in the GIS database. Information on which lifts are open and the length of any queues is similarly transmitted to the piste database by wireless link or land-based telephone line. Customers using the Web to access piste maps see current information, displayed in real time. Clicking on any of the pistes and ski lifts gives access to more detailed information, including condition histories and seasonal summaries. Customers can also download these maps as KML files for use with Google Earth.

## WEB CAMS

Several web cameras (web cams) are placed around the resort. The location of web cams is marked on the piste maps as 'hot spots'. Customers clicking on the hot spots can pull up a real-time video image of the view from the web cam. They can see the weather, the condition of the pistes, length of the lift queues and whether there are any spaces left in the car parks. This service is also provided to hotels in the resort via cable TV.

## RESORT MONITORING

It is essential for the smooth running of the resort that managers have instant access to information about conditions in the resort. In addition to the information provided by the dynamic piste maps and web cams, resort managers have access to data from electronic monitoring devices around the resort that are linked via landline or telemetry to the central database. These data include:

- the volume of people going through ski lift turnstiles;
- the volume of traffic on the resort's road network;
- the location and movement of ski buses;
- the number of spaces left in car parks;
- the location of ski ranger;
- the weight of the snow pack;
- data from strain gauges located on avalanche-prone slopes; and
- data on meteorological conditions around the resort.

All data can be displayed dynamically on maps of the resort. Meteorological data are used by the resort's met office and linked to regional and national weather data to provide short-, medium- and long-term weather forecasts.

## NAVIGATION AND TRAFFIC MAPS

The Happy Valley website contains a link to a third-party route manager and traffic monitoring website. This allows customers to plan the quickest route to the resort and also check current traffic conditions.

## BOOKING SERVICE

A booking service is available from the Happy Valley website. Customers can check the availability and cost of hotels, ski school places and ski hire. A map of services in the Happy Valley resort helps customers choose the services they want. Detailed information about a particular hotel or ski school can be viewed simply by clicking on its location. If a service provider has its own website, a link to this is provided. For most hotels, virtual tours of facilities are available. The map of services is customizable by the user, so that only those services of interest are displayed. Bookings and payment can be made online. Once an online transaction is made, the information is passed directly to the central database and to the service providers.

Despite the cost of setting up the monitoring systems and Internet and intranet access tools, the Happy Valley resort management are confident that the benefits of these developments will be better resort management, increased revenue from service commissions and increased visitor numbers.

Datasets and activities relating to the Happy Valley Case Study can be found online at **www.pearsoned. co.uk/heywood**.

## ■ GIS AND SPATIAL DECISION SUPPORT

At an organizational or strategic level the most important aspect of GIS output is its use in spatial decision support. Central to this is that GIS provides a much-needed framework for approaching, supporting and making spatial decisions. Decision making as a scientific discipline has a much longer history than GIS. As such, much of the decision-making theory is well established and has its own specialized literature. Within the wider decision research field, computers have been used to develop decision-support systems (DSS). Geoffrion (1983) identifies six distinguishing characteristics of DSS.

1  Decision-support systems are used to tackle ill-structured or semi-structured problems that may occur when the problem and objectives of the decision maker cannot be coherently specified.

2  Decision-support systems are designed to be easy to use, making sophisticated computer technology accessible through a user-friendly graphic user interface (GUI).

3  Decision-support systems enable users to make full use of the data and models that are available, perhaps by interfacing external routines or database management systems.

4  Users of decision-support systems develop a solution procedure using models as decision aids to generate a series of decision alternatives.

5  Decision-support systems are designed for flexibility of use and ease of adaptation to the evolving needs of the user.

6  Decision-support systems are developed interactively and recursively to provide a multiple-pass approach.

GIS has been referred to as a special kind of decision-support system dealing with problems that involve a high degree of spatiality (Densham, 1991). This is perhaps not entirely an accurate view, as GIS *per se* does not meet all the six characteristics of a decision-support system as defined by Geoffrion (1983). However, GIS can provide a framework for the development of 'spatial decision-support systems' (SDSS), particularly when either loosely or tightly coupled with other modelling software (**Chapter 7**).

SDSS are an extension of DSS as defined by Geoffrion (1983). They share the same characteristics, but in addition provide spatial data models, the means of inputting and displaying spatial data and additional spatial analysis tools. Some of the requirements for an SDSS are met by standard GIS, but others, such as advanced spatial analysis tools and the user-friendly GUI (specified for DSS by Geoffrion, 1983) are not. One of the most important rules governing the use of GIS for SDSS is that GIS themselves do not make decisions – people do. In SDSS the emphasis is on the use of spatial data and in GIS it is on supporting people and the decisions they need to make.

### Maps as decision tools

Maps have a long history of being used to support decision making. Ever since maps were first used as a means of navigation, they were used as a form of decision-support tool. Which is the best route to take between one place and another? What is the best means of transport? What will I need to take with me? Military planners and explorers have all relied on maps. Good maps often meant the difference between success and failure, even life and death. It is not unusual, therefore, to find that maps play a very important role in modern decision making, be it for retail planning, government policy making, business management, facilities management or, indeed, military planning and exploration. In two of our case study examples, the output from GIS has been used for decision support in one way or another. This is illustrated in Box 8.6. In many cases digital spatial information is being used through either GIS or custom-designed software to create spatial decision-support systems (SDSS). Recent developments have linked Web-based GIS methods with SDSS principles to create online PPGIS (Public Participation GIS) aimed at encouraging and enabling public participation in decision making. Box 8.7 provides an example of Web-based PPGIS. GIS is also being used for front-line, real-time decision support in disaster relief operations. The case study in Box 8.8 on p. 282 describes how GIS and qualified personnel are being drafted in to help UN relief operations on the ground in disaster areas around the world.

# BOX 8.6 Examples of GIS used in a decision-support role

PRACTICE

### ZDARSKE VRCHY

In the Zdarske Vrchy GIS output was used to help determine new areas for consideration as nature conservation zones. GIS was used to generate maps showing a range of different scenarios for new protected areas given a range of environmental considerations. These maps were then used to structure discussions about nature conservation plans for the Zdar region. The GIS output acted as a catalyst for informed debate between a range of interested parties, including landowners, conservationists, government officials and international organizations. In this respect the GIS helped bring professionals from disparate backgrounds together to exchange ideas and information. The focus for their discussions was the maps produced from the GIS. The ease and flexibility with which new maps could be produced showing changes to the spatial distribution of protected areas following changes in environmental criteria offered additional benefits to the decision-making process.

### RADIOACTIVE WASTE DISPOSAL

In the case of radioactive waste disposal in the UK, GIS has been used to help decide the best or near-optimum sites for nuclear waste disposal. Carver (1991b) suggests that the combination of GIS and multi-criteria evaluation (MCE) represents a major contribution towards the development of effective SDSS. It is suggested that GIS/MCE-based systems may be used to assist decision makers in site identification and to plan evaluation, since radioactive waste disposal is a typical example of an ill-structured or semi-structured and complex spatial problem of the type referred to by Geoffrion (1983). Site-search procedures could significantly benefit from the use of SDSS from the initial survey stage to the final siting decision and public inquiry. It is further envisaged that a stand-alone or networked GIS-based SDSS in a committee room could create significant improvements in the way siting decisions are made. In addition, these types of SDSS may have an important role to play in providing more effective means of public participation and consultation throughout the site search process by allowing for greater public input and feedback to decision makers. Later work suggests that this may best be achieved by SDSS installed on the web to ensure as large an audience as possible (Carver *et al.*, 1997).

# BOX 8.7 Public Participation GIS

CASE STUDY

### Richard Kingston

Public Participation GIS (PPGIS) tools help planners, policy makers and citizens of varying backgrounds build consensus about the design, development and future direction of a community. If used properly, these tools can help citizens make better planning decisions by enabling improved communication, dialogue, design and analysis of ideas and proposals. If the extent to which citizens are involved is enhanced, the decision-making process will be improved (Healey, 1997; OECD, 2001), and assessment of the manner in which citizens are engaged provides a useful way of assessing the improvement of the public participation process.

In England all local authorities must produce a Statement of Community Involvement (ODPM, 2004). The United Nation's Aarhus Convention (UNECE, 1998) requires the public to have access to information and participation in decision-making processes. GIS offer tools which can engage the public in this type of spatial policy making. For example, GIS can be used to present data and information to the public which were previously available as hard-copy maps. Access to GIS over the Internet gives the public the additional possibility to manipulate data, visualize alternative scenarios and develop their own proposals.

Researchers have been developing PPGIS tools to allow access to spatial data and fuller

BOX 8.7

public participation in spatial decision-making processes (Ramasubramanian, 1996; Kingston *et al.*, 2000). The concept of PPGIS was developed at an NCGIA specialist meeting in 1998 (Craig *et al.*, 1998) which focused on the use of GIS by community groups and the public. The term has deliberately remained open and flexible to allow for a wide interpretation although some (for example, Tulloch, 2003) have attempted to define it as use and application of geographic information and/ or geographic information systems technology by members of the public (both as individuals and grass-root groups) for participation in the public processes (data collection, mapping, analysis and/or decision making) affecting their lives.

Ongoing work by many people (including Carver, 2001; Craig *et al.*, 2002) has been investigating the development of new methods and techniques for building GIS tools to help improve public participation in spatial planning. A recent research programme funded by the European Commission (IntelCities) has been investigating the potential of new technology to manage and plan cities of the future using new technology such as PPGIS. One of these tools developed as part of this project is the *e-Participation* system.

### THE E-PARTICIPATION SYSTEM

The e-Participation system engages citizens in the planning and regeneration process. An online PPGIS allows participants to discuss new proposals or identify issues relating to the development of their community. This type of system encourages citizen-led participation and allows individuals and community groups to start discussions and debate issues, and in a sense, create a geo-referenced community database.

The e-Participation tool is based on the OGC-compliant Web Map Server (WMS) technology and supports a range of spatial data formats including Esri Shape files, MapInfo and various raster data formats such as GeoTIFF. It has functions to allow the inclusion of additional geospatial data layers for relevant spatial policy data. An SQL database is used to store geo-referenced comments linked to a discussion forum (Figure 8.14). Some of the features of the e-Participation tool are:

- map navigation by zooming and panning;
- display or removal of selected map layers;
- queries;
- discussions created by clicking on the location about which the user wishes to comment. Users can respond to other users' comments in these discussions;
- a discussion can be based on a particular theme (for instance. transport, open space or derelict land); and
- discussion issues can also be initiated by the local authority to focus discussion on current policy proposals.

**Figure 8.14** PPGIS integrated with a geo-referenced threaded discussion forum

BOX 8.7

The development of online PPGIS and the provision of tools such as the e-Participation tool support the vision of delivering public services via e-Government. They also provide access to data and information which, until recently, were not easily available to the wider public. These types of systems are beginning to improve government decision making and have the potential to increase government efficiency and enhance the role of citizens in decision making.

PPGIS may also assist the management of cities to help achieve long-term physical, social and economic sustainability. For example, the 'Environment on Call' system stores georeferenced complaints and enquiries. It allows policy officers to monitor the spatial dispersion of service use across the city. This allows real-time monitoring, assessment of clusters of problems and appropriate targeting of resources when problems arise.

PPGIS is a relatively new sub-field of Geographic Information Science and its potential is not yet fully understood. In particular, a number of key operational questions for PPGIS have still to be answered:

- should a moderator be present for discussions?
- what spatial data layers should be included for public use?
- how should links to information showing 3D environments be dealt with?
- should facilities be provided to allow users to link to their own text and graphics and provide their own material?
- should users be able to upload their own spatial data?

### FURTHER INFORMATION

PPGIS experiments, publications, software and other resources: www.ppgis.manchester.ac.uk/
The Urban Regional Information Systems Association PPGIS pages: www.urisa.org/ppgis/

*Source*: Richard Kingston, University of Manchester

## BOX 8.8 MapAction: a personal view

CASE STUDY

### Chris Ewing

MapAction is a UK-based charity which uses GIS and GPS technology to help in disaster situations worldwide. If a natural disaster occurs the chances are that MapAction will be there providing mapping support. As a MapAction volunteer I have attended natural disasters in Jamaica (Hurricane Dean) and Bolivia (severe flooding), and a capacity-building mission in Lesotho in partnership with an HIV/AIDS charity. This case study presents some of my personal experiences.

Becoming a volunteer for MapAction is not an easy process. There is a three-stage interview process with two face-to-face interviews and a GIS test (the 'GISquisition') on your Esri ArcGIS capabilities. The 45-minute test included making a situational overview map for a mock earthquake. Ten injections and vaccinations were required as part of the induction process since working for MapAction may require travel anywhere in the world, at very short notice, to respond to a typhoon, cyclone, earthquake or other natural disaster.

Training sessions for MapAction take place monthly, and include disaster-related exercises, landmine awareness training, and firstaid. The training sessions are an opportunity to meet fellow volunteers and they help us prepare for real-life disasters. Knowing each other before going to a disaster area is very important – when doing 18 hour days in a small team and communicating in a foreign country it is good to be with people you know.

Combining volunteering with a full-time job is a juggling act – last year I volunteered about 25 days of my time which included weekends, days off, and evenings. The company I work for, Halcrow Group Limited, are very supportive of MapAction and my colleague Helen Wood is also a volunteer. A lot of the skills learnt as a MapAction volunteer can be applied in the workplace, such as team-working and interpersonal skills, thinking and problem-solving in high-pressure situations, working in unfamiliar environments, and working with people of different languages and cultures to help those in need. MapAction volunteers also have many other

**BOX 8.8**

technical skills and experience – for example in GIS, GPS, telecoms and IT, as well as experience of humanitarian working and overseas travel.

If help is requested following a natural disaster MapAction aims to deploy within 48 hours. In order to provide maps of the disaster, situation data are collected from a variety of sources both primary (for example, GPS locations of points of interest such as airfields) and secondary (for instance statistics, data and reports from other non-governmental organizations (NGOs). At MapAction headquarters an archive of map-base data is available for disaster prone countries. In the field maps are made of the current situational overview. These are distributed electronically (via www.mapaction.org) and printed to allow different NGOs, UN agencies and government officials to see the maps and take to meetings or presentations. The maps provide a clear operational picture of sometimes complex emergency situations.

There are two types of missions: disaster response and planned missions. MapAction has sent teams to some 15 humanitarian emergencies since 2004. 'Light missions' have ranged from delivering GIS training in Tajikistan to working with landmine NGOs in Angola and Iraq. As well as working in the field, there are a number of support-base activities to help with. Tasks can include uploading digital maps to the catalogue or helping with complex GIS analysis.

In early 2008, along with another volunteer, Emese Csete, I attended flooding in Bolivia caused by the El Niño weather system. Ongoing flooding had led the government to declare a state of emergency. After a journey of more than 20 hours from London, we arrived at La Paz airport just before 7am Bolivian time. The first task was to join up with the United Nations Disaster Assessment Team (UNDAC) in the On-Site Operations Co-ordination Centre (OSOCC) at the UNICEF offices in the city. The problems brought by the floods were wide ranging: thousands of displaced people, contaminated water supplies, destroyed crops, impassable roads and transport links. With as many as 750,000 people affected, and transport problems, the delivery of aid was difficult.

**Figure 8.15** MapAction team with UN OSOCC in Haiti
*Source*: Courtesy of MapAction

With the worry of more rain to come, the aim of the UNDAC team was to obtain a clear operational picture of the situation with the help of maps from MapAction, as an aid to planning for further flooding should it happen. Emese went to Santa Cruz in the flood-affected area, where the National Guard and military were also based. I stayed in La Paz to provide a country overview and help with secondary information about the disaster coming into the UNICEF office (see Figure 8.15). For ten days we provided mapping, GIS and GPS support to the different aid agencies working in the country. We also provided a much needed situation overview for the many international donors and agencies dealing with the disaster. To enable MapAction to provide a continued presence during the period of flooding, after ten days we were replaced by two more volunteers. The experience was very demanding but ultimately rewarding. Providing maps of the situation gave the organizations who were responding to the emergency the ability to plan aid deliveries and reach those affected much more effectively.

**FURTHER INFORMATION**

MapAction: www.mapaction.org/

*Source*: Chris Ewing, MapAction

## REFLECTION BOX

- The map presented in the Box 2.6 is an example of a 'linked display'. Outline some possible uses for linked displays in your own discipline or area of interest.
- What are the advantages and disadvantages of providing GIS output as hard and soft copy? What do you encounter most often? Why?

- What do you think will be the dominant form of GIS output in the future? Will maps remain important?
- What is an SDSS? How could SDSS be applied in your own discipline or field of interest?

## CONCLUSIONS

It is difficult to beat the map as a method for communicating spatial information. It is an accepted part of our culture and has been refined and developed over a far greater time than any of the new visualization tools being developed by today's computer scientists. However, it is important to recognize that maps come in many guises, each with its own subtle distinctions, and, furthermore, that each map tells a different story.

The application of GIS to spatial problem solving has meant that many more people are producing maps as part of their day-to-day jobs. No longer is a professional cartographer required to produce a map; anyone can do it. Maps have become a throwaway commodity generated by the GIS analyst in response to the changing needs of decision makers. The automated nature of much of the GIS map production process – with the software automatically adding default legends, keys and scales – means that maps can be produced in minutes rather than the weeks or months associated with the traditional process. There is often a price to pay for this freedom – and poor-quality output that does not communicate its message effectively to users abounds. An interesting example is shown in Figure 8.16. This is an example of GIS output and yet break several cartographic rules on numbers of classes, choice of colour scheme and missing classes. Some of the interactive map composition tools available on the Internet will only serve to compound this problem by making it easier for people without cartographic training to design and produce inappropriate map products. It would be too much to expect the GIS analyst or the decision maker using a desktop GIS to become a professional cartographer overnight – such skills have

to be learnt. In this chapter we have tried to provide some basic pointers for map design as well as recommend sources of further study. Maps are powerful tools and users should acknowledge the responsibility they take on when they choose to use them for communicating ideas and information.

**Figure 8.16** Example of poor quality output. Spot the mistakes (answers appear at end of chapter).

**CONCLUSIONS**

Besides maps, there are a number of other output options available to the user of GIS. These include a range of other cartographic presentation techniques such as cartograms, computer movies and animated map sequences. GIS output may also take the form of individual numbers, statistics or tables of values. In addition, GIS output can now be integrated with other media to produce multimedia presentations. For future developments it is suggested that the Web will play an increasingly important role in the dissemination of GIS outputs to a wider audience.

However, it is important to remember that GIS output, be it as a paper map or as part of a multimedia presentation, or a virtual reality experience, is rarely an end in itself. Usually, it is a catalyst that facilitates some decision-making or problem-solving situation.

This chapter concludes Part 1 of the book on the 'fundamentals' of GIS. We have now dealt with definitions of GIS, spatial data, data models and data management, data input, data analysis and spatial modelling, and output. The second part of the book deals with a selection of major GIS 'issues' including the historical development of the technology, data quality, human and organizational issues and project management. To conclude Part 2 we give some thought to where the future of GIS may lie.

## REVISION QUESTIONS

- Why is map design an important consideration when creating maps using a GIS?

- What are the main elements of a map?

- What are cartograms and how do they differ from traditional map output?

- When may non-cartographic output be more appropriate than drawing a map?

- What is the difference between permanent and ephemeral output?

- What mechanisms are available for the delivery of GIS output?

- What are the six characteristics of a DSS and an SDSS?

- Describe how the use of the Internet has changed GIS output.

## FURTHER STUDY – ACTIVITIES

- Find some Web mapping sites and create your own maps. Can you create maps that include all the elements suggested earlier in the chapter?

- Draw a mental map of the town or city where you live and work. Use appropriate symbolism. Ask friends or colleagues to draw their own mental map of the same place and compare the results.

- Draw a routed line cartogram for a rail or bus network that you know well.

- Find some data for a set of spatial units such as employment rates or population figures for districts or counties. Use a GIS (or pen and paper) to create choropleth maps of the data, experimenting with different class interval methods. Imagine you are a politician or journalist trying to make the data look as bad or as best you can. How different do the patterns look?

- Find examples of alternative forms of cartographic output on the Web, including cartograms, 3D visualizations, photorealism and animated 'fly-throughs'.

- Access and use examples of Web-based PPGIS.

## FURTHER STUDY – READING

Whilst this chapter has given a summary of the subject of GIS output, it is not possible within the confines of this book to cover all the relevant material in depth. This is especially true for many of the cartographic principles of map design. Therefore to develop your skills as a cartographer you should read further texts. An excellent starting point is Robertson (1988).

Wood and Keller (1996) provide a more detailed look at how computer technology can be used to advantage in cartographic design, and Brown and

Feringa (2003) cover relevant aspects of colour science and technology. Brewer (2005) addresses issues of map design specifically for GIS users.

Other specialist texts cover the developments mentioned in this chapter. These include Peterson (1995), McMaster and Shea (1992) and MacEachren and Taylor (1994). MacEachren has also produced books on symbolization and design (MacEachren, 1994) and representation and visualization (MacEachren, 1995). Recent developments in geovisualization are covered by Dyke *et al.* (2005)

When considering the use and misuse of maps as tools communicating spatial information, there is an excellent series of books by Monmonier (1993, 1995 and 1996) entitled *Mapping it Out*, *Drawing the Line* and *How to Lie with Maps*.

There are several texts available about the Internet and GIS. These include Kraak and Brown (2000) on web cartography and Plewe (1997) *GISOnline*.

Brewer C (2005) *Designing Better Maps: a Guide for GIS Users*. Esri Press, Redlands

Brown A, Feringa W (2003) *Colour Basics for GIS Users*. Pearson Education Limited, Harlow

Dyke J, MacEachren A M, Kraak M-J (2005) *Exploring Geovisualization*. Pergamon, Oxford.

Kraak M-J, Brown A (eds) (2000) *Web Cartography: Developments and Prospects*. Taylor and Francis, London

MacEachren A M (1994) *Some Truth with Maps: a Primer on Symbolization and Design*. Association of American Geographers, Washington.

MacEachren A M (1995) *How Maps Work: Representation, Visualization, and Design*. New York: The Guilford Press

MacEachren A M, Taylor D R F (1994) *Visualization in Modern Cartography*. Pergamon Press, New York

McMaster R B, Shea K S (1992) *Generalization in Digital Cartography*. Association of American Geographers, Washington DC

Monmonier M (1993) *Mapping it Out*. University of Chicago Press, Chicago

Monmonier M (1995) *Drawing the Line*. Henry Holt, New York

Monmonier M (1996) *How to Lie with Maps*. University of Chicago Press, Chicago

Peterson M P (1995) *Interactive and Animated Cartography*. Prentice-Hall, Englewood Cliffs, NJ

Plewe B (1997) *GISOnline: Information Retrieval, Mapping and the Internet*. Onword Press, Santa Fe.

Robertson B (1988) *How to Draw Charts and Diagrams*. North Light, Cincinnati, Ohio

Wood C H, Keller C P (eds) (1996) *Cartographic Design*. Wiley, Chichester

## WEB LINKS

### Online maps and mapping:

- USGS National Map Viewer: nationalmap.gov/nmjump.html

- Digimap: edina.ac.uk/digimap/

- David Rumsey Historical Maps: www.davidrumsey.com/

- World Mapper: www.worldmapper.org/index.html

- Google Earth: earth.google.com/

- Google Maps: maps.google.com/

- Bing maps: www.bing.com/maps/

- NASA's World Wind: worldwind.arc.nasa.gov/java/

- Wikimapia: www.wikimapia.org/

- MapTube: www.maptube.org/

- Second Life: www.secondlife.com

### Cartographic techniques and case studies:

- Case Studies of visualization in Social Sciences: www.agocg.ac.uk/reports/visual/casestud/contents.htm

- CSISS 'Recipe' for creating a choropleth map in Arc/GIS: www.csiss.org/cookbook/recipe/36

- Cartogram types: www.ncgia.ucsb.edu/projects/Cartogram_Central/types.html

- Worldmapper: www.world mapper.com

Visit our website **www.pearsoned.co.uk/heywood** for further web links, together with self-assessment questions, activities, data sets for practice opportunities and other interactive resources.

## Answers to Figure 8.16:

- missing title
- no north arrow
- no indication of map projection or grid
- poor colour choice for DEM — looks like high is low and vice versa
- no indication of units on DEM in key
- poor choice of background colour for key
- items in legend that don't appear on map
- inappropriate symbolism for point features
- use of file names in legend
- poor choice of symbolism for line features — rivers look like roads and vice versa
- no indication of units in scale bar
- clutter and superfluous detail in form of contours
- useless information in form of 'My House'
- lacks inset map or any indication of general location
- no metadata information (source of data, date, author, etc.)

# PART 2

# Issues in GIS

# 9

# The development of computer methods for handling spatial data

## Learning outcomes

By the end of this chapter you should be able to:

■ Describe methods of handling spatial data before GIS

■ Outline why computer methods for handling spatial data have developed

■ Identify disciplines which have contributed to the development of GIS

■ Provide examples of some of the important events in the development of GIS

■ Outline the characteristics of GIS in the early 21st century

# INTRODUCTION

Understanding the history of GIS helps to provide some context for the systems and concepts we work with today. GIS may be a mature technology in your area of interest, widely used in business organizations, research and teaching, or it may be a relatively novel technique in your discipline, still under development and testing in new applications. Utilities management and environmental planning are perhaps two of the earliest and most established areas of application. Areas involving location-based systems, and Web-based access to geographical information are newer fields of development. Both established and new application areas can learn a lot from each other, and an appreciation of how the technology has developed can help us develop a better understanding of the strengths and limitations of the techniques we use today.

Summarizing the important developments in GIS is difficult as there are no GIS archives and little critical examination of the history of GIS (Pickles, 1995). Many of the developments in GIS have been technology- or application-led, rather than guided by the development of theory (Martin, 1995). Particular personalities, companies and projects have played a major role in the pace and nature of developments in different countries. The influence of policy constraints and different commercial settings has also been important. Many different disciplines have been involved, including those related to the technology: computer mapping, databases, computer science, geography, remote sensing, data processing, mathematics and statistics; and those concerned with methodological and institutional issues such as systems theory and information management (Tomlinson, 1990; Coppock and Rhind, 1991). Developments have often been the result of cooperation and integration, matching the integrating nature of the technology.

This chapter considers the way GIS have developed. We do not attempt to present a comprehensive history of GIS, but aim to give some context for the systems and concepts we use now. First, methods of handling spatial data that were used before computers were commonly available are examined, since these give an insight into what we require computers to do, and how they can help

(or hinder) existing practice. Computer methods for handling spatial data existed before GIS, so these are reviewed, then developments in GIS are discussed together with relevant examples from a selection of complementary disciplines. A brief decade by decade review of the main features of GIS and key development issues is also provided.

# HANDLING SPATIAL DATA MANUALLY

Imagine a world before computers were commonly available – maybe back in the 1940s or 1950s. Spatial data were encountered most often as maps. Maps then, as now, could be found in most households, shops and businesses. Maps were used for a plethora of applications: identifying areas for new developments, routing delivery vehicles, planning weekend walks, targeting services or conducting warfare.

Maps were important for sieve mapping, a technique that combines several map layers to identify sites meeting a number of criteria (Burrough, 1986). The concepts of sieve mapping were introduced in an important book *Design with Nature* (McHarg, 1969). Each map layer is placed in turn on a light table and areas of interest highlighted. The layers produced are then placed on top of each other to identify areas of overlap. Sometimes the areas of importance on each map are shaded differently to indicate the relative importance of resulting maps. In the early days McHarg and his colleagues used carefully selected colour pens and shading patterns to represent different values which would be meaningful when overlaid (Anon., 1995). The techniques of sieve mapping were used by planning agencies, nature conservation agencies and many others. An early example of the use of manual sieve mapping was to map land capability (Bibby and Mackney, 1969). Manual sieve mapping is inaccurate and can cope with only relatively coarsely mapped data because fine details are obscured in the overlaying process. Problems also arise if there are more than a few map layers involved, since even with transparent media and a good light source, it is difficult to see through layers of different colours.

Another common use of maps was for route finding, for planning deliveries or an emergency evacuation from a disaster zone. Measuring shortest distances on a map by hand is a tedious process involving the use of a piece of string. The string is used to trace the route, then the length of string used is measured and converted to distance using the map scale. The use of a manual or digital map measurer can speed up the process. There are many problems with these methods: they are inaccurate, difficult to repeat, and time-consuming. It is also difficult to check alternative routes quickly or to take changing local conditions (for example, road repairs, speed limits or one-way streets) into account. Knowledge of local conditions and main routes is necessary.

Calculating areas from maps is even more difficult without help from computers. Without access to digitizers and GIS or similar technologies users relied on tracing areas onto graph paper and then 'counting squares' to obtain area estimates. An innovative technique is said to have been used by the US Bureau of Census for area estimation before their acquisition of GIS software. Their method involved cutting out polygons from a paper map and weighing them to gain a measure of relative area (Tomlinson, 1988).

These manual techniques of spatial data handling were in widespread use until computers became generally available and they share many problems. These include the slow speed of update of data, the slow speed of access to data, difficulties in extracting relevant data, difficulties in analysing data, inaccuracies in data and analysis, and the inflexibility of the data-storage media.

There are other problems associated with the media used. Paper maps stretch and shrink (due to folding or surrounding climatic changes), are always out of date, require complete redrawing for even minor changes, contain inconsistencies in the way features are represented, are easily destroyed, require considerable space for storage and can be difficult to transport.

Other spatial data – for example, printed census data – also have features that prevent efficient use. For those researching historical events, population census records contain a wealth of information. However, these need to be combined with birth and death records, marriage certificates and other documents for effective reconstruction of the lives and places of the past. Locating such records may be difficult since data sets may have been dispersed or even destroyed. It may be necessary to wade through files, books or microfiche to locate the exact details required. Moreover, without training, the old-fashioned script may be almost impossible to decipher. Other problems relate to inconsistencies and changes in spellings, and errors in records.

With all these problems associated with spatial data handling and analysis, it was hardly surprising that cartographers and other spatial data users were keen to exploit and explore computer technology.

## ■ THE DEVELOPMENT OF COMPUTER METHODS FOR HANDLING SPATIAL DATA

It is primarily advances in computer technology that have enabled the development of GIS. The decreasing cost of computer power over the last few decades has been one stimulus for wider use of GIS. There are other developments that have had an important impact: improvements in graphics technology, data access and storage methods, digitizing, programming and human–machine interfaces, and developments in systems theory. Martin (1995) considers that GIS 'have grown out of a number of other technologies and a variety of other application fields, and are thus a meeting point between many different disciplines' (Martin, 1995: 2). Some of the areas of prime importance are databases, remote sensing, global positioning systems, computer-aided design and computer graphics. Hardware developments in screens, printers and plotters and input devices such as digitizers and scanners have also had a major impact on GIS developments. However, it is the areas of computer cartography and spatial statistics that have produced many of the features of GIS. Most, but not all, early developments were in North America and the UK (Coppock and Rhind, 1991). In the USA the Public Health Service, the Forestry Service, the Bureau of Census and the Harvard Graphics Laboratory all played a role (Coppock and Rhind, 1991). In the 1960s there was little commercial development in the field, but groups such as the British Oxford Experimental Cartography Unit, as well as a number of individual researchers, were working with digital technology. Only after advances had been made in computer cartography and spatial statistics could the systems we now know as GIS emerge. In other areas of IT, such as computer-aided design (CAD) and automated mapping/facilities management (AM/FM), developments led to the development of hybrid systems. These take some elements from CAD or AM/FM and others from GIS. In addition, the development of the systems approach as a theoretical framework for the development of GIS concepts and techniques has had an important influence on GIS.

## The systems approach

General systems theory, one component of the systems approach, suggests that to understand the complexity of the real world, we must attempt to model this complexity. Unlike other scientific methods, which involve breaking down systems into component parts for study, general systems theory attempts to put back together the pieces of reality that science has 'dismembered' for analysis (Huggett, 1980). A macroscopic, rather than microscopic, view is taken (Odum, 1971). This holistic systems approach to modelling complexity developed from the ideas of writers such as Boulding (1956), Von Bertalanffy (1968), Churchman (1968) and Ackoff (1971).

Systems theory is the conceptual approach to applying general systems theory to problem solving. Using systems theory, all aspects of the real world can be modelled as human activity or physical systems, or as hierarchies of systems. The avalanche prediction model presented in **Chapter 8** is a complex system, which could be viewed as an element of a much larger system examining the hydrology and meteorology of the Happy Valley region.

Systems theory considers how a process works. Thus it is a meta-discipline, taking ideas from other areas and applying to many fields, in a similar manner to mathematics (Reeve, 1997). In the same way GIS is a meta-discipline, linking fields of study such as computer science, remote sensing, cartography, surveying and geography (Goodchild, 1995). There are many parallels between the way in which systems theory and GIS view the world. GIS is all about the integration of data from a wide variety of sources (Burrough, 1986; Aronoff, 1991); these data are frequently the distillation of ideas from different disciplines.

There are many examples of the application of systems analysis, the methodology for implementation of systems theory. These include models of the global hydrological cycle (Ward, 1975) and models of social systems and of the universe (Huggett, 1980). Perhaps more important for GIS is the use of systems analysis as a methodological framework for the development of computer software and systems that can handle users' information requirements. Two approaches, hard and soft systems analysis, have been used for the development of computer systems, although the boundary between them is fuzzy. Both methods may be used to help improve our understanding of a problem or to develop computer systems (**Chapter 12**).

## Computer cartography

Computer cartography (also known as digital mapping, automated cartography or computer-aided cartography) is the generation, storage and editing of maps using a computer (Mather, 1991). The development of computerized GIS is closely linked with the development of computer cartography. Pioneering developments in computer cartography, such as those outlined in Box 9.1, were initiated by map producers, not map users. Their interests were in speeding up and improving the process of map production. Therefore areas such as name placement, plotting, the use of different fonts and the production of colour separations for printing were important (Tomlinson, 1988).

# BOX 9.1 Developments in computer cartography

**CASE STUDY**

### THE ATLAS OF BRITAIN

This atlas (Bickmore and Shaw, 1963) is widely recognized as one of the catalysts for the development of computer methods for handling spatial data in the UK. Although published in 1963, work on this traditional atlas had started in the 1950s. Many cartographers and designers were involved and by the time of publication there was both

→

## BOX 9.1

praise and criticism for the work. The cartography was praised, but criticisms were levelled at the out-of-date material included and the unwieldy format of the atlas, and the book made a loss for its publishers (Rhind, 1988). The problems faced during production led Bickmore and his team at the Oxford Experimental Cartography Unit to consider using computers to check, edit and classify data and to experiment with graphics. Experiments were started in the late 1950s into the automatic placement of names on maps and methods of digitizing. These investigations led to the development of the first free-cursor digitizer and programs for line measurement, projection changing and data editing (Rhind, 1988).

### SYMAP

SYMAP was probably the first commercially available automated cartography package (Goodchild, 1988).

It was first marketed in the USA in about 1965 and could produce crude but useful choropleth maps. Production of maps was quick, but the quality of the output was poor since this was dictated by the output device of the time – the line printer. The line printer achieved area shading by repeated printing of the same character or number. Mather (1991) describes time-consuming early versions of SYMAP. Use involved the production of decks of punched cards held together with elastic bands that were read by a card reader. A deck of cards was submitted to an operator and output collected later. Later, however, file input was possible and other important developments included the implementation of previously subjective cartographic procedures such as interpolation (Goodchild, 1988) and SYMVU, the first three-dimensional mapping package. Both SYMAP and SYMVU were still in widespread use well into the 1980s. Examples of SYMAP and SYMVU output are shown in Figure 9.1.

**(a)** SYMAP

**(b)** SYMVU output

**Figure 9.1** Examples of (a) SYMAP and (b) SYMVU output

## Spatial statistics

Spatial statistics and geographical analysis methods were an area of much research and development in the late 1950s and early 1960s as computers became available to help the statistician. Initially, developments were in areas such as measures of spatial distribution, three-dimensional analysis, network analysis and modelling techniques (Tomlinson, 1988). These developments took place remote from any GIS or computer cartography developments primarily because the methods were developed using synthetic data sets and analyses were generally conducted only once. This is in marked contrast to the typical GIS 'analysis', which uses real-world data, and where analysis may be repeated many times.

Many of the methods developed in the middle of the twentieth century by spatial statisticians are still not available in GIS. Despite the 'quantitative revolution' in geography during the 1960s and the huge increase in the application and teaching of spatial statistics, these methods are still not appreciated or required by many users, and are not included in new GIS products and releases. Some of the techniques, notably those used in geodemographics and location planning (gravity models and location-allocation models), appear in separate packages. However, these frequently remain in the academic domain or as tools for independent experts to use for the provision of information to clients.

## AM/FM and CAD

Automated mapping/facilities management (AM/FM) and computer-aided design (CAD) are two areas of technology closely related to GIS. Early AM/FM systems were used to store information about utility facilities and to link external files to electronic maps. The maps and their related databases contained information about entities such as power lines or transformers (Reina, 1997). With the realization that location was important to the operation of most utility organizations, the use of AM/FM took off in the 1970s. In common with GIS, it took the enthusiasm of individuals to encourage its widespread use (Reina, 1997).

CAD is used to enter and manipulate graphics data, whereas GIS is used to store and analyse spatial data (Henry and Pugh, 1997). Traditionally CAD was used (in architectural design, engineering and facilities management) for the production, maintenance and analysis of design graphics. CAD provided no functions for the production or analysis of maps. However, many current CAD packages do permit the creation of spatial relationships between graphical entities (such as site plans and design drawings) and include basic spatial analysis tools.

## ■ THE DEVELOPMENT OF GIS

GIS and map analysis developments began around the same time as related developments in computer cartography and spatial statistics. They were prompted by the limitations of hard-copy maps, problems with overlaying data sets and the increasing size and number of available data sets (Tomlinson, 1988). The initial developments (in the 1960s) were technical and were aimed at developing a set of spatial data handling and analysis tools that could be used with geographical databases for repeated problem solving (Tomlinson, 1990). The technical issues for early consideration included digitizing, the development of data structures to allow updating and analysis, the compacting of line data to ease storage and the development of computer methods that would operate on large or small sets of data. Unlike the cartographers, initial developments were not focused on the production of high-quality output, but on the analysis of data. One of the first systems called GIS appeared in Canada in 1964 (Box 9.2).

Goodchild (1995) considers that the roots of current GIS lie in the 1960s. He suggests that the main developments of the decade were the adoption of a layer approach to map data handling in CGIS (Box 9.2), and pioneering work by the US Bureau of Census that led to the digital input of the 1970 Census (Box 9.3).

In turn, these developments influenced work at the Harvard Graphics Laboratory, which led directly to the production of some of the first commercial GIS software, including Esri's package ARC/INFO. Other technical developments of the 1960s should not be forgotten. These included improvements in the conversion of scanned data into topologically coded data, and the implementation of computer-based methods for efficient point-in-polygon analysis (Tomlinson, 1988).

## BOX 9.2  CGIS: an early GIS

**CASE STUDY**

The Canadian Geographic Information System (CGIS) used data collected for the Land Inventory System and was developed as a result of the requirements of the Canadian Agriculture and Development Act. It was designed to produce maps of the crops that areas of land were capable of producing and to map land capability for forestry (based on soil, climate, drainage and physical land characteristics). Land within the survey area was classified according to wildlife potential, taking into account all factors that were favourable or unfavourable for certain species. The land inventory also classified land for recreation. The problem of measuring and comparing all this information necessitated a technological solution. The approach adopted was also influenced by the volume of data required, the huge size of Canada, the need for a consistent spatial referencing system and the need for map linkage methods (Crain, 1985). The search for a computer solution led to the

development of CGIS. The date usually quoted for the beginning of CGIS is 1964, but the system was not fully operational until 1971. CGIS was capable of tackling the 1650 'person-year' task of comparing 6000 maps in days, or selecting areas that were suitable for forage crops but not suitable for grain growing or forestry in a few hours (Symington, 1968). CGIS has been modified significantly since then, but it is still a significant component of the Canada Land Data System. Many currently accepted and used GIS concepts, terminology and algorithms are originally from CGIS. It was the first general-purpose GIS to go into operation; the first system to use raster scanning for efficient input of large numbers of maps; the first GIS to employ the data structure of line segments or arcs linked together to form polygons (which has now become a standard); and the first GIS (in 1975) to offer remote interactive graphics retrieval on a national basis (Crain, 1985).

**Figure 9.2** Land use and land capability for agriculture

**BOX 9.2**

Over the years CGIS has been modified and improved to keep pace with technology and the requirements of users. For example, standard PCs instead of old mainframe systems are now used for data input and analysis. However, the overall components of CGIS have remained constant. There is a subsystem for scanning input and raster editing that allows editing and verification of scanned images and some auto-processing (for example, line thinning, closing of small gaps and raster to vector conversion). Additional subsystems include the interactive digitizing and editing subsystem and a cartographic output subsystem.

CGIS has been applied to a wide variety of areas and topics. The land inventory data, which took 12 years to gather, has been used in land-monitoring programmes focusing on changes in fruit-growing areas near to urban centres, acid rain monitoring and toxic waste siting. Considerable benefits have been obtained from the system, including cost savings, cost avoidance, improved administrative decisions, wiser land use, better-managed national parks and more effective agriculture and forestry industries (Symington, 1968; Crain, 1985).

The Canadians recognized from the beginning that good, accurate and relevant information was vital for the success of the GIS (Symington, 1968). CGIS was still being used until the late 1980s to help with land-use planning, pollution incidents and resource management (Heywood, 1990) and data sets are downloadable via the Internet. An example of CGIS and Canadian Land Use Inventory mapping is shown in Figure 9.2.

## BOX 9.3 Topology for the US Census – the development of DIME

PRACTICE

In the late 1960s the US Bureau of the Census needed an efficient method of storing the detailed information contained on US street maps. This led to the development of the GBF-DIME (Geographic Base File, Dual Independent Map Encoding) data format, a major step forward in GIS data models. GBF-DIME files are computerized representations of detailed street maps for the major urban areas of the USA that have proved an invaluable aid to census taking (see Figure 9.3). The list below outlines some of the major developments during the development of GBF-DIME:

1965   Census Small Area Data Advisory Committee is established
1966   A Census Use Study (CUS) is planned and begins in New Haven, Connecticut
1967   CUS begins testing procedures for the 1970 Census, and starts computer mapping experiments. Problems are faced with the conversion of analogue maps into digital format. Each street intersection in a normal US rectangular street grid is being digitized eight times.

Mathematician James Corbett presents the principles of map topology to CUS and these are used to overcome the problem. Street intersections and blocks are numbered. Street segments are encoded by reference to the areas to the left and right of them, as well as to the nodes that they connect (to node and from node). This model gives some redundancy that assists automated error checking

George Farnsworth of CUS names the new process DIME, and details of the process are presented to URISA and published.
1970s   GBF-DIME files are digitized for all US cities

DIME was a practical innovation – it assisted efficient digitizing and error removal, and it supported choropleth mapping of census results. It is at the heart of many vector data models, including that used by ArcInfo. The topological edit is perhaps the most significant contribution of the model allowing the creation of a perfect replica of the elements on the original analogue map (US Bureau of Census, 1990).

→

BOX 9.3

**Figure 9.3** Example of GBF/DIME data for Montgomery County, Maryland
*Source*: From the Montgomery County, Maryland website, www.montgomerycountymd.gov/gistmpl.asp

DIME files are also a key component of the current TIGER (Topologically Integrated Geographic Encoding and Referencing) system that is, in turn, a critical part of the National Spatial Data Infrastructure. The DIME files, and the organizations formed to produce them, also became the foundation for the geodemographics industry in the USA.

*Sources*: Mark *et al.* [no date]; US Bureau of Census (1990); Sobel (1990)

## REFLECTION BOX

- Write some step-by-step instructions for someone wishing to carry out *either* sieve mapping *or* measurement of distance from a paper map by hand. Make a list of the problems they might encounter.
- Produce a table summarizing the development of computer methods for handling spatial data. Use two columns, headed 'technological development' and 'methodological developments'.
- What do systems theory and GIS have in common?

- Which developments in GIS do you think came from map producers, and which from map users?
- Why do you think CGIS remained in use for so long?
- Goodchild (1995) considers that the roots of current GIS lie in the 1960s. Provide details of important developments from the 1960s to support this statement.

## The 1970s

Developments in GIS and related disciplines in the 1970s were many and varied, assisted by rapid advances in computing technology and other technological developments. The decade also saw important advances in remote sensing, with the launch of NASA's first three Landsat satellites. During the mid-1970s, developments were still occurring in two separate areas: in GIS, which were now commercially available with limited analysis capabilities; and in computer cartography, where good quality graphics and plotting and easy data editing were still the focus of improvements.

New computer cartography products entered the market (for example, GIMMS, MAPICS and SURFACE II). The Geographic Information Mapping and Manipulation System (GIMMS) may have been the first globally used GIS. Rhind (1988) reports that GIMMS was available from 1973 at a cost of $250. He considers GIMMS a 'prime antecedent of contemporary GIS', and notes that it employed a topographic data structure, user command and macro languages, and provided user control of graphics.

In the UK the first multi-colour standard series map to be produced digitally appeared in 1971. Soon after this the Ordnance Survey began the routine digitizing of topographic maps to speed up the production of conventional paper maps (Rhind, 1987). With digitizing a common pursuit, another important technical development in the 1970s was the Douglas–Peucker algorithm, used for reducing the number of points required to represent a line after digitizing (Douglas and Peucker, 1973) and still used in many GIS.

Computer technology advanced rapidly in the 1970s, with memory sizes and processing speeds increasing and costs falling. Interactive graphics became available, and blind digitizing (digitizing without a graphical display) could be avoided (Tomlinson, 1990). These technological developments were advantageous to all fields using computers. The decade was one of 'lateral diffusion' of the technology, with more users and more subject areas being introduced to GIS (Tomlinson, 1988).

The 1970s saw the first conferences and published work on GIS. Conference proceedings and other publications began to appear around 1974 (Coppock and Rhind, 1991). The first UK meeting of academics to discuss GIS was in 1975 (Unwin, 1991). This was part of an increasing awareness of and communication about GIS, which included the first texts on GIS published by the International Geographical Union (Tomlinson, 1990).

Another major contribution of workers in the 1970s was the recognition that the problems facing GIS were not just technical. Tomlinson *et al.* (1976) were among the first to recognize that there are as many problems associated with the management side of implementing an information system as with the technical side, and probably more.

## The 1980s

By the 1980s the demand had grown for good graphics and data analysis, real-time querying of databases and true topological overlay. Computers had by now found their way into all aspects of cartographic production (Taylor, 1991a) and computer cartography, and GIS and spatial analysis converged. The term 'GIS' was considered to be in common usage (in the UK) by the mid-1980s (Rhind, 1987).

One of the major events of the decade, in computing terms, was the production of the first microcomputers using the Intel 80386 chip in 1986. This paved the way for a new generation of microcomputer-based GIS products (Taylor, 1990). Significant technical developments in the 1980s included the increased use of the raster data model as the cost of computer memory decreased, and the development of the quadtree data structure (Goodchild, 1988). Spatial modelling methods began to be included in software, but GIS developments and demand for this aspect of GIS have lagged far behind the developments in spatial analysis. Developments were, in general, still related to improving technology, but, owing to price decreases, hardware was no longer such a limiting factor (Maguire, 1989).

Some developments were assisted by new GIS research initiatives. The USA's National Center for

Geographic Information and Analysis (NCGIA), spread over several university sites, and the UK's Regional Research Laboratories (RRLs) were perhaps the two largest initiatives to begin in the 1980s (Goodchild and Rhind, 1990; Masser, 1990). Other research programmes began in countries such as Australia, the Netherlands and France. Frequently, the remit for these centres included raising awareness of GIS and training, as well as the development of new techniques and applications. Training was facilitated by the appearance of the first textbooks, including Burrough (1986); by the establishment of the *International Journal of GIS* in 1987; and by discussions accompanying the development of GIS curricula and teaching materials such as the AUTOCARTO syllabus (Unwin *et al.*, 1990) and the NCGIA's core curriculum (Goodchild and Kemp, 1990).

A series of government reports charts the development of GIS in the UK. These include a review of the Ordnance Survey in 1978 (the Serpell Report); the review of remote sensing and digital mapping published in 1981 led by Lord Shackleton; and, perhaps most important, the 1987 Report of the Committee of Inquiry into the handling of geographic information, often referred to as the Chorley Report after the committee's chairperson (Department of the Environment, 1987). The Chorley Committee made recommendations relating to digital topographic mapping, data availability, data integration, awareness raising in GIS and the role of government in GIS development (Department of the Environment, 1987; Rhind and Mounsey, 1989). One of the recommendations of the Chorley Report was that a Centre for Geographic Information should be established. The Association for Geographic Information (AGI), launched in January 1989, has gone some way towards fulfilling this role. It is a '. . . multidisciplinary organization dedicated to the advancement of the use of geographically related information. . . . It aims to increase awareness of the benefits brought by the new technology, and assist practitioners . . .' (Shand and Moore, 1989: 9). In addition to the promotion of the technology, the AGI is active in co-ordinating and organizing interest groups, collecting and disseminating information about GIS, developing policy advice on GIS issues, and encouraging GIS research.

The decade saw a massive increase in interest in the handling of geographical data by computers, and this led to the rapid evolution of systems in response to user needs rather than any clear theoretical structure (Martin, 1991). Tomlinson (1990) cites an estimate that over 1000 GIS and computer cartography systems were in use in the USA by 1983, and forecasts then were for 4000 systems by the end of the decade. Developments during the 1980s were not necessarily restricted to the players who had been involved thus far. Developing, as well as developed, countries were seeing applications for GIS and international and continental projects were emerging. For the developing nations a steady expansion in applications took place, particularly after 1986 (Taylor, 1991b). The major areas of application were still the natural environment, national and regional government and the utilities. Many new projects relied on inter-organization cooperation and exchange of data (for example, CORINE, Box 9.4). This initiated new debate on data issues such as transfer format, costs and copyright and began the general trend towards information being regarded as a commodity (Openshaw and Goddard, 1987).

Besides CORINE there are other environmental examples of inter-organizational projects, including WALTER, the terrestrial database for rural Wales (Martin, 1991), and socio-economic examples such as REGIS (the Regional Geographic Information Systems project) in Australia (O'Callaghan and Garner, 1991). The utilities in the UK also investigated GIS, realizing that there were benefits to be gained from the exchange of data for pipes and cables. Savings could be made if repeated 'digging up' of road surfaces for repairs and maintenance could be avoided. The Dudley joint utilities digital mapping trial, initiated in the UK in 1982, aimed to build one database on one computer that could be accessed from the offices of the utilities and local government (Mahoney, 1991). The project highlighted discrepancies in the boundaries of the operating units of each utility, and led to the development of a standard format for the exchange of map data (NJUG, 1986).

## BOX 9.4  CORINE: An international multi-agency GIS project

One example of an international, multi-agency GIS project is the European Commission's CORINE (Coordinated Information on the European Environment) programme, initiated in 1985 (Rhind *et al.*, 1986). This GIS was designed as a general environmental database that would encourage the collection and co-ordination of consistent information to aid European Community policy (Rhind *et al.*, 1986; Mounsey, 1991a and 1991b). It was used initially to address the setting up of an inventory of sites of scientific importance for nature conservation, acid deposition and the protection of the Mediterranean environment. Mounsey (1991b) summarizes the nature of the project and reviews the lessons learnt. She considers that some of the constraints on the development of environmental

databases at that time were the availability of data, education of users and need for substantial resources. In addition, Mounsey suggests that the establishment of environmental databases should be pro-active (that is, not as a reaction to existing problems), and conducted according to systematic methodologies (such as systems design) with the full support of all interested parties.

The main achievements of CORINE were:

- the standardization of practices and methods for recording environmental data;
- the demonstration of the feasibility of a single database to meet a variety of end-user needs; and
- the development of similar activities at national level.

## GIS in the 1990s

In the early 1990s GIS were general-purpose software packages. Automated computer cartography products were widely used, and many users were beginning the long task of establishing digital databases for large applications in the utilities, local government and environmental research.

With many applications in the process of being established, data encoding was an important topic. Data encoding was a drawn-out and expensive process because of the limited availability of digital data. In the UK, for example, the national mapping agency (Ordnance Survey) had digitized only part of the national base data that are essential for many applications. Data for many areas, particularly rural areas, were not available although emerging technologies such as GPS (which was made freely available worldwide from 1993) were creating new opportunities for data collection. Widespread use of manual digitizing techniques made the data-conversion process slow. Large volumes of data were generated, but storage devices were still relatively limited. PCs were widely available, but CD-ROMs, disks with storage capacities of over a few hundred megabytes and the use of electronic communications for data transfer and exchange were relatively rare. Large applications needed serious computing power

– either workstations or mainframe computers. Data sets were being digitized by many different organizations, but it was difficult to find out if someone else had already digitized the data that you needed. If you could find digital data, there were questions of format, price, copyright and confidentiality, which all made data exchange a difficult process.

At a more conceptual level, the layer model of the world was dominant in GIS. Some packages offered a vector view of the world, and others offered only raster-based functionality. Analysis was restricted to one data model or the other, with little or no integration of raster and vector data sets. Even the inclusion of a raster background map was relatively unusual. Modelling was often conducted outside the GIS package, with the results being imported for map display. Alternatively, GIS was used to generate variables for models run elsewhere.

Issues of software usability were being discussed. The GIS of the early 1990s was not Windows-based – most packages were command line driven. Programs on PCs were DOS-based; those on mainframes used other non-intuitive operating systems. Users had to become expert at remembering complex commands. These commands had to be typed in, without errors, to operate software. If data transfer or editing were required, GIS users would write programs, using programming languages such as FORTRAN, to undertake reformatting, as spreadsheets and database software were unable to perform many of the tasks required. A high level of skill was required, and generally GIS expertise was at a premium. Training and awareness of GIS were important, and GIS courses were becoming widely available. In academia the distinction between GIS and GIScience was a topic of interest. In the commercial world many organizations that could benefit from GIS were unaware that it existed, and in others an individual 'champion' had promoted the case for GIS to get it into their workplace. The case for GIS generally was hampered by a lack of good case studies of successes and problems, and lack of any evaluated cost–benefit analysis to justify investment in systems. On the other hand, a huge amount of hype was beginning to emerge, with large-scale conferences, vendor offers to education and the emergence of new products. Very little, if anything, was really known about the impact of GIS on organizations, although the potential of GIS to add value to an organization's data was clear, and research had begun to improve data issues and help organizations with the implementation of systems. For example, research was undertaken on the impacts and diffusion of GIS, and attempts made to try to identify 'critical factors' for success (Campbell and Masser, 1995; Masser *et al.*, 1996).

At the beginning of the 1990s Maguire *et al.* (1991) considered that GIS had not found their full potential. Instead of being used for decision support, and for exploration and analysis of our world, the systems in use were often limited to mapping, inventory and information management-type applications. By the end of the decade the situation had altered with the emergence of Web-based applications a particularly important development.

## GIS in the 2000s

GIS became routine and mainstream in the early years of the twenty-first century. All the major organizations of the world, both in the government and commercial sectors, were using GIS in some form or another. Most people's lives were in some way touched by GIS, either directly or indirectly. GIS-based technologies appeared in navigation systems and online mapping systems and became accessible using everyday devices such as mobile phones, home PCs and wireless laptops. Most users of GIS-based services were consumers with little or no knowledge of GIS or GIScience.

Most GIS systems could now handle rasters and vectors (not just one or the other), and data exchange was standardized. Most systems worked with common data formats such as Esri Shapefiles, MapInfo MIF and TAB files. However, the development of new products for map viewing and querying, and for specific market sectors, meant that functionality between products diverged. Some of the companies producing 'high-end,' multi-purpose GIS for large-scale organizational applications even failed. The market moved away from this type of product and towards out-of-the-box solutions, targeted applications with specific functionality perhaps even including pre-loaded data. GIS was often a set of interoperable applications and APIs. This tied in with the near saturation of the market for systems in organizations such as utilities, local government and large commercial organizations, and the predicted expansion of markets in service and business sectors. There was also a growth in 'value-added services', particularly consultancy services to meet one-off applications. Market analysis is an area where this was particularly prevalent (Birkin *et al.*, 1996) with companies such as GMAP seeing considerable growth servicing a demand for advanced spatial analysis tools to support location decisions in the retail sector.

Despite developments in the field of spatial data modelling, GIS data models were still limited to raster and vector, and some hybrids of these basic modelling approaches. The third and fourth dimensions were still poorly catered for, and there was an overemphasis on the map and image presentations of spatial data. The object-oriented data model had undergone significant development in the 1990s and commercial systems using this model were now available. However, most GIS were still firmly fixed in the raster and vector views of the world, and there were no widely used GIS with established methods for handling true three-dimensional or temporal data, despite advances in related fields such as geological modelling and signs that true 3D data models were starting to creep into mainstream GIS as add-on functions (Figure 9.4). The lack of tools for handling temporal data, an issue important in many environmental applications, restricted the integration of process models with GIS in the 1990s and continued into the 2000s (Goodchild, 1992).

(a) Environmental variables above Sydney airport

(b) Voxel model of gas concentrations over Sydney airport

(c) Extruding 2D to 3D object

**Figure 9.4** Example 3D object modelling in MapInfo using Encom's Engage3D add-on
*Source*: Encom Technology Pty Ltd

There was widespread development of analytical functionality in the 2000s. Links to other software were often provided in GIS products, and model and menu builders existed in some packages to allow the integration of models from elsewhere, or the addition of new functions programmed by the user. Visualization of both 3D and time data was also an area of research and development (Batty *et al.*, 2001; Llobera, 2003; Batty and Hudson-Smith, 2005).

Data lineage tracking functions in some software assisted quality control; however, there was less progress on the widespread integration of error-handling tools. Apart from Monte Carlo simulation techniques, it seems that to most software developers the inclusion of error-handling tools would have been an admission that their software was faulty. However, quality issues were still being considered. Data quality was often examined and documented by the data supplier, and data standards for supply and transfer of data were widely adopted (see **Chapter 2**).

Improvements in accuracy, miniaturization and affordability meant that GPS-equipped devices were now commonplace. These included in-car navigation and vehicle tracking devices, GPS-equipped PDAs and handheld devices including mobile phones. There were also advances in expensive high-end surveying equipment capable of sub-centimetre accuracy through a steady improvement in the quality of the service and the ability of GPS receivers to reduce errors either in real time or via post processing. This was achieved by using more channels to process more signals, by averaging and by using differential methods (see **Chapter 2**). In addition, Selective Availability (SA) was switched off by the US government in 2000, thus removing the intentional errors from civilian GPS signals. Developers also produced hardware and software capable of using signals from both US GPS and Russian GLONASS satellites, further improving accuracy, whilst receivers capable of using WAAS (Wide Area Augmentation System) led to further improvements for civilian users. Another development set to revolutionize satellite navigation will be the introduction of the European GALILEO system of 30 satellites that will be fully operational by 2013 (ESA, 2008).

Conversion routines and standard data formats allowed direct input of GPS field data into GIS packages. GPS receivers could collect attribute as well as positional data, and field-based computing systems combining GIS and GPS technology in one sealed and rugged unit allowed direct plotting of data onto a map base whilst in the field. GPS chips in devices such as mobile phones and digital cameras made it possible to record the location of photographs, video and sound clips and associated attribute data in multimedia meta-data files which could then be downloaded to create point data sets with linked attributes and multimedia. As GPS became routinely used, data duplication was a problem, and standards and quality difficult to ensure in field data integration – for example, with remotely sensed data. This is an area in which there is now a wide published literature (for example, Dominy and Duncan, 2001 and Gao, 2002).

Mass digitizing was largely over in countries such as the USA and UK by the late 2000s, as national mapping agencies had completed the initial encoding of their resources, and many of the major utilities completed their data capture programmes. The new challenges were updating and revising these data. In a new twist, the digitizing of UK population census boundaries was put out to tender and subcontracted to more than one organization to ensure quality and competitiveness. However, there was still a great deal of data capture to be completed in many areas of the world, and in some cases the production of a useful digital base topographic data set was hampered by the limitations and incompleteness of available paper maps. The development of automatic digitizing tools and the increased resolution and availability of good aerial photography and satellite imagery offer alternatives which continue to be explored.

Despite the decreasing cost and increased capacity of data storage media there were still some constraints on the size of GIS applications, especially those dealing with global data sets and high-resolution, remotely sensed imagery. The size of individual images and attribute files could make processing difficult, and the amount of memory demanded in some cases exceeded that available.

In the USA spatial information was regarded as a public good (Muller, 1993), with access to many data sets available to every citizen for a nominal charge. The situation in other countries was sometimes very different, often due to government policy on commercial value or national security. In the UK the Ordnance Survey (OS) were required to recover costs, so some users were excluded from GIS due to their need for digital base data from the OS, and their inability to afford them. Reported difficulties in accessing data collected at taxpayers' expense prompted two journalists to start a campaigning blog (Arthur and Cross, 2006) which continues to address issues of access. Privacy and confidentiality of data remain issues for users, particularly in connection with freely available data such as that provided online by Google Maps Street View.

The customers for GIS were also changing. The market outside the USA expanded, as indicated by the presence of major vendors in countries such as the Netherlands, the UK and Australia. Another indication of the growth of markets elsewhere was the translation of products into languages such as Russian and Chinese, and the growth of user communities in Eastern Europe, Latin America, Africa and the Far East. However, there were still problems for the development of GIS in large parts of the developing world where limited availability and poor quality of data represented a major obstacle. Further developments in GPS and remote sensing will help ease these data problems, and data availability is being assisted by the provision of more and more data freely over the Internet (such as through Google Earth).

As the number of business applications of GIS has expanded, there was been a renewal of interest in making the business case for GIS with focus on issues such as the return on investment that can be achieved (Maguire *et al.*, 2008). In many commercial applications GIS became part of wider management information systems and an important component of an asset-management system, or business information system. Users might be unaware that they were working with GIS technology – a trend that is likely to continue.

Training and educational opportunities also expanded. In the UK GIS appeared in the National Curriculum for schools. In the USA the NCGIA's core curriculum was widely adopted in universities and colleges, and vendors addressed the needs of the K–12 (kindergarten to year 12) GIS curriculum. Undergraduate, postgraduate and professional development programmes expanded to the point where there were courses available at technician level, analyst level and manager level; by traditional study or distance learning; and online. A number of professional GIS qualifications emerged, such as the GISCI (GIS Certification Institute) and in the UK the Association for Geographic Information offered a programme of Continuing Professional Development for GI practitioners. Resources for those wishing to update their skills independently were also abundant and included general and specialist resources, online learning materials and instruction books for different software products and different types of user – from 'Dummies' (DeMers, 2009) to the end user in a particular application area (e.g. Gregory and Ell, 2007), or the technical expert (for example, Davis, 2007).

## REFLECTION BOX

- Identify the contribution of the following individuals to the history of GIS:

    D P Bickmore, Lord Roger Chorley, Michael Goodchild, Ian McHarg, Peter Burrough.
- What barriers to the adoption and diffusion of GIS exist in your own country or discipline?
- Produce a table summarizing the characteristics of GIS in the 1990s and 2000s. Use 'applications', 'data issues' and 'software characteristics' as your column headings.

- In which of the following areas do you think GIS advances have been particularly rapid over the last two decades? Why?
    - Data encoding
    - Data models for GIS
    - Data analysis
    - System usability

## CONCLUSIONS

Coppock and Rhind (1991) identify three 'ages' of GIS: the pioneering age, the age of experimentation and practice and the age of commercial dominance. These stages in the development of GIS, together with their characteristics, are summarized in Table 9.1. Additional 'ages' have been added to give a picture of the more recent development of GIS.

There has not been space in this chapter to cover all of the disciplines contributing to GIS developments or likely to influence the field in the future. However, there are some particularly important developments that deserve a brief recap. The first is the closer integration of remote sensing and GIS. With the emergence in the 1990s of GIS really

TABLE 9.1 **The stages of GIS development**

| Stage and date | Description | Characteristics |
|---|---|---|
| 1 Early 1960s–1975 | Pioneering | ■ Individual personalities important<br>■ Mainframe-based systems dominant |
| 2 1973–early 1980s | Experiment and practice | ■ Local experimentation and action<br>■ GIS fostered by national agencies<br>■ Much duplication of effort |
| 3 1982–late 1980s | Commercial dominance | ■ Increasing range of vendors<br>■ Workstation and PC systems becoming available<br>■ Emergence of GIS consultancies and bureaus<br>■ Launch of trade journals such as *GIS World* (USA) and *Mapping Awareness* (UK) |
| 4 1990s | User dominance, vendor competition | ■ Embryonic standardization<br>■ Systems available for all hardware platforms<br>■ Increasing use of PC and networked systems<br>■ Internet mapping launched |
| 5 2000s | Mainstream embedded GIS | ■ GIS within corporate information systems<br>■ Data rich, with focus on updating and extending the range of data available<br>■ Mobile and online GIS for everyone |

*Sources*: Adapted from Tomlinson (1990); Coppock and Rhind (1991)

**CONCLUSIONS**

capable of handling and analysing both vector and raster data, the use of remotely sensed data in GIS has grown. In addition, developments in remote sensing, including the launch of new satellites, increased data resolution and access to new data (such as those from the former Eastern Bloc) are helping to promote the further integration of the two disciplines. Second, developments in surveying technology, particularly in global positioning systems, also have a big part to play in the future of the technology. Finally, the ease of access to spatial information afforded by the Internet and mobile devices has brought GIS into the mainstream, where it has become a tool for use by the general public. At the same time business and research applications have grow in complexity and sophistication to help address current issues and global concerns. **Chapter 13** provides an insight into some of the directions GIS may take in the future.

## REVISION QUESTIONS

■ How were map data used before the introduction of computer methods? What were the problems with these methods of data handling?

■ Briefly list the major events in the development of GIS during the 1960s, 1970s, 1980s,1990s and 2000s.

■ How have application areas for GIS changed over the last 40 years? In which disciplines were GIS first applied, and in which areas are new applications being developed today? Can you suggest any reasons for these trends?

■ What have been the major developments in IT over the last three decades? How far do the developments in GIS over the same period reflect general developments in information processing and handling?

■ Describe briefly the nature of current GIS. What developments will affect the nature of GIS in the next decade?

## FURTHER STUDY – ACTIVITIES

■ Visit and explore some of the GIS timeline websites listed below.

■ Research the development of GIS policy and applications in your own country, or another country of interest. Have there been initiatives similar to those identified for the UK?

■ Find out about the history and development of GIS companies such as Esri, Clark Labs (Idrisi) and MapInfo, or others with which you are familiar. How does their history fit with the developments outlined in the chapter?

■ Identify current trends in GIS that you think might (in time) be recorded as key elements in the history of GIS in the early twenty-first century?

## FURTHER STUDY – READING

The history of GIS has, in general, been poorly documented. There are, however, a number of useful sources of information, including some highly personal and very interesting reviews of developments. Coppock and Rhind (1991) offer a good overview of the history of GIS. Tomlinson (1990) outlines the development and applications of CGIS, and considers general GIS developments in the 1970s and 1980s. A special issue of the journal *American Cartographer* in 1988 presented a number of papers by eminent GIS professionals, and gave them the opportunity to reflect on the development of GIS. Coppock (1988), Rhind (1988), Goodchild (1988) and Tomlinson (1988) offer personal views from different sides of the Atlantic on the major developments during the early years of GIS. In an article in *GIS World* magazine (Anon., 1995), McHarg reflected on methods introduced in his book *Design with Nature* (McHarg, 1969), and offered comment on the development of GIS. Roger Tomlinson was interviewed by the same magazine in 1996 (Anon, 1996) and more recently for *GEOconnexion International* (Baumann, 2009)

Foresman (1997) offers interesting 'perspectives from the pioneers' written first-hand by many of those involved in early developments. Major projects, including CGIS and the development of the TIGER data structure, are described and developments in a range of agencies are reviewed. Contributions are mainly from US authors, but there are chapters that outline developments in Canada, Europe, Australia and international applications.

For chronological accounts of GIS development, there are a number of timelines that can be consulted. National Research Council (1997) presents a timeline that was constructed by participants at a workshop on the future of spatial data and society. The participants included many famous GIS personalities, and the unedited timeline they produced can be seen online. Rana *et al.*'s (2001) timeline, viewable online, can be searched and includes some interesting graphics, such as output from SYMAP.

Rhind *et al.* (1991), Rhind (1992), Maguire *et al.* (1991) and Goodchild (1992) all provided predictions on the nature of GIS in the twenty-first century which it is interesting to revisit now that the first decade has passed. The proceedings of research conferences such as GIS Research UK, which are published in a series of books entitled *Innovations in GIS* (e.g. Wise and Craglia, 2008; Drummond *et al.*, 2007), give a feel for current research directions. Reading in other areas will also give you ideas for the development of GIS. For example, journals and magazines such as *Personal Technologies* and *Wired*, cover a range of technological areas.

Anon. (1995) *GIS World* interview: Ian McHarg reflects on the past, present and future of GIS. *GIS World* 8 (10): 46–9

Anon. (1996) *GIS World* interview: Roger Tomlinson the father of GIS. *GIS World* 9 (4): 56

Baumann J (2009) Roger Tomlinson on GIS history and future. *GEOconnexion International Magazine.* February 2009. http://www.geoconnexion.com/uploads/roger_tomlinson_intv8i2.pdf

Coppock J T (1988) The analogue to digital revolution: a view from an unreconstructed geographer. *American Cartographer* 15 (3): 263–75

Coppock J T, Rhind D W (1991) The history of GIS. In: Maguire D J, Goodchild M F, Rhind D W (eds) *Geographical Information Systems: Principles and Applications.* Longman, London, vol. 1, pp. 21–43

Drummond J, Billen R, Joao, E, Forrest D (2007) (eds) Dynamic and mobile GIS: investigating changes in space and time. *Innovations in GIS.* CRC Press, Taylor and Francis Group, Boca Raton, FL.

Foresman T (ed) (1997) *The History of GIS.* Prentice Hall, New Jersey

Goodchild M F (1988) Stepping over the line: the technological constraints and the new cartography. *American Cartographer* 15 (3): 277–89

Goodchild M F (1992) Geographical Information Science. *International Journal of Geographical Information Systems* 6 (1): 31–45

Maguire D J, Goodchild M F, Rhind D W (eds) (1991) *Geographical Information Systems: Principles and Applications.* Longman, London

McHarg I L (1969) *Design with Nature.* Doubleday, New York

National Research Council (1997) *The Future of Spatial Data and Society.* National Academies Press, Washington DC

Rana S, Haklay M, Dodge M (2001) GIS Timeline. www.casa.ucl.ac.uk/gistimeline

Rhind D (1988) Personality as a factor in the development of a discipline: the example of computer-assisted cartography. *American Cartographer* 15 (3): 277–89

Rhind D (1992) The next generation of Geographical Information Systems and the context in which they will operate. *Computers, Environment and Urban Systems* 16: 256–68

Rhind D W, Goodchild M F, Maguire D J (1991) Epilogue. In: Maguire D J, Goodchild M F, Rhind D W (eds) *Geographical Information Systems: Principles and Applications.* Longman, London, vol. 2, pp. 313–27.

Tomlinson R F (1988) The impact of the transition from analogue to digital cartographic representation. *American Cartographer* 15 (3): 249–63

Tomlinson R F (1990) Geographic Information Systems – a new frontier. In: Peuquet D J, Marble D F (eds) *Introductory Readings in Geographical Information Systems.* Taylor and Francis, London, pp. 18–29

Wise S, Craglia M (eds) (2008) *GIS and Evidence-based policy making* (Innovations in GIS). CRC Press, Taylor and Francis Group, Boca Raton, FL

## WEB LINKS

### GIS timelines:

- CASA's GIS timeline: www.casa.ucl.ac.uk/gistimeline/
- GIS Development timeline: www.gisdevelopment.net/history/

### Systems and organizations:

- US Census Bureau: www.census.gov/geo/www/tiger/index.html
- NCGIA: www.ncgia.ucsb.edu/
- Idrisi/Clark Labs: www.clarklabs.org
- Esri: www.esri.com/about-esri/about/history.html

### Online mapping services:

- Google Earth: earth.google.com/index.html
- Bing maps: www.bing.com/maps/
- Mapquest: www.mapquest.com/
- Streetmaps (UK): www.streetmap.co.uk/

### Other:

- GISCI: www.gisci.org/
- AGI: www.agi.org.uk
- Free our data blog: www.freeourdata.org.uk/index.php

Visit our website **www.pearsoned.co.uk/heywood** for further web links, together with self-assessment questions, activities, data sets for practice opportunities and other interactive resources.

# 10

# Data quality issues

# Learning outcomes

By the end of this chapter you should be able to:

■ Explain key concepts and terminology associated with data error and quality

■ Describe errors in spatial data

■ List types of errors that arise in GIS

■ Outline typical sources of error in a GIS project

■ Explain how GIS errors can be modelled and traced

■ Describe how errors in GIS can be managed

# ■ INTRODUCTION

In many industries, whether they are in the manufacturing or service sectors, quality control is everything. Without a quality product that is properly designed, well made and reliable, customers will go elsewhere for the goods and services they require. GIS users strive for quality products from their data and systems, and hope to produce good quality results and output. Whether users are in local government, business, research or any other sector, quality is always an issue. But mistakes and errors can enter a GIS in many different ways, leading to difficulties for many different users. For example, a mistake in digitizing a field boundary may result in inaccurate calculation of potential crop yields. A mistake in the transcription of a grid reference could result in emergency personnel being sent to the wrong incident location. Misinterpretation of satellite imagery may lead to inaccurate flood plain definition and in turn inappropriate planning decisions. And ill-considered selection of interpolation operations may lead to the incorrect calculation of wind turbine visibility. Errors may come from source data, human error, incorrect selection of analysis operations, computational errors and inappropriate presentation of final results.

The computing saying 'garbage in, garbage out' (GIGO) recognizes that if you put poor-quality data into your program, you will output poor quality results. The saying applies to GIS, since the results of analysis are only as good as the data put into the GIS in the first place. Previous chapters have shown how easy it is to combine maps, create buffer zones, interpolate point data sets and perform dozens of other operations with a GIS to produce complex and attractive maps. However, errors in input data may be compounded during these analyses, and errors in classification or interpretation may create misleading final output for end-users. Maps produced by a computer have an inherent feeling of quality and authority. Naive users have a tendency to believe what they see on the screen, and complexity, clever colours and intrinsic visual appeal may fool users into thinking that GIS output is of a high quality. But what is a high-quality product? What exactly are good-quality data? How can we describe and recognize poor-quality output?

Two issues are particularly important in addressing quality and error issues: first, the terminology used for describing problems and, second, the sources, propagation and management of errors. Describing data problems in GIS is difficult since many of the words used are also common in everyday language. Words such as quality, accuracy and error not only mean different things to different people but also have precise technical definitions. The terms used for data error and quality are introduced at the start of this chapter, since the first step to solving problems is to be able to recognize and describe them. The remainder of the chapter outlines the types and sources of errors in GIS to help you recognize and deal with problems at the appropriate stage of a GIS project. Techniques for modelling and managing errors are also considered. These techniques are important, as few data sets are error-free. GIS users must learn to manage and live with the errors their data and systems produce. Perhaps more importantly, GIS users should document the limitations of their source data and the output generated. Such documentation will ensure that output will be of value to others and will permit appropriate future use of data. Returning to our car analogy, you might be less likely to buy an old vehicle from an auction (which might have high mileage, several owners and no service history) than a smart little car with quality assurance documentation from an established dealership (with guaranteed low mileage, one careful owner and a full service history). The same principles apply to GIS data. Quality control should be impeccable because confidence in GIS analysis and results relies heavily on having access to reliable, good-quality data.

Earlier in the book we compared the fuel in a car to data in a GIS. In the same way that a poor quality fuel may cause problems with the running of the car, poor-quality data will introduce errors into your GIS.

# ■ DESCRIBING DATA QUALITY AND ERRORS

GIS users must describe data problems before they can solve them. So, before discussing the sources of data problems and their implications and management, this chapter considers the terms used to describe data problems in a GIS context. These terms will be used later in discussions of the sources of error in GIS.

'Quality' is a word in everyday use that is very difficult to define. It has different meanings for different people and in different contexts. There is a long philosophical debate about the meaning of quality in Robert Pirsig's book *Zen and the Art of Motorcycle Maintenance* that is worth reading; however, a simple dictionary definition of quality is 'degree of excellence' (Collins, 1981). In GIS, *data quality* is used to give an indication of how good data are. It describes the overall fitness or suitability of data for a specific purpose or is used to indicate data free from errors and other problems. Examining issues such as *error*, *accuracy*, *precision* and *bias* can help to assess the quality of individual data sets. In addition, the *resolution* and *generalization* of source data, and the data model used, may influence the portrayal of features of interest. Data sets used for analysis need to be *complete*, *compatible* and *consistent*, and *applicable* for the analysis being performed. These concepts are explained below.

Flaws in data are usually referred to as *errors*. Error is the physical difference between the real world and the GIS facsimile. Errors may be single, definable departures from reality, or may be persistent, widespread deviations throughout a whole database. In the Happy Valley database a single error in a point data set would occur if, for example, a co-ordinate pair representing a ski lift station were entered incorrectly, perhaps with two digits in the wrong order. A more systematic error would have occurred if the co-ordinates for all the ski lift stations in the data set had been entered in (y,x) order instead of (x,y).

Accuracy is the extent to which an estimated data value approaches its true value (Aronoff, 1991). If a GIS database is accurate, it is a true representation of reality. It is impossible for a GIS database to be 100 per cent accurate, though it is possible to have data that are accurate to within specified tolerances. For example, a ski-lift station co-ordinate may be accurate to within plus or minus 10 metres.

Precision is the recorded level of detail of your data. A co-ordinate in metres to the nearest 12 decimal places is more precise than one specified to the nearest three decimal places. Computers store data with a high level of precision, though a high level of precision does not imply a high level of accuracy. The difference between accuracy and precision is important and is explained in Box 10.1.

## BOX 10.1 Accuracy and precision

**THEORY**

It is perfectly feasible to have a GIS data set that is highly accurate but not very precise, and vice versa. The difference between accuracy and precision is best illustrated using diagrams. Imagine the Happy Valley biathlon has just taken place. Four contestants in the shooting have produced the results illustrated in Figure 10.1. Contestant A was accurate but not precise; contestant B was inaccurate but precise; contestant C was neither accurate nor precise; and contestant D, the winner, was both accurate and precise.

Extending this illustration to GIS, imagine the target is geographical space. The 'bull's eye', or central cross on the target, is the true geographical location of a feature we are trying to locate and the athletes' 'hits' on the target represent points collected by surveying or positioning techniques.

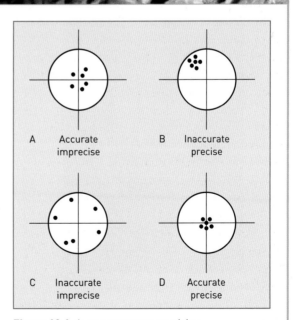

A    Accurate imprecise

B    Inaccurate precise

C    Inaccurate imprecise

D    Accurate precise

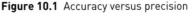

**Figure 10.1** Accuracy versus precision

Bias in GIS data is the systematic variation of data from reality. Bias is a consistent error throughout a data set. A consistent overshoot in digitized data caused by a badly calibrated digitizer, or the consistent truncation of the decimal points from data values by a software program, are possible examples. These examples have a technical source. Human sources of bias also exist. An aerial photograph interpreter may have a consistent tendency to ignore all features below a certain size. Although such consistent errors should be easy to rectify, they are often very difficult to spot.

Bias, error, precision and accuracy are problems that affect the quality of individual data sets. Data quality is also affected by some of the inherent characteristics of source data and the data models used to represent data in GIS. Resolution and generalization are two important issues that may affect the representation of features in a GIS database.

'Resolution' is the term used to describe the smallest feature in a data set that can be displayed or mapped. In raster GIS, resolution is determined by cell size. For example, for a raster data set with a 20-metre cell size, only those features that are 20 × 20 metres or larger can be distinguished. At this resolution it is possible to map large features such as fields, lakes and urban areas but not individual trees or telegraph poles, because they measure less than 20 × 20 metres. Figure 10.2 allows comparison of a 25-metre resolution vegetation map with a 5-metre resolution aerial photograph of the same area. Notice how the vegetation map only picks out the main pattern in vegetation, and generalizes the detailed mosaic of different vegetation types on the aerial photograph. Vector data can also have resolution, although this is described in different terms. Resolution is dependent on the scale of the original map, the point size and line width of the features represented thereon and the precision of digitizing.

Generalization is the process of simplifying the complexities of the real world to produce scale models and maps. Cartographic generalization is a subject in itself and is the cause of many errors in GIS data derived from maps. It is the subjective process by which the cartographer selectively removes the enormous detail of the real world in order to make it understandable and attractive in map form. Generalization was introduced in **Chapter 2** and discussed further in **Chapter 8**. Further examples of how cartographic generalization may influence data quality are given in Box 10.2. These cartographic practices, whilst understandably necessary for clear map production, cause problems when using paper maps as data sources for GIS. Positional inaccuracies are introduced through generalization of real-world detail, representation of area features as points, exaggerated line thickness and lateral dislocation of adjacent features.

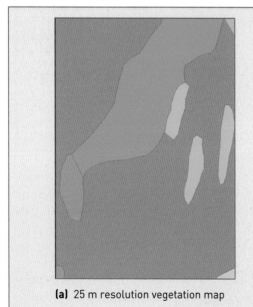

**(a)**  25 m resolution vegetation map

**(b)**  5 m resolution colour aerial photograph

**Figure 10.2** Resolution and generalization of raster datasets

# BOX 10.2 The influence of cartographic generalization on data quality

PRACTICE

**1** A ski piste has two distinct edges and follows a sinuous course. On a map, there is limited space to represent this. The cartographer may opt to draw the piste as a single coloured line without many of the smaller bends, whilst still retaining the general trend of its course, thus producing a generalized representation.

**2** Area features on a map may be generalized to points. This extreme form of generalization is common in small-scale maps, where important features such as towns and villages can be too small to be drawn as area features. They are effectively reduced to a single annotated point to communicate their name and location.

**3** The cartographer may draw features larger than they actually are in order to emphasize their importance. This is useful to highlight important communication routes. For instance, a road may be drawn on a 1:50,000 map as a broad red line approximately 0.8 mm wide. If we multiply this by the map scale, it suggests the road is 40 metres wide, whereas in reality it is in the order of 10 metres wide or less.

**4** Linear features that run close together, such as a road running beside a river, may have the distance separating them on the map considerably exaggerated to avoid drawing them on top of each other.

**5** Scale is very important in determining the level of generalization present on a map. This can be illustrated by considering the cartographic portrayal of line and area features at various map scales. In Figure 10.3 the generalization of a village is shown at three different scales. At 1:2,500, the village is drawn as a series of points, lines and polygons representing individual features such as trees, fences and buildings. At 1:25,000, the village may be reduced to a few lines and polygons representing the roads and main groups of buildings. At 1:250,000, the village has become a single point at the intersection of a few main roads.

**(a)** 1:2,500          **(b)** 1:25,000          **(c)** 1:250,000

**Figure 10.3** Scale-related generalization

There are further data quality considerations if data sets are to be used in a GIS. *Completeness, compatibility, consistency* and *applicability* are introduced here.

A complete data set will cover the study area and the time period of interest in its entirety. The data are complete spatially and temporally, and should have a complete set of attribute information. Completeness of polygon data is relatively easy to determine. Errors in attribute data might be seen as polygons lacking attribute information, and errors in spatial data may be seen as polygons with two sets of attribute data. Completeness of line and point data is less obvious as features may be missing without showing up in the database. The only way to check is by comparison with another source. Sample data, whether taken via a random, regular or systematic survey, are by definition incomplete. In this instance, completeness is a matter of having a sufficiently large sample size to represent adequately the variation in the real world. In the case of time series data, completeness is very difficult to define, let alone assess. Almost all time series data in GIS are referenced to discrete moments in time. However, time is continuous. A set of maps relating to a series of discrete times is therefore incomplete, and requires broad assumptions to be made regarding change in the intervening periods.

Compatible data sets can be used together sensibly. With GIS it is possible to overlay two maps, one originally mapped at a scale of 1:500,000 and the other at 1:25,000. The result, however, is largely worthless because of incompatibility between the scales of the source documents. Maps containing data measured using different scales of measurement cannot be combined easily. Overlaying a map of snow depths in Happy Valley (in metres – ratio scale) with a map of piste and non-piste areas (binary classification – nominal scale) using a union or multiplier operation would produce a meaningless output map.

To ensure compatibility, ideally data sets should be developed using similar methods of data capture, storage, manipulation and editing. Consistency applies not only to separate data sets but also within individual data sets. Inconsistencies can occur within data sets where sections have come from different source documents or have been digitized by different people. This will cause spatial variation in the error characteristics of the final data layer. One area of the final data set may contain more errors than another. Problems of inconsistency also come from the manner in which the data were collected. For example, in Happy Valley there are meteorological stations that record snowfall. The equipment used to measure snowfall is of different ages and designs and thus the accuracy of the measurements obtained varies from station to station. If meteorological stations are included in a GIS as points, with snowfall records as their attributes, inconsistencies will be introduced to the database. Another example is that of national and international boundaries. These are subject to constant revision as a result of international disputes and internal reorganization, thus positional inconsistencies may occur where boundary revisions have taken place.

Applicability is a term used to describe the appropriateness or suitability of data for a set of commands, operations or analyses. For example, interpolating height data using Thiessen polygons (**Chapter 6**) may be an unsuitable match between data and technique, as height data vary continuously (not abruptly

as assumed by the Thiessen polygon method). Alternatively, applicability can be used to describe the suitability of data to solve a particular problem. It is inappropriate to use height data to estimate some socio-economic variables, such as car ownership, but height data may be applicable to estimating the number of days in a year with snow cover.

The criteria introduced above provide some pointers for gauging the overall quality of GIS databases. However, in the same way that different terms can be interpreted in different ways, the quality criteria regarded as important by data creators and users are likely to be different. Suppliers of digital topographic maps to the Happy Valley GIS team may regard their data as of high quality if they are complete and up to date at a range of scales, and if they deliver a known and measurable degree of accuracy. Their reputation depends on their ability to supply a product of consistent standard. However, different criteria may be important for the data users. The GIS team may require precise data on particular themes, and data that are compatible with their GIS system. They want all slopes and cliffs to be represented consistently. They may be unwilling to use the data for decision support if they know it contains errors.

## ■ SOURCES OF ERROR IN GIS

Having the terminology to describe flaws in data is a good first step to providing quality GIS. Awareness of the sources of problems comes next. Spatial and attribute errors can occur at any stage in a GIS project. They may arise during the definition of spatial entities, from the representation of these entities in the computer, or from the use of data in analysis. In addition, they may be present in source data, arise during conversion of data to digital format, occur during data manipulations and processing, or even be produced during the presentation of results. The best way to explore how and where errors may arise is to view them within the context of a typical GIS project.

### Errors arising from our understanding and modelling of reality

Some of the errors arising in GIS have a source far away from the computer. Errors can originate from the ways in which we perceive, study and model reality. These errors can be termed *conceptual* errors, since they are associated with the representation of the real world for study and communication.

The different ways in which people perceive reality can have effects on how they model the world using GIS. Mental mapping (Box 10.3) provides an illustration of how varying perceptions can manifest themselves in a geographical context.

Our perception of reality influences our definition of reality, and in turn our use of spatial data. This can create real errors and often gives rise to inconsistencies between data collected by different

surveyors, maps drawn by different cartographers and databases created by different GIS users.

Different scientific disciplines have developed their own particular ways of explaining and studying reality. Compare the reductionist approach of the physical sciences, such as physics and chemistry, with the wider, more holistic approaches of the environmental sciences such as geography and ecology. The reductionist view attempts to explain and model parts of the world system by isolating individual physical, chemical and biological processes at

## BOX 10.3 Perceptions of reality – mental mapping

**THEORY**

Your mental map of your home town or local area is your personal, internalized representation that you use to help you navigate on a day-to-day basis and describe locations to other people. You could commit this to paper as a sketch map if required. Figure 10.4 presents two sketch maps describing the location of Northallerton in the UK. They would be flawed as cartographic products, but nevertheless are a valid model of reality. Keates (1982) considers these maps to be personal, fragmentary, incomplete

and frequently erroneous. Consider the mental maps that a group of residents and visitors in Happy Valley might draw. The residents might stress their homes, place of work and community facilities. Visitors would include their hotel, ski school, ski runs and the après-ski facilities. In all cases, distance and shape would be distorted in inconsistent ways and the final product would be influenced by the personality and experience of the 'cartographer'.

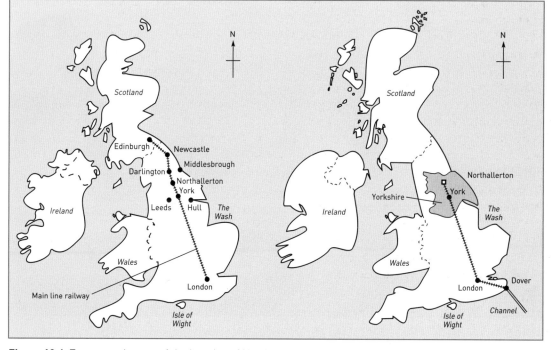

**Figure 10.4** Two mental maps of the location of Northallerton in the UK

a micro scale, whereas the holistic view attempts an explanation on the basis of related and interdependent macro- and meso-scale systems. For example, a biologist is more likely to be interested in photosynthesis as a biochemical process, taking place at the level of an individual plant cell or leaf. An ecologist is no less interested in the process of photosynthesis but takes a much wider view of its role. Given such different approaches to science it is understandable that difficulties (such as bias and compatibility) arise when different people's models of space and process are compared.

In GIS we use spatial models to reflect reality. The main models in use are raster, vector, object-oriented and layer-based (**Chapter 3**). All of these spatial models have limitations when it comes to portraying reality. For instance, the raster model assumes that all real-world features can be represented as individual cells or collections of cells. This is clearly not the case. The vector model assumes that all features can be given a single co-ordinate or a collection of Cartesian co-ordinates. The world is actually made up of physical and biological material, which is, in turn, made up of molecular and sub-molecular matter grouped into complex systems linked by flows of energy and materials (solids, liquids and gases). Consider the problem of defining a common geographical feature such as a mountain in a GIS. First we need to define exactly what we mean by 'mountain'. This in itself becomes problematic since different people have different ideas as to what a mountain exactly is, where it begins, where it ends and how it is linked to other morphological features such as valleys and passes. Does Mount Everest (Figure 10.5) begin at sea level in the Bay of Bengal to the south, or does it rise out of the Tibetan Plateau to the north? From a mountaineer's perspective, the base of the mountain may be base camp below the Khumbu icefall — the place from which most climbing expeditions start. Everest base camp lies at 5500 metres above sea level and is itself several hundred metres higher than any of the peaks in the European Alps or mainland USA; however, the peaks of Europe and the USA are still mountains and no less spectacular (Figure 10.6).

**Figure 10.5** Terrain model of Mount Everest and its surrounding area based on photogrammetric survey data.
3D vizualization of Mount Everest
*Source*: Armin Gruen/Institute of Geodesy and Photogrammetry, ETH Zurich

**Figure 10.6** The Finsteraarhorn, Switzerland (4273 m)

The summit of Mount Everest (8849 m) is part of a larger massif and is joined to Lhotse (8501 m). From these two peaks are several ridges or arêtes that link satellite peaks including Nuptse (7861 m) and Shartse II (7590 m). All might be considered part of the bulk of Mount Everest, but could also be considered separate mountains in their own right, and this still leaves us with the problem of defining where Everest begins and ends (Figure 10.5). Box 6.9 also discussed the problem of identifying mountains.

Next we need to decide what feature type is best used to show the location of the mountain at the scale being used. Is it a point with attributes of name and height? Is it a series of nested contour lines and spot heights with height attributes? Is it a polygon around the base of the mountain or series of polygons showing the individual morphological units that make up the mountain? Is it a surface feature shown as a TIN or DEM? Or is it, as Fisher and Wood (1998) suggest, actually a fuzzy object whose extent is dependent on personal perception, scale and resolution? Whatever GIS model we adopt, it is a simplification of this reality, and any simplification of reality will include errors of generalization, completeness and consistency. Figure 10.7 overleaf attempts to visualize this problem for Tryfan, a mountain in north Wales.

## Errors in source data for GIS

Our models of reality in GIS are built from a variety of data sources including survey data, remotely sensed and map data (**Chapter 2**). All sources of spatial and attribute data for GIS are likely to include errors. Three examples of data sources are examined below to identify possible errors.

Survey data can contain errors caused by mistakes made by people operating equipment or recording observations, or by technical problems with equipment. For example, if sites of damage to ski pistes were being recorded using a GPS, errors in spatial data might occur if there were technical problems with the receiver. Errors in attribute data could occur if features were recorded incorrectly by the operator. A visitor survey of holidaymakers in Happy Valley could include spatial errors if postal codes for the visitors' home addresses were wrongly remembered or recorded. Attribute errors would occur if the characteristics of the respondents were wrongly allocated, or incorrectly noted.

Remotely sensed and aerial photography data could have spatial errors if they were spatially referenced wrongly, and mistakes in classification and interpretation would create attribute errors. For example, clouds on remotely sensed imagery

**(a)** Photograph of Tryfan

**(b)** Ordnance Survey 1:50,000 topographic map showing area above 750m

**(c)** Contour map

**(d)** DEM

**Figure 10.7** Multiple representations of Tryfan, north Wales
*Source*: (b) Reproduced by permission of Ordinance Survey on behalf of HMSO. © Crown Copyright 2011. All rights reserved. Ordnance Survey Licence number 100030901.

**(e)** TIN

**(f)** Topographic features

**Figure 10.7** *Continued*

**Figure 10.8** Problems with remotely sensed imagery: (left) example satellite image with cloud cover (A), shadows from topography (B), and shadows from cloud cover (C); (right) an urban area showing a building leaning away from the camera
*Source*: Ian Bishop (left); Google UK (right)

**Figure 10.9** Land use change detection using remotely sensed imagery. Example from IDRISI
*Source*: Screenshot shows IDRISI interface. Reproduced by permission of Clark Labs, Clark University

commonly obscure features of interest on the ground. Shadows from mountains, buildings, trees and other tall structures can lead to problems of misclassification and further obscure interesting detail. Examples of these problems are shown in Figure 10.8. It should also be remembered that these data sources are 'snapshots' of reality, showing the location and nature of features at a particular moment in time. Thus, for any changeable features the date of the image used is extremely important. Remotely sensed imagery can be very useful as many earth-observing satellite platforms provide regular repeat coverage of the same area making their data ideal for identifying change. An example using SPOT imagery for land use change detection is shown in Figure 10.9.

Maps are probably the most frequently used sources of data for GIS. Maps contain both relatively straightforward spatial and attribute errors, caused by human or equipment failings, and more subtle errors, introduced as a result of the cartographic techniques employed in the map-making process. Generalization is one cartographic technique that may introduce errors, and this has been discussed earlier in this chapter. There are other cartographic difficulties, some of which affect other types of data as well as maps. These include the representation of continuous data and features with indistinct (or fuzzy) boundaries. Problems may also be introduced because of sampling and measurement difficulties. These are explained in Box 10.4.

## BOX 10.4 Examples of errors from cartographic data sources

### CONTINUOUS DATA

Continuous data require special cartographic representation. For example, height is usually represented by contours on paper maps. Annotated lines indicate land of a particular height above a datum (usually sea level). Limited additional information is provided by spot heights. Apart from this, no information is given about the height of the land between contours, except that it lies somewhere in the range denoted by the bounding contours. Since height varies continually over space, the vector-based contour and spot height model is always a compromise for the representation of height data.

### FUZZY BOUNDARIES

Where changes between entities are gradual in nature, such as between soil types, vegetation zones or rural and urban areas of settlement, indistinct boundaries exist. These 'fuzzy' boundaries are difficult to represent cartographically. As a result, features are often made distinct by drawing sharp boundaries around entities. This will ultimately lead to positional errors and attribute uncertainty in digital map data.

### MAP SCALE

Whether the 'fuzziness' of natural boundaries is important depends largely on the scale at which the data are mapped and viewed. For instance, if a soil boundary is mapped at a large scale (for example 1:1250) then the representation of a fuzzy boundary can result in a significant amount of error. However, if the boundary is mapped at a much smaller scale (such as 1:25,000) then the error resulting from the representation, although still there, is no longer so significant. The thickness of the line used to draw the boundary on the map at this scale may well account for much of the fuzziness between the soil types on the ground.

### MAP MEASUREMENTS

Measurements on maps and other data sources are rarely 100 per cent accurate. Generalization aside, measurements such as contour heights, feature locations and distances have limited accuracy because of human and mechanical error. Human measurement error can result from personal bias (for example, in the interpretation of aerial photographs), typographical errors, equipment reading errors and mistakes in the use of equipment. Examples of machine error may include limitations of surveying equipment, limited accuracy of GPS receivers and equipment calibration errors.

## Errors in data encoding

Data encoding is the process by which data are transferred from some non-GIS source, such as the paper map, satellite image or survey, into a GIS format (**Chapter 5**). The method of data encoding and the conditions under which it is carried out are perhaps the greatest source of error in most GIS. Digitizing, both manual and automatic, is an important method of data entry, so this section will focus on errors that can arise during digitizing.

Despite the availability of hardware for automatic conversion of paper maps into digital form, much of the digitizing of paper maps is still done using a manual digitizing table (**Chapter 5**). Manual digitizing is recognized by researchers as one of the main sources of error in GIS (Otawa, 1987; Keefer *et al.*, 1988; Dunn *et al.*, 1990); however, digitizing error is often largely ignored. Beyond simple checks in the editing process, practical means of handling digitizing errors do not exist in proprietary GIS software.

Sources of error within the digitizing process are many, but may be broken down into two main types: source map error and operational error. Source map errors include those discussed in the previous section and in Box 10.4. Operational errors are those introduced and propagated during the digitizing process. Human operators can compound errors present in an original map (Poiker, 1982) and add their own distinctive error signature. Some of the operational errors that can be introduced during manual digitizing are outlined in Box 10.5.

# BOX 10.5 Operational errors introduced during manual digitizing

**PRACTICE**

Following the exact course of a line on a map with a digitizer cursor is a difficult task, requiring skill, patience and concentration. Hand wobble, a tendency to under- or overshoot lines, strain and lack of hand–eye co-ordination are all sources of human error in the digitizing process. Jenks (1981) categorizes human digitizing error into two types: psychological and physiological. To these can be added errors resulting from problems with line thickness and the method of digitizing which is used.

**1** *Psychological errors*. Psychological errors include difficulties in perceiving the true centre of the line being digitized and the inability to move the cursor cross-hairs accurately along it. This type of error results in lateral offset between the true and digitized line and under- or overshoots at corners. Under- and overshoots and lateral displacement of features can be difficult to spot and correct.

**2** *Physiological errors*. Physiological errors result from involuntary muscle spasms (twitches and jerks) that give rise to random displacements or 'spikes' and polygonal knots or 'switchbacks' (Figure 10.10). Some of these errors, such as spikes and knots, are easy to spot and correct.

**3** *Line thickness*. The thickness of lines on a map is determined by the cartographic generalization employed. All lines are drawn so as to be visible to the user and thickness may in some instances reflect the perceived importance of the feature being depicted rather than its actual size on the ground. A major road may be shown on a map by a line wider than the cross-hairs on the digitizer cursor. Common sense suggests that the true course of the line is along its mid-point, but it is a difficult task for the operator to follow the centre of the line exactly. Some displacement of the cursor either side of the centre line is inevitable, leading to positional errors in the resulting data.

**4** *Method of digitizing*. There are two basic methods of manual digitizing: point mode and stream mode (**Chapter 5**). Point mode digitizing allows careful selection of sample points to create a faithful representation of a line. The method involves a certain amount of generalization by the operator, who decides both the number and location of sample points. The greater the sample size, the greater the detail, but the positional error associated with each point remains much the same. In stream mode digitizing, sampling is determined by the frequency set by the user at the start of the digitizing session. The number of points sampled is controlled more by line complexity, since the operator can move relatively quickly along straight sections but needs to slow down significantly to navigate the cursor along more convoluted sections.

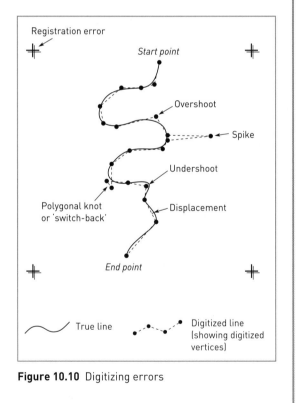

**Figure 10.10** Digitizing errors

The errors outlined in Box 10.5 occur during the digitizing process. However, errors can arise earlier. For example, before starting to digitize, the map needs to be placed on the digitizing table and registered to enable later conversion from digitizer to geographic co-ordinates. If this is done incorrectly or carelessly, then the whole set of co-ordinates digitized during that session will contain the same error or bias. Care when setting up the map document on the digitizer is therefore essential.

Automatic digitizing, like manual digitizing, requires correct registration of the map document before digitizing commences, but there the similarity ends. By far the most common method of automatic digitizing is the use of a raster scanner. This input device suffers from the same problems regarding resolution as the raster data model. Cell size is important. The smaller the cell size, the greater the resolution of the resulting image and the greater the quantity of the data produced. Cell size is determined by the resolution of the machine being used; a fine resolution may produce data with less positional error than manual digitizer output. Some of the other practical problems faced when using an automatic digitizer are discussed in **Chapter 5**.

Data input by either manual or automatic digitizing will almost always require editing and cleaning. This is the next source of errors in GIS.

## Errors in data editing and conversion

After data encoding is complete, cleaning and editing are almost always required. These procedures are the last line of defence against errors before the data are used for analysis. Of course, it is impossible to spot and remove all the errors, but many problems can be eliminated by careful scrutiny of the data. As a potential source of error, cleaning and editing do not rank high; they are positive processes. Some GIS provide a suite of automated procedures, which can lead to unexpected results. For example, it is common for vector GIS software to contain routines to check and build topology. The closing of 'open' polygons and the removal of 'dangling' lines (overshoots) are controlled by tolerances set by the user. If the tolerances are set too low, then the gaps in polygons will not be closed and overshoots will not be removed. Alternatively, if the tolerances are set too high, small polygons may disappear altogether and large dangling lines, which may be required on networks, may also disappear (Figure 10.11a).

A different problem occurs when automated techniques are used to clean raster data. The main problem requiring attention is 'noise' – the misclassification of cells. Noise can be easy to spot where it produces a regular pattern, such as striping. At other times it may be more difficult to identify as it occurs as randomly scattered cells. These noise errors can be rectified by filtering the raster data to reclassify single cells or small groups of cells by

matching them with general trends in the data. The 'noisy' cells are given the same value as their neighbouring cells. Filtering methods for this process are discussed in **Chapter 6**. Choosing an appropriate method is important as the wrong method may remove genuine variation in the data or retain too much of the noise.

After cleaning and editing data it may be necessary to convert the data from vector to raster or vice versa. During vector to raster conversion both the size of the raster and the method of rasterization used have important implications for positional error and, in some cases, attribute uncertainty. The smaller the cell size, the greater the precision of the resulting data. Finer raster sizes can trace the path of a line more precisely and therefore help to reduce classification error – a form of attribute error. Positional and attribute errors as a result of generalization are seen as classification error in cells along the vector polygon boundary. This is seen visually as the 'stepped' appearance of the raster version when compared with the vector original (Figure 10.12). Box 10.6 on p. 325 provides further examples of errors created during the vector to raster conversion process. The results of this process are often more generalized data with more uncertainty attached. The conversion of data from raster to vector format is largely a question of geometric conversion; however, certain topological ambiguities can occur – such as where differently coded raster cells join at corners as in Figure 10.11b. In this case it is impossible to say, without returning to the original source data, whether the vector polygons should join. Where vector maps have been derived from raster data, conversion may result in a stepped appearance on the output map. This can be reduced, to some extent, by line-smoothing algorithms (**Chapter 5**), but these make certain assumptions about topological relationships and detail that may not be present in the raster source data.

Most GIS packages provide data conversion software so that their files can be read by other GIS. For example, a GIS program may not be able to read data from a CAD package, so a conversion program is supplied to convert to another format, such as DXF, which both can handle. Users naturally assume that these conversion routines work without introducing new errors into the data. However,

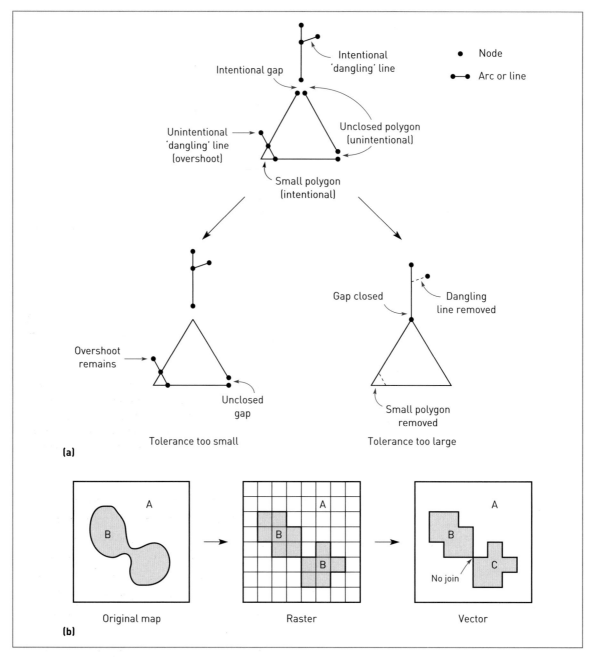

**Figure 10.11** Topological errors in vector GIS: (a) effects of tolerances on topological cleaning; (b) topological ambiguities in raster to vector conversion

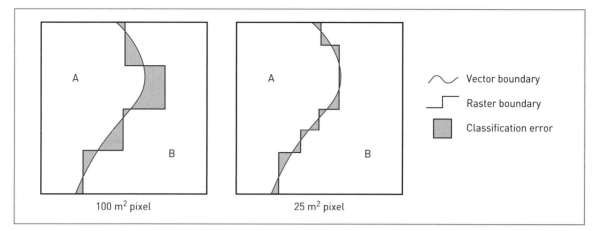

**Figure 10.12** Vector to raster classification error

**BOX 10.6 Rasterization errors**

THEORY

Vector to raster conversion can cause an interesting assortment of errors in the resulting data. For example:

1 *Topological errors*. These include loss of connectivity, creation of false connectivity (Figure 10.13a), loss of information, where individual vector polygons are smaller than the chosen raster cell size, and ambiguities due to changes in raster orientation and datum. Connectivity problems occur when rasterizing vector polygons are separated by narrow gaps or connected via long narrow corridors. Significant features such as these are easily lost if they are narrower than an individual raster cell.

2 *Loss of small polygons*. When individual vector polygons are less than half the area of the raster cell or small enough to be missed by adjacent cell centroids, they may be lost in the conversion process.

3 *Effects of grid orientation*. If two identical grids are placed over the same vector map but at different angles (such that the difference between the angles is not a multiple of 90), the resulting raster maps will be different from each other. It is usual when rasterizing vector data for the grid to be oriented parallel to the co-ordinate system of the source data. However, certain circumstances may dictate that a different orientation is required (such as when integrating rasterized vector data with satellite or aerial imagery where the satellite or aircraft track is not parallel to the vector co-ordinate system or when using data digitized on a different projection system). This is illustrated in Figure 10.14.

4 *Variations in grid origin and datum*. Using the same grid but with slightly different origins (where the difference is not a multiple of raster size) will result in two different raster maps in much the same fashion as with grid orientation. This is illustrated in Figure 10.14.

**Figure 10.13** Topological errors in vector GIS: (a) loss of connectivity and creation of false connectivity; (b) loss of information

**Figure 10.14** Effect of grid orientation and origin on rasterization

the transfer of a GIS database from one software package to another can lead to errors through changes in format and definitions. These database transfer errors are a kind of 'mechanical' error that is often overlooked. To overcome this problem standard formats have been developed such as those being promoted by the OpenGIS Consortium.

## Errors in data processing and analysis

Errors may be introduced during the manipulation and analysis of the GIS database. GIS users must ask themselves questions before initiating a GIS analysis. For example: are the data suitable for this analysis? Are they in a suitable format? are the data sets compatible? Are the data relevant? Will the output mean anything? Is the proposed technique appropriate to the desired output? These questions may seem obvious, but there are many examples of inappropriate analysis. These include the inappropriate phrasing of spatial queries, overlaying maps which have different co-ordinate systems, combining maps which have attributes measured in incompatible units, using maps together that have been derived from source data of widely different map scales and using an exact and abrupt method of interpolation (such as Thiessen polygons) to interpolate approximate and gradual point data (height data collected using GPS).

GIS operations that can introduce errors include the classification of data, aggregation or disaggregation of area data and the integration of data using overlay techniques.

The way in which we classify and reclassify data affects what we see and therefore how we interpret the data. Choropleth mapping is an example (**Chapter 8**). Much has been written in the cartographic literature on how the definition of class intervals in this sort of map can radically affect the pattern produced (Monmonier, 1996). Take, for example, a choropleth map of unemployment by region drawn from census data. How the class intervals are defined can affect perception of unemployment across the country. Similarly, how the areas are shaded also influences what we see in terms of pattern and magnitude. Dark colours in this instance can be used for the top two classes to create a more depressing feel.

Classification errors also affect raster data. Classified satellite images provide a reflectance value for each pixel within a specific wavelength range or spectral band (for example, red, near infra-red or microwave). Raster maps of environmental variables, such as surface cover type, are derived by classifying each pixel in the image according to typical reflectance values for the range of individual cover types present in the image. Error can occur where different land cover types have similar reflectance values (for example, asphalt car parks may have similar reflectance to mature spruce forest) and where shadows cast by terrain, trees or buildings reduce the reflectance value of the surface. Careful choice of classification method can help to reduce this type of error.

Where a certain level of spatial resolution or a certain set of polygon boundaries are required, data sets that are not mapped with these may need to be *aggregated* or *disaggregated* to the required level. This is not a problem if the data need to be aggregated from smaller areas into larger areas, provided that the smaller areas nest hierarchically into the larger areas. Problems with error do occur, however, if we wish to disaggregate our data into smaller areas or aggregate into larger non-hierarchical units. The information required to decide how the attribute data associated with the available units aggregates into the larger but non-nested units or disaggregates into a set of smaller units, rarely exists. This problem is normally approached by assuming that the data in the source units are evenly distributed by area (for example, population within an enumeration district is evenly distributed across its total area). The data for the new set of units or areas are in this case simply calculated as a function of the area of overlap with the source units. However, such a homogeneous distribution of data within the source units cannot always be correctly assumed. In the case of population, distribution will depend on urban morphology, which can vary over small areas. For example, even population distribution may be assumed across an area of similar housing type, but in an area of mixed land use (for example, housing, parkland and industrial estates) it cannot. Aggregation, disaggregation and classification errors are all combined in the modifiable areal unit problem (MAUP) (**Chapter 6**).

Error arising from map overlay in GIS is a major concern and has correspondingly received much attention in the GIS literature (Chrisman, 1989; Goodchild and Gopal, 1989; Openshaw, 1989). This is primarily because much of the analysis performed using GIS consists of the overlay of categorical maps (where the data are presented in a a series of categories). GIS allows the quantitative treatment of these data (for example, surface interpolation or spatial autocorrelation), which may be inappropriate. Map overlay in GIS uses positional information to construct new zones from input map layers using Boolean logic or cartographic modelling (**Chapter 6**). Consequently, positional and attribute errors present in the input map layers will be transferred to the output map, together with additional error introduced by multiplicatory effects and other internal sources. Data output from a map overlay procedure are only as good as the worst data input to the process.

Perhaps the most visual effect of positional error in vector map overlay is the generation of sliver polygons. Slivers (or 'weird' polygons) occur when two maps containing common boundaries are overlaid. If the common boundaries in the two separate maps have been digitized separately, the co-ordinates defining the boundaries may be slightly different as a result of digitizing error. When the maps are overlaid a series of small, thin polygons will be formed where the common boundaries overlap (Figure 10.15). Slivers may also be produced when maps from two different scales are overlaid. Of course, sliver polygons can and do occur by chance, but genuine sliver polygons are relatively easy to spot by their location along common boundaries and their physical arrangement as long thin polygonal chains.

Also shown in Figure 10.15 is the attribute error that can result from positional error. If one of the two maps that are overlaid contains an error (such as the missing 'hole' feature in map A that is present as an island in map B), then a classification error will result in the composite map (polygon BA). Chrisman (1989) identifies these two cases as extremes and points out that one type of error (positional or

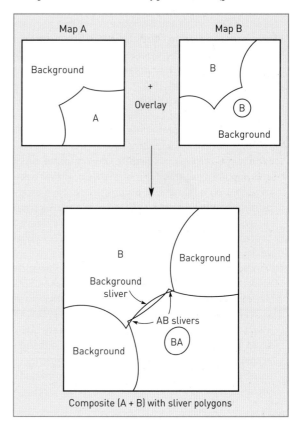

**Figure 10.15** Generation of sliver polygons

attribute) will inevitably result in the other type of error in the output map and vice versa. Classification errors occur along the common boundary where the slivers are produced as a result of positional error and attribute error.

## Errors in data output

From the preceding discussion it should be clear that all GIS databases will contain error. In addition, further errors will be introduced during manipulation and analysis of the data. Therefore, it is inevitable that all GIS output, whether in the form of a paper map or a digital database, will contain

inaccuracies. The extent of these inaccuracies will depend on the care and attention paid during the construction, manipulation and analysis of the databases. It is also possible that errors can be introduced when preparing GIS output. For example, if we choose the map as a means of communicating the results of our analysis, it may be necessary to generalize data for the sake of clarity. This will add errors and assumptions into the database.

Box 10.7 provides a case study of the use of viewshed analysis in archaeology, which considers the range of errors that can arise in a GIS project, including errors arising from our understanding of reality, errors in source data and those that arise during data processing.

---

# BOX 10.7 Viewshed analysis in archaeology

**CASE STUDY**

## Graham McElearney

Archaeologists have been interested in viewshed analysis (see **Chapter 6**) from a relatively early stage of GIS usage within the discipline. Viewshed analysis has been applied in fairly functional applications (for example, looking at visibility between Iron Age hill forts to assess their defensive and/or territorial functions), to what might be considered more symbolic forms of analysis in which archaeologists analyse locations of ceremonial monuments such as burial mounds and stone circles in an attempt to model other aspects of past perceptions of the landscape. However, despite the relative ease and popularity of this form of analysis, it has been criticized on a number of grounds. This case study outlines some of the problems with viewshed analysis in archaeological applications, reviewing conceptual and pragmatic problems as well as those which are technical and procedural.

Viewshed analyses are based on Digital Elevation Models (DEMs) derived in the current day. These DEMs pay no attention to current or past vegetation such as trees or scrubland, and this can have a profound effect on visibility, especially at locations very close to an observer's viewpoint (Figure 10.16). Although archaeologists can reconstruct past environments to some extent, the approach they take tends to be quite broad-brush and would, for example, rarely identify a

stand of trees around a burial mound. When looking at viewsheds calculated over longer distances, the issue of object-background clarity can arise and this can result in objects which fall within a calculated viewshed not actually being visible with the naked eye.

In addition to the effects of distance, factors such as seasonal and diurnal variations will affect visibility over differing timescales. Over and above normal differences between day and night, seasonal differences in vegetation and weather have a significant effect on both distance and clarity of view. For example, field observations suggest that summer haze can attenuate view just as much as a winter mist, and stands of deciduous trees will present a more permeable barrier to visibility in winter than in summer.

Further problems exist in assuming that 'views from' a monument automatically equate to 'views to' the same monument from the surrounding landscape, yet this is implied in GIS derived viewshed maps. GIS based viewsheds present also a binary view, with features simply being either in view or not, and for many reasons above this can be an over-simplification.

The DEM is the primary data source from which viewshed analysis is derived, so any factors that affect the quality of the DEM will affect the quality of the results based upon it. For example, the resolution (grid-cell size) of the original DEMs and early data sets that were available for GIS analysis were quite coarse.

**BOX 10.7**

**(a)**

**(b)**

**Figure 10.16** A composite image showing a GIS-generated viewshed (a) and a photograph taken from the same location (b). Note how the vegetation in the near field obscures the view of the landscape behind, which is indicated as visible by the viewshed map
*Source*: Courtesy of Dr Graham McElearney, University of Sheffield. Digimap download data © Crown Copyright Ordinance Survey. All rights reserved.

Generally speaking, the higher the resolution of the DEM, the more accurate the results of the analysis should be, so early analyses were relatively crude.

Related to resolution is the quality of the interpolation of the DEM from the source data. Artefacts can be created during interpolation, typically resulting in spikes or sink holes in the data. Whilst these can be filtered out, the process may also smooth out naturally existing features such as hillcrests or outcrops, the features that often delineate viewshed boundaries, so the erosion of these is problematic.

Any errors in the underlying data set will be propagated in analytical techniques that combine multiple viewshed maps, such as Cumulative Visibility Analysis (Wheatley, 1995). Figure 10.19 provides another example of the difficulties for visibility analysis that may be due to errors in a DEM.

The choice of GIS software will also have a bearing on viewshed analysis, as different packages may use different algorithms and implementations of the basic analytical principle. Tests for robustness have resulted in as much as a 50 per cent difference in viewsheds derived by different analytical routines, or based on different DEM interpolation methods.

Over and above these technical limitations, archaeologists face their own theoretical and interpretive issues when applying this analysis technique. Perhaps the most significant of these is assessing the importance of the viewshed phenomena to those living in the past. Just because we may be able to demonstrate some potential visible association between certain kinds of monuments in the present, this does not necessarily mean that these relationships had any significance to those who built them in the past. It is certainly very difficult to suggest that any relationships were important factors when past populations were choosing to locate themselves or the sites they left behind in the landscape. So, whilst analysis of a defensive structure such as Hadrian's Wall in Northern England might seem relatively straightforward, it is much more problematic when looking at archaeological monuments whose functions are far less tangible.

Despite these limitations, and if analysis is undertaken with awareness of possible conceptual, technical and interpretative issues, GIS-based viewshed analysis still presents archaeologists with a very powerful technique which enables a complexity and speed of analysis that is currently impossible to achieve by other means.

**FURTHER INFORMATION**

For further discussion of these and related issues, see:

Wheatley, D (1995) Cumulative Viewshed Analysis: a GIS-based Method for Investigating Intervisibility, and its Archaeological Application. In: G Lock and Z Stančic (eds) *Archaeology and Geographic Information Systems: a European Perspective*. Taylor and Francis, London, pp. 171–85

Wheatley D, Gillings M (2000) Vision, Perception and GIS: Developing Enriched Approaches to the Study of Archaeological Visibility. In: G R Lock (ed) *Beyond the Map: Archaeology and Spatial Technologies*, Amsterdam: NATO Science Series A: Life Sciences, 321:1–27

*Source*: Graham McElearney, University of Sheffield

**REFLECTION BOX**

- Think about the digital topographic base data for your country or region. How well do these data represent reality? Write brief notes on the completeness, compatibility, consistency and applicability of the data for an application of your choice.

- Summarize the errors that arise from adoption of the raster or vector data models. What additional errors might occur during conversion from one data model to the other
- Create a table summarizing the errors that might occur in a GIS for a project in your area of interest. Use the template below.

| Brief description of error | Is this a spatial or attribute error? | What is the source of this error? |
|---|---|---|
| 1. | | |
| 2. | | |

## ■ FINDING AND MODELLING ERRORS IN GIS

Much research effort has focused on identifying, tracking and reducing error in GIS. One American GIS research organization, the National Center for Geographic Information and Analysis, even placed the problem of error in GIS first on its list of research priorities (NCGIA, 1989). Despite this research effort, little has been done to incorporate error identification, modelling and handling techniques within proprietary GIS packages. This has had the effect of keeping awareness of the general error problem at a low level amongst non-specialist GIS users. However, there are methods for detecting and modelling errors that users can employ to improve confidence in their data and results. Methods for checking data include visual inspection, double digitizing, the examination of error signatures and statistical analysis.

### Checking for errors

Probably the simplest means of checking for data errors is by *visual inspection*. This should be included as a matter of course in the data cleaning and editing process. Once in a GIS format, data can be plotted on paper or on the graphics screen for visual comparison

with the original source document. Obvious oversights in the digitizing process and missing data will be revealed and can be amended. The attributes of spatial features can be checked by adding annotations, line colours and patterns. Again, comparison with the original map should reveal major errors.

Double digitizing is an error-checking method used by large digitizing operations (for example, map producers such as the UK Ordnance Survey). The same map is digitized twice (possibly by different people) and the resulting copies compared to identify inconsistencies. Polygon areas, length of lines, polygon perimeters and the number of sliver polygons produced are calculated and compared. If the two versions are deemed significantly dissimilar, then the map may be re-digitized and the comparison repeated. However, this is a costly and time-consuming method of error checking.

Work on the psychological and physiological aspects of digitizing (Box 10.5) suggests that each operator has a different error signature that describes how they digitize. An operator may, for example, consistently over- or undershoot changes in line direction or show lateral displacement relative to position on the digitizing table. If different digitizer operators do have recognizable and repeatable error signatures, these could be used to effectively cancel out the errors introduced in digitizing.

Various statistical methods can be employed to help pinpoint potential errors. Attribute errors in GIS data sets may stand out as extreme data points in the distribution of data values; or outliers in scatter plots if plotted against a correlated variable (for example, total annual rainfall against number of rain days). A statistical measure of the error attributed to generalization and line thickness can be gained by estimating the total area of the map covered by lines. An example given by Burrough (1986) shows that a 1:25,000 scale soil map measuring 400 × 600 mm may have as much as 24,000 mm of drawn lines, covering 10 per cent of the map's total area.

## Error modelling

There are relatively few tried and tested approaches to modelling of errors in spatial data and these tend to focus on the effects of quantifiable positional errors. Two such methods are epsilon modelling and Monte Carlo simulation. Epsilon modelling is based on an old method of line generalization developed by Perkal (1956). Blakemore (1984) adapted this idea by using the model to define an error band about a digitized line that described the probable distribution of digitizing error about the true line. The shape of a graph of the error distribution is a matter of debate, but the graph is probably a bell-shaped curve. A true normal distribution is unlikely since very large errors represented by the tails of the distribution are rare and easily spotted and removed in the data cleaning and editing process. Some authors believe a 'piecewise quartic distribution' to be most realistic. This has a flatter central peak and tails that quickly run out to zero (Brunsdon *et al.*, 1990). Others have suggested a bimodal or platykurtic distribution on the grounds that digitized points are very unlikely to fall exactly on the line.

In an empirical example based on overlaying employment offices (points) with employment office areas (polygons), Blakemore (1984) found that he could unambiguously assign only 60 per cent of the points to any polygon. The remaining 40 per cent were found to lie within the specified epsilon band width for the employment office area polygons. From this work Blakemore defined the

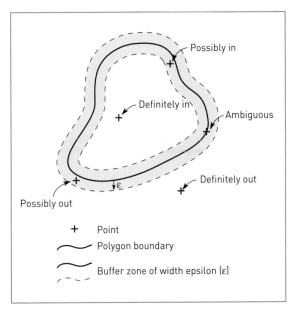

**Figure 10.17** Point-in-polygon categories of containment
*Source*: Blakemore (1984)

five different categories of containment illustrated in Figure 10.17.

The epsilon model has been developed in various ways and with various error distributions to produce a seemingly robust means of error modelling. Given information on the distribution of error in the input data layers, this technique can be used to handle error in GIS operations by allowing for the epsilon band in GIS analyses to improve the confidence of the user in the results (Carver, 1991a).

A Monte Carlo simulation approach has also been used to model the effects of error in GIS overlays (Fisher, 1991; Openshaw *et al.*, 1991). This method simulates the effect of input data error by the addition of random 'noise' to the line co-ordinates in map data and as such is a kind of 'jack-knifing' (a technique often used in the statistical analysis of error and uncertainty based on re-sampling data based on random models). Each input data source is assumed to be characterized by an error model that represents reasonable estimates of positional error thought to be present in a similar manner to the epsilon approach. The random 'noise' added to the map co-ordinates simulates the combined effects of source map and digitizing errors by slightly varying the course of the original shape of the line. The randomized data are then processed through the required series of GIS operations and the results saved. This entire process is repeated many times. If the output is numeric, then the distribution of the results gives some indication of the effects of input data error. If the output is in the form of a map, then the total set of saved maps can be used to draw confidence intervals. A simple example would be the use of the D8 algorithm to derive stream networks from DEMs. In **Chapter 7** it is described how results from the D8 algorthm are more uncertain in areas of lower relief. This is illustrated in Figure 10.18 where the effects of uncertainty in the DEM and D8 algorithm have been simulated using a Monte Carlo simulation approach. Note how the uncertainty envelopes are much wider in the lower relief areas where adding random noise to the DEM has greatest effect.

Fisher (1993 and 1994) describes how visibility analyses are susceptible to errors in digital elevation models. He uses a Monte Carlo simulation approach to investigate uncertainty in the results that are caused by errors in the DEM and algorithms used. Because many of the angles of lines of sight are so acute (see Figure 6.32), even very small errors in the height values within the DEM can produce comparatively large errors in the visible area polygons. In Happy Valley line of sight is important for ski rangers out on patrol because the range of their radios is significantly affected by the steep terrain. Although reception can be maintained in the shadow of valleys and peaks as the radio signals are diffracted around and bounced off the mountains, the best reception is always maintained when there are no intervening hills or mountains between receivers and the base station. Figure 10.19 gives an example for selected radio 'viewpoints' located in the mixed terrain of the Happy Valley ski resort and its surrounding area. A Monte Carlo simulation is used to describe the certainty of the predicted radio coverage for each location. Note how uncertainty in the DEM used has the greatest effect in areas of relatively flat relief.

**Figure 10.18** Simulating effects of DEM error and algorithm uncertainty on derived stream networks

In both these examples the results of these simulations can be overlaid on the normal, unvaried results as a visual means of communicating the likely effects of input data error. The main drawback with this method is the large number of simulations required to achieve a statistically valid result. For GIS pro-cesses where the effects of error need to be estimated, up to 100 extra sets of processing may be required. Monte Carlo simulation is only likely to be used for very important analyses requiring a high level of confidence in the results. Although initial research on Monte Carlo simulation focused on its use for modelling positional error in vector map overlay (Openshaw *et al.*, 1991), the approach is equally applicable to modelling attribute or positional error in raster or vector GIS operations. Another example of the use of Monte Carlo simulation modelling to prepare multiple surfaces representing uncertainty in terrain data is provided in Box 10.8.

**Figure 10.19** Simulating the effects of DEM error and algorithm uncertainty on radio communications in the Happy Valley area

Monte Carlo simulation could also be applied to the modelling of avalanche risk zones. Because the models used are by no means perfect and there are known levels of error in the data used, and because the implications of making mistakes in the drawing up of 'avalanche safe' zones that are suitable for constructing ski pistes, hotels and other developments, the Happy Valley GIS team need to be absolutely certain that they are confident in their model results. The higher overheads involved in running the avalanche model 100 times within a Monte Carlo simulation are in this case easily outweighed by the potential costs to life and property should the wrong decisions be made.

In combinatory models such as sieve mapping, the sensitivity of the resulting map to various input layers can be investigated using a technique known as 'bootstrapping'. This is a kind of leave-one-out analysis applicable to overlay analyses where the analysis is repeated $n$ times, where $n$ is the number of input data layers, leaving one of the input layers out of the analysis on each successive model run. The results of the $n$ analyses can be

## BOX 10.8 Uncertainty and terrain data in ice-sheet modelling

CASE STUDY

### Ross Purves and Felix Hebeler

Ice-sheet modelling seeks to represent the behaviour of ice sheets under past, present or future climates. It is often based on some form of proxy record: for example, the temperature signal extracted from an ice core record. The data obtained can help us explore concerns about future climate change such as the impact of climate on ice extents and thus potential changes in sea level and the influence of melt water inputs to ocean systems.

Ice-sheet modelling, in common with all numerical modelling, seeks to improve our understanding of the real world through an abstraction of reality. Any abstraction cannot completely represent a system and thus differences between a modelled

scenario and 'reality' must exist. It is important to understand these differences, or uncertainties, in modelled results to make use of models as tools for understanding and predicting the behaviour of natural systems. One source of uncertainty arises because ice-sheet models are typically run at low resolutions, so certain processes are not well resolved. For example, inception (the formation of new ice sheets) can only take place in regions where the annual mass balance is positive (that is, more snow accumulates over a year than melts). Such regions are typically at high elevations – for example, in the snow-covered European Alps. However, when we use low-resolution data, the value of elevation for a single grid cell encompassing a large area may be derived in a number of different ways from

→

BOX 10.8

higher-resolution data. Figure 10.20 shows high-resolution data collected for the Zermatt region by the Shuttle Radar Topography Mission for the Swiss Alps, and a corresponding generalization of these terrain data to 20 km (a resolution typically used in an ice sheet model).

This case study illustrates the use of Monte Carlo simulation to explore the dependence of model results on uncertainties in elevation values. Monte Carlo simulations require that a model is run many times with one variable being varied according to some probability distribution, in order to explore the model's sensitivity to that variable. In the following example the use of standard GIS techniques to prepare multiple surfaces representing uncertainty in elevation is explained. The approach adopted involved four tasks.

### TASK 1: PREPARE TOPOGRAPHIC DATA FOR INPUT TO MONTE CARLO SIMULATIONS

The first task in preparing topographic data for input to Monte Carlo simulations is to characterize the probability distribution which will be applied to the low-resolution data. For example, regions where elevation varies little will have low standard deviations in elevation – for example, the Prairies of Mid-West America. Mountainous regions such as the Alps have high standard deviations in elevation.

Resampling is a standard technique in GIS to derive rasters of lower resolutions. For example, to resample from a resolution of 100 m to 1 km, the origin of the new raster must first be selected. Then, values for the new raster at a spacing of 1 km must be identified. Typically, a neighbourhood function is applied to the source resolution raster to derive a value at the new target low resolution. The neighbourhood function may, for example, take the mean of all 100 cells inside the 1 km cell, or may apply some neighbourhood function to a subset of these cells.

To calculate standard deviations for 20 km cells, a 20 km raster with unique values for each cell was created. These cells were used as the zones in calculating zonal means and standard deviations with the 100 m data as an input. Figure 10.20a shows four cells representing the topography of the Zermatt valley at a resolution of 20 km, alongside the same region represented at a resolution of 100 m. The smoothing of the topography is dramatic. By resampling high

**Figure 10.20** Resampling SRTM data from 100 m to 20 km resolution. (a) A view of Zermatt from the east at 100 m resolution (top) and 20 km resolution (bottom). (b) DEM of the European Alps (top) and graph of standard deviation in elevation with respect to elevation for resampled data (bottom)

**BOX 10.8**

resolution data with a range of origins and functions it is possible to describe statistically what we see in this picture. The standard deviation of all the cell values in the source data set (the 100 m data) gives a measure of the roughness of the surface. By plotting the standard deviation for the whole of the European Alps, it is possible to see that roughness varies with altitude (Figure 10.20b). This in turn implies that the uncertainty in an elevation value in a low-resolution raster will also vary with altitude.

### TASK 2: CREATE TOPOGRAPHIC SURFACES FOR INPUT TO MONTE CAROL SIMULATION

The next step in creating a set of topographic surfaces for input to Monte Carlo simulations is to generate multiple surfaces that represent the uncertainty in elevation shown in Figure 10.21a. In our case, only low-resolution data were available for Scandanavian topography, and we hypothesized that the relationships described in Task 1 for the European Alps, would also hold true for Scandanavia – in other words, that standard deviation in elevation varies with altitude in both cases.

In order to create the error surfaces, altitude is sampled at 250 random points on the original raster of Scandinavia (sampling approximately every 50th cell). At each location $i$, a standard deviation $\alpha$ is derived from the trend line shown in Figure 10.21b. A random number is then drawn from a normal distribution of mean $\mu = 0$ and standard deviation $\alpha$, which is set as the value $\varepsilon$ at location $i$. The result is a set of points with values of $\varepsilon$ that have an altitude dependent and a random component. Between these points, a surface is interpolated using inverse

(a)

(b)

(c)

**Figure 10.21** DEM error and ice sheet modelling: (a) An example error surface (one of 150 generated) drapped over the Scandinavian topography. (b) Standard deviation in ice sheet extent through time.(c) A snapshot of minimum and maximum ice sheet extents after 12,000 model years

## BOX 10.8

distance weighting (Figure 10.21b). In total 150 error surfaces are generated in this way and added to the original topography.

### TASK 3: RUN THE ICE SHEET MODEL

Having produced the 150 error surfaces, the ice sheet model (GLIMMER) can be run with identical parameters for each surface. In this case, the ice sheet model is driven by temperature data derived from an ice core record between 120 and 60 thousand years before present. Figure 10.21c shows the resultant variations in ice sheet extent and volume through time and space.

### RESULTS

The aim of this research was to explore uncertainties in ice sheet models as a result of uncertainties in low-resolution rasters. However, lessons can also be drawn in a more general sense when we think about how we use a GIS. For example, Figure 10.20a portrays very starkly what we are doing when we reduce resolution by resampling. Often we do this over large areas and the general form of relief is maintained. However, such a zoomed-in view of

the new data is a useful way of seeing what really happens. Second, the statistical relationship plotted in Figure 10.20b shows that, in general, lower places tend to be flatter. However, by expressing this statistically we can use the relationship to explore the implications in a real modelling problem.

In Figure 10.21 we can see what these results mean for the question we are researching. Relatively small absolute uncertainties (of the order of 50 m) in elevation can cause large differences in the final configurations of the derived ice sheet volumes. These differences are of course largest when the ice sheets are smallest, but are over the whole model run of the order of 10 per cent. These initial results from our research show that the influence of uncertainty in terrain is important in considering the results of ice-sheet models. However, much more important, deriving these results makes us think about how we use GIS to better understand the world around us, and of course it is also fun.

### FURTHER INFORMATION

GLIMMER: forge.nesc.ac.uk/projects/glimmer/

*Source*: Ross Purves and Felix Hebeler, University of Zurich

compared to identify their sensitivity to each of the input layers. For example, in sieve mapping analysis, the number of times a polygon or cell appears in the final decision set is indicative of its

robustness as a suitable location. This is illustrated in Box 10.9 which provides an overview of the different error types that might occur in the nuclear waste cases study.

# BOX 10.9 Errors in the nuclear waste case study

CASE STUDY

### SOURCE DATA ERROR AND THE EFFECTS OF GENERALIZATION

Just as with any other GIS application, the source data used in this study are subject to problems of data error. The four siting factors used by NIREX present different problems:

1 *Geology.* These data were derived from 1:625,000 scale maps of deep hydrogeological environments produced by the British Geological Survey (BGS). The requirement of any suitable geological environments is that water flow through the rock strata is either non-existent or so slow that it would take thousands *of years for it to return radionuclides leached*

*out of* the waste to the surface. Although the surface geology of the UK is accurately mapped, very little is known about geological structures 500–1000 metres underground. The maps showing potentially suitable geological environments were interpolated and extrapolated from a very limited number of deep boreholes, seismic records and 'intuitive guesswork'. Given the complexity of geological structures and the smoothness of the areas identified by the BGS, these data are subject to a high degree of generalization and therefore positional and attribute uncertainty. Given the element of guesswork involved it is very difficult to quantify the level of error involved, but it can only be assumed to be large.

## BOX 10.9

**2** *Population*. The population data were 1-kilometre-square gridded population counts derived from the 1981 OPCS Census of Population. The census data themselves can be assumed to be reasonably accurate for the night on which the census was taken. The derivation of gridded population counts from ED-level data for studies of this kind makes certain assumptions about the data and the distribution of people within the enumeration districts. This is referred to as the modifiable areal unit problem (**Chapter 6**).

**3** *Accessibility*. Accessibility data were derived by buffering the road and rail networks at predetermined widths. The road and rail data were taken from Ordnance Survey 1:625,000 digital data. The relatively small scale of this data source means that a significant amount of scale-related cartographic generalization is present. Also, if digitizing error is assumed to be in the order of 0.8 mm either side of the mapped line (Dunn *et al*., 1990), then at this scale this would account for possible lateral displacement of the digitized line by up to 500 metres.

**4** *Conservation*. The location of conservation areas was derived from 1:250,000 scale maps provided by the Nature Conservancy Council (NCC). Again, the effects of scale-related cartographic generalization need to be recognized as well as errors introduced during the digitizing process. At this scale these could account for up to 200 metres lateral displacement.

The effects of source data error and generalization may be investigated through jack-knifing or Monte Carlo simulation. Figure 10.22 shows the effects of adding random noise to data inputs for each of the four siting factors in the nuclear waste siting example.

### CHOICE OF CRITERIA

The choice of the criteria used in performing categorical map overlay is critical to the outcome of the site search. Guidelines produced by the IAEA (1983) suggested that geology, population distribution, accessibility and conservation be taken into account. However, no specific criteria specifying how these were to be mapped were included. The criteria used in this case study to produce the maps shown in Figure 1.8 mimic those chosen by NIREX using the best available digital data sets at the time.

**1** *Geology*. Suitable hydrogeological environments were mapped by the BGS and no further processing of these maps was required after digitizing.

Definitive final areas output from standard sequence of GIS operations

99% probability areas defined using epsilon algorithm

90% 'credibility regions' defined using Monte Carlo simulation model

99% 'credibility regions' defined using Monte Carlo simulation model

**Figure 10.22** Simulating uncertainty in the siting of nuclear waste facilities

BOX 10.10

**2** *Population*. Areas of low population density were defined by adapting the Nuclear Installations Inspectorate's (NII) relaxed nuclear power station-siting guidelines. These state that there should be a maximum population of 100,000 people within five miles of the reactor site (Charlesworth and Gronow, 1967; Beale, 1987), and this was converted into a population density threshold of 490 persons per square kilometre. Here the same criterion is used, but is mapped using the 1-kilometre gridded population data.

**3** *Accessibility*. Local accessibility to the national network of roads and railways is considered more important than 'strategic' accessibility to waste-producing sites based on Weberian location analysis. Good local accessibility is defined as a maximum straight-line distance to a motorway or railway line of 3 km, or 1.5 km to a 'primary route'. Proximity to local transport networks would keep the costs associated with building new access roads and rail sidings to a minimum. These criteria are not based on any formula, but are deemed to be 'reasonable' figures by the analyst.

**4** *Conservation*. As with geology, the location of existing conservation areas was provided on maps by the NCC and, after digitizing, no further processing of these was required.

The criteria that are used to map population and accessibility clearly affect the results shown in Figure 1.8. 'What if?' analyses can be carried out to investigate the effects of specifying different criteria. However, the question of the appropriateness of the criteria still remains. For example, why is a population density of 490 persons per square kilometre considered to be the most appropriate threshold? What makes a site with a population density of 490 better than a site with a population density of 491? Why not use a density threshold of 49 persons? Why not use 4.9? The answer is probably that a lower population density threshold would have excluded too much of the UK land area from consideration, making the task of finding an acceptable site far harder for NIREX. Choosing appropriate threshold criteria and being able to justify them is critical when defending an analysis of this kind (particularly if the siting decisions made are likely to provoke a hostile reaction from local residents).

## ANALYTICAL ERROR

Any analytical error affecting this case study is likely to come from within the GIS software via the analyst. The opportunity for gross errors in the choice of appropriate analytical tools is minimal since the analysis is limited to categorical mapping (reclassification functions), buffering and map overlay. All the source data are at comparable scales and resolutions. It is conceivable that errors may occur in the processing of the data due to hardware or software faults.

## CONCEPTUAL ERROR

The above discussion of data error and choice of criteria reveals the difficulties of conceptualizing what at first appears to be a map overlay problem. In the first instance, it is not safe to assume that the IAEA's suggested siting factors (geology, population, accessibility and conservation) are correct for every situation. From a personal point of view you may suggest 'as far away from me as possible!' as an additional factor. In terms of the data used and the criteria chosen we can see that the site search pursued was driven to a certain extent by economic and political considerations. By specifying a relatively high population-density threshold they were able to keep their options open. Sites they already owned were identified. By defining good accessibility they could keep transport costs to a minimum. The last site investigated was near to the Sellafield reprocessing plant in Cumbria (the main source of the UK's nuclear waste). This was clearly a choice driven more by economic factors such as low transport costs and political pragmatism than by other geographical considerations. The conceptual error associated with choice of siting factors could usefully be investigated using the bootstrapping approach by generating four comparative suitability maps from four separate overlay procedures, each one leaving out one of the four siting factors. The results of such an analysis are shown in Figure 10.23. It should be noted that these do not show the effects of any uncertainty or error in the definition of the siting factors from positional error, attribute uncertainty or reclassification criteria, rather they show the relative impact of the exclusion of selected factors in the sieve-mapping process.

→

BOX 10.9

**(a)** Suitable-deep geology

**(b)** Population

**(c)** Accessibility

**(d)** Conservation areas

**(e)** Comparative suitability map resulting from overlay of siting factors a, b and c

**(f)** Comparative suitability map resulting from overlay of siting factors a, b and d

**(g)** Comparative suitability map resulting from overlay of siting factors a,c and d

**(h)** Comparative suitability map resulting from overlay of siting factors b, c and d

**(i)** Comparative suitability map resulting from overlay of all siting factors a, b, c and d

**Figure 10.23** Bootstrapping or 'leave one out' analysis

# ■ MANAGING GIS ERROR

We must accept that no matter how careful we are in the preparation of our data and how cautious we are in our choice of analysis, errors will find their way into the GIS database. To manage these errors we must track and document them. The data quality parameters discussed earlier in this chapter – accuracy, precision, resolution, generalization, applicability, bias, compatibility, completeness and consistency – provide a useful checklist of quality indicators. Documenting these for each data layer in a GIS, and for any new layers produced by analysis, will help subsequent users of the GIS to keep track of data quality. Data quality information is frequently used to help construct a data *lineage*.

Lineage is a record of data history that presents essential information about the development of data from their source to their present format. Lineage information should provide the data user with details of their source, method of data capture, data model, stages of transformation, editing and manipulation, known errors and software and hardware used. In a way, lineage is like the service history in our car example. A good knowledge of lineage gives the user a better feel for the data and is indicative of its overall quality. Above all (as in the car's service history) it lets the user know where problems with the data are likely to occur. Unfortunately, just as with our car example, full service histories are

something of a rarity in GIS. Depending on the context, documentary evidence may be available. Some digital data suppliers provide good lineage information in their accompanying materials. Users are less likely to maintain good lineage records. They are not helped by the fact that few GIS packages provide any means of storing lineage information.

Together, lineage and consistency give an indication as to the *repeatability* of the methods being used. Could the same set of results from a GIS analysis be obtained if the data were reconstructed from scratch using the same methods and processes as described in the lineage? If consistency in the data sources, methods of capture and data processing is low it is unlikely that the results will be the same. Inevitably this leads to a lack of confidence in GIS data and analysis. Thus, it is essential that all possible means are used to ensure as high a level of consistency and overall data quality as is practical. This is especially important in those cases where the results of a GIS analysis are to be used for real-life decision making where lives, property and money are at stake.

A number of organizations involved in the creation, processing and use of digital map data have attempted formalization of lineage requirements. Both the Association for Geographical Information (AGI) in the UK and the National Committee on Digital Cartographic Data Standards (NCDCDS) in the USA define basic lineage requirements (see Box 10.10).

## BOX 10.10 Basic lineage requirements

PRACTICE

**1** *Description of source data.* This should contain information relevant to the data source, including its name, date and method of production, date of last modification, producer, reference, map scale and projection. It should also include the necessary backward links so that the user can see how the source data were generated.

**2** *Transformation documentation.* Transformation information contains details of the operations and actions that have led to the production of the data in their present form. This includes the commands, parameters and context of use.

**3** *Input/output specifications.* This records details of the data files used and the products generated. This information should include descriptions of file formats, transfer formats, input/output procedures, media specifications and any interaction the user makes with the process.

**4** *Application-dependent information.* This category has been proposed to record information about the purpose for which a particular data set was generated. This can help the user determine for which applications the data are useful.

Lineage information can be recorded manually or automatically. The user may manually create the lineage by recording the information needed directly on paper or in a computer text file. A paper-based scheme involves the user keeping a record in a log book or card system as operations are carried out on the data. Wholly computerized systems of manually recording lineage are perhaps the best option. This is possible if a database system into which the user records lineage information runs concurrently with the GIS. Automatic systems remove the need for user input altogether. These are rare in GIS, but some GIS do produce log files automatically which can then be converted to lineage information when combined with data-specific information. The benefits of lineage in quality control are summarized in Table 10.1.

**TABLE 10.1  Benefits of lineage**

| | |
|---|---|
| Error detection | Lineage helps recreate analysis processes in which previous data sets containing errors have been used |
| Management accountability | Lineage provides information from which accounting can be undertaken (Aronoff, 1991). It allows the assessment of workloads and efficiency |
| External accountability | Lineage records the work of each GIS user, which allows others to assess the validity of the work undertaken |
| Quality reporting | The AGI in the UK and the NCDCDS in the USA define lineage as a requirement for GIS data quality reports. These should include information on data history, positional accuracy, attribute accuracy, completeness and logical consistency |

## REFLECTION BOX

■ Consider a data encoding technique which you commonly use, or which is in common use in your discipline. List the errors which might result from this technique and identify ways to avoid or minimize these during the data encoding process.

■ Describe how you would visually check a data set digitized from a paper map. What errors would you be looking for?

■ Briefly outline the main elements of epsilon and Monte Carlo modelling. What are these techniques used for?

■ Try to document a data lineage for a data set you are familiar with. Include details of the source, method of data capture, data model, stages of transformation, editing and manipulation, known errors and software and hardware used.

## CONCLUSIONS

Error in GIS is a 'fact of life' but not an intractable problem. Adopting good practice in terms of data capture and analysis should, in most instances, be sufficient to ensure errors are kept to a minimum. Conscientious documentation and the incorporation of lineage information and quality statements on GIS output will help ensure that end users are aware of the limitations of GIS products. Box 10.9 outlines some of the known errors in the nuclear waste case study by way of an example of the types of errors that can be documented.

Despite the existence and documentation of errors such as those presented in Box 10.9, some GIS users, like car drivers, press on with their work despite the GIS equivalent of the flashing of the oil warning light, the clouds of steam coming from the radiator and the clanking noise from the rear axle. Like (petrol engine) motor cars, GIS require the correct data (petrol not diesel), data which are fit for the purpose and system (unleaded or leaded petrol), appropriate choice of analysis method (careful driving) and maintenance of correct lineage (good service history).

→

As GIS become widely used in everyday decision making, the issue of legal liability raises its head. If a GIS is used to support a decision, which subsequently results in loss or injury where the GIS analysis has been proven to be at fault, who or what is to blame? Is it the decision maker who decided to use GIS? Is it the GIS analyst who directed the GIS work? Is it the GIS technician who pushed the keys? Is it the GIS or the vendor who supplied the software? Is it the data or the data supplier? Chances are that the 'buck' stops with the GIS analyst, for it is they who will have made the crucial decisions regarding what data and methods of analysis to use. Precedents exist in the USA, where people have been successfully sued over faulty GIS analyses – think back to the example of the Happy Valley GIS team and the risk zones produced using their avalanche model. Openshaw (1993) gives an interesting account of these and other issues in his editorial 'GIS crime and criminality'.

With expertise in GIS comes responsibility. Responsibility to ensure that the data and the chosen method of analysis are appropriate to the problem in hand. Responsibility to ensure that possible alternative solutions have been considered and are presented to the decision maker. Responsibility to ensure that the best available data are used and any potential errors are accounted for and communicated clearly to the end user. As suggested in **Chapter 8**, GIS output never fails to impress management and decision makers; it is colourful, it has been produced using a computer, it is often complex. Is it correct? Any doubt in the mind of the GIS analyst, whether due to potential data error or to uncertainty about the analysis, should be clearly stated in the output. Be careful in your choice of data. Be careful in your chosen analysis procedure. Communicate any doubts clearly in the results and output.

## REVISION QUESTIONS

- What is the difference between accuracy and precision? Give examples of both in a GIS context.

- Write a brief explanation of the following terms: applicability, bias, compatibility, completeness, consistency.

- Conceptual error is related to how we perceive reality. How does conceptual error affect GIS?

- What are the main sources of error in GIS data input, database creation and data processing?

- The modifiable areal unit problem (MAUP) is commonly cited as a difficulty faced when attempting to overlay data layers in a GIS. What is the MAUP, and why does it cause problems for data integration?

- Describe the main sources of error in digital terrain models. What implications might they have for subsequent analyses?

- Discuss some of the ways in which errors in GIS data can be avoided.

## FURTHER STUDY – ACTIVITIES

- Draw a mental map for either your home town or a journey that you are familiar with. Compare your map with a traditional paper map of a suitable scale. How do they differ in their portrayal of space and place?

- For a GIS problem you are working on, try to identify all the likely sources of error and the probable impacts they may have on your analysis. Can you quantify the error?

- Draw a flow chart illustrating a Monte Carlo simulation model of a GIS analysis you are working on.

- Run a Monte Carlo simulation on a simple model in Excel or other spreadsheet software. Use the RAND function or equivalent to add random noise to one or more of the input variables and re-run the model several times. Save the results and plot a frequency distribution graph.

# FURTHER STUDY – READING

The problems of data quality and error propagation in GIS saw a flush of research papers in the 1980s and 1990s. The NCGIA research initiative on the accuracy of spatial data started much work in the USA and produced one of the few books dedicated to the subject (Goodchild and Gopal, 1989). Similarly, in the UK the NERC/ESRC-funded Geographical Information Handling project did much to stimulate research on error propagation and generalization in GIS (see Mather, 1993). Prior to these initiatives, Burrough (1986) was one of the first authors to recognize the problem of error in GIS and dedicated a whole chapter in his book to the issue. A good section on data quality can be found in Longley *et al.* (2005).

Many studies have been carried out into digitizing error (for example Jenks, 1981; Otawa, 1987; Keefer *et al.*, 1988; Bolstad *et al.*, 1990; Dunn *et al.*, 1990). Errors associated with map overlay operations have been addressed by several authors. Sliver polygons are addressed by McAlpine and Cook (1971) and Goodchild (1978); whilst Newcomer and Szajgin (1984), Chrisman (1989) and Walsh *et al.* (1987) consider the cumulative effects of map overlay and map complexity on output errors. The paper by Walsh *et al.* (1987) addresses the problem of errors propagated through vector to raster conversion; this work has been followed up by Carver and Brunsdon (1994). Key papers in the development of techniques of error handling include Brunsdon *et al.* (1990), Openshaw *et al.* (1991), Fisher (1991 and 1993) and Heuvelink *et al.* (1990).

Bolstad P V, Gessler P, Lillesand T M (1990) Positional uncertainty in manually digitized map data. *International Journal of Geographical Information Systems* 4 (4): 399–412

Brunsdon C, Carver S, Charlton M, Openshaw S (1990) A review of methods for handling error propagation in GIS. *Proceedings of 1st European Conference on Geographical Information Systems,* Amsterdam, April, pp. 106–16

Burrough P (1986) *Principles of Geographical Information Systems for Land Resources Assessment.* Clarendon Press, Oxford

Carver S, Brunsdon C (1994) Vector to raster conversion error and feature complexity: an empirical study using simulated data. *International Journal of Geographical Information Systems* 8 (3): 261–72

Chrisman N R (1989) Modelling error in overlaid categorical maps. In: Goodchild M F, Gopal S (eds) *The Accuracy of Spatial Databases.* Taylor and Francis, London, pp. 21–34

Dunn R, Harrison A R, White J C (1990) Positional accuracy and measurement error in digital databases of land use: an empirical study. *International Journal of Geographical Information Systems* 4 (4): 385–98

Fisher P (1991) Modelling soil map-unit inclusions by Monte Carlo simulation. *International Journal of Geographical Information Systems* 5 (2): 193–208

Fisher P F (1993) Algorithm and implementation uncertainty in viewshed analysis. *International Journal of Geographical Information Systems* 7 (4): 331–47

Goodchild M (1978) Statistical aspects of the polygon overlay problem. In: Dutton G H (ed) *Harvard Papers on Geographical Information Systems,* vol. 6, Addison-Wesley, Reading, Mass., pp. 1–21

Goodchild M F, Gopal S (eds) (1989) *The Accuracy of Spatial Databases.* Taylor and Francis, London

Heuvelink G B, Burrough P A, Stein A (1990) Propagation of errors in spatial modelling with GIS. *International Journal of Geographical Information Systems* 3: 303–22

Jenks G F (1981) Lines, computer and human frailties. *Annals of the Association of American Geographers* 71 (1): 142–7

Keefer B K, Smith J L, Gregoire T G (1988) Simulating manual digitizing error with statistical models. *Proceedings of GIS/LIS '88.* ACSM, ASPRS, AAG and URISA, San Antonio, Texas, pp. 475–83

Longley P A, Goodchild M F, Maguire D J, Rhind D W (eds) (2005) *Geographical Information Systems: Principles, Techniques Management and Applications.* Wiley, Chichester

Mather P M (ed) (1993) *Geographical Information Handling – Research and Applications.* Wiley, Chichester

McAlpine J R, Cook B G (1971) Data reliability from map overlay. *Proceedings of the Australian and New Zealand Association for the Advancement of Science 43rd Congress, Section 21 – Geographical Sciences.* Australian and New Zealand Association for the Advancement of Science, Brisbane, May 1971

Newcomer J A, Szajgin J (1984) Accumulation of thematic map errors in digital overlay analysis. *American Cartographer* 11 (1): 58–62

Openshaw S, Charlton M, Carver S (1991) Error propagation: a Monte Carlo simulation. In: Masser I, Blakemore M (eds) *Handling Geographic Information: Methodology and Potential Applications*. Longman, London, pp. 78–101

Otawa T (1987) Accuracy of digitizing: overlooked factor in GIS operations. *Proceedings of GIS '87*. ACSM, ASPRS, AAG and URISA, San Francisco, pp. 295–9

Walsh S J, Lightfoot D R, Butler D R (1987) Recognition and assessment of error in Geographic Information Systems. *Photogrammetric Engineering and Remote Sensing* 53 (10): 1423–30

## WEB LINKS

- NCGIA I-1 Closing Report: www.ncgia.ucsb. edu/Publications/Closing_Reports/CR-1.pdf

- The Geographer's Craft – Error, accuracy and precision: www.colorado.edu/geography/ gcraft/notes/error/error_f.html

- DTM error: www.geocomputation. org/1998/80/gc_80.htm

Visit our website **www.pearsoned.co.uk/heywood** for further web links, together with self-assessment questions, activities, data sets for practice opportunities and other interactive resources.

# CHAPTER 11

# Human and organizational issues

# Learning outcomes

By the end of this chapter you should be able to:

■ Describe the types of GIS applications that exist in businesses and other organizations

■ Explain how investment in GIS can be justified

■ Identify different types of GIS users

■ Outline how an organization can choose an appropriate GIS

■ Explain how successful implementation of GIS can be ensured

■ List some of the organizational changes that may result from GIS adoption

# ■ INTRODUCTION

This chapter moves away from the technical and conceptual issues that underpin GIS and takes a look at some of the human and organizational issues surrounding commercial and business applications. In utilities, local government, commerce and consultancy, GIS must serve the needs of a wide range of users and should fit seamlessly and effectively into the information technology strategy and decision-making culture of the organization. The technical issues for such systems will be the same as those discussed in earlier chapters. For instance, the most appropriate data model must be chosen, the most effective data input method selected and suitable analysis procedures designed. However, successful application of large-scale GIS projects depends not only on technical aspects, but also on human and organizational factors relating to the implementation and management of systems. This has been recognized since the 1970s (Tomlinson *et al.*, 1976) and there has been considerable research into the organizational and human factors that may impede the implementation of GIS (Department of the Environment, 1987; Onsrud and Pinto, 1991; Medyckyj-Scott and Hearnshaw, 1994; Campbell and Masser, 1995; Hernandez *et al.*, 1999, Nasirin *et al.*, 2003, Roche and Caron, 2009). This chapter examines some of the human and organizational issues associated with the choice and implementation of GIS for business and other large organizations.

The issues discussed in this chapter are, in many cases, not unique to GIS, but are also faced by organizations seeking to use any information technology for the first time. In terms of our car analogy, you might be faced with similar issues when deciding whether or not to use a car, bicycle or train as your main means of transport. If you decide to buy a car, you have to decide which car to purchase. If you are to be the main user you must carefully assess your needs and identify your requirements from the vehicle. This will help you set parameters to allow comparison of different makes and models that are available and within your price range. Good fuel consumption, the reputation of the vendor for after-sales service, the insurance cost, the colour or the dimensions may be important. You can test some of the models, perhaps by seeing demonstrations or taking a test drive. If you go ahead with a purchase you may find that the car brings major changes to your lifestyle. Your routines and relationships may change as you can visit friends further away more often, use out-of-town shopping facilities and work further away from home. You may also find you have less disposable income as the costs of insurance, servicing and fuel eat into your budget! In summary, you need to justify your investment in a car, review the range and functions of models on offer, formulate a list of your requirements, take advice on purchase options and be prepared for the changes that will occur once you acquire the car. The questions you need to ask are similar to those posed by organizations considering adopting GIS:

- *What applications of GIS exist?* What GIS are available? What are the options for hardware and software?

- *What are the needs of potential users?* Who are the users of a GIS? How can users be involved in setting up a new GIS? How will GIS impact on their work tasks and jobs? How can their commitment to a GIS project be ensured?

- *Can investment in GIS be justified?* Is GIS a proven technology in the organization's sphere of activity? Will GIS bring benefits to the organization? How can potential benefits be assessed?

- *Which system is appropriate, and how should it be implemented?* How can an appropriate system be chosen? How can a system failure be avoided? Can resistance to a new system be prevented?

- *What changes will GIS cause in the organization?* How can implementation of a new GIS be undertaken so that the new methods of working have minimum impact on the work culture of the organization?

The importance of each of these issues in the process of choosing and implementing a GIS varies depending on the nature of the organization and its function. Awareness of the characteristics of GIS applications, and hardware and software choices that have to be made, are important first steps in appreciating the significance of these issues.

# GIS APPLICATIONS

From the early days of custom-designed and specially built systems such as CGIS (**Chapter 9**), the GIS market has grown to accommodate general-purpose and tailored packages for a wide range of application areas. In the 1990s, according to *GIS World* (Anon., 1994), the most frequent users of GIS were in government (over 20 per cent) and education (over 15 per cent). Today, the growth of Web-based GIS applications, such as Google Earth and Google Maps, has meant that many people are using GIS technology every day. GIS applications can now be found everywhere and support a very diverse range of activities. With such an enormous user group it is not surprising that GIS applications vary considerably in their scale and nature. They can serve the whole organization, several departments, one project or just one user (Box 11.1).

GIS applications tend to fall into one of three categories: *pioneering*, *opportunistic* or *routine*. Pioneering applications are found in organizations that either are at the cutting edge of their field, or have sufficient financial reserves to allow them to explore new opportunities. Pioneering applications are characterized by the use of unproven methods. As such they are considered high-risk and often have high development costs (Figure 11.1). However, successful pioneering GIS applications can provide high paybacks by giving considerable advantages over competitors.

Opportunist applications are found in organizations that kept a careful eye on the pioneers and quickly adopted the technology once they saw the benefits. Following this approach, opportunist organizations let the pioneers bear more of the cost and risk of application development. Those who wait longer become users of routine GIS

## BOX 11.1 Corporate, multi-department and independent GIS applications

**THEORY**

**Corporate GIS** is developed across an entire organization. GIS is usually implemented using a 'top down' approach to promote data sharing, reduce data duplication and create a more informed decision-making environment. Corporate GIS is appropriate for a utility company where all departments – customer support, maintenance, research and development, logistics, sales and marketing – could benefit from the sharing of data and access to GIS. Local government agencies also benefit from corporate GIS since all departments work within the same geographical area.

**Multi-department GIS** involves collaboration between different parts of an organization. GIS is implemented in a number of related departments who recognize the need to share resources such as data, and the benefits of working together to secure investment for a system. Multi-department GIS may be appropriate in a retail organization. Marketing, customer profiling, store location planning, logistics management and competitor analysis may all make use of GIS. Other activities, such as product design and planning, brand management or financial services, may be completed without the use of GIS.

**Independent GIS** exists in a single department and the GIS will serve the host department's needs. These systems can be adopted quickly and are generally task specific. GIS is usually adopted in response to a specific clearly identified need within the department. In a telecommunications company an independent GIS may be used to assist the siting of masts for a mobile phone network.

Corporate and multi-department GIS share many benefits, although the benefits will be at a corporate level where a corporate system is implemented, and only found in cooperating departments in the multi-department case. These benefits include integration of data sets, generation of shared resources, more informed decision making, increased data sharing and improved access to information. Improved control over priorities, information and staffing can be benefits for independent systems.

Corporate and multi-department systems can suffer where individual departments have different priorities, and where there are different levels of skills and awareness of GIS and spatial data. Independent systems may be hampered by lack of support, isolation and limited lobbying power.

*Source*: Hernandez *et al.*, 1999

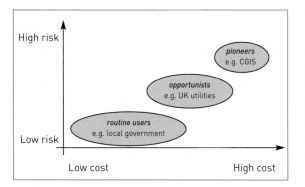

**Figure 11.1** Development of GIS applications

applications. Routine users adopt a tried, tested and refined product with lower risk and cost.

Whilst there are now plenty of examples of routine and opportunistic GIS applications, truly pioneering applications are rare. This is not surprising, since few organizations are prepared to be the first to apply GIS to new areas because of the high risks and costs that may be involved. For example, in our nuclear waste disposal example,

the organization concerned was unlikely to risk any adverse reaction or publicity to such a sensitive problem by application of what was, in the mid-1980s, a new technology. Despite being, by today's standards, a relatively straightforward GIS application, the use of GIS to perform site search and evaluation analysis was, at the time, the realm of academic research (Openshaw *et al.*, 1989; Carver, 1991b).

Pioneering, opportunistic and routine applications of GIS have permeated the entire range of business activities from operational, through management to strategic (Box 11.2).

In many organizations GIS applications cross the boundaries between the three types of activity introduced in Box 11.2. For example, in a utility organization GIS may be used to help operational maintenance of pipelines, improve the management of marketing services to customers and plan the strategic development of the organization. In many organizations a GIS will initially provide

---

## BOX 11.2 Business applications of GIS

PRACTICE

### OPERATIONAL APPLICATIONS

Operational GIS applications are concerned with managing facilities and assets. For example:

- a utility company may use GIS to identify assets such as pylons and masts in need of routine maintenance;
- a waste management company may use GIS to route waste collection vehicles; or
- a property management company may maintain hotel buildings and grounds with the help of maps and plans from its GIS.

### MANAGEMENT/TACTICAL APPLICATIONS

Management (or tactical) GIS applications are concerned with distributing resources to gain competitive advantage. For example:

- a ski holiday company may use GIS to identify appropriate potential customers to receive direct mailings;

- an education department might use GIS to produce options for school closures when deciding how to distribute limited resources; or
- a telecommunications company might use GIS to identify and evaluate possible communications mast sites to serve the maximum possible population.

### STRATEGIC APPLICATIONS

Strategic GIS applications are concerned with the creation and implementation of an organization's strategic business plan. For example:

- a ski equipment retailer may decide which geographical areas to target over the next five years after geodemographic analysis using GIS;
- a local government organization may decide on budget reallocations after analysis of population growth or decline in certain areas using modelling and GIS; or
- a catering business may decide to expand a restaurant chain to 100 outlets after analysis of the location of its competitors with GIS.

information in support of one activity and, if it is successful, develop so that information can be provided to support other activities. The evolution of CGIS, the Canadian GIS, illustrates this development.

CGIS underwent three major stages of development (Crain, 1985). The system was used first in an operational role for data inventory, with data collection and querying the dominant functions. Then CGIS became more of a management tool with new data being produced through the use of spatial analysis techniques, complex queries and data manipulation. Now the system is used in a strategic role as a decision-support tool for modelling and simulation. The evolutionary path followed by CGIS is typical of many GIS applications where the need to integrate, collate and manage data drives users to become more sophisticated and explore, through analysis, spatio-temporal patterns within the database. In turn, users begin to develop decision-support applications.

Therefore, when evaluating whether or not to develop a GIS for a particular application much can be learnt from the experiences of other organizations. An appreciation of the nature of the application (pioneering, opportunistic or routine) and the type of business activity being supported (operational, management or strategic) is important. However, in the business world issues of secrecy, power and commercial confidence often prevent the sharing of information and experiences.

## GIS hardware and software

The existing information technology (IT) infrastructure of an organization will affect the nature of any GIS implemented. There may be organizational standards for hardware to which new purchases must adhere. Any new installation may have to fit in with existing IT networks. Only specific 'company-approved' software products may be permitted, and there may be restrictions on the use of Internet services. The physical location of personnel may be important in determining the nature of a GIS installation. The GIS and analysts may all be located in a single office or department, or spread widely across an organization. Whether the GIS application needs to be accessible over the Web or using mobile devices needs to be considered.

Given the pace of developments in IT, evaluating and classifying GIS installations according to the hardware and software used may not be the best place to start for the development of future applications. However, it is useful to review the nature of current technology to give a clear picture of the wide range of technology used in GIS installations.

In 1996 *GIS World* published the results from a survey that suggested that the type of computer hardware used in an organization was related to the size of the user community (Anon., 1996b). In local and state or provincial governments, where there were small numbers of GIS users, the PC platform was prevalent. In similar organizations with more GIS users, more powerful, larger and more costly workstations were commonly used. It was suggested that larger organizations regarded high-speed networking, sharing of databases and improved processing power as important to their GIS applications. This trend in hardware use may be reflected in other GIS user sectors. In recent years, the use of Web-based and mobile GIS systems has also complicated this picture (Pick, 2008).

The range of GIS and related software can be reviewed by examining annual surveys and directories produced by the trade press. Longley *et al.* (2005) classify the main GIS software packages into six groups as summarized in Table 11.1.

The distinctions between packages are often difficult to discern, and are becoming more and more blurred. In all cases it is 'fitness for purpose' that is most important – the package which is chosen must meet the needs of the users it will serve. Table 11.2 offers a list of topics that it would be useful for any prospective GIS user to consider when evaluating hardware and software.

In addition to the proprietary software systems outlined above, many organizations have opted for specially designed customized systems. At the simplest level customization may be adaptation of proprietary software by changing the interface and including a few special functions. On the other hand, customization may require the writing of completely new software to perform specific tasks, or to allow the integration of GIS with other databases and software in use in the organization. The GIS developed for Network Rail (Box 2.6) is an example of a customized system. At the core is Esri software but the interface has been customized to

**TABLE 11.1 GIS software types**

| | |
|---|---|
| Professional | High-end GIS products that offer the full range of modelling and analysis functions – these are the 'true' GIS that fit the definitions of GIS offered in **Chapter 1**. These are the most expensive software products and are likely to have the least users |
| Desktop | Desktop mapping systems have become popular GIS products. These do not offer the full range of GIS analysis or data input functions, but are easy to use and geared towards users who need to display and query data. Less expensive than professional products, these offer a wide range of functions |
| Hand-held | GIS software specifically for use in the field on portable and hand-held hardware devices has emerged in the last few years. These systems are used for data capture, navigation and the provision of information to field staff. Many of these products work together with Internet products. Hand-held GIS are likely to be cheaper than desktop systems and relevant to more users. They provide a more limited range of functions |
| Component | Tool kits of functions (or components) can be used by an experienced programmer to 'build their own' GIS. Component GIS are particularly relevant to highly specialized applications requiring only a subset of the full range of GIS functions. Some of the resulting systems are embedded within other packages so end users may be unaware that GIS has been employed |
| GIS Viewers | A GIS Viewer provides functions that allow the display of files in common formats. Simple query and mapping functions may also be available, but sophisticated analysis will not. These are inexpensive, entry level products designed to encourage the widespread use of a particular vendor's products or specific standards for data exchange |
| Internet GIS | Longley *et al.* (2005) consider that Internet GIS have the highest number of users. These products offer a very limited range of functions at low cost to the widest range of users. These are 'end user' products that are used to develop particular applications – such as online location finding, or data exchange. Tools such as MapTube (www.maptube.org) and the mapping functions embedded within products such as Google Docs are increasing the analytical power of Internet GIS applications |
| GIS Apps | With the arrival of Apple's iPhone in 2007 a whole new type of GIS applications became possible – the GIS App. Some of these are single purpose applications with a spatial element that allow the user to perform one function. An example is Groundspeak's geocaching application (www.groundspeak.com) which allows users to identify and seek geocaches (geocaching is a modern form of treasure hunting in which 'treasure' is hidden in a small container and the co-ordinates of each 'cache' are uploaded to the web so that it can be found using a GPS). Other examples of Apps with a wider range of GIS functionality include Esri's ArcGIS |

*Source*: Adapted from Longley *et al.* (2005)

meet users' needs, and additional software functions were written to meet specific organizational requirements. Cresswell (1995) offers examples of customization of GIS in the retail sector but also considers that, whilst customized systems would seem to offer an ideal solution for an organization investing in GIS, there are a number of reasons for initially purchasing a proprietary system. Proprietary systems are likely to be cheaper, able to serve a large number of users, easy to use and well supported, and flexible in dealing with different data and problems. More recently the development of Web-based GIS tools such as MapTube (www.maptube.org), a free resource for viewing, sharing, mixing and mashing maps online, created by University College London's Centre for Advanced Spatial Analysis, have meant that customized GIS applications are now available to an even wider range of organizations.

Grimshaw (1999) summarizes the choices available to an organization considering GIS as:

■ to develop its own software in-house;

■ to purchase a standard software package;

■ to purchase a ready developed software system from a third-party user;

**TABLE 11.2 Hardware and software questions for organizations considering GIS**

| Hardware questions | Software questions |
| --- | --- |
| ■ What hardware is necessary for the application? | ■ What functions does the software offer? |
| ■ How many computers are needed? | ■ Will the functions available meet the requirements of the organization? |
| ■ Which computers are needed? PCs, workstations or mainframe? | ■ Are all the functions of a full GIS really necessary, or would a more easily used desktop mapping package be more suitable? |
| ■ What storage capacity, memory, speed, etc., must the computers have? | ■ Is the package user-friendly? |
| ■ What graphic display devices are needed? | ■ Can the package be customized? Does the organization have to do the customizing? |
| ■ Which peripherals are needed – printers, plotters, digitizers, scanners, data loggers? | ■ Are there extra modules/functions for more experienced users – for example, is it possible to write macros? |
| ■ Will the GIS be run over a network or the Internet? Will additional hardware be required for this? | ■ What standards does the software use – for instance, database query language, data transfer formats, handling of address standards? |
| ■ Will any new hardware be compatible with existing hardware in the organization? | ■ Can the software exchange data with other software in use in the organization? |
| ■ What about the reliability and maintenance of the hardware? | ■ What documentation is provided? |
| ■ How much will it cost? | ■ What training is available? Will new training need to be developed? |
| ■ Is it possible to expand or upgrade the hardware once it is in place? | ■ Are upgrades to the software going to be provided? |
| ■ How long can the hardware be expected to last? | ■ How much will it cost? |
| ■ Will new staff be required to install/run/maintain the hardware? | ■ Will the software continue to be developed and supported by the vendor? |
| ■ Will GIS be run over the Internet? Does the GIS need to run on mobile devices? | ■ Which operating system will be required? |

■ to employ consultants to write new software; or

■ to use Internet GIS and GIS for mobile devices.

This list needs to be updated to include the option of using Internet GIS and GIS Apps for mobile devices.

Awareness of the range of GIS applications, and the options for hardware and software, is an essential first step in the GIS implementation process. Comparison with similar organizations can give information about whether the organization is a GIS pioneer or a routine user following an accepted trend for the business sector. Understanding and learning from successful applications of the same nature is one way of attempting to justify investment in GIS, but more formal methods are often required. However, first in any project come the users.

## ■ GIS USERS

Before GIS adoption and implementation it is important to consider just who the users of a GIS will be. Consider, for example, a large sports equipment retailer. It has many stores throughout Europe (including one in Happy Valley), and offers a range of services to customers, including mail order, financial services and a store card. This company, SkiSupplies Inc., has a GIS that it uses for a variety of applications, including:

■ location analysis for the siting of new stores;

■ target marketing of store card customers;

■ managing distribution of stock from warehouses to stores; and

■ producing maps of individual stores to help customers find their way around.

The SkiSupplies GIS, therefore, has a number of different users. At one end of the spectrum are the company directors and managers. These decision makers require strategic information from the GIS, but they are unlikely ever to use the system in a 'hands-on' way. They are more interested in hard-copy output, such as maps of stores that are performing well, or maps that justify the decision about where to locate a new store. They have a good knowledge of the context of the applications – the retail environment – but little GIS knowledge or experience. They rely on a group of GIS analysts to perform appropriate GIS tasks to provide them with the information they require. These analysts have an understanding of retailing and GIS. They are able to translate the managers' requirements into real GIS analysis. They also have a role in designing and maintaining the GIS, including the development of new applications for the system, so that it will continue to meet the needs of other users and justify the amount of money that has been spent on it. A third group of users are the computer technicians. These users have other responsibilities in addition to the GIS. They are responsible for the wages and personnel computer system, which has no GIS element. They are not GIS experts, but computer specialists who assist with data formatting and input, hardware maintenance and system upgrading. Finally, the customers are also users of the system. For example, on-screen maps are available in larger stores to help customers navigate around the store and obtain additional information on products and services. There is also a GIS App available for customers to download to their mobile phone. This allows them to find the location of the SkiSupplies store closest to their home and pre-order equipment. Although these are not 'true' GIS they are providers of spatial information. Customers also receive targeted mailings produced as a result of the GIS analysts' manipulations of the customer databases. Such users do not need to know that it is a GIS they are interacting with. They are unlikely to have any retailing, GIS or computing expertise. They are end users of the products and services offered by the company.

Eason (1988, reported in Grimshaw, 1999) has a similar view of users of information systems. He classified the four groups of users discussed above as 'occasional professional' (with high task expertise and low system expertise); 'application specialists' (with high task and high system expertise); 'computer specialists' (with low task expertise and high system expertise); and 'the public' (with low expertise in both tasks and system). These ideas are shown in Figure 11.2. For most GIS applications a similar classification of users can be achieved or even expanded. Brown (1989, reported in Huxhold, 1991) suggested 10 roles that might be required in a GIS application. These have been included in Figure 11.2.

Understanding the range of potential users is important in the 'user-centred approach' advocated by many researchers for successful GIS implementation (Eason, 1988, 1994; Medyckyj-Scott, 1989). Different types of users have different requirements

|  | High application skills | Low application skills |
|---|---|---|
| **High GIS skills** | **GIS analysts**<br>*applications specialists*<br>e.g. system manager<br>analyst<br>cartographer | **Computer technicians**<br>*computer specialists*<br>e.g. programmer<br>data processor<br>database administrator<br>digitizing technicians |
| **Low GIS skills** | **Managers**<br>*occasional professionals*<br>e.g. end-users<br>decision makers | **Customers**<br>*the public*<br>e.g. customers |

**Figure 11.2** GIS users
*Source:* Adapted from Brown (1989); Eason (1994) and Grimshaw (1999)

from GIS and will need different functions. For instance, the managers may be interested in the data output functions and the ability to reproduce the company logo, whereas the analysts may be interested in the analysis functions and the ability to link the GIS with appropriate retail modelling packages. Different users have different skill levels with computers and GIS, and different levels of awareness about what GIS can achieve. Thus, the questions they will ask the GIS will differ. Different users also have varying time available to work with the GIS. For example, the analysts may interact with the system every day, whereas an interested manager, who likes to find out the answers to her own queries from time to time, will need a system that is easy to use, and quick to relearn after a period without using it. To ensure that the GIS will meet the needs of all its users a range of issues must be considered:

- education, awareness raising and training;
- the adoption of standards; and
- GIS usability.

Educating users not only in how to use the new system, but also in the reasons why it has been necessary to develop it in the first place, is an important, but often overlooked, element of introducing a new GIS application into an organization. This education may take many different forms, from the awareness raising generated by articles in a corporate newsletter, to formal training in software development and use.

Adopting standards can help new users to learn and accept a new innovation more quickly. There are many standards associated with GIS for software, hardware and data. Some of the organizations responsible for these are shown in Table 11.3. Adoption of standards with which users are already familiar will reduce the learning curve associated with a new system, and the implementation of standards that are required by legislation or convention will help ensure commitment to applications. Further information on GIS standards can be found in **Chapters 5** and **10**.

System usability is affected by many factors: ergonomic aspects such as hardware design and location; training; support factors such as documentation; and software characteristics such as the interface design. The human–computer interface can be considered to be more than just the design of the graphics on the screen; it is the way in which the users' requirements are translated into actions in the computer, and the way in which the user interacts with the software. The effects of a selection of interface characteristics are reviewed in Box 11.3.

**TABLE 11.3 A selection of national and international standards development organizations**

| | |
|---|---|
| AFNOR | Association Française de Normalisation |
| AGI | Association for Geographic Information (UK) |
| ANSI | American National Standards Institute |
| BSI | British Standards Institute |
| CEN | European Committee for Standardization |
| DGIWG | Digital Geographic Information Working Group |
| EUROGI | European Umbrella Organization for Geographical Information |
| EUROSTAT | Organization responsible for developing standards for the dissemination of geostatistical data within the European Commission |
| ISO | International Standards Organization |
| NMA | Norwegian Mapping Authority |
| NNI | Netherlands Normalisation Institute |
| OGC | Open GIS Consortium |

*Source*: Adapted from Harding and Wilkinson (1997)

## BOX 11.3 GIS interfaces

**THEORY**

Some of the characteristics of the human–computer interface assist users with the adoption of a new product. Two examples are the visual characteristics of the software, and help and error messages.

### SOFTWARE APPEARANCE

Different screen interfaces may be appropriate for different types of user. For occasional users of the SkiSupplies Inc. GIS, or the general public, a Web-style GIS interface, or even a touch-screen providing access to a limited range of functions would be appropriate. This would enable users to adapt to the GIS quickly, and menus and prompts would remind them which functions to use if they had forgotten. For more frequent GIS users, touch-screen buttons might be inappropriate and limited, and a Windows-style, menu-based interface most likely, possibly with 'wizard' style help for less frequently used functions. Some GIS programmers still prefer to use the 'old style' command-line driven systems that were typical of DOS and Unix GIS platforms in the 1990s. Therefore, it may be necessary to select a GIS, or a range of GIS, which will allow different interfaces to be used by different users.

### HELP AND ERROR MESSAGES

Windows-based interfaces have adopted a common format for help and error messages. Information about commands is usually displayed at the bottom of the screen, and context-sensitive help can be obtained using the function keys. Error messages normally appear in 'pop-up' windows that can be closed once remedial action has been taken. These features are helpful for users, and if they are already familiar with the way in which Windows software offers help and advice, a new system can be learnt quickly. However, not all advice from software is helpful. A message to 'ring technical support for help' whilst working with GIS at a field site may not be very helpful. For a single user, the only GIS user in an organization working on a stand-alone system, the message to 'contact the system administrator' is also unhelpful. A single user may be the system administrator, as well as analyst, data entry technician and cartographer. So, problems can remained unsolved. However, today virtually all GIS software products and applications will provide a direct link to support via the Internet. This usually takes one of three forms: answers to a list of frequently asked questions; the ability to search online discussion forums where users provide advice and support to each other; and the option to chat 'online' with a technical support person. Comprehensive training materials may also be available online to allow 'just in time' training to support users.

---

Davies and Medyckyj-Scott (1996) conducted a study of GIS usability. Whilst they found that most users were happy with the usability of their system, they were critical of error messages, documentation and feedback, and infrequent users of GIS were found to encounter more problems. From their observations, Davies and Medyckyj-Scott estimated that around 10 per cent of the time spent in front of a GIS screen was spent on unproductive tasks, such as sorting out 'snags' and waiting for the system to process. Over a decade later there are still challenges to be met to produce user-friendly software to suit all users.

Standards and interfaces are just two issues that can have a huge effect on users. If users are not kept happy throughout the implementation of a new GIS project, signs of resistance to a project may emerge. Stern and Stern (1993) suggest that some of the ways in which employees could resist the implementation of a new information technology system include inserting or deleting data, copying or destroying files or installing 'bugs'. Some examples of resistance to GIS were identified by a survey, conducted in 1990, of GIS users in the UK (Cornelius and Medyckyj-Scott, 1991). Unwillingness to use the systems, reluctance to change to new methods, and finding faults with improved output were uncovered as problems from unhappy users. Unhappy users may be one reason for under-use of systems, but there are others. In particular, a poorly researched and constructed business case may lead to misplaced investment in GIS (Pick, 2008; Roche and Carron, 2009).

## REFLECTION BOX

- Describe, using examples from your reading or experience, the characteristics of pioneering, opportunistic and routine GIS applications.
- Imagine a parcel distribution business, or alternatively a company producing environmentally friendly products such as cleaning products for domestic users. What operational, tactical or strategic applications of GIS could be used by the organization? Give brief details of the applications and explain why they fit into the categories above.

- For GIS software with which you are familiar, think about how it fits into the classification of software types presented in Table 11.1. Can you provide examples of software to fit into the other categories?
- Ten roles for GIS users are suggested in Table 11.2. Which of these roles are most important in applications in your discipline or area of interest? Are all of the roles needed?

## ■ JUSTIFYING THE INVESTMENT IN GIS

A business case must be made for almost all GIS projects in commercial organizations, to persuade sponsors to support the project. Clear objectives must be set for the GIS, and the advantages and disadvantages of introducing a system need to be considered and evaluated. Questions that need to be addressed include:

- What will be the benefits of introducing a GIS?
- How will GIS help to improve the organization's effectiveness?
- What will be the costs of the new GIS?
- Will the expected benefits outweigh the anticipated costs?

One of the most frequently used methodologies at this stage of a GIS implementation is cost–benefit analysis.

### Cost–benefit analysis

Cost–benefit analysis is a methodology that is not unique to GIS as it has applications in many other areas. The method involves a thorough assessment of all the costs and benefits expected in association with a new project. Each cost or benefit is given a monetary value, and a graph, sometimes known as a *baseline comparison chart*, may be drawn up that predicts how costs and benefits will vary in relation to each other over a period of a few years (Huxhold, 1991). An example of such a graph is provided in Figure

11.3 (after Bernhardsen, 1992). The point on the cost–benefit graph where the cost and benefit lines cross is known as the break-even point, and this gives an indication of when a new project may start to become profitable – when the initial investment will start to produce returns.

The results of cost–benefit analysis can also be expressed as a payback period (in years) where:

$$\text{Payback period} = \frac{\text{total cost of investment}}{\text{estimated annual revenue}}$$

In GIS projects cost–benefit analysis can be used to justify a project in its entirety from the outset, or to compare the costs and benefits associated with differing GIS solutions or products. For example, Huxhold (1991) compared the costs of implementing and operating a GIS with the costs of not implementing and operating a GIS between 1984 and 1993 for an American local authority. Without a

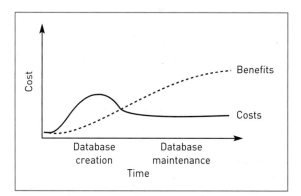

**Figure 11.3** Cost–benefit graph
*Source:* Adapted from Bernhardsen (1992)

GIS, costs were set to increase fairly steadily over the period under consideration, whereas analysis showed that with a GIS considerable savings would be made. Smith and Tomlinson (1992) offer a methodology for cost–benefit analysis that they applied to the city of Ottawa in Canada. They consider that cost–benefit analysis is the best way of assessing the value of GIS, but admit that there are problems with the technique. Further examples of the application of cost benefit in a GIS context can be found in Pick (2008).

In GIS the main problem with cost–benefit analysis is the difficulty of identifying and quantifying the costs and benefits associated with a new project.

Table 11.4 suggests some of the costs and benefits that may be associated with a GIS project. Direct benefits, such as staff savings or cost savings, are easy to quantify. However, other benefits (intangible or indirect benefits) are hard to quantify. These may include, for instance, improved information flow, better decision making and improved access to information (see, for example, Openshaw *et al.*, 1989). Another problem with applying cost–benefit analysis to GIS is the difficulty in identifying initial costs. Particularly important is the high cost of the creation of a digital map base. Burrough (1986) estimated that 46 per cent of the cost of a GIS project

**TABLE 11.4 Costs and benefits of GIS**

| Direct costs | | Direct benefits | |
|---|---|---|---|
| 1 Hardware / software | ■ Hardware<br>■ Fewer staff required<br>■ Software<br>■ Software development<br>■ Customization<br>■ Consumables (printer supplies, etc.)<br>■ Software upgrades<br>■ Maintenance and support contracts<br>■ Communications networks | 1 Savings | ■ Reduced cost of information production and provision<br>■ Fewer staff required<br>■ Less space required for, e.g., map storage<br>■ Time savings for routine and repetitive tasks |
| 2 Data | ■ Database creation<br>■ Database maintenance<br>■ Data updating | 2 Increased effectiveness | ■ Faster provision of information<br>■ Greater range of information services provided<br>■ Information more readily available<br>■ Up-to-date information available |
| 3 Human/admin. | ■ Insurance<br>■ Administration<br>■ Security<br>■ Training<br>■ Rent | 3 New products | ■ New range of output – maps, tables, etc.<br>■ Better-quality output |
| 4 Method for choosing system | ■ Pilot project<br>■ Benchmarking<br>■ Cost–benefit analysis<br>■ Consultancy | | |

| Indirect costs | Indirect benefits |
|---|---|
| ■ Increased reliance on computers – vulnerability to failures, changes in software/hardware, etc.<br>■ Poorer working environment – noise, heat, tedious tasks for users<br>■ Higher-skilled workforce required | ■ Improved information sharing and flows<br>■ Better-informed decision making<br>■ Stronger competitive ability<br>■ Better-motivated workforce – more career options, less tedious tasks<br>■ Greater analysis and understanding of problems<br>■ Justification for decisions made<br>■ Improved visualization of data |

was in base map conversion, and Huxhold (1991) estimated that these costs could rise as high as 80 per cent. The situation may be different with respect to more recent business applications, where data may be obtained from secondary sources rather than digitized in-house. The expected lives of hardware, software and data compound this problem. Bernhardsen (1992) considers that whilst hardware and software can be expected to be replaced after two to five and three to six years respectively, the data for a GIS may last for 15 or 20 years. This means that cost–benefit analysis should be extended for at least this period of time.

A further difficulty with cost–benefit analysis is that there is, to date, no standard methodology for GIS. Mounsey (1995) even considers that the traditional methods of cost–benefit analysis are inappropriate for some business applications of GIS, where timescales are short, implementations are small and there is unpredictability and change in potential applications, people, business processes, data and technological requirements. She suggests an alternative 'short cut' method for justifying GIS investment in such organizations that involves four stages:

1  Determine the business drives for GIS. This will help to clarify the *strategic benefits* of a new system and interest potential sponsors.

2  Identify the range of applications to assess the scope and practical application of the GIS. This should consider the users of the system, provide a basis for comparing the systems being considered and identify the *operational benefits* of a GIS.

3  Identify the technological framework within which the GIS will fit. Consideration of software, data, hardware and implementation issues will help to identify potential costs.

4  Develop the plan for implementation.

Finally, having performed a comprehensive cost–benefit analysis there is no guarantee that the benefits identified will actually be realized, so the predicted break-even point and savings may not be achieved. However, if little or no attempt is made to establish the business case for GIS using a methodology such as cost–benefit analysis, there is a greater chance the application will be poorly received within the organization.

## ■ CHOOSING AND IMPLEMENTING A GIS

Once the users of a GIS application have been identified and a business case for development established, the organization must choose and implement a system. There are many formal models of information system design and implementation that can be used as a starting point (see Applegate *et al.*, 2009). A variation on one of these models is presented in **Chapter 12** as a framework for the design of a GIS project, and others are introduced in Box 11.4.

---

### BOX 11.4 Methods for information system design

THEORY

Four of the most commonly used types of method for information system design are (after Avison and Fitzgerald, 1988; Grimshaw, 1999; Reeve, 1997; Applegate *et al.*, 2009):

**1** *Structured systems methods.* Structured systems analysis and design (SSADM) is an example of a structured systems design methodology. In SSADM, an understanding of the current working system is developed in order to create a logical model. A second logical model of the required system is developed after consultation with users, and, following implementation of this users develop a specification of the system they require. A sequence of design phases follows. LBMS (Learmouth Burchett Management Systems) is another widely used example of a structured systems method.

**2** *Soft systems methods.* One of the major soft systems methodologies, soft systems analysis (SSA), was developed as a general-purpose methodology for investigating unstructured management problems (Checkland, 1981). These approaches place emphasis on the human aspects of problems, →

and have been used to investigate organizational aspects of information systems development. An adapted soft systems method is presented in **Chapter 12**.

3 *Participative methods*. ETHICS (Effective Technical and Human Implementation of Computer-based Systems) is a participative information systems design methodology. The method aims to develop systems that will improve the job satisfaction of those who use them, and thus improve system

use and acceptance. All those affected by the system – users, managers, customers and suppliers – are involved in the design process.

4 *Hybrid methods*. MULTIVIEW is a hybrid design methodology, incorporating ideas from soft systems analysis and data analysis techniques. A number of different perspectives of system development are taken to allow the incorporation of broader human issues as well as technical issues into the system design process.

---

However, formal methods such as those introduced in Box 11.4 are not always implemented, and use of these methods is not in itself a guarantor of a successful project. Therefore, this section considers how organizations choose and implement their GIS by reviewing the research findings of surveys (Anon., 1993; Medyckyj-Scott and Cornelius, 1991; Campbell and Masser, 1995) and evidence from case studies. First, the way organizations choose their GIS systems is examined, then how users can be involved in GIS selection and implementation by adoption of a 'user-centred approach' is considered. Some of the methods available for testing and supporting purchase decisions (benchmarking and pilot studies) are also introduced.

### Involving users

There have been a limited number of surveys undertaken to try to find out how organizations choose their GIS. For example, Woodcock *et al.* (1990) considered factors which users in

environmental management should take into account when choosing a system. Top of their list was cost, second was functionality (the range of operations offered by the software). They considered that users should opt for the widest range of functionality offered. In practice this may not be a sensible decision, since the organization may pay for functions that will never be used. A *GIS World* survey in 1993 put functionality at the top of a list of users' considerations when buying a GIS (Anon., 1993). Other factors were compatibility with existing hardware and software within the organization, technical support, the reputation of the vendor and cost. The issue of compatibility with other systems also came top of the list of criteria that were important to utilities (Battista, 1995).

In almost all cases the key to success is considered to be the adoption of a *user-centred* approach to GIS purchase decisions (Burrough, 1986; Eason, 1988; Medyckyj-Scott and Cornelius, 1991; Roche and Caron, 2009). Details of one user-centred approach are given in Box 11.5.

## BOX 11.5 User-centred GIS implementation

THEORY

A user-centred approach to GIS implementation involves all users from the outset in any decision making related to the GIS. This helps to overcome the problems faced where managers and designers make assumptions about users' requirements. In addition, involving users helps to:

■ specify functionality requirements;

■ ensure user commitment and cooperation;
■ identify areas where organizational change may be required; and
■ limit resistance from individuals to any new system.

Medyckyj-Scott and Cornelius (1991) suggest a four-stage approach to achieving a user-centred implementation:

**BOX 11.5**

**1** Establish an appropriate management structure, with time for meetings to discuss human and organizational factors as well as technical issues.

**2** Explain the benefits and rewards that the new GIS will offer. Employees who are aware of the benefits are less likely to offer resistance.

**3** Allow time for human and organizational changes to occur. Training and extra time to adjust to new working practices may be necessary.

**4** Ensure that there is a carefully planned implementation process, backed up by proper resources, appropriate staffing and a positive attitude to the changes expected.

There are various ways in which a user-centred approach can be implemented (Eason, 1988; 1994). First, users can be directly involved in the choice of system, using methods such as user-needs surveys and data audits. Then, users can be involved in benchmarking exercises and pilot projects. They must be aware of the implications and changes that will result from the implementation of the system. They should also have a direct input into the strategy for implementation that is adopted. In all cases, the users must be involved and care must be taken to evaluate their needs, raise their GIS awareness, train them and provide appropriate support (Medyckyj-Scott, 1989).

---

However, a user-centred approach can be difficult to implement, since, as Campbell and Masser (1995) found, users do not always know what they want. They stress that defining user needs is one of the hardest, yet most critical aspects of the GIS implementation process. Evidence of other implementation problems is found in reports of failed applications (one example is presented in Box 11.6) and in the reasons why organizations do not implement GIS.

Apart from the financial constraints associated with adopting GIS, there are other reasons why an organization may decide not to implement GIS:

■ There may be a lack of senior management support.

■ There may be concerns about the technology and data conversion problems.

■ There may be a lack of knowledge and awareness about GIS.

■ There may be internal political disputes or constraints.

■ There may be concerns about the time needed to implement a system.

This list clearly shows that the issues that prevent the uptake of GIS are predominantly human and organizational in nature. The reasons why GIS implementations fail are not dissimilar. Eason (1994) identifies three reasons for GIS failure:

1 Organizational mismatch (where the system introduced does not meet the needs of the organization).

2 User acceptability (where the system has negative implications for user groups, which may result in resistance to the system).

3 Non-usability (where users may face technical or other difficulties when trying to use the system).

---

**BOX 11.6 A problematic implementation**

**CASE STUDY**

In 1992 the London Ambulance Survey updated its computerized command and control system by including amongst other things some level of GIS functionality. According to newspaper reports at the time, the new computer system cost well over £1 million and involved 40 networked PCs together with a number of servers. It incorporated vehicle tracking, GIS and other off-the-shelf software (Hobby, 1992; Mullin, 1992). It was considered to be a 'pioneering' application.

As soon as the system was implemented, problems started to occur. Calls from those needing an ambulance were lost in the system, and the response time for ambulances to reach

→

**BOX 11.6**

accidents and emergencies increased. Several deaths were initially blamed on the failure of the new information system. The staff had no back-up system to turn to and problems continued until the old system could be reinstated. A number of causes for these problems were identified. It was suggested that there had been insufficient testing of the system and insufficient training of the staff (some new users had limited IT experience), and that the system was overcomplicated (Arthur, 1992). The method of selecting the system also came in for criticism. It was suggested that the Ambulance Service had not examined applications in other similar organizations before making its choice, and that no competitive tendering had been carried out (Mullin, 1992).

Page (reported in Arthur, 1992) implied at the time that the failure of the system might have been due to incomplete specification of requirements, and that the system had not been designed from the 'people outwards' – a user-centred approach had not been adopted. He also suggested that the failure was due not to the computer system (as there was little evidence of widespread technical problems), but rather to the method by which the system had been integrated into the working practices of the organization. Commenting on the official report on the incident in 1993, Arthur (1993) reported the findings that management had made mistakes in connection with procuring, installing and operating the system. It was estimated that it would take five years to properly introduce the system, including testing and training.

Again, Eason advocates a user-centred approach to help avoid these problems, and suggests that users should be involved in user-needs surveys, benchmarking and pilot studies to help select an appropriate system (Eason, 1994).

## User-needs surveys

Clear articulation of user needs (through a user-needs survey or a user-requirements analysis) is essential to assist the development of an appropriate list of requirements for a GIS. This requires talking and listening to all potential users, from management to technicians. Goodwin (1995) reports the process used by the retailers 'Boots the Chemist' for its choice of GIS. The first phase of its GIS development project was to talk. Talking went on within the information systems department and with all those who might have business applications for the proposed GIS. These discussions resulted in a list of all of the potential users and applications for the GIS. In turn, this allowed the writing of a technical specification for the system that included details of the functionality required.

Evaluation of users' needs may be difficult if users are new to GIS. As stated previously, users do not always know what they want. Awareness-raising activities, such as demonstrations or seminars, may be helpful, although it is probably best to start with an evaluation of potential users' current roles and information requirements. Clarke (1990) considers that user-requirements analysis can be conducted

through interviews, documentation reviews and workshops. Topics that need to be considered are workflows, the characteristics of spatial data used in these workflows and the information products required (hard copy, on-screen displays, reports and data in various output formats). Assessment of these topics should result in an understanding of what data are being used, by whom, and how they are being collected, processed, stored and maintained. As well as giving a basis against which to evaluate the acquisition of GIS, this will help to develop data requirements for issues such as accuracy, update frequency and classification. In addition, issues such as the number of users, response times required and data volumes will help with the identification of hardware requirements. Performance indicators developed during user-requirements analysis can be evaluated during benchmarking or pilot studies.

## Benchmarking

Benchmarking is a technique often employed to help with the decision about which software package to select. The prospective buyer organizes a benchmark test. Three or four system vendors will be invited for a day or more to the buyer's site and will be provided with a list of tasks that the buyer would like their GIS to be able to perform. The buyers may also supply data, and possibly some expertise from within the organization to assist and advise. The vendors will then attempt to demonstrate that their system will meet the buyer's requirements. Outputs from the benchmarking process may include measures of processing and operator time, as well as products such as graphics and statistics (Clarke, 1990). The user interface and system documentation can also be assessed. This is an excellent method of choosing between systems, and more effective than viewing vendors' demonstrations and listening to presentations on the experiences of other users. In a benchmarking situation, some of the problems of data entry and conversion, as well as the limitations of analysis operations, may be uncovered. However, the whole exercise requires a fair level of expertise and awareness on the part of the potential buyer, particularly with respect to the identification of the tasks to be performed by the vendors.

## Pilot systems

A pilot project or system may be defined as the limited-term use of GIS, using data for a small geographical area to test the planned application and demonstrate the capabilities (Huxhold, 1991). It is a test run on a small-scale system that may involve an investment of 5–10 per cent of the total system costs. The system may have been leased from the vendor, with an option to purchase should the pilot project go well. Such a project provides the opportunity for an organization to test each potential application, test user acceptance, demonstrate facilities to improve awareness and acceptance, and improve cost–benefit analysis. The pilot system may be selected after benchmarking. Overall, a pilot system will decrease the financial risks involved in a large-scale GIS project since it does not require the full commitment of a new full system initially. However, time and resources are still needed to ensure that the pilot system will be a successful test of the GIS application.

There are many examples of pilot projects in the GIS literature. For instance, Grimshaw (1999) reports the findings of pilot GIS studies carried out by the Woolwich Building Society, a financial organization in the UK. Its study raised the question of whether a single system to meet all needs, or several systems to deal with specific tasks, would be appropriate, as well as questions about the data to be used. The pilot study was an important step in the design and acquisition of a system that would meet all users' requirements. Bernhardsen (1992) outlines the conclusions of a pilot project undertaken in Oslo for municipal management. This project

involved cooperation between a wide range of user groups. The benefits to these user groups were projected forward over 25 years, and the findings led to a change to the originally suggested collective system. A distributed system was developed, and since the pilot project showed that no one system would meet the needs of all users, a total of four or five different software packages were included in the final GIS configuration.

In practice, most GIS users do not appear to use methods such as pilot projects and benchmarking to help with their choice of system. Medyckyj-Scott and Cornelius (1991) found that the most common methods adopted to help with the choice of a new system amongst UK users were visits to and from suppliers, visits to other GIS users and discussions with other GIS users. More formal methods such as feasibility studies, pilot periods, the use of systems analysis techniques, benchmark tests and cost–benefit analysis were far less common. In the London Ambulance Service case study described in Box 11.6, there appears to have been limited use of any of these methodologies, and little attempt to learn from the experiences of other similar users. It remains a matter for speculation whether any of these more formal methods would have helped to prevent the problems that were experienced with the system.

## Implementation strategies

As Campbell and Masser (1995) recognize, even if there is nothing technically wrong with a GIS system chosen by an organization, problems and failure may occur. One of the causes of problems can be the way in which the GIS is implemented. Eason (1994) and, more recently, Roche and Caron (2009) review the methods for implementing GIS. Stern and Stern (1993) and Applegate et al., (2009) discuss similar approaches for IT more generally. Implementation methods fall into four main categories:

- *direct conversion* from the old system to the new;
- *parallel conversion*, where both old and new systems run alongside one another for a short time period;
- *phased conversion*, where some of the functions of the old system are implemented first, then others follow; and
- *trial and dissemination*, for example the running of a pilot system, then appropriate conversion to a new system.

Eason (1994) reflects on the level of user involvement with each of these approaches. With a direct conversion, also known as a 'big bang' conversion, the old system is switched off one day, and immediately replaced by the new. Users must adapt immediately. This was the approach used in the case of the London Ambulance Service, and the London Stock Market computer system, which also crashed in 1986. In such an approach the speed of change is rapid, leaving little time for users to learn and adjust to new working practices and methods. However, such an implementation can be successful if managed well and prepared for properly. Parallel running of an old and a new system requires extra resources, but decreases the risks of failure (Grimshaw, 1999). Extra demands are placed on users, who need to operate both systems. An incremental implementation system, with limited conversion to new methods at any one time, gradually results in a fully functional new system. Here the rate of change is much reduced, giving users more time to learn and adapt to the new situation.

Grimshaw (1999) considers organizational characteristics that can help the choice of implementation. If the system needs a 'large critical mass', then a big bang or parallel running approach would be appropriate. If normal business would be placed at risk during the implementation, parallel running or phased implementation would be appropriate. If users are willing and quick to learn a new system, a big bang may be possible, and if the needs of specific groups of users must be considered, pilot projects may be necessary.

The method of implementation is one factor that can affect the success or failure of a GIS project. Gilfoyle (1997) considers that other factors identified by the Department of the Environment (1987) are still applicable a decade later. These suggested that GIS would succeed if:

- the organization can afford some experimental work and trials;
- there is a corporate approach to geographic information and a tradition of sharing and exchanging information;
- there is strong leadership and enthusiasm from the top of the organization with a group of enthusiasts at the working level; and
- there is experience of and commitment to IT and the use of digital data.

Other researchers have examined success factors in particular application areas. Campbell (1994) studied UK local government. She identified success factors as simple applications, an awareness of resource limitations, and innovative and stable organizations or those with an ability to cope with change. The importance of these factors has been recognized in other areas. A study of a US Army GIS by Peuquet and Bacastow (1991) revealed slow development with a two-and-a-half-year delay. The reasons for this were felt to be the vague requirements and newness of GIS technology but also the inability of the organization to deal with change, and the non-involvement of the whole organization.

Research has confirmed that the factors identified by Campbell and Masser (1995) apply to other sectors of GIS users (Sieber, 2000; Nasirin et al., 2003). The example of international retailing is looked at in more detail in (Box 11.7). It also remains to be investigated whether GIS is actually any different from other areas of IT in this respect. For example, are there additional problems associated with the implementation of GIS as opposed to a large command and control system, or a stock management system?

## BOX 11.7 Keys to successful GIS in the retail sector

**PRACTICE**

Following research in retail organizations in Canada, the Netherlands and the UK, it has been possible to identify factors that are critical to the success of GIS in this sector during the planning, use and development of systems.

**PLANNING**
- use GIS intelligently to gain competitive advantage;
- identify business issues and then look for specific GIS solutions, not vice versa;

---

**BOX 11.7**

- involve end-users and make sure that everyone who may be affected by the system can make some input in the early stages;
- get support from someone at executive level who will 'sign-off' on the investment;
- quantify possible benefits and costs at the planning stage, but recognize that some will be unknown;
- keep the project manageable, and place it within broader IT and data strategy framework.

**USE**

- keep initial applications simple;
- allocate resources to training and education;

- secure early and visible quick results in critical area.

**DEVELOPMENT**

- get the GIS champion to promote and market the technology within the organization;
- accept that GIS is likely to be seen as just a part of a complex process – whilst you may view it as central to decision making, others may not;
- maintain momentum. Plan development to involve more datasets and exploit new analytical methods.

*Sources*: Hernandez *et al.* (1999); Hernandez and Verrips (2000)

---

## ORGANIZATIONAL CHANGES DUE TO GIS

In most organizations the introduction of a GIS brings with it a period of change. This is in common with other IT applications and other innovations that change working practices. A new GIS system can have a number of impacts: on the organization, on relationships with external suppliers and bodies, and on individuals. For example, a new system for the Happy Valley retailer SkiSupplies Inc. may:

- require the organization to undergo some internal restructuring (for example, a new business analysis department may be created to house the GIS);

- require redeployment and retraining of staff (staff may be redeployed to the new department and retrained in IT, or digitizing, or GIS);

- require revisions to administration arrangements and working practices (delivery vehicles may be supplied with printed or in-vehicle, on-screen maps in addition to written instructions of routes to be followed); and

- require implementation of new data sharing and exchange policies along with appropriate standards and guidelines (the distribution department will need to use the same base map data as the marketing and location analysis department; all address data held will need to be updated to a common format).

Betak and Vaidya (1993) summarize the impacts of GIS, and in particular increased data

sharing, at Conrail, the American Consolidated Rail Corporation. These included changes in the way employees related to customers, competitors and suppliers. At a more general level, changes included the consolidation of business functions, a reduction in departmental barriers and the de-layering of the management structure.

Changes that influence the success of a GIS project may also be initiated outside the organization. External factors that affect GIS could include the stability of the GIS supplier organizations, political and institutional restructuring, and changes to law and policy in associated areas.

As in any business sector, there are long-established GIS companies that have a broad customer base and stability. Others might be subject to take-over, rapid staff turnover or even closure. Even within a long-established company, particular product lines may be withdrawn or phased out, leaving customers unsupported or unable to upgrade. GIS users often use equipment and software from a range of vendors – the GIS software from one company, the computer hardware from a second, the printer from a third and the digitizer from a fourth – and this can compound problems associated with the stability of suppliers.

Political and institutional restructuring can affect any business, but can cause particular problems when imposed on GIS users. An example is British local government, where many GIS applications have been established and then restructuring of the local government system has been imposed. This has caused job insecurity and uncertainty about the future, and has resulted in the abolition of some local government organizations. The future of the

GIS projects was not a major consideration for the decision makers in this process. Some of the newly created local government organizations have been left with two GIS systems, which will need to be rationalized and integrated.

Changes to law and policy can also affect GIS, in both a positive and negative manner. For instance, legislation in the UK to ensure that the register of waste disposal sites was made available to the public resulted in a host of vendors offering GIS-style applications for the management and communication of the waste disposal register. Other policy changes, such as the adoption of standards for data, can also be beneficial in the long term, but disruptive and time-consuming in the short term.

In short, the introduction of a GIS is not simply a matter of choosing the product, ordering the equipment and getting it running, but a complex interplay of technical and human and organizational factors that may reflect change and uncertainty. During and after implementation there are likely to be considerable organizational changes due to the introduction of a GIS. Different users and stakeholders will be affected in different ways. Individuals' jobs and roles may be affected, relationships between departments may change and the nature of the services provided to outsiders may alter. It is worth addressing issues of organizational change from the outset. For example, what impact will a new GIS have on jobs? What difference will it make to the roles of individuals using it? What changes will be required to training programmes? How will the way information flows between departments change? Will the structure of the organization have to change to accommodate the GIS? Consideration of these issues may facilitate any changes required, identify areas of potential conflict and avoid problems of resistance and under-use.

## REFLECTION BOX

- What is cost–benefit analysis? What problems might be encountered while trying to apply cost–benefit analysis to a new GIS application?
- Why is user needs analysis an important part of the GIS selection process? How can user needs analysis be carried out?

- What are benchmarking and pilot projects? Why are they often carried out prior to full system implementation?
- Think of a GIS in an organization with which you are familiar. How was the GIS implemented? Which implementation strategy was adopted?

## CONCLUSIONS

The motor car has led to profound changes in society. Increasing the mobility of the population has influenced the work and social habits of individuals and the widespread use of vehicles has led to changes in the way society is structured, and the way settlements are planned and constructed. Veregin (in Pickles, 1995) considers that the most significant impact of technology tends to occur when the technology becomes indistinguishable from everyday life. Cars, for most of us in Europe, North America and other parts of the world, are indistinguishable from everyday life. In some areas, however, the motor car is less common, and life can be very different. The same is true for GIS. Some users are now at the stage where they do not even know they are using a GIS. GIS has become part of their everyday work, and is seamlessly integrated with other IT in their organizational information strategies (Pick, 2008). In many cases GIS has had a profound impact on the way individuals work and departments interact (Roche and Caron, 2009). For some organizations, GIS is still a technology they intend to adopt when it can be justified, and for others, the benefits of adopting a system still need to be evaluated.

This chapter has attempted to ask questions that need to be answered during the implementation of a GIS. Some answers have been proposed, but the best methods of introducing systems, and assessing the impacts of GIS, are areas requiring more research for clear generic recommendations appropriate to all organizations to be made. Perhaps it will be impossible to develop generic guidelines, since all organizations differ in their nature, in the scope of their GIS applications and the range of users. However, many authors have attempted to generalize methodologies for IT implementations, and the question remains whether GIS is really any different from other IT implementations.

## REVISION QUESTIONS

- Discuss the nature of GIS users. Why is a user-centred approach to GIS required for successful implementation?

- What factors can be considered when selecting a GIS for an organization? Which of these do you consider to be the most important, and why?

- How can techniques such as cost–benefit analysis, benchmarking and pilot projects be incorporated in the GIS procurement process?

- Describe the strategies available for GIS implementation and discuss their relative advantages and disadvantages.

- Outline the problems which organizations implementing GIS may face. Are these problems any different from those associated with other large IT applications? What makes GIS special?

## FURTHER STUDY – ACTIVITIES

- Use the literature (including trade magazines) and web to research organizations relevant to your own discipline or field of interest. How have they selected and implemented their GIS?

- Having seen what has been achieved in Happy Valley, another ski resort has employed you as a consultant to prepare a proposal for their GIS. Write a checklist, or produce a mindmap, of questions you need to ask to help prepare the proposal.

- Find out about GIS users in an organization of your choice. Identify all the users of GIS and try to classify them into the scheme shown in Figure 11.2.

## FURTHER STUDY – READING

For an overview of GIS in particular applications areas it is necessary to consult a wide range of sources. There have been some published surveys of users, their applications and difficulties faced when adopting GIS. Examples include Gould (1992), who surveyed UK health authorities and considered their use of GIS and computer-assisted cartography; Campbell and Masser (1992, 1995) and Campbell (1994), who review the applications and nature of GIS in UK local government; and Medyckyj-Scott and Cornelius (1991, 1994), who conducted a more general review of GIS in the UK, looking at the nature of systems being used and problems faced when setting up a GIS.

Medyckyj-Scott and Hearnshaw (1994) offer interesting chapters on various aspects of GIS in organizations. The chapter by Eason (1994) entitled 'Planning for change: introducing a GIS' is particularly recommended. Eason (1988) is also an interesting book, offering a more general picture of IT in organizations, but with many ideas that are directly applicable to GIS.

An entire section in Longley *et al.* (1999) addresses GIS management issues. The chapter by Bernhardsen (1999) offers comments on choosing a GIS that includes consideration of hardware and software issues. Longley *et al.* (2005) includes a useful chapter on GIS software that expands on the types of systems and provides examples of products in each category.

Obermeyer (1999) takes an in-depth look at cost–benefit analysis for GIS. Campbell (1999) examines the institutional consequences of the use of GIS and discusses different managerial approaches to GIS. Further information on the research into the retail sector by Hernandez can be found in Hernandez *et al.* (1999) and Hernandez and Verrips (2000).

More recently, Roche and Caron (2009) have brought the debate up to date in their book entitled *Organizational Facets of GIS* in which they look at the developments of the geographical information society from a historical perspective and reflect on its broader connection to the information technology field in order to explore issues and problems associated with the implementation, diffusion and use of GIS by organizations.

Bernhardsen T (1999) Choosing a GIS. In: Longley P A, Goodchild M F, Maguire D J, Rhind D W (eds) *Geographical Information Systems*. Wiley, New York, 589–600

Campbell H J (1994) How effective are GIS in practice? A case study of British local government. *International Journal of Geographical Information Systems* 8 (3): 309–26

Campbell H J (1999) Institutional consequences of the use of GIS. In: Longley P A, Goodchild M F, Maguire D J, Rhind D W (eds) *Geographical Information Systems*. Wiley, New York, 621–31

Campbell H J, Masser I (1992) GIS in local government: some findings from Great Britain. *International Journal of Geographical Information Systems* 6 (6): 529–46

Campbell H J, Masser I (1995) *GIS and Organizations: How Effective are GIS in Practice?* Taylor and Francis, London

Eason K D (1988) *Information Technology and Organizational Change.* Taylor and Francis, London

Eason K D (1994) Planning for change: introducing a Geographical Information System. In: Medyckyj-Scott D J, Hearnshaw H M (eds) *Human Factors in Geographical Information Systems.* Belhaven, London, pp. 199–209

Gould M I (1992) The use of GIS and CAC by health authorities: results from a postal questionnaire. *Area* 24 (4): 391–401

Hernandez T, Verrips A (2000) Retail GIS: more than just pretty maps. *GeoEurope 9* (4): 16–18

Hernandez T, Scholten H J, Bennison D, Biasiotto M, Cornelius S, van der Beek M (1999) Explaining retail GIS: the adoption, use and development of GIS by retail organizations in the Netherlands, the UK and Canada. *Netherlands Geographical Studies* 258, Utrecht

Longley P A, Goodchild M F, Maguire D J, Rhind D W (eds) (1999) *Geographical Information Systems: Principles, Techniques, Management and Applications.* Wiley, New York

Longley P A, Goodchild M F, Maguire D J, Rhind D W (2005) *Geographical Information Systems and Science.* 2nd edn. Wiley, Chichester

Medyckyj-Scott D J, Cornelius S C (1991) A move to GIS: some empirical evidence of users' experiences. *Proceedings of Mapping Awareness 1991*, conference held in London, February, Blenheim Online, London

Medyckyj-Scott D J, Cornelius S C (1994) User viewpoint: a survey. In: Hart T, Tulip A (eds) *Geographic Information Systems Report.* Unicom Seminars, Uxbridge, UK

Medyckyj-Scott D J, Hearnshaw H M (eds) (1994) *Human Factors in Geographical Information Systems.* Belhaven, London

Obermeyer N J (1999) Measuring the benefits and costs of GIS. In: Longley P A, Goodchild M F, Maguire D J, Rhind D W (eds) *Geographical Information Systems.* Wiley, New York, 601–10

Roche S, Caron C (2009) *Organizational Facets of GIS.* Wiley, New Jersey.

## WEB LINKS

### Cost–benefit analysis:

- Cost–benefit analysis and justifying investment in GIS: www.nysgis.state.ny.us/ coordinationprogram/reports/cost/index.cfm

- EU GISIG Guidelines for best practice in user interface in GIS: www.gisig.it/best-gis/Guides/ chapter9/nine.htm

### GIS implementation:

- Implementing GIS: lagic.lsu.edu/gisprimer/ phased_implementation.asp

Visit our website **www.pearsoned.co.uk/heywood** for further web links, together with self-assessment questions, activities, data sets for practice opportunities and other interactive resources.

# GIS project design and management

## Learning outcomes

By the end of this chapter you should be able to:

- Outline how a GIS project can be designed and managed

- Explain what is meant by a *rich picture*

- Describe the difference between a physical and a conceptual data model

- Explain what cartographic modelling is, and how it can be used

- List a range of approaches available for managing the implementation of a GIS project

- Describe problems that may be encountered when implementing a GIS

- Plan for the implementation of your own GIS project

## ■ INTRODUCTION

The theoretical and technical knowledge you have gained from earlier chapters provides a solid foundation from which to create your own GIS applications. The aim of this chapter is to build on this foundation and provide a framework for you to carry out your own GIS project. The emphasis is on the practical aspects of designing and project managing the implementation of a GIS application. The project design and management approach outlined in this chapter is suitable for small-scale GIS projects in any application area – for example, the type of project which you may be required to complete as part of a GIS course or research project. To deliver such a project you will need to develop, alongside your knowledge of GIS, skills to enable you to:

- design your GIS application;

- project manage its implementation; and

- evaluate whether your GIS application has been successful.

The skills required to design a GIS application involve understanding how to: identify the nature and scope of a problem; define the system to be built; quantify the amount and type of data necessary; choose the data model and the establish the analysis required. Project management skills are required to ensure a project is delivered on time and is of appropriate quality. Evaluation skills are necessary to help you critically appraise whether or not your project has delivered what you set out to achieve and helped you learn lessons for the future.

Good project design, management and evaluation skills are essential to produce a useful and effective GIS application. In the context of the car analogy you are now ready to use your car to go on your first real journey. Before you set out you must plan, or design, the excursion to ensure that you have everything you need – for example, fuel, food and maps. You must also manage your journey to make sure you arrive on time, and in the right place. An evaluation afterward will help you look back to assess whether your journey was successful, and whether you need to implement any changes – for example, to your route or the supplies you take with you, in future.

The project design and management approach outlined here does not embrace any specific design methodology or project management process; rather it is an integration of a number of ideas. Various elements of the approach, when scaled up, could provide a methodology for the implementation of larger GIS applications. However, it should be remembered that many organizations have corporate approaches to managing and co-ordinating the design and delivery of projects. Therefore, you should check if there is any guidance or procedure you must follow. It should also be remembered that there is no generic blueprint for GIS success. Any design and project management approach adopted should be able to be adapted to meet the needs of the application, the available technology, the users of the system and the organizational culture in which the GIS must reside.

This chapter starts by considering the design of a GIS application. A simple three-stage approach is introduced, which looks at how to scope out the problem to be addressed; design and choose a data model and establish an appropriate analysis framework. To help with implementation we then take a closer look at several project management approaches and techniques.

Few if any GIS projects are implemented without any problems and unexpected difficulties are common. To help you plan for the unexpected, some of the main things to look out for are outlined. Tips on how to identify and manage any risks associated with a project and guidance on how to evaluate your project are also provided.

To conclude, a checklist is provided to help you keep track of the design and implementation of your own GIS project. The house-hunting case study is used to illustrate the approach throughout the chapter.

## ■ GIS DESIGN

One of the key elements in delivering a successful GIS project is to get things right from the start. This is often referred to as *Project Design*; a process which involves working out what it is you are trying to do and how best to go about it. The following three-stage process may be helpful:

- *Step 1: Problem identification.* Every GIS application is designed to address a specific challenge or problem. For a GIS application to be successful, you must make sure that you fully understand the nature of the problem you are seeking to address. To help you scope out the problem two methods are introduced below: the *rich picture* and *root definition*.

- *Step 2: Design and choose a data model.* Successful GIS applications model the part of the real world which is of interest to the problem under investigation. To create an appropriate model an understanding of the distinction between the *conceptual data model* and the *physical implementation* of this model in the computer is necessary. These two elements, including tools to help you design your own data model are discussed further below.

- *Step 3: Design the analysis.* All GIS applications are designed to allow users to explore a problem

and your own GIS project will be no exception. The key to successful analysis of a problem is the design of an appropriate framework for this analysis. One technique which can provide a structure for setting out the analysis required by a GIS application is *cartographic modelling*.

## ■ STEP 1: PROBLEM IDENTIFICATION

Before developing a GIS application the problem that the GIS will address must be identified. There are two techniques that can be used to assist problem identification: creating a *rich picture* (a schematic view of the problem being addressed), and developing a *root definition* (a statement of an individual's or group's perspective on the problem). Both these techniques are drawn from the soft systems approach to system design (Box 12.1).

## BOX 12.1 The soft systems approach

THEORY

The original soft systems ideas were developed by Checkland (1981) and have been added to by other researchers (Wood-Harper *et al.*, 1995). The soft systems approach to problem identification provides a method for addressing unstructured problems (Skidmore and Wroe, 1988; Checkland and Poulter, 2006). This is useful in a GIS context because many GIS problems are unstructured and often difficult to define (Balram and Dragićević, 2006). To understand the difference between structured and unstructured problems it is helpful to revisit the house-hunting case study.

- *Structured problem.* Identifying the location of the properties an estate agent has for sale is an example of a structured problem. To develop a GIS to address this problem, the system developer needs to know where houses are and whether or not they are for sale.

- *Unstructured problem.* Choosing a neighbourhood in which to live is an example of an unstructured problem. In this situation all home buyers will have different opinions on where they want to live and

different priorities influencing their choice of property. They may use some common data sets, such as proximity to public transport or workplace, but how they use these data to make an informed decision will vary. Developing a GIS to meet their needs is much more problematic than developing a GIS that just shows the location of houses currently for sale.

To formulate a problem, users should appreciate the context, or world view, from which the problem is being considered. This is the key to the soft systems approach. For example, the house-hunting GIS was constructed from the perspective of a home buyer searching for a place to live. An alternative perspective would be that of the estate agent trying to sell property. From the soft systems perspective it is not models of real-world activities, which are created, but models of people's perception of an activity. How people feel about and view the activity are included (Checkland and Poulter, 2006). Therefore, soft systems models are abstract logical models that help with our understanding and structuring of a problem.

## The rich picture

A rich picture is a schematic view of the problem your GIS project will address. It presents the main components of the problem, as well as any interactions that exist. The rich picture for the house-hunting GIS (Figure 12.1) adopts the conventions of other authors, in particular Skidmore and Wroe (1988) Avison and Wood-Harper (1991) Reeve (1996) and Curry *et al.* (2005). These include the use of:

■ *Crossed swords.* A crossed swords symbol expresses conflict. It is used to indicate the differences between the home buyers and the estate agents. There is conflict since the motives of the two groups for system development are different. The

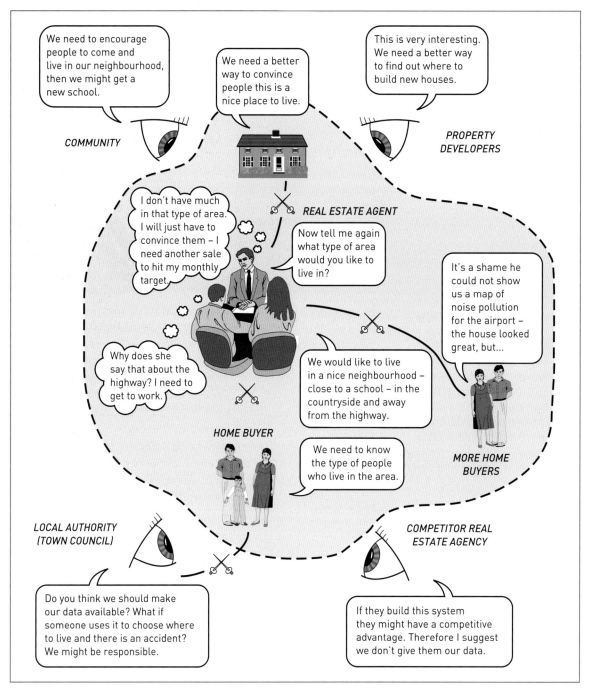

**Figure 12.1** Rich picture for the house-hunting GIS

home buyer wishes to find the house that best suits their needs whereas the estate agent wishes to sell a house to gain commission.

- *Eyes*. Eyes are used to represent external observers. Property developers interested in identifying new areas for housing development may be external observers.

- *Speech bubbles*. Personal or group opinions are indicated in speech bubbles. The different priorities home buyers see for the system may be included in the rich picture in this way.

Drawing the rich picture records various thoughts on paper and helps to organize ideas. For a small-scale application, one individual may draw the rich picture. A rich picture drawn by a project team will represent a consensus view of a problem reached by all the project participants. A single composite rich picture can be achieved by asking all members of the team to draw their own rich pictures. These are then discussed and combined to create a single picture that reflects the views of all parties. Skidmore and Wroe (1988) suggest that rich pictures are particularly useful when considering the design of computer systems within organizations because:

- they focus attention on important issues;

- they help individuals to visualize and discuss the roles they have in the organization;

- they establish exactly which aspects of the information flows within the organization are going to be covered by the system; and

- they allow individuals to express worries, conflicts and responsibilities.

The development of a rich picture should not be rushed, particularly if it is trying to reflect an unstructured problem. A poorly defined rich picture may translate into a poor GIS application. An additional check to ensure that the problem is well understood is to develop a root definition.

## The root definition

Like rich picture, the term root definition also comes from the soft systems approach (Box 12.1). The root definition is a view of a problem from a specific perspective. Different users have different views of a problem. In the house-hunting GIS, the views of groups involved in the design process might be quite different and lead to a degree of conflict. For example, home buyers may see the GIS as 'a system to help identify and rank possible homes', whereas the estate agents may see it as 'a system to help maximize house sales'. These two statements are the root definitions of these particular groups. The system developer must get these two groups to agree on a common root definition: for example, 'a system that identifies properties for sale which meet the requirements of individual home buyers'.

Establishing a common root definition for a problem will help others to evaluate and understand why a GIS has been constructed in a particular way. Likewise, understanding that others may view a problem from a different perspective will ensure a GIS application is designed to address a range of needs. If a single root definition can be agreed upon, there is a greater chance the GIS will meet the requirements of all concerned. Once a rich picture and root definition exist the main aims and objectives for a project can be identified and a GIS data model can be created.

If it proves difficult to draw a rich picture or formulate a root definition then the problem being addressed may be unstructured (Box 12.1). Unstructured problems are the most difficult to address with GIS. However, the rich picture method can still be used; typically it will start with only a few elements of the problem clearly defined. Additional elements are added after talking to potential users of the GIS, consulting the literature and discussing the project with others working in a similar field.

As the rich picture is developed and the root definition formulated the resources available to the project must also be considered. In some cases the only resource will be one individual's time and commitment. In larger projects there may be access to several members of staff and a budget. It is important to consider, given the resources available, whether it is possible to address the whole problem that is unfolding, or whether it will be necessary to break the problem down into smaller parts. Breaking the problem down into more manageable pieces may allow quicker results, which may be important where the GIS activities are taking place in a large organization. Small but useful results, produced relatively quickly, will gain recognition and respect for a project. This may result in further support and resources being allocated to the project. Therefore, many system designers use pilot projects to produce results quickly. These results

## BOX 12.2 Problem identification: questions

**1** What is the problem you wish to address with GIS? Is the problem structured or unstructured?

**2** Can the problem be summarized in a rich picture? Does the rich picture include the views of all those who will use the GIS and all those who will be affected by it?

**3** Can a single root definition be defined for the problem? Is this root definition acceptable to all potential users and those who will be affected by the GIS?

**4** Can the problem be tackled within the resources available? Would it be better to break the problem into smaller component parts and tackle one of these first?

can be disseminated widely throughout the organization to encourage support for the GIS. Box 12.2 offers key questions to be addressed at this stage of the GIS project design process.

## ■ STEP 2: DESIGN AND CHOOSE A DATA MODEL

The rich picture and root definition help to identify the main characteristics of the problem for which a GIS application is required. The next step is the development of a model of the part of the real world which is of interest to the problem – the data model.

The term *data model* has different meanings in different contexts. In **Chapter 3** it was used to describe the method for representing spatial entities in the computer. Peuquet (1984) and Frank and Mark (1991), however, use data model as a collective term for the process of identifying all the design elements used in the construction of a GIS. Worboys and Duckham (2004) offer a useful solution to this confusion by distinguishing between conceptual and physical data models. The conceptual data model is a high-level view that is independent of the computer system. This is the user's view of a problem and its elements. This is close to the way that Peuquet (1984) and Frank and Mark (1991) use the term. The physical data model, on the other hand, describes the organization of data in the computer. This is how the term has been used in **Chapter 3**.

Therefore, it is important to understand that the data model used by your GIS application will consist of two parts: a conceptual data model which captures the main elements and aspects of the problem;

and a physical data model which reflects how you have actually chosen to represent these elements in the computer. Consider these ideas in the context of the development of a computer game. The conceptual data model is the context for the game – what it is you are going to do? For example, are you going to hunt for treasure, build a city or shoot monsters? The conceptual data model should also reflect how the elements in your game interact. For instance, you may decide that your monsters will morph into something new when they are shot. The physical data model is how you are going to represent features in the computer. For example, would your monsters be represented by vector graphics or blocks of pixels?

In GIS the conceptual data model also adds spatial detail to the rich picture by including elements of spatial form and spatial process. The physical data model is concerned with how to represent the conceptual model within the computer. Details about the spatial data model (raster or vector), the appropriate data structure and the analysis scheme are included in the physical data model.

Unfortunately, the data-modelling stage is frequently neglected in the design and development of GIS projects, often with disastrous consequences. Insufficient attention to data modelling may lead to the failure of the GIS to meet the expectations of users.

### Creating a conceptual and physical data model

One way to create a conceptual data model is to borrow heavily from the ideas of hard systems analysis. Hard systems analysis advocates the clear identification of the elements of the data model: the entities, their states and their relationships to each other (Box 12.3). One method of presenting this is to use a flowchart. In systems analysis, flowcharts use a range of symbols to communicate different aspects of the model. This is illustrated in Figures 1.6, 1.11 and 1.13 in **Chapter 1**, where flowcharts represent conceptual models of the three case studies. A box has been used to identify entities and their states, a diamond represents decisions, a hexagon indicates procedures and lines with arrows show relationships and linkages. The aim of the flowcharts is to illustrate all the stages involved in modelling the problem from identifying the data requirements, through defining the analysis, to determining the output requirements. Figures 1.6, 1.11 and 1.12 give an idea of the stages that must be gone through to use a GIS

to address the problems of siting a radioactive waste dump, identifying areas for nature conservation or locating a neighbourhood in which to live.

GIS terminology can be avoided when constructing the conceptual data model. This is a good idea as the resulting flowchart will then explain what it is the GIS application will do in a way that is clear to all interested parties. It will give those with little GIS experience the opportunity to provide feedback on the approach. Moreover, if the conceptual model is jargon-free, it can be given to GIS programmers with different software backgrounds. This may be an advantage in large-scale GIS projects where an organization wishes to compare how well different software products can address a task.

Bell and Wood-Harper (1992) provide a useful checklist for the development of a conceptual model:

## BOX 12.3 Hard systems analysis and GIS

**THEORY**

Hard systems analysis advocates trying to understand reality by rebuilding part of it. The link to GIS is clear, as GIS data models attempt to reconstruct parts of reality for specific purposes. During the 1970s and early 1980s the hard systems approach was the dominant methodology used for the design of computer systems. It is possible that the early developers of GIS software used a hard systems approach to design.

There are four phases in hard systems analysis. These are outlined below (after Huggett, 1980). However, before discussing these phases in detail, it is necessary to explain some of the terminology of the hard systems approach. Three key terms are entities, states and relationships. The entities, or elements of a system, are either physical objects or concepts. In GIS terms entities are points, lines, areas, surfaces and networks (**Chapter 2**). Entities also possess properties known in hard systems terms as *states*. The states associated with an entity give its character. In GIS terms states are attributes (**Chapter 4**). In addition, relationships exist between entities. In GIS this relationship could be the topological links between features (**Chapter 3**).

The four stages in hard systems analysis, in a GIS context, are:

1 *The lexical phase*. The objectives of the lexical phase are:
- to define the problem;
- to define the boundaries of the problem;
- to choose the entities that define the components of the problem; and
- to establish the states of these entities.

In GIS this involves:
- identifying the nature of the application;
- selecting the study area;
- defining the real-world features of interest; and
- identifying associated attributes.

2 *The parsing phase*. In the parsing phase the relationship between entities and groups of entities are defined. The entities and knowledge about their states are used to create a computer model (**Chapter 3**).

3 *The modelling phase*. In this phase the GIS is used to address the problems identified during the lexical phase. The way in which entities and their states will interact and respond under differing situations is expressed. This may involve linking GIS software to other software.

4 *The analysis phase*. This phase is the validation of the modelling phase. Testing occurs to find out how closely the GIS model (of both form and process) fits what is observed in reality.

1 *Develop a rich picture and root definition.* Everyone associated with the problem should agree upon these. They are used to focus the aims and direction for the project.

2 *Create a list of actions the system must be able to perform.* In the house-hunting example these actions may include permitting users to select neighbourhood characteristics such as proximity to schools, railway stations and shops, and allowing users to weight these characteristics in terms of their relative importance. These actions are known as *activities*.

3 *Identify a list of system inputs and outputs.* In GIS terms system inputs are data sources and outputs are products such as maps. In the house-hunting example, the data sources would include street networks, public transport routes and the location of properties for sale. Outputs might be a list, or map, of properties meeting home buyers' criteria.

4 *Group activities, inputs and outputs into a logical, chronological order.* Arrows symbolizing some form of action are used to join activities together. For example, in the house-hunting GIS the combination of data from different sources could be effectively represented in this way.

The physical data model requires additional detail that describes how to model the spatial entities, their associated attributes and the relationships between entities in the computer. **Chapters 3** and **4** examined the detail of spatial and attribute data models in detail and these chapters should be consulted before reading any further, if you have not reviewed them already. A physical data model which is most appropriate for your purpose should be chosen. You should choose a physical data model that is closest, in terms of representation, to the aspect of the real world feature which you wish to model. Sometimes this will be a relatively straightforward decision – for example, the use of connected vectors to represent a road network to allow traffic flows and routes to be mapped. In other circumstances decisions may be more difficult. For example, a forest could be represented as a discrete vector polygon (to which a descriptive attribute could be added to reflect vegetation type), or it could be represented as a series of raster pixels (classified to reflect vegetation type).

Your choice of data model will be influenced by how you want to use your GIS model. If you have a clear picture of the type of analysis you wish to perform, your choice of physical data model may be easier. Thus, it is helpful to undertake Step 3 of the design process before finally selecting a physical data model. However, sometimes the choice of physical data model will be constrained by the GIS software available for a project. In this case it is helpful to remember that the physical data model is just that – a model – and there will usually be another way to explore your problem.

## ■ STEP 3: DESIGNING THE ANALYSIS

There are a number of different approaches to designing an analysis scheme; here we concentrate on an approach known as *cartographic modelling*.

The origins of cartographic modelling are difficult to establish. As Tomlin (1991) states, they are derived from a collection of old ideas that have been organized, augmented and expressed in terms amenable to digital processing. However, it is the work described by Tomlin (1983) as 'Map Algebra' and Berry (1987) as 'Mapematics' that established cartographic modelling as an accepted methodology for the processing of spatial information.

Cartographic modelling, at its simplest, is a generic way of expressing and organizing the methods by which spatial variables, and spatial operations, are selected and used to develop an analysis schema. Tomlin (1991) considers that the fundamental conventions of cartographic modelling are not those of any particular GIS. On the contrary, they are generalized conventions intended to relate to as many systems as possible (Davis, 2001). The truth in this statement is illustrated by the number of GIS software products that use the concepts of cartographic modelling in their approach to spatial analysis.

The concepts that underpin cartographic modelling borrow heavily from mathematics. Cartographic modelling is a geographic data-processing methodology that views maps (or any spatial data layer) as variables in algebraic equations. In algebra, real values are represented by symbols, such as $x$, $y$ and $z$. In map algebra these symbols may represent numeric attributes of map elements (for example, pH values associated with a given soil type) or even whole maps. Numbers assigned to symbols in an equation interact to generate new numbers using mathematical operators such as *add*, *subtract*, *multiply* and *divide*. In the same way, in map algebra maps are transformed or combined into new maps by the use of specific spatial operations. Figure 12.2 shows how a GIS overlay procedure could be represented as a cartographic model and as an algebraic expression.

There are four stages in the development of a cartographic model (after Burrough, 1986 and Burrough and McDonnell, 1998):

1 Identify the map layers or spatial data sets required.

2 Use natural language to explain the process of moving from the data available to a solution.

3 Draw a flowchart to represent graphically the process in step 2. In the context of map algebra this flowchart represents a series of equations you must solve in order to produce the answer to your spatial query.

4 Annotate this flowchart with the commands necessary to perform these operations within the GIS you are using.

To explore these stages it is helpful to consider the house-hunting case study. In **Chapter 1** we completed stage 1 of the cartographic modelling process by identifying the different data layers to be used in the model (Figure 1.12). Figure 1.12 also shows how natural language was used to describe what happens at various stages in the model (stage

**Figure 12.2** A simple map algebra equation

2). The use of natural language in the development of a cartographic model is important because a user who is able to express in words the actions that he wishes to perform on the geographical data should be able to express that action in similar terms to the computer (Burrough, 1986). Tomlin (1983) recognized the role for natural language to express the logic of spatial analysis when developing the analogy between cartographic modelling and algebra. The result was GIS software with a natural language interface: the Map Analysis Package. In this package each spatial operation is a verb which acts on subject nouns, representing map layers, to create object nouns, or new map layers. Table 12.1 provides a selection of key verbs of spatial operations used in Tomlin's approach.

**TABLE 12.1 Selected examples of natural language keywords**

| Keyword | Operation | Description | Example of use |
|---------|-----------|-------------|----------------|
| SPREAD | To create a corridor from a linear data set, or zone of influence around a point | Calculate the distance of all geographical positions in the data set from a given starting point or line | To create buffer zones representing proximity to schools, main roads or stations |
| OVERLAY | To find the intersection of two different sets of area entities covering the same geographical region | Lay two different sets of area entities over each other to produce a new, more complex set of area | To create a map of accessibility from an overlay of proximity to stations and proximity to main roads |
| EXTRACT | To extract a new data set from an existing data set | Select specified values and/or ranges of values from one overlay to make a new overlay | To create a new data layer showing an existing data set only those houses that are for sale within the area defined by the accessibility map |

*Source*: Adapted from Tomlin (1991)

Stage 3 of the cartographic modelling approach involves adding detail to the flowchart to represent the logic of the analysis required by the GIS project. Figure 12.3 provides an example of how this might be done for part of the house-hunting GIS. Table 12.2 presents four of the equations required by the analysis scheme shown in Figure 12.3. The equation numbers refer to those spatial operations identified by a number in Figure 12.3.

The final stage in the cartographic modelling process is to annotate the flowchart with the appropriate commands from the GIS package in which it is intended to perform the analysis. Many GIS software packages have a graphical modelling interface which allows a visual picture of the required analysis to be manipulated by the user, negating the need for complex list of commands.

From the above example, it should be apparent that the power of cartographic modelling lies in the structure it provides to the GIS designer. This structure allows the designer to tackle a complex spatial problem by breaking it down into its components. Simple statements, sections in a flowchart or solvable equations can then express these. However, it may not be possible or sensible to do all your analysis in the GIS. In certain cases it may be necessary to couple the GIS with other applications to obtain results (**Chapter 7**).

If care has been taken over the construction of the rich picture and root definition and their subsequent translation into a conceptual and physical data model, then, at least in theory, the computer implementation of a GIS data model should be relatively straightforward. In reality, it is likely that building an application will start whilst detail is still being added to the GIS data model. There is nothing wrong with this approach as long as the enthusiasm for implementation does not take over, leaving missing details forgotten. Trial and error with systems development is an accepted approach, but often individuals working alone may spend many hours developing a solution only to find that they have not documented their work and are therefore unable to explain to others how the result was reached. How you implement a physical data model will depend upon the nature of your problem and the organizational setting in which you are working. However, good project management will help ensure that goals are met. Box 12.4 on p. 378 suggests questions that should be addressed at this stage of the data modelling process.

**TABLE 12.2 Simple equations for Figure 12.3**

| Equation 1. From land use 'extract' countryside | Equation 2. From road map 'spread' road by distance |
| --- | --- |
| $a - b = c$ | $(d - e) + f = g$ |
| where: | where: |
| $a$ = land use map | $d$ = road map |
| $b$ = urban land use map | $e$ = portion of map which is not road |
| $c$ = countryside | $f$ = zone of spread for a specified distance either side of the road |
|  | $g$ = proximity to road map |
| **Equation 3. From house status 'extract' houses for sale** | **Equation 4. 'Overlay' houses for sale in countryside away from roads** |
| $h - i = j$ | $j + k = l$ |
| where: | where: |
| $h$ = house status map | $j$ = houses for sale |
| $i$ = houses not for sale | $k$ = countryside away from roads |
| $j$ = houses for sale | $l$ = houses for sale and countryside away from roads |

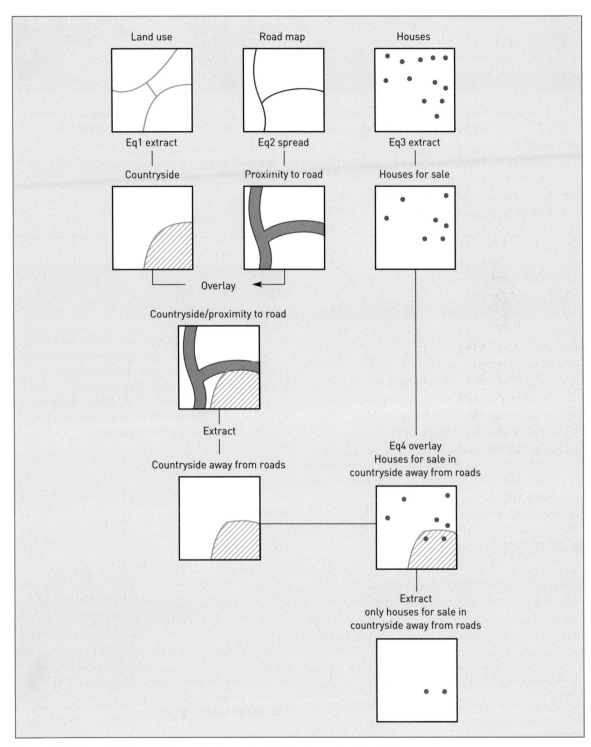

**Figure 12.3** Example of part of the analysis required by the house-hunting GIS

## BOX 12.4 Designing a data model: questions

**1** What are the elements of your conceptual data model: entities, attributes and relationships?

**2** What are the four stages in the design of a conceptual model? Can they be represented in a flowchart?

**3** What spatial and attribute data models and structures are required?

**4** What approach to designing an analysis scheme is appropriate? Can the ideas of cartographic modelling be applied?

**5** Can GIS offer all the functions required for the analysis?

## REFLECTION BOX

- Write brief explanations of the following terms and phrases:
  - Soft systems model
  - World view
  - Rich picture
  - Root definition
  - Hard systems modelling
  - Cartographic modelling.
- Create a rich picture for a GIS project with which you are familiar. What are the benefits of using a rich picture?

- With reference to an application from your own discipline or area of interest, explain the difference between a conceptual and a physical data model.
- Review Bell and Wood-Harper's checklist for the development of a conceptual model presented above (Bell and Wood-Harper, 1992). Can you apply this list to a project of your own?

## ■ PROJECT MANAGEMENT

This chapter has considered techniques for identifying the character and extent of a spatial problem and techniques for helping with the design of GIS data models. However, as discussed in **Chapter 11**, accurate problem definition and good data model design are only part of the story. Once the data model is constructed, the GIS must be implemented and in many cases integrated into the wider information strategy of an organization. To ensure success during implementation good project management is essential.

There are many different approaches to managing information technology (IT) projects. In this text we focus on the System Development Life Cycle (SDLC) approach. The SDLC is a framework for planning, managing and controlling the development of an information system, which can also be applied to GIS projects (Curry *et al.*, 2005). Figure 12.4 provides an overview of the SDLC process. We have already considered the planning, design and analysis stages presented in Figure 12.4. In this section we introduce two of the most common approaches used to implement a GIS project: the waterfall approach and prototyping.

### The waterfall approach

The waterfall approach offers a linear process for the management, development and implementation of an IT system (Skidmore and Wroe, 1998). The waterfall analogy is used because the outputs from the first stage of the process inform the second phase, and the outputs from the second phase affect the third phase, and so on. There are many variations

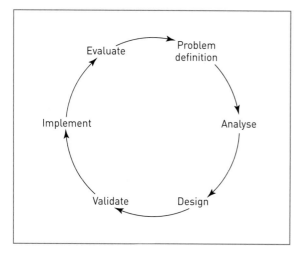

**Figure 12.4** The System Development Life Cycle

on the general approach. The house-hunting case study provides an example of how a GIS project could be managed using the waterfall approach.

1 *Feasibility study.* This would involve asking the real estate agents and home buyers questions about whether they would make use of the system being proposed for development and what the costs and benefits of developing a GIS would be. If the feasibility study is positive then the project moves to the second phase.

2 *System investigation and system analysis.* The GIS designer would try to establish the current way in which home buyers and real estate agents interact to identify houses for sale in appropriate neighbourhoods. This would include identifying the data and analysis requirements as well as the preferred output types. A soft systems approach could be used to help with this phase.

3 *System design.* The GIS data model is constructed using information collected in the previous phase. In the house-hunting example cartographic modelling techniques might be used to help structure the analysis requirements of the GIS.

4 *Implementation, review and maintenance.* Now the house-hunting GIS is built and provided to users. This may be the first opportunity for users to comment on, or interact with, the system since their involvement in the feasibility study. Users' experiences inevitably require changes to the system. These may include the addition of new data layers, new analysis techniques or new ways of visualizing the output.

The main advantage of the waterfall approach is that it provides a very structured framework for the management of a GIS project. This can be extremely important when good time management is an essential aspect of the project. In addition, it is often easier to budget for the resources required by a waterfall approach because the requirements of the system are established at an early stage in the project.

Despite its popularity as a project management tool for IT projects there are a number of problems with the approach:

- Designers who use the waterfall method often fail to address the context of the business for which the system is being developed. The approach encourages the designer to focus on only a part of the information problem of an organization. In the case of the house-hunting example the approach might prevent some of the external factors that influence where people choose to live being included in the model.

- The timescale and linear nature of the process do not allow for change in the scope and character of the problem. By the time the system is implemented it may be out of date. In the house-hunting case there may be a government policy review, which affects the way houses are bought and sold between the feasibility study and implementation.

- The waterfall approach does not put the user at the centre of the system design. It emphasizes the identification of flows of information rather than understanding why they are required. This creates problems, because it only allows a system that reflects the current way of doing things. This may be a problem for GIS design as a new system may radically change the way information is managed.

- Finally, the waterfall approach is often considered to favour hierarchical and centralized systems of information provision. It offers a very technocentric view of system development.

## The prototyping approach

The prototyping approach to IT project management developed as a response to the criticisms of the waterfall method, particularly in response to the lack of consideration of users (Checkland and Howell, 1998). Figure 12.5 shows the main stages in the method. The user first defines the basic requirements of the system. This could be achieved by using the rich picture and root definition techniques described earlier in this chapter. The system designer takes these basic ideas to construct a prototype system to meet the needs identified by the user. In GIS projects such systems are often described as *demonstrators*. The users who identified the original requirements for the system then experiment with the demonstration system to see if it is what they expected. Other potential users of the final system may be brought in at this stage to see if the system is of wider value. The system designer uses their recommendations to improve the system.

The prototyping approach has a number of advantages over the waterfall method:

■ Users have a more direct and regular involvement in the design of the system.

■ It is easier to adapt the system in the face of changing circumstances which were not identified at the outset of the project.

■ The system can be abandoned altogether after the first prototype if it fails to meet the needs of users. This reduces the cost of developing full systems.

■ If money and time are available, a number of prototypes can be built until the user is satisfied.

The prototyping approach, however, does have problems:

■ Prototyping can be difficult to manage. There may be large numbers of users with large numbers of ideas and opinions.

■ The resource implications may change following the development of the first prototype.

■ Knowing when to stop development can also be a problem. However, some GIS designers argue

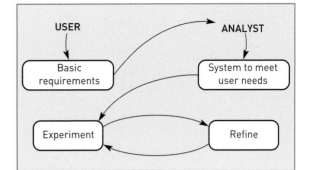

**Figure 12.5** The prototyping approach

that this is a positive aspect of the approach since few, if any, GIS systems are ever finished.

A development of the prototyping method is Rapid Application Development (RAD) (Stair and Reynolds, 2010), in which a GIS project is broken down into smaller parts and individual prototypes developed quickly to meet the needs of specific groups of users. This approach is particularly suited to large-scale or institutional GIS applications where there are many users with diverse needs.

The waterfall, prototyping and RAD approaches are just a few of many available for the management of a GIS project and there are many variations on the basic methods outlined above. It is also possible to pick and mix aspects of the three approaches to develop a management style that suits the development environment. In addition, there are a wide range of project management techniques and tools, which can be used to help with various phases in the different approaches. Those already considered – rich pictures, root definitions and flowcharts – are summarized alongside other techniques in Tables 12.3 and 12.4. General questions about project management that need to be addressed in relation to a GIS project are presented in Box 12.5.

## BOX 12.5 Project management: questions

**PRACTICE**

**1** Which management approach is appropriate for your application – waterfall, prototyping or RAD?

**2** What techniques will you use to ensure that your project is adequately documented?

**3** Will it be possible for others to recreate how the application has been constructed if things go wrong?

**4** Can someone else use the documentation you are producing as part of your system development to tackle the project?

**TABLE 12.3  Project management techniques**

| Technique | Purpose |
|---|---|
| SWOT analysis (strengths, weaknesses, opportunities and threats | This technique is used to establish the strengths, weaknesses, opportunities and threats associated with the development of the GIS. It is often used as part of the feasibility study in the waterfall approach |
| Rich pictures and root definitions | These techniques come from a methodology known as 'soft systems' and are used to help system designers determine the scope of a problem. They can be used both in the feasibility phases of the waterfall approach and to help users identify the basic requirements of a system in the prototyping approach |
| Demonstration systems | These are demonstration GIS applications, which are designed to help users evaluate whether or not the full system being proposed by the system designer will work. They are most frequently used when a prototyping approach to project management is used |
| Interviews and data audits | These techniques are used to help problem definition and to establish the current information and analysis needs of an organization. They are more structured than the rich picture and root definition techniques described above. Data audits can be particularly valuable in a GIS context, as many organizations are unaware of the spatial data to which they have access |
| Organization charts, system flowcharts and decision trees | These three techniques are all variations on the flowcharting theme. The organization chart maps out the flows of information within the organization. The system flowchart describes how the system will model these information flows. The decision tree shows the problem from a decision-making perspective and focuses on showing how different decisions cause information to be used in different ways within the organization. The technique used will depend on the experience of the system designer and the character of the problem |
| Data flow diagrams and data flow dictionaries | These techniques are drawn from hard systems analysis and represent a much more structured approach to system design than some of the techniques described above. They can be of immense value in GIS for tracking what happens to a data layer through the analysis process. This is extremely valuable in monitoring data quality and providing lineage information |
| Cartographic models and entity relationship diagrams | These techniques are of most value in structuring the analysis schemes used in GIS. In this respect they help plan the functional requirements of each stage of the analysis |

**TABLE 12.4  Project management tools**

| Project management tool | Description |
|---|---|
| GANTT charts | GANTT charts are time management tools that establish when and where, over the life of the project, a particular task will take place. Figure 12.6 shows a GANTT chart for the house-hunting GIS |
| PERT (program evaluation review techniques) | These are graphical tools for managing a project by showing how a task depends on the completion of others before it can be undertaken. One of the main values of a PERT chart is that it shows which tasks can be undertaken in parallel (Figure 12.7) |
| CASE (computer-assisted software engineering) tools | These are software tools that help reduce program development time. They can be used in a number of ways. They may help with project management, for example in the construction of GANTT and PERT charts. They can be used to assist software development and the construction of conceptual models and analysis flowcharts. And in the later stages of a project they may be used, like 'wizards' that allow automatic construction of documents in a word processor, to help create prototypes |
| PRINCE2 | PRINCE2 (Projects IN Controlled Environments) is a process-based method for effective project management. This method is in the public domain and offers guidance on project management. The key features of the approach are:<br>■ a focus on business justification;<br>■ a structure for the project management team;<br>■ emphasis on dividing the project into manageable and controllable stages; and<br>■ flexibility for application to projects of different sizes.<br><br>Many of the tools and techniques described above can be used as part of a project managed using the PRINCE2 approach |

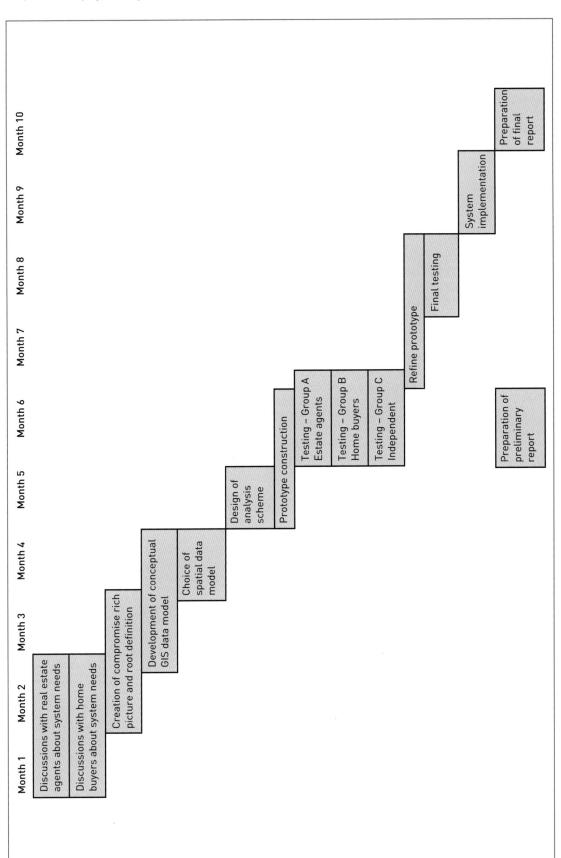

**Figure 12.6** GANTT chart for the house-hunting GIS

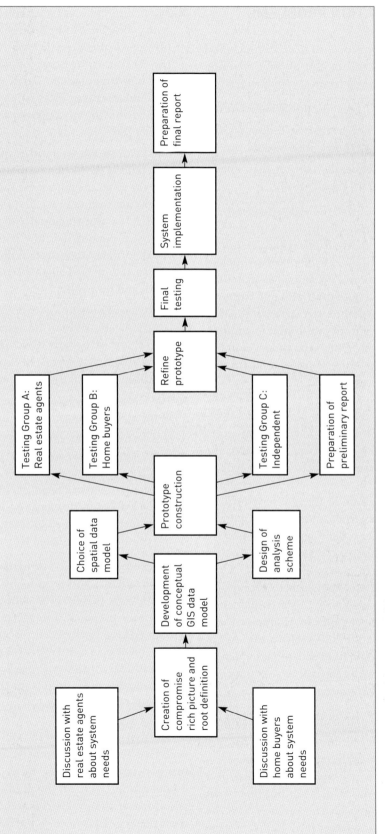

**Figure 12.7** PERT chart for the house-hunting GIS

# ■ IMPLEMENTATION PROBLEMS

There will always be problems for GIS design and development which no amount of prior planning can prepare for. Three of the most common are:

- data in the wrong format for the GIS software;
- a lack of GIS knowledge imposing technical and conceptual constraints on a project; and
- users of the GIS frequently changing their mind about what they want the GIS to do.

In many GIS projects the data required are unavailable in a format compatible with the GIS software or analysis needs. If this is the case, there are two options: to look elsewhere for a supplier, or to convert the data into the desired format. Both these approaches have been dealt with in **Chapter 5**. However, in the case of the latter, errors may creep in as data are changed from one format to another. Alternatively, the conceptual data model could be revisited to assess the importance of the data, and evaluate alternative data options.

It is inevitable, at some point, that applications will be limited by users' technical or conceptual knowledge about which spatial data model, data structure or analytical operation is most appropriate for the task required. Much can be learnt from other applications and other users. Colleagues working in similar areas, other organizations, or the Internet can all be sources of help. For many organizations the solution is to employ an independent expert to undertake application development or specific analysis.

A risk-management plan, which includes a risk register, may be a useful tool to help prepare for any implementation problems (DeMarco and Lister, 2003). A risk-management plan, in its simplest form, is a list of the main anticipated problems, the scale of impact these issues could have on a project and how they could be dealt with. The risk register is a practical tool for keeping track of these problems as the project progresses and is usually designed as a spreadsheet which can be updated at project meetings. Templates are available for the creation of a risk register (see the Web links at the end of the chapter).

The creation of a risk-management plan is a straightforward process which can be done by an individual for a small-scale GIS project, or with the input from others for a larger project. A five-stage process is necessary:

- *Step 1: Identify the risks.* Either independently or with colleagues list the main risks to which your project is exposed. At this stage the likelihood of the risks occurring is not considered. In a team situation it may be appropriate to ask individuals to compile separate lists first, then combine these, removing any duplicate risks, to create a master list.
- *Step 2: Rate the risks.* Each risk is given two scores. One reflects the likelihood of the risk occurring and the other the impact the risk would have if it did occur. A numerical scale of 1 to 5 is normally used. A low score indicates a low probability of the risk occurring and thus minimal impact. A high number suggests that there is a greater chance that the risk might occur and have a big impact on the project.
- *Step 3: Establish triggers.* Few risks develop into full-scale problems without any warning. In most cases there will be early signs that things are not going according to plan. For each of the risks identified potential triggers should be identified.
- *Step 4: Identify mitigating action.* This is a description of the action necessary to put things right, should a particular risk actually occur.
- *Step 5: Identify the owner of the risk.* A named person should be given responsibility for keeping track of the status of a risk and implement any mitigating action if necessary. In a small-scale GIS project an individual may 'own' all the risks, but in a larger project these will be shared across the project team.

When developing a risk-management plan, it is important to capture as many risks as possible as it is always better to be over-prepared, rather than under-prepared, for anything that may go wrong. To keep track of the risks identified in a risk-management plan a risk register is necessary. A table or spreadsheet is used with rows representing the identified risks and columns for the characteristics and status of the risk. Box 12.6 provides an example of a risk register.

The dynamic nature of the GIS design process is such that the information needs of users are often in a constant state of flux. By the time a GIS data model is implemented, the needs of the users and the scope of the problem may have moved away from the original defined by the rich picture. This is a major issue in the development of GIS applications

## BOX 12.6  Risk register

A risk register contains information about all the potential risks that may impact on a GIS project. Risk registers are usually set up in a spreadsheet or table with rows completed for each risk. The main items normally found in a risk register are:

- *Risk number*: a unique identifier to allow the risk to be tracked.
- *Risk description*: a short description of the risk.
- *Likelihood of occurrence*: This will be a score, usually from 1 to 5 (where 1 is a low likelihood of occurrence and 5 is high).
- *Scale of impact*: This will be a score, usually from 1 to 5 (where 1 indicates a low impact and 5 high potential impact).
- *Risk score*: This is calculated by combining the scores for likelihood of occurrence and scale of

impact, usually by multiplying them together. The larger the risk score, the greater the risk and the more closely it should be watched.

- *Mitigating action*: This should describe what action will be taken if things go wrong.
- *Owner*: The name and contact details of the person who owns the risk
- *Status*: This provides the current status of the risk. 'Traffic light' colours (red, amber and green) are often used. Red indicates a risk is 'live' and has actually occurred. Amber indicates that there is a reasonable chance that a risk may occur. Green indicates that there is no problem at present.
- *Action taken*: A brief description of any action taken to address the risk.

---

for larger organizations, where applications development may take considerable time, and the awareness of key players about GIS may increase in the meantime. The solution is to gain frequent feedback from the individuals who will be the end users of the GIS. They should reconfirm that the scope of the project has not changed, allows changes to be brought into the design process, or potential risks to be re-assessed. Even in a small-scale GIS project, as the knowledge of the GIS analyst grows there may be changes in the aims and scope of a project. Box 12.7 presents some of the questions that should be addressed during the implementation of a GIS.

## ■ PROJECT EVALUATION

After a GIS application has been constructed, some problems may be just about to start. It is important that the output produced by the system is usable, valid and meets the goals set at the beginning of the project. Validation of results is often difficult to achieve, particularly if results are in the form of predictions. However, if you are working with an organization, testing the GIS and validating output will be a crucial part of the design process. In many cases, this may well result in adjustments to the rich picture and the GIS data model. In extreme cases

## BOX 12.7  Implementation problems: questions

1 Are data available? What format are they in? Can they be entered into the GIS?

2 Can an appropriate attribute database be created?

3 Is appropriate software available? Does the software impose restrictions on the analysis that can be conducted? If so, can the analysis scheme be revised or the software changed?

4 Will the output produced still meet users' needs? Have the users' needs changed? Has the root definition changed in the light of increased knowledge and awareness of GIS and the problem to be addressed?

5 Are there any risks associated with the implementation of the project? How will these be handled if they occur?

the GIS may have to be abandoned, and the project restarted. This feedback process can be very costly and often explains why many organizations adopt the prototyping approach to project management described earlier in this chapter. Prototyping should prevent a system from being inappropriate, as frequent testing and evaluation should be taking place.

There are three tests that can be used to check whether a GIS application meets the goals set for it at the start of the design process (see Box 12.8). First, all the parties involved in the design and development of the GIS can be asked if they are using the application for the purpose for which it was designed. If they are not, or have even reverted to using the old methods, it is a sure sign that something has gone wrong somewhere. The goals originally used to guide them in helping to identify the scope of the problem may have changed as time and work priorities have altered. Alternatively, users may find the application difficult to use, or without a key feature. In such cases, further training, or adaptations to software, may be all that is required to ensure that the GIS are used effectively.

Second, the GIS output can be checked against reality. This type of test would be appropriate for the avalanche prediction model as the location of avalanches could be predicted before the start of a season and then compared against actual avalanches at the end of the season. For the house-hunting GIS the best check would be to see if home buyers found a house that met all their requirements.

Third, the adaptations and changes that had to be made when moving from the rich picture through the GIS data model to the GIS implementation can be evaluated. These changes and adaptations can be assessed to establish whether they were the result of knowledge deficiencies, poor problem definition or system adaptations required because the software or data would not permit implementation of the model as planned, can be assessed. If the system adaptations have dominated the development of the application, it may be that a technical solution has been provided that bears little resemblance to the reality of the initial problem.

---

## BOX 12.8 Project evaluation: questions

**PRACTICE**

**1** Does the GIS help to address the problem identified at the outset?

**2** Has the project been implemented as planned?

**3** Have there been many system adaptations or modifications?

**4** Are the users of the GIS satisfied with the product?

---

## REFLECTION BOX

- Try to apply the ideas of the waterfall model to a GIS project in your own area of interest. What are the problems of using this approach?
- What are the differences between prototyping and Rapid Application Development? Produce a table summarizing the advantages and disadvantages of each approach.
- Think of a novel application of GIS in your own area of interest. Produce a SWOT analysis to help present your thoughts about the development of this application.

- Make a few notes on how you would evaluate a GIS project of your own. Consider:
  - whether project goals have been met;
  - whether output is valid; and
  - what adaptations were made to original project design during implementation.
- Are there any risks common to all GIS applications? How relevant are they for your GIS project?

## CONCLUSIONS

This chapter has reviewed the stages involved in the design and management of a GIS application. Details of a methodology that will encourage successful design, development and implementation of a GIS have been presented. The methodology described draws heavily on the ideas of both hard and soft systems analysis, as well as ideas from the System Development Life Cycle (SDLC) including the waterfall, prototyping and rapid application development approaches. The ideas and techniques presented in this chapter provide a 'toolbox' from which to select techniques and approaches suitable for a particular project. Different design methods have advantages in different situations depending upon the type, scale and nature of the GIS application being developed.

A key element of developing a successful GIS application is the time and importance attached to problem identification, the design of the GIS data model, the development of an appropriate analysis schema and careful project management. In addition, there is a need for all individuals concerned to be involved in the design and validation of the GIS.

Whilst this chapter has presented a framework for the design and implementation of a GIS project, in practice the design issues need to be addressed concurrently with a whole set of practical matters including:

- data availability;
- data encoding;
- analysis methods available;
- output requirements;
- software availability;
- time available; and
- resources available.

Some of these issues have been touched upon, and in reality these may dictate the stages in the design and development of a project. For example, in the Zdarske Vrchy case study a lack of remotely sensed data for a part of the study area meant that this area had to be excluded from some of the analysis, despite its being included in the project design. Thus, results of analysis were unavailable for this region. However, good project design and management mean that less will be left to chance and a GIS application will be more likely to meet the needs of its users. Box 12.9 offers a final checklist to facilitate implementation of the ideas in this chapter. This checklist, together with the questions from boxes earlier in the chapter, provides a useful framework for project management.

## BOX 12.9 Checklist of details to be included in a GIS project log

PRACTICE

1 GIS project title
2 Aim and scope of the project
  2.1 Root definition
  2.2 Rich picture
3 Geographical area concerned
  3.1 Dimensions of the area concerned
  3.2 Resolution required
4 Numbers of users of the GIS
  4.1 Who are the users?
  4.2 What kind of knowledge do they have?
5 List of digital thematic maps required
  5.1 Feature
  5.2 Entity type (point, line or area)
  5.3 Raster or vector?
  5.4 Accuracy
  5.5 Source

6 List of attribute data requirements
  6.1 Feature
  6.2 Description
  6.3 Character or numeric?
  6.4 Accuracy
  6.5 Source
7 Database requirements
  7.1 Database queries required
  7.2 Frequency of updating
8 Analysis requirements
  8.1 List of analysis functions required
  8.2 Simple language flowchart of the GIS analysis model
  8.3 System specific flowchart of the GIS analysis model
9 Description or picture of interfaces required for user interaction

---

**BOX 12.9**

10 Choice of GIS software
   10.1 Principal requirements
   10.2 PC/workstation/network and number of installations
   10.3 Availability over the Internet
11 Choice of hardware
   11.1 PCs/workstations
   11.2 Networking
   11.3 Peripherals
12 GANTT chart

13 Resources and costs
   13.1 Fixed investments (PCs, workstations, other hardware, software, data)
   13.2 Continuous costs (manpower, GIS specialists, system manager, project manager, maintenance, data updating)
14 Risks
   14.1 Risk-management plan
   14.2 Risk register

---

## REVISION QUESTIONS

■ What is a rich picture and how is it used to assist GIS project design?

■ What are the differences between conceptual and physical data models?

■ What is cartographic modelling? How can cartographic modelling be used to develop a GIS analysis scheme?

■ Review the main problems that might occur during the implementation of a GIS application.

■ Compare and contrast the waterfall, prototyping and RAD approaches to managing a GIS project.

■ Discuss the role of flowcharts in the management and design of GIS projects.

■ How can a GIS application be evaluated?

## FURTHER STUDY – ACTIVITIES

■ Produce a flowchart summarizing a GIS analysis that you have carried out.

■ Write some cartographic modelling equations for a GIS application or project with which you are familiar. If you can, use a GIS to which you have access to try out the equations.

■ Research the implementation of GIS in an organization of your choice. What project management approaches were adopted? What problems (data, technical and human and organizational) were faced during implementation?

■ Produce a GANTT or PERT chart for a project of your own. Keep it alongside you as you work through the project and assess the usefulness of the approach.

■ Produce a Risk Management Plan and a Risk Register for a project of your own. Keep it up to date as your project progresses.

## FURTHER STUDY – READING

There are several sources of further reading which provide details of designing and managing the development of a GIS application. DeMers (2009) offers a very practical and straightforward overview of GIS technology and how to use it to deliver a project. Reeve and Petch (1998) put an emphasis on soft systems and prototyping methodologies. Chrisman (2002) discusses several ideas for assessing data quality and defining the analysis requirements of a GIS project. Although a little dated perhaps, one of the most comprehensive overviews of GIS project management is provided by Huxhold (1991) in his chapter titled 'The model urban GIS project'. This chapter provides a good overview of the practicalities of developing a large-scale GIS project, and also considers how organizational factors influence the process. More recently, the chapter entitled 'Managing a GIS' in Longley *et al.* (2010) offers much good advice.

From a technical perspective, the tutorial material accompanying specific GIS software may provide a good starting point and many of the tips presented for one system often have generic application.

Outside the GIS literature there is an extensive literature on project design and management. Checkland and Poulter (2006) provide an excellent introduction to the Soft Systems approach and Checkland and Howell (1998) introduce the field of information systems management. Curry *et al.* (2005) provide an overview of the development and management of IT systems from a business perspective. Other texts on IT project management will also cover the range of techniques and tools introduced in this chapter.

Checkland P B, Howell S (1998) *Information, Systems and Information Systems: Making Sense of the Field.* Wiley, London

Checkland P B, Poulter J (2006) *Learning for Action: A Short Definitive Account of Soft Systems Methodology and its use for Practitioners, Teachers and Students,* Wiley, London

Chrisman N R (2002) *Exploring Geographic Information Systems.* 2nd edn. Wiley, New York

Curry A, Flett, P, Hollingsworth, I (2005) *Managing Information and Systems: The Business Perspective.* Routledge, Oxford

DeMers M N (2009) *GIS for Dummies.* Wiley, New York

Huxhold W E (1991) *An Introduction to Urban Geographic Information Systems.* Oxford University Press, New York

Longley P A, Goodchild M F, Maguire D J, Rhind D W (2010) *Geographical Information Systems and Science.* 3rd edn. Wiley, Chichester

Reeve D E, Petch J R (1998) *GIS Organizations and People: a Socio-technical Approach.* Taylor and Francis, London

## WEB LINKS

### Cartographic modelling in practice:

- PCRaster: pcraster.geog.uu.nl/ documentation/pcrman/c1194.htm

- Cartographic modelling lab: cml.upenn.edu/

### Project management tools:

- GANTT charts: www.ganttchart.com/

- Smartdraw: www.smartdraw.com/specials/ projectchart.asp?id=15063

- SWOT analysis: www.quickmba.com/strategy/ swot/

- RISK Management Plans: www.wikihow.com/ Develop-a-Risk-Management-Plan

Visit our website **www.pearsoned.co.uk/heywood** for further web links, together with self-assessment questions, activities, data sets for practice opportunities and other interactive resources.

# CHAPTER 13

# The future of GIS

## Learning outcomes

By the end of this chapter you should be able to:

■ Appreciate the difficulties associated with predicting the future for GIS

■ Explain how the relationship between GIS and other technologies might impact on future developments

■ Speculate on what GIS might be in 10, 20 or even 50 years' time

## ■ INTRODUCTION

Making predictions for any field involving rapidly developing and changing technology is a difficult task, as the quotations provided in Box 13.1 illustrate. In spite of this, there are many authors who have been prepared to predict the future. Sometimes they look forward as an intellectual exercise, using their knowledge and imagination to try to anticipate the world in a few years' or decades' time. They also look forward to assist businesses that depend on staying one step ahead and meeting the next, perhaps as yet unidentified, customer need. Managers and decision makers contemplating spending thousands or even millions of pounds on a GIS also need to look to the future. They need to 'future proof' a new system, and make sure that it will last, or at least be upgradeable when the next set of technological developments comes along. Some of the well-known techniques for justifying investment in new technology, such as cost–benefit analysis, involve an element of crystal ball gazing. Costs and benefits are predicted to justify a project, despite the fact that the benefits may never actually be achieved. Strategic planning exercises, such as writing development or research plans for a business or company, also require a look into the future.

Imagine that you are writing a report that assesses the current state and future potential of the motor car. What do you need to consider? Your adventures in previous chapters have helped to establish many of the technical and practical issues of car ownership and use. Now you must consider the future. First you may examine the predictions of others for the next 20 years. If these are not available you could make your own predictions. To do this you must understand the nature of the current industry, and identify strengths and weaknesses. For example, perhaps you might be concerned cars use too much fuel, create too much pollution, or that the roads are too congested for efficient car use. Having identified and evaluated current problems, you might suggest areas where more research is necessary and begin to make predictions for the future.

This chapter revisits the predictions that were made for GIS in the third edition of this book to show just how much has happened in recent years (see Box 13.2). Some of the issues for GIS development in the twenty-first century are then explored, and finally, we set you the challenge of predicting what might happen next and present a few of our own ideas. The chapter concludes with an 'Epilogue' which attempts to highlight the role of GIS in our dynamic world.

---

## BOX 13.1 Some famously incorrect predictions for technological innovations

**PRACTICE**

Some famously incorrect predictions for technological innovations

- 'One day there will be a telephone in every major city in the USA.' – Alexander Graham Bell, c. 1880
- 'The horse is here to stay, but the automobile is only a novelty.' – President of Michigan Savings Bank, 1903, advising Henry Ford's lawyer not to invest in the Ford Motor Company.
- 'I think there is a world market for as many as 5 computers.' – Thomas Watson, head of IBM, 1943.

- 'There is no reason anyone would want a computer in their home.' – Ken Olsen, founder of Digital Equipment Corporation, 1977.

*Source*: www.audiencedialogue.org/predict.html

And Bill Gates, the founder of Microsoft, almost missed the opportunity provided by Internet. Initially he considered it to be a passing fad.

*Source*: www.internet-story.com/microsoft.htm

# BOX 13.2 Revisiting predictions for the twenty-first century

PRACTICE

In 2006 we presented our predictions for GIS which addressed some issues which had been around for a while, some which were not yet solved, and some arising from contemporary technical and market developments. These are revisited below with comments on recent developments:

**1** *New and improved data models will be developed, especially for true 3D modelling, global data sets and temporal modelling.* Global energy companies such as BP and Shell now use 3D models on a regular basis to model what is happening in oil reservoirs around the world (see Manning *et al.*, 2009). Other developments include Google's pioneering work on Google Earth as a global database and work to monitor the temporal flows of traffic to ease traffic congestion in some of the world's major cities (Lenhart *et al.*, 2008).

**2** *Developments in related fields will influence developments in GIS (for example, artificial intelligence, neural computing, geocomputation, spatial statistics, multimedia, virtual reality, the Internet and office packages).* The last few years have seen a proliferation of GIS applications as part of the growth in mobile computing. Apple and Google have been the pioneers – allowing developers to embed mapping and spatial analysis routines into a wide range of applications from finding and navigating to restaurants, to keeping track of the location of contacts in your address book. In addition, intelligent, pattern-matching algorithms are now being used by a growing number of social networking sites to connect you to people with similar interests who live near to where you live and work.

**3** *There will be a continued repackaging of products to meet sector-specific needs, and, with continued dominance of the PC, further expansion of the desktop market and a reduction in the workstation/mainframe market.* This has gone further than anticipated with Internet tools and mobile GIS applications becoming mainstream. Most leading GIS software providers now offer Internet and mobile versions of their products and, in some cases, have teamed up with data providers to deliver simple applications for the customer. Examples include the recent mobile GIS application developed by the British Geological Survey which allows smart-phone users to find out about the rocks beneath their feet (British Geological Survey, 2010).

**4** *GIS tools will be downloadable as Java applets for specific tasks, negating the need to buy a complete GIS system.* The launch of the iPhone by Apple in 2007, and the subsequent emergence and growth of the 'downloadable app' has made this a reality for simple GIS tasks. Esri's Arc/GIS for the iPhone and iPad is an example of this growing trend.

**5** *There will be further additions to GIS functionality: for example, much-needed exploratory data analysis and error analysis.* Research in this area continues to be a major focus with a wide range of new and innovative techniques being developed to analyse, visualize and interrogate spatial data. For example, Nakaya and Yano (2010) have used kernel-density estimation techniques and scan statistics to create a 'Space-Time Cube' to visualize the proliferation of snatch and run crimes in Kyoto City, Japan.

**6** *Data collection and exchange will continue to be influenced by GPS development, coupled with future developments in satellite communications and data collection methods.* Once again the integration of GPS and mobile phone technology has driven forward this area of development. In 2007 it would have seemed futuristic to expect that, less than three years later, Internet photo-sharing sites like Flickr (www.flickr.com) would allow you to upload and map the location of photographs taken on a GPS enabled mobile phone.

**7** *Peripherals such as high-quality printers will become widely available. In fact, the need for hard copy will be reduced, as new forms of output, such as animation, fly-through and 3D models, can be easily exchanged over communications networks and more effectively presented in digital format.* A standard home colour printer is now able to produce output of a quality suitable for the majority of GIS applications. Output as fly-throughs can be produced by packages like Google Earth and there are screen-capture and animation packages available which allow these to be captured, annotated and shared via sites like YouTube (www.youtube.com).

**8** *Data storage will cease to be an issue, although the development of real-time systems will challenge even the largest storage devices.* The availability of affordable terabyte drives and data storage in the 'cloud' is keeping pace with current demands. However, as the resolution at which data are captured increases further, it is unlikely that there will

**BOX 13.2**

be a reduction in the demand for even more storage capacity.

**9** *The ideas of interoperability, allowing the use of the most appropriate software tool for the solution of a particular problem, will extend to influence other areas of GIS, including data and education.* Whilst free Internet data services (such as Google Maps) have pioneered public access to GIS, the National Mapping agencies have been much slower to open up access to their data. North America has always led the way in promoting access to spatial information, and this remains the case. However, there are now signs that this trend may be spreading For example, in 2010 the UK's Ordnance Survey opened up free access to many of its data sets. The Internet now provides opportunities for sharing of GIS teaching resources with numerous online texts, courses and other resources available.

**10** *Location-Based Services will continue to develop with new generations of GPS-equipped mobile phones and PDAs, enabling large amounts of location specific information to be at our fingertips 24/7 anywhere in the world.* As discussed above, it is perhaps this area more than any other where the progress in recent years has been most striking. The mapping and navigational capabilities of mobile devices have created new opportunities for individuals and organizations to use spatial data.

**11** *As GPS chips become smaller and cheaper, everything of value will come with its own GPS chip,* *resulting in the concept of 'lost' becoming a thing of the past.* We are not quite there yet, however. GPS chips and RFID chips (Radio Frequency IDentification) capable of tracking the object they are embedded within are now found in all sorts of items including computers, phones, cameras, trains, planes and cars, with new items being geo-tagged daily.

**12** *New education and training products will be developed to address skill shortages for both potential users and the research and development needs of the GIS industry.* The basic core of GIS education and training has remained relatively static in recent years. However, GIS is now taught across a wider range of disciplines. In the USA, there has been further growth in GIS teaching in schools, and a wide range of supporting resources are now available.

**13** *Crowd sourcing and NeoGeography will continue to grow and develop into an important data source, providing a fundamental information window on people, communities and social behaviour.* The principal data sources of maps, remote sensing and surveys cannot provide detailed information about people, whether as individuals or in groups. New social geographies facilitated by spatially referenced tweets, blogs, twitters and mash-ups allow hitherto impossible applications and analyses about everyday social behaviour and how people interact with their world through enhanced ICT.

## ■ WHERE NEXT FOR GIS IN THE TWENTY-FIRST CENTURY?

What is GIS? Is it a computer system, a method of handling geographic data for spatial problem solving or a way of thinking? Definitions can change rapidly and they vary according to who is giving them. The definitions of the early 1990s were functional in nature. They focused on what a GIS could do and what it would produce. Now definitions tend to address issues like information strategies, service provision, and adding value to data. The range of views of GIS has increased in recent years, in response to the expansion of the market for GIS and the applications to which it has been applied. Is there a danger, as with the term 'management information systems', that 'GIS' could become a meaningless term, or one which means different things to different people? In

the case of GIS the answer is 'possibly not', because it is the G (for the geographical or spatial data) which is arguably the most important component. The geographical perspective on information distinguishes GIS from other types of information systems.

Goodchild (1992) considered the unique features of spatial data which make special data handling tools necessary, and his ideas are still relevant. First, geographical data has to be accessed using a spatial reference. These make up a continuous referen-cing system in two dimensions, so that any location can be visited or studied, and attribute data can be retrieved for any location. Second, locations close together are likely to have similar characteristics. This is Tobler's first law of geography, and justifies, for example, the assumption that all the people with the same postcode have similar income levels, professional status and family characteristics. Third, geographical data are distributed continuously, in three dimensions, all over the surface of the Earth. To date, most analysis has been performed on a projected set of these data. There are still few possibilities for the analysis of geographical data in their correct spherical configuration. However, this is changing as more use is made of GPS-enabled data loggers which capture data and transmit these data, and their position, in real time.

As long as we continue to have a requirement for the analysis and management of geographical data, tools that can cope with the special characteristics of these data will be essential. In fact, Goodchild (1995) suggests that this is a 'philosophy' on which GIS is based. The range of geographical data, and the fact that they can be described, structured and handled in similar ways, gives a justification for GIS.

Goodchild (1995) also looked to the future of GIS and identified a number of limitations which still hold true today. These are that GIS:

- are two-dimensional with limited abilities to handle the third dimension;

- are static, with limited abilities to cope with temporal data;

- are good at capturing the physical positions of objects, their attributes and their spatial relationships, but have very limited capabilities for representing other forms of interaction between objects;

- offer a diverse and confusing set of data models; and

- are still dominated by the idea of a map, or the view of a spatial database as a collection of digital maps. The development of visualization tools such as Google Earth and Google Street View has changed this slightly, but the map still dominates.

These observations prompt questions that we can consider when looking at the future of GIS in the twenty-first century. For example, can GIS be truly three-dimensional? Can real-time GIS become a reality? Can computers be used to accurately model the interactions between features? Is there a better way of modelling geographical data than the current raster/vector/object-oriented paradigms? Are there other ways of representing space that do not rely on maps? Will GIS end up as part of a much wider database approach to managing our

ever-more complex world? These questions give us topics to address to consider the future of GIS. The following section considers four aspects of the future – possible applications, the future of the technology, the future of geographic data and the future for users and organizations.

## GIS applications

Maguire (1991), perhaps rather rashly, suggested that by the end of the twentieth century everyone in the developed world would use GIS every day for routine operations. He was of course right (although it took a little longer than he anticipated) and GIS is now well on the way to infiltrating other areas of IT, such as spreadsheet, database and graphics packages; maps and mapping functions are available over the Internet to anyone who has a networked PC; and GIS-type functionality is appearing in everyday use in other types of applications such as in-car navigation systems (Figure 13.1). One of the most radical developments in recent years has been the addition of mapping tools to Web-based products such as Google Documents, which allow users to plot their own spatial data over a map background. Virtual globes, such as NASA's World Wind and Google Earth (Figure 13.2), now permit everyday users to explore any location on the Earth's surface using intuitive interactive interfaces that combine maps, terrain models, satellite imagery and attribute information with powerful search engines.

There have been other exciting developments that include elements of GIS and involve the handling of geographic information. Touch-screen interfaces (such as that of the Apple iPad) offer

**Figure 13.1** In-car navigation system
*Source*: Courtesy of Gordon Sell Public Relations, www.gspr.com with permission

**Figure 13.2** Google Earth
*Source:* Google UK.

direct and intuitive ways of exploring spatial data. Many other mobile devices now have embedded positioning technology and allow you to view your geographical location when on the move. For example, you can obtain your real-time location overlaid on a satellite image or map whilst on a train journey. Additional data from the train operator can provide real-time updates on your progress against the published timetable. Whilst still on the train you can also check for hotels, places of interest and even contacts in your address book who are close to your current location or destination. You could take geo-tagged photographs which can be immediately uploaded to an online photo-sharing site or social network or used to create a 3D model (for example, with Microsoft's Photosynth – www.photosynth.net) Other people may also be able to track your progress (if appropriate permission is allowed). And should you accidently leave your mobile device behind on the train, a GPS receiver and mobile broadband connection will allow you to keep track of its location. If necessary a tracking code can be passed on to the police to allow them to track a stolen device.

Another example of the integration of GIS and GPS technology is the development of GPS audio guides for major cities and other tourist destinations. One example, the Culloden Battlefield Guide developed for visitors to this historic site in the North of Scotland, is described in Box 13.3.

---

## BOX 13.3 The Culloden Battlefield Guide

**CASE STUDY**

### Steve Bone

The course of British, European and world history was changed at Culloden Moor, Scotland on 16 April 1746 when the Jacobite army was finally defeated in its attempt to reclaim the throne of Britain from the Hanoverians for a Stuart king. The Culloden Battlefield, near Inverness in the Highlands of Scotland, is now a major visitor attraction managed by the National Trust for Scotland (NTS). In December 2007 the NTS opened a new visitor centre at the Battlefield and commissioned an outdoor battlefield audio guide to complement the technology-led approach to interpretation taken within the centre. This case study describes the innovative guide, which was one of the world's first GPS triggered electronic guides in use at a tourism and heritage site.

The Culloden Battlefield Guide (CBG) includes specially created audio, video and supporting illustrations. This multimedia content is triggered automatically as visitors explore Culloden Moor, at their own pace, allowing them to understand and appreciate what actually happened on 16 April, 1746. As they soak up the atmosphere of the site, they receive unobtrusive, location specific multimedia content (Figure 13.3). The guide was initially provided in English and Gaelic and is now available in eight other languages from Spanish to Japanese.

A key objective of the Culloden project was that the battlefield site should be restored, as close as possible, to that seen by the forces on that fateful day in April 1746. Satellite technology was used, as more traditional techniques, such as interpretative panels or markers, were not felt to be in keeping with the sanctity of the Moor as a war grave. As well as the multimedia content that is automatically triggered as the visitor explores the site, additional information is available on a menu that changes depending on the visitor's position, allowing them to gain a more complete understanding of the historic events (Figure 13.4).

**BOX 13.3**

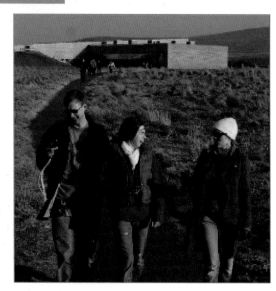

**Figure 13.3** Using the CBG at the Culloden Battlefield

**Figure 13.4** The CBG menu

This project was innovative technologically and in terms of the content provided to visitors. For example:

- On being given the CBG, the audio instruction suggests visitors put it in their pocket and forget about it – a warning 'bleep' will sound shortly before the audio content triggers. If the core audio is supported by visual material, an audio prompt is given for the user to look at the screen. In this way, the visitor needs to interact with the device only when they choose.

- Satellite technology automatically triggers information relevant to the visitor's position as they explore the site.
- The default screen displays the site map – a moving icon shows the visitor's real-time location.
- The information menu changes depending on the visitor's position on the battlefield – if they wish, visitors can use the screen options, such as a 'Find out more' and 'Fact File', to access further interpretation material.
- Audio accounts from genuine contemporary participants are vividly brought to life, Combined with careful use of sound effects, these create an immersive and emotive tour of the site. The experience of hearing their accounts as you stand where they stood is truly memorable, and often poignant.
- Complementing the character accounts, the latest historical and archaeological evidence is also incorporated, explaining how the battle was fought and how the landscape influenced its outcome. Artefacts discovered in the ground – such as fragments of ballistics and touching personal items – are shown on screen, triggered at the very location they were discovered. Visitors frequently return to the displays in the exhibition to look with fresh eyes at the real items, which have taken on a new significance through the CBG interpretation.

Overall the CBG offers engaging, user-driven, multi-layered interpretation of the battlefield without physical intrusion on to this sensitive site.

**FURTHER INFORMATION**

Further information about the Culloden Battlefield: www.nts.org.uk/culloden
Other GPS audio tour guides:
South Africa: springwise.com/tourism_travel/greatguide/
Yellowstone National Park: georoameryellowstone.com/cms/
New York:
itunes.apple.com/us/app/5th-avenue-new-york-gps-audio/id367293064?mt=8

*Sources*: Steve Bone, Consultant; www.nts.org.uk/culloden

The examples above show how GIS has become mainstream, but what about the suggestion by Gelernter (1992) that one day you would be able to look into your computer screen and see reality: a 'mirror world'. Automated web cameras and live data feeds from scientific instruments scattered around the world help us get part way towards this vision, as do projects such as virtual field trips. However, perhaps the closest we have come is in the simulation engines that power computer games. Some current online multiplayer games could be regarded as complex GIS, as they allow gamers to interact with huge volumes of spatial data, albeit for virtual rather than real worlds. The application of gaming technology to 'real-world' scenarios is less common, although some augmented reality games and applications are emerging, such as the augmented reality browser Layar (www.layar.com) which allows the overlay of real-time digital information on images of the real world from your camera or mobile phone. As an avatar in a virtual world it is also possible to interact with virtual objects, and replicas of those from the real world: for example, David Rumsey's map library, which can be visited in Second Life (www.davidrumsey.com/).

GIS are also being used in new ways. Beyond the original decision-support and problem-solving role suggested by early definitions and applications, GIS are now being used as tools for generating ideas, and for improving the participation of all stakeholders in the decision-making process. So, from a situation where a planning authority would present a set of authoritative, computer-produced maps to gain support for its actions, now the local community can be involved in the setting of important criteria, the modelling of alternative scenarios and in decision making that will affect them (Carver *et al.*, 2001, Stevens *et al.*, 2007). This is beginning to be influenced by developments in social computing such as blogs and Twitter. Social and personal map mash-ups facilitated through free Web-mapping tools such as Google's API are beginning to put the power of the map and GIS into the hands of an increasingly technologically savvy public with potentially far-reaching effects. For example, applications could be developed for calculating a taxi fare from A to B, finding out the location of police speed traps, mapping disease incidence and plotting the potential effects of climate-induced, sea-level rises.

In the face of these developments will GIS continue to exist? What will be the implications of this increased access to GIS technology? What are the implications for society, for organizations and for individuals, and for the field of GIS as a whole? GIS is an integrative discipline involving those who produce the computer tools, collect and prepare the data, and interpret the output. Computer scientists, surveyors, photogrammetrists, remote sensors, spatial analysts, database technicians, cartographers, planners, utility managers and decision makers are all involved, and new applications for the technology will be found in all these disciplines, and in areas involving more than one of these disciplines.

## GIS technology – how will we handle spatial data?

In 1995 Barr (1995a) suggested that GIS as software products would disappear from sight. He has, in part, been right. In the marketplace today GIS is sold, not so much as a versatile 'toolkit' as it was in the past, but as straightforward applications for solving single problems. There are, for instance, routing packages and geodemographic data-analysis packages. There are also an increasing number of custom-built applications to address specific problems such as traffic management, see Figure 13.5.

GIS as a recognizable and separate category of software products is starting to fade from sight and is being replaced by a series of applets that are downloaded on demand for a particular task (as, for example, the growing number of map-based and GPS-enabled applications appearing as Apple or Google 'Apps'). GIS is also gradually being subsumed into a much larger suite of general-purpose, office productivity software.

The proliferation of GIS applications is driving a demand for geographical data. Indeed, data are one commodity we are not going to be short of in coming years; in fact we are more likely to be overwhelmed by data, as suggested in **Chapter 4**. So the need for tools to handle geographical data effectively and reliably is growing ever greater.

Whatever happens, it is likely that everyone will use more spatial data and GIS-type tools (whether they are aware of them or not) as they draw on Geographical Information Services, but that specialists in spatial analysis will still be required for high-end applications such as modelling and forecasting.

Future users of GIS will expect systems with tools for multimedia for real-time modelling of complex phenomena and for better representation of the real world. At present GIS can provide, for example, a range of options for the siting of a new ski piste, based on data provided and criteria chosen and applied by the end-user. Information is output: for example, the

**Figure 13.5** Example real-time traffic map, Houston, Texas
*Source*: From the Houston Transtar website, http://traffic.houstontranstar.org/layers/layers/aspx, with permission of the Houston Transtar Consortium

routes for possible new ski pistes. However, human interpretation is still required to assess the appropriateness of the options presented, to assess any errors, and to evaluate whether there are additional criteria that need to be applied to the problem. In the future, GIS applications may be called upon to provide answers too, by drawing on knowledge collated over many years and stored within GIS as 'spatial intelligence.' Whilst this may sound like an excellent idea, such developments must always be placed back into the cultural context within which they will operate. Decision making has always been the focal application of GIS. However, it could be argued that decisions are never made solely on the basis of GIS output, and indeed should not be. In many businesses GIS results are used to confirm or authenticate decisions that have been made in other ways.

Finally, a growing trend in the technology sphere has been that of cooperation between vendors. A number of major organizations have, for example, come together to produce common standards to allow the development of open systems that will support real-time access to geographic data across diverse systems and the use of the most appropriate software tool for a particular task. Open systems can also be the product of collaboration between individual enthusiasts, and further development of open systems is likely to be a major theme in coming years.

## Emphasizing the information

The major data entry projects of the 1990s have been completed and the national utilities and national mapping agencies have their data digitized and ready to use, and maintenance programmes in place to keep their data up to date. A shift is taking place away from the development of tools that help with the input, correction and structuring of data to those that add value to data and provide information to improve our knowledge and understanding of the world.

None of the expected future applications and technical developments will be possible without access to appropriate data. Data access issues have been a subject of debate for nearly thirty years. In the UK the Chorley Report (Department of the Environment, 1987) highlighted barriers to the development of GIS, which focused in the main on data issues such as referencing, standards and confidentiality. Many of these issues remained unresolved during the twentieth century (Heywood, 1997), and despite major strides forward, including access to data over the Internet, the development of commercial data providers and the development of standards for address referencing, data formats and data exchange, there is still plenty of work to be done. In Europe many of the challenges, such as the standardization of the European Address system (Barr, 1995c) remain. Barr noted that, for many GIS applications, address is the most important spatial reference. Individuals know their home address whereas they are unlikely to know the grid reference of their home, or even the name of their electoral district. In the USA the Census Bureau keeps the TIGER addressing system up to date, including referencing, assisted by appropriate legislation. However, in Europe no such system exists. Standardization of address formats at an international level has yet to be considered. In reality, the need for consistency in address referencing may decline in significance as advances in GPS, mobile devices and geo-tagging mean that the location of any object can be tracked.

Using the huge amounts of data we are now producing presents some exciting opportunities and challenges for the future. Real-time data, such as those from CCTV pictures, car monitoring, store cards and other automatic data-capture techniques, provide huge resources of data to exploit to improve our understanding of the world. Although often collected for other purposes, much of this data includes a spatial reference. For example, if you speed past a police camera, the number plate identifying your car may be captured and the location of your misdemeanour recorded. If you purchase an item using a store loyalty card, information about your purchases will be added to a database already containing your address and location details. If you telephone a company for information on its services you will be asked for your postcode so that the company database can be updated with your location. Thus, the data mountain continues to grow, and there is a need for new techniques to manage, mine and analyse this resource. Data issues are still foremost amongst problems and challenges for the future.

## People and GIS

It is unlikely that the human and organizational issues associated with GIS will ever be completely overcome. Consider the training and updating of users. Even now, if you disconnect yourself from the Internet (for example during a holiday), you will have to catch up on your return with new software releases and upgrades, Internet updates, and tweets, blogs and email from members of your social and professional networks. The pace of change and development is exciting, but also frightening at an individual level. How can you remain an expert in your field? Lifelong learning, retraining and professional development are necessary for everyone in a society where change is the norm rather than the exception. Tools and aids to learning and professional updating are required. Software agents, which search the Internet on topics in which you are interested, are now commonplace, but other tools are needed. Individuals in the developed world are already adapting to a world where jobs for life are rare, where working from home is becoming more common and where major career changes are frequent. Organizations too are adapting and changing.

Two areas of skill and knowledge shortage in GIS were identified by Marble (1997). These were in the basic understanding of geographical phenomena (spatial thinking) and high-level technical skills. Skill shortages in these areas are critical. A lack of understanding of spatial issues and topics such as geodesy and cartography will prevent GIS from being used effectively. Users will be unable to recognize the potential of and the possible applications of GIS. Some would argue that this danger still exists today as mobile and Internet GIS applications put GIS into the hands of a wider audience, many of whom have limited knowledge about how to interpret and apply geographical information. However, this can also been seen in a positive light, since at the same time more people are being exposed to a geographical perspective. It is important that we address the need for widespread geographical understanding.

At the other end of the skill spectrum, a lack of highly skilled technical personnel with skills in programming, software research and product development still has the potential to limit the development of the tools and programs used to handle geographical data. However, as the technology becomes easier to use, so the focus can move from

## REFLECTION BOX

- Alan Kay, an American computer scientist, is credited with the idea that 'technology is anything that was invented after you were born'. This statement implies that, if something was invented before you were born, then you tend to take it for granted. For the authors of this book, calculators, video recorders, the PC and the mobile phone are all 'technologies'. But for younger readers, some of these technologies will be 'taken for granted'.

- Think about GIS in the context of this idea. Do you think the users of the future, those who were born after GIS was invented, will take GIS for granted?
- Geographical information services are predicted to be an area of growth and development over the next few years. What services would you like to have available on your mobile phone, PDA, MP3 player or other portable devices?

---

the development of software tools to adding value to geographic information. For users to 'add value' effectively they must appreciate how to work with, manipulate and analyse spatial information.

## ■ EXPLORING THE FUTURE FOR GIS

When we began revising this chapter we considered asking people who are working in leading research institutes and technology companies to gaze into their crystal balls and tell us what they could see. They, better than anyone, should know what the future holds for GIS. However, we have decided to be bolder and ask everyone who reads this book to share their thoughts on what the future holds for GIS. We want to know what you think will happen next. How will GIS technology develop? What new technologies will emerge? What will they look like? What will they do? How will they be different? Who will use them and what will they use them for? Where will technologies converge, as they have with the GPS and the mobile phone? What challenges will the geographical information scientists of the future face? Who will provide the data needed to fuel these new applications? And what global challenges will GIS help us address?

Box 13.4 explains how we would like to capture your ideas about the future of GIS, using a blog available on the book's companion website. To help you get started, we provide below three of our own 'tales from the future' and from time to time we will be asking GIS practitioners to share their own predictions and will add these to the website. We hope that as well as adding your own ideas, you will help evolve and shape the predictions made by others.

## BOX 13.4 Contributing to our vision of the future for GIS: how to create your story

**PRACTICE**

Please visit the companion website for this book (**www.pearsoned.co.uk/heywood**) where you will find a blog in which we hope to collect your ideas about the future of GIS. You are welcome to contribute to this blog individually, or as part of a group or class project, and from time to time we hope that experts from the field of GIS will also add their ideas about the future.

Your contribution can be in whatever form you choose – as a picture, story, cartoon, podcast, short video or any other appropriate media.

To help you start planning your contribution let's revisit the car analogy we have used throughout the book. Imagine that you have been asked to write a report that assesses the future potential of the motor car. To do this you must understand the nature of the current industry, and identify strengths and weaknesses. For example, you might be concerned that cars use too much fuel, create too much pollution, or that the roads are too congested. Having identified and evaluated current problems, you might be able to suggest areas where more research is necessary.

**BOX 13.4**

Another way to get started might be to look at how cars are used today and think about how they could be improved in the future, for example by using different fuel or accommodating more or fewer people. Try also to think differently – for example, what would happen if cars could fly, or if they merged with other kinds of transport, such as trains? Don't be limited by your own imagination: feel free to draw on books, movies, cartoons, computer games or whatever else helps your creativity.

We look forward to seeing your ideas.

The three stories which follow present some of our ideas about what the future may hold for GIS. The first story explores the potential for 'true' 3D modelling; the second considers the implications of ubiquitous geo-tagging; and last explores possible developments in geo-marketing. To create our stories, which speculate on just three possible areas of future development, we have drawn inspiration from a number of different sources, from real research projects to movies.

## Natural Disasters

'There's got to be a solution. I mean, if this rain keeps up I don't think the dam will hold much longer. Let's take another look.' I swivel round in my chair.

Jill and Ed sit on the other side of the table, looking at me. I tap the 'launch' button. The desk in front of us lights up and within seconds we are staring at a miniature version of the dam and the surrounding landscape. No matter how many times I work with these miniature replicas of the real world, I am always amazed by the quality of the three dimensional rendering. These models are the real deal. They are nothing at all like the holograms which they replaced several years ago. You can see every detail, from the curvature in the dam wall to the forested slopes in the north, and the river to the east. What's more, you can peel back the layers of the landscape, like the skin of an onion, to reveal what lies below – the soil, water table and geology. By switching to simulation mode, you can play with the model, try out different scenarios, and see what would happen if things were to change. That's what Jill is doing now – she's increasing the rate of the predicted rainfall to four times the normal amount.

'OK – let's re-check the current level of the water table and see just how much more this baby can hold.' Jill adjusts the slider bars on the control panel, and the top layer of the model lifts away from the underlying structure, revealing just how much of the soil remains unsaturated. The floating slivers of landscape look very thin indeed.

'We've been here before,' says Ed. 'We know there is not enough capacity left to accommodate the expected rainfall. The soil will be fully saturated in a few more days, and then everything is going to start moving – vegetation, soil, rocks. The whole place will be one big mud bath. And when that lot hits the reservoir it is going to displace so much water – the dam will never hold.'

'Do you want me to run the scenario again?' asks Jill.

I don't think we need to put ourselves through that again, I reply.

When we'd run the scenario the day before, the three of us had watched in horror as the miniature world on the desktop ripped itself apart. Mud slides cascaded down the valley sides taking everything with them, until they reached the reservoir, where they displaced a wall of water. This surged over the dam wall, arching like a renegade wave into the valley below. The dam had then collapsed, releasing an even greater torrent of water into the valley. The closest towns and villages weren't shown on the model, but it didn't need much imagination to guess what the impact on them would be. The first wall of water would do most of the damage, and anything left would be obliterated by the wave that followed.

'So we know that the soil can't hold the expected rainfall – and that if the soil cover becomes fully saturated, it will only be a couple of days before the whole valley starts to slide. So, what are the options?' I ask, thinking out loud.

'Other than switch off the rain...,' says Ed.

'If only we could,' says Jill. 'But there's only one solution. We have to get those villages evacuated.'

'But to do that, we need to convince the politicians that this is really going to happen. Jill, have you got the movie of the simulation?' I ask.

'…And the data to back it up?' adds Ed. 'We're going to need it.'

'Sure.'

Thirty minutes later the three of us are sitting in front of the office video wall; a floor to ceiling communications portal which allows us to share data, images and ideas with teams anywhere in the world.

'OK Sam, what have you got for us? Good news I hope,' says Bill Williams, in his usual upbeat manner. Bill is the head of our regional emergency response team and based at HQ. His team is charged with managing everything from the impact of natural disasters through to plane crashes.

'Jill, let's show Bill the movie of the 3D simulation.' I turn back to the screen. 'Bill, we're going to switch over to 3D projection mode now, so you might want to put on the glasses.'

Ten minutes later, with the show over, Bill isn't looking quite as relaxed as before.

'OK, you win', he says in a quiet voice. 'You've got your evacuation.'

## Daddy, What Does Lost Mean?

I sit in the basement looking at the screen of our main home computer. A floor plan of the upstairs of the house flashes into view. I can clearly see that the offending item, my son's favourite space hero toy, is in the bathroom. Its location is clearly visible as a flashing red icon.

'In the laundry basket,' says the computer in a quaint, rather pompous, English accent.

Just at that moment my son appears.

'I bet you thought you'd lost that,' I say.

He looks puzzled.

'Daddy, what does "lost" mean?' he asks, staring at the screen.

You see, my son has been brought up in a world where nothing is ever lost any more. Things can be misplaced for a short while maybe, but only until you can look up their whereabouts on the Internet. It all started about ten years ago, when scientists created a locational nano chip, small enough to put inside absolutely anything. The new geo-chips, as they became known, can be used to track everything. It doesn't matter whether the object is: indoors, outside, beneath the ground or under water. Wherever the object is, it can be located, and if it moves, it can be tracked. The geo-chips are tracked by a combination of enhanced GPS and local

positioning technology which is a modern variant of the RFID chips from the early years of the twenty-first century. These technologies are a world away from the early GPS systems, when accuracy was limited unless you could afford the most expensive devices. Geo-chips also have the advantage that they can report their vertical position, as well as their (x,y) location. Great news if you lose your keys in a 25-storey tower block!

Locating misplaced toys is only one of the many everyday tasks that geo-chips help with. A bit like the printing press, affordable domestic electricity, and the Internet, geo-chips have had an impact far beyond the wildest dreams of their inventors. There is no more lost property: a hat, a cat, a book or anything that's misplaced can always find its way back to its owner. Fly tipping, where people dump their unwanted possessions, has become a crime of the past. Parents always know where their children are. Stores know where the products they've sold have gone, so you can't steal anything. Goods can be tracked from the point when they were created until the time they are finally disposed of. Food can be traced from farm and factory to the restaurant and the home… and so it goes on.

But geo-chips haven't been all good news. In the early days there was quite a lot of opposition to the idea of putting them in some items. A significant number of people were not happy with the idea that they could be tracked through their possessions. However, they were won over when a privacy function was added to the chips, which allows the owner of the chip to govern who can track them.

Like my son, I'm glad they were invented, and that 'Lost' is a word from the past.

## The Ad Man's Dream

'It'll be an ad man's dream,' I say. 'A billboard that knows who you are and what you want, at any time on any day. Actually, it already exists, let me show you.'

Ten minutes later, we are in the underground car park of the Institute of Geo-marketing. The car park is full as the Institute's annual conference is in progress. The great and the good from the world of marketing are all here to catch a glimpse of the future. They don't always like what they see, but I think these new billboards will attract some interest.

One of the innovations that did get some opposition at a recent conference was the 'pay as you go' service that started with Google. This was a clever idea in which the advertisers only got paid when potential customers actually clicked on an ad, rather than being paid for simply making the ad and sticking it up on the Web. This was liberating for both the customer and the service provider. The customer could ignore the stuff they didn't want to see and the service provider only paid the ad agency if the ad encouraged potential customers to find out more. You could even track which ads worked and which didn't. Some of the more traditional advertising companies who couldn't respond to this change in the way of doing things went out of business. It was not longer necessary to pay upfront for an expensive marketing campaign, which might miss the target audience, when you could pay for ads that worked. The 'responsive billboard' we are on our way to see is perhaps the next step in individually targeted advertising.

We reach the car and I unlock it using my phone. All new cars now use mobile phones instead of the traditional 'key'. To be frank it's a much more secure way of letting the car know that you are the person who should be driving it. Databases can be checked, routes updated, seats adjusted, in-car entertainment set... quite smart really, so long as you haven't misplaced your phone!

'Plug in your phone, make sure locational services are activated and set data sharing to public,' I say to Chris as we reach the car. 'OK, input the postcode for this address and let's go.'

The car pulls out of the underground car park, and within a few minutes we are moving quickly through the streets behind the Institute. It starts to rain heavily.

'OK, on your left, what do you see?'

'It's a billboard,' Chris answers.

'Yes, and what's on the board?' I ask.

'An ad for cat food.'

'And do you have a cat?' I ask smiling, knowing what the answer will be.

'No. Now is this going to take much longer? I need to get back – I have another session to chair this afternoon.'

'Five more minutes,' I say, just as the in-car navigation system prompts us to take a left turn. In front, on the left-hand side, is another billboard, except this time something is different.

'Stop the car,' Chris says, looking at the billboard in amazement.

'But, how did it know that? It's just like you said in your presentation! It is almost as if it knows what I am thinking. How did it know I wanted that?' the questions tumble out of him, lots of them, in no coherent order.

Several minutes later, he is still looking at the sign, transfixed, mesmerized. He can't take his eyes off it.

'Do you see the same thing?' he asks.

'Yes,' I answer. 'But this is only a prototype and I set it to focus on you and your needs. The real version will analyse what we both want and show us something we're both interested in – like the location of a good Italian restaurant!'

'And if there are a lot of people on the road, it will analyse all our needs and look for common threads, depending on the time of day, the weather and what day of the week it is ... so an advert for warm, dry, coat wouldn't go amiss today!' I say, smiling, as I look at the cold grey sky and the rain.

Two years later, they are everywhere – on every highway, in every tube station, airport and the back of every plane, train and tram seat. They come in every shape and size you could imagine. Intelligent billboards, connected to the Internet – they know who you are, what you want and, most importantly, when you are going be driving towards them, or standing in front of them. They know where you are, where they are, and where the things you need can be found. A global connection of live, intelligent, geographically aware personal adverts which all started with one intelligent billboard!

## CONCLUSIONS

As we said at the outset, predicting the future is a dangerous occupation. What is clear is that the trend in recent years has been away from GIS being seen as a specialist tool, to greater widespread use of GIS in everyday life. To conclude, let's consider what might happen in the case-study scenarios that have been used throughout this book. The radioactive waste-siting GIS could become a system allowing widespread public participation in the decision-making process. Using the Internet, via their TV or mobile devices, members of the public could influence the siting criteria from their armchairs. In the house-hunting scenario an online house sales service could offer multimedia presentations of houses for sale, virtual reality displays of houses yet to be built and automatic selection of houses meeting buyers' criteria. Maps and details available on mobile phones would help them find the properties. Direct links to land registry systems (as are already being developed) could then speed up the house-buying process. In the Zdarske Vrchy case study there are opportunities for automatic data collection and the generation of new data to help model building and predictions. GPS and satellite data collection facilities could be coupled with bar-coded visitor passes to find out about the resources of the area and the pressures on them.

In Happy Valley the possibilities are endless! Perhaps the valley will have its own real-time man-agement information system. This could be used to enhance services, target potential clients, provide effective avalanche prediction, and for real-time management of resources. Information on avalanche risk, weather forecasts and special events could be continuously provided in multiple languages to guests' mobile devices, and on hand-held computers used by emergency services and service providers.

These are possible scenarios for the future. At the heart of them all are a number of challenges. The first is data. Automatic data collection, data exchange and real-time modelling using a range of data types all need to be tackled. Another challenge is the techno-logy. It is likely that the technological challenges, as in the past, will be easier to address than the organizational and human issues, and many new opportunities will come from areas outside the field of GIS. A third challenge is how to present geographical data and the results of geographical modelling to a whole new audience. New ways of visualizing the results of modelling and analysis are needed to help decision making and resource management. New interfaces are needed to make it possible for people to interact with geographical data and decision-support systems. Finally, there is the challenge for education and training. A new audience of users must be prepared to work and live with geographical information.

## ■ EPILOGUE

Before you finish reading and continue on to explore the use of GIS in your work or study, you may like to consider one more time the place of GIS in the wider world. If you read the newspapers, listen to the radio or watch TV news, you will appreciate that the world is a very dynamic place. Things change and they seem to be changing faster than ever before. Understanding the way we humans interact with other humans, respond to and alter our environment is fundamental to effecting and managing that change. In this context information truly is power and GIS can help us use that information effectively. Consider some of the dramatic events that have taken place in recent years. There have been many human events and natural disasters ranging from political upheaval, terrorist attacks and wars, to famine and extreme natural events such as earthquake, tsunami and hurricane. Some of these could have been predicted, and indeed, some were planned. In other cases, such as earthquakes, we may never be able to accurately predict them before they occur. In all instances GIS has and is being used to help co-ordinate the human response using many of the information sources, tools and methods described in this book. For example, GIS was used to help co-ordinate the response to Hurricanes Katrina and Rita that hit the southern coastal states of the USA in August and September 2005, including mapping flooding depths (Figure 13.6), predicting wind speeds, precipitation and other weather information (Figure 13.7), and providing before and after satellite images

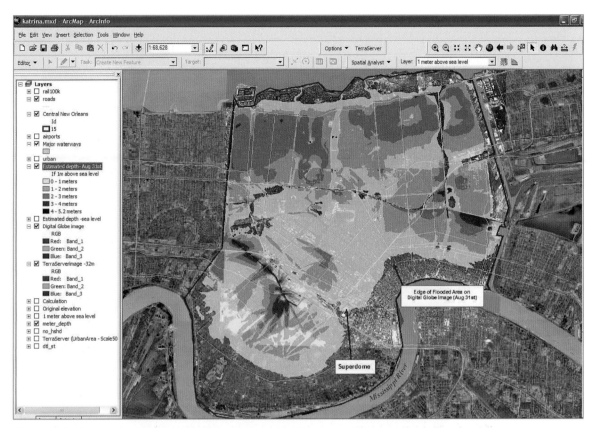

**Figure 13.6** Mapped flooding depths in New Orleans after Hurricane Katrina
*Source*: From Esri website, www.esri.com. Image Copyright © DigitalGlobe, used with permission. The ArcMap/ArcView/ArcInfo
Graphical User Interface is the intellectual property of Esri and is used herein with permission. Copyright © Esri

that have been used to estimate the extent and cost of the damage (Figure 13.8). There is now an online Hurricane Disaster viewer (Figure 13.9) provided by Esri using its Arcweb mapping services. Another example, where GIS has been used in emergency response planning was in the 2004 Boxing Day tsunami that affected large areas of the Indian Ocean. In this instance, data on which coastlines were affected and damage estimates were immediately available using high-resolution satellite imagery (Figure 13.10). This helped target immediate disaster relief efforts, generate casualty figures from population and census data, and has subsequently been used to help inform re-building efforts.

Climate change is another area that is creating new challenges and new opportunities for GIS. Much of this work relates to modelling what changes might occur, monitoring changes in human and environmental systems and predicting how these will affect us. Examples include mapping land areas that are likely to be inundated if sea levels rise in response to predicted global temperature increases, modelling changes in the spatial extent of ecosystems and wildlife habitats as regional climatic zones shift horizontally and vertically in response to changes in temperature and precipitation, and predicting changes to critical environmental cycles such as the hydrological, carbon and nitrogen cycles as climatic patterns change (see Figure 13.11). Other examples include monitoring the break-up of the Antarctic ice shelves and changes in the major ice streams (Figure 13.12), through to major changes in human land use such as deforestation in the Amazon basin for beef and soya farming (Figure 13.13).

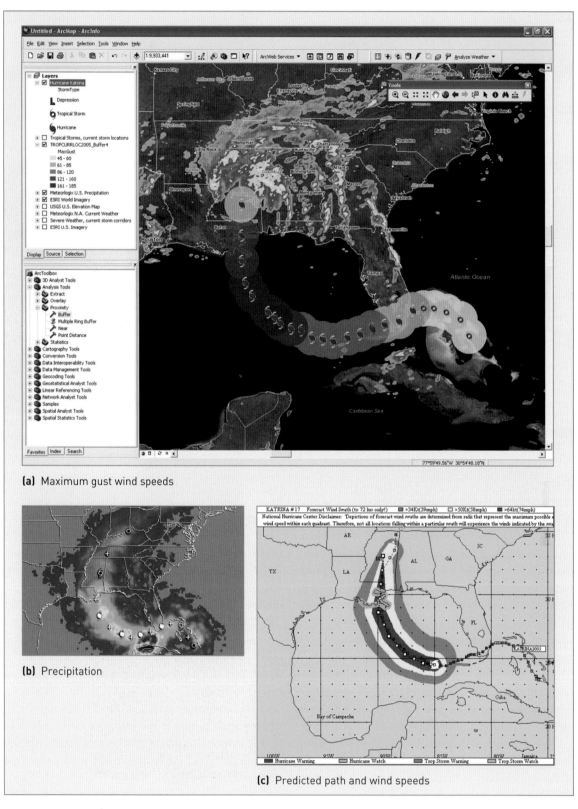

**(a)** Maximum gust wind speeds

**(b)** Precipitation

**(c)** Predicted path and wind speeds

**Figure 13.7** Predicted weather patterns in Hurricane Katrina

*Sources:* (a) From the Esri website, www.esri.com. Copyright © 2005–2006 Data Transmission Network Corporation, d/b/a Meteorlogix. The ArcMap/ArcView/ArcInfo Graphical User Interface is the intellectual property of Esri and is used herein with permission. Copyright © Esri; (b) NASA; (c) From Hurrevac website, www.hurrevac.com with permission of Sea Island Software

(a) Before                                    (b) After

**Figure 13.8** Hurricane Katrina, before and after satellite images of Biloxi
*Source*: DigitalGlobe (www.digitalglobe.com/Katrina_gallery.html), used by permission

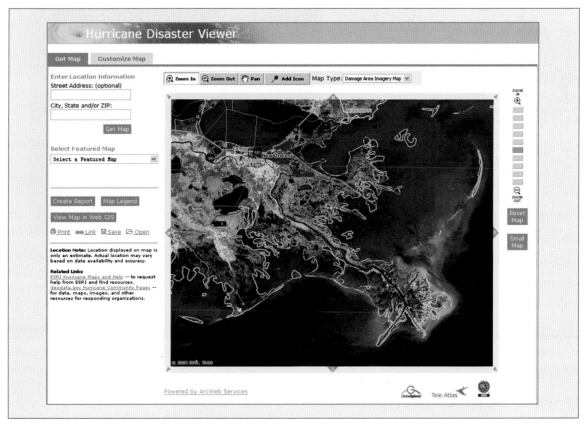

**Figure 13.9** Esri's Hurricane Disaster viewer
*Source*: Reproduced from http://arcweb.esri.com/sc/hurricane_viewer/index.html. Copyright © 2005 Esri, TANA. Used with permission

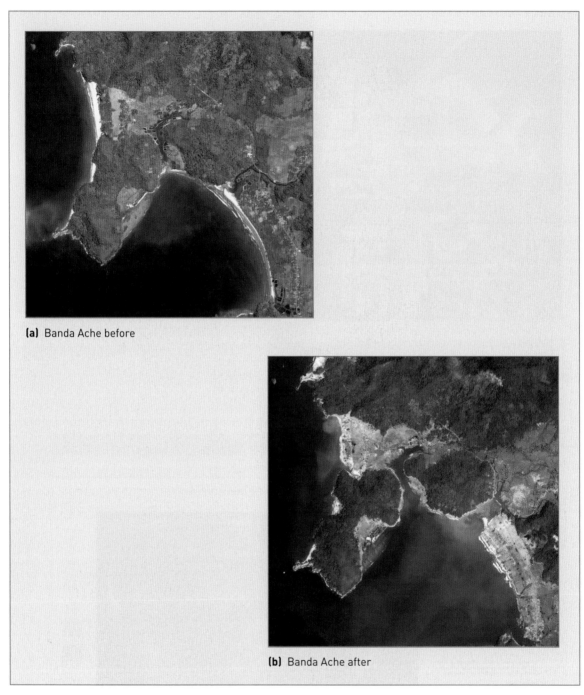

(a) Banda Ache before

(b) Banda Ache after

**Figure 13.10** Before and after images of areas hit by 2004 Boxing Day tsunami
*Source*: DigitalGlobe (www.digitalglobe.com/ tsunami_gallery.html), used by permission

'May you live in interesting times' is a saying oft, if incorrectly, attributed to an old Chinese curse though it is more likely to be of more recent origin. Wherever it comes from it implies that 'interesting times' are those most likely to be characterized by periods of rapid change, strife and upheaval. It would seem apparent that we do live in such times as these. Whatever happens over the next few years and decades, however, it is certain that GIS and its associated technologies, together with the people that use it and those that require the answers it provides will be there, working to either effect the change or respond to it. Since you are reading this book, it is likely that you will be one of them. We wish you luck in your endeavours.

**Figure 13.11** Inundation models for predicted global sea level rise

**(a)** Break up of the Larsen B ice shelf     **(b)** Ice flow rates (after NERC, 2000)

**Figure 13.12** Monitoring ice conditions in the Antarctic
*Sources*: (a) From the Earth Observatory / NASA website: http://earthobservatory.nasa.gov; (b) From the Natural Environment Research Council website, www.nerc.ac.uk/ publications/latestpressrelease/2000-13-1icemelt.asp. Copyright © Jonathan L Bamber, used with permission

**(a)** Deforestation in the Napo river valley, Ecuador 1986

**(b)** Deforestation in the Amazon Basin: before 1975 (left) and after 1986 (right)

**Figure 13.13** Deforestation in the Amazon Basin
*Sources*: (a) MECAA project website, www.unites.uqam.ca/mecaa; (b) LANDSAT Pathfinder satellite images

## REVISION QUESTIONS

- Have the predictions of GIS writers over the last 10 years been realized?

- What are the characteristics of GIS in the early twenty-first century?

- What are the problems facing future developments of GIS?

- How will developments in related fields, such as hardware, communications and multimedia, influence the future of GIS?

- Give examples of the use of GIS-type products in daily life, and suggest other areas where GIS use might become more widespread amongst the general population.

## FURTHER STUDY – ACTIVITIES

- Try to find your home, college or place of work using some online mapping services from the list below. Do you think that users without training in geographic information can make effective use of these products?

- Survey friends and family to find out what use they currently make of geographic information. Make some predictions for them – how will GIS impact their lives at home and work in the next few years.

- Think about some of the dramatic events that have happened recently and how GIS could be used to help. Use the Web to find examples of where GIS is being used.

- Investigate the future of areas and technologies related to GIS such as GPS, remote sensing or aerial photography. How will future developments in these areas impact GIS?

- Share your ideas about the future of GIS with us via the companion website for the book.

## FURTHER STUDY – READING

For further details of the predictions of others as to the character of GIS today, consult the full text of articles by Rhind *et al.* (1991), Rhind (1992), Maguire *et al.* (1991) and Goodchild (1992). For predictions for the future of GIS, you need to read up-to-date literature and keep your eye on the growing number of GIS-related blogs. Baumman's interview with Roger Tomlinson provides a short, but thought-provoking reflection about how the history of GIS has helped shape the future direction for GIS (Baumann, 2009). For a comprehensive overview of future challenges see Berry (2010). In addition, GIS magazines often contain the predictions of key individuals and annual reviews of the state of the GIS industry. Other information about developments will come from the research literature: for example, by reading up-to-date articles by innovative researchers. The proceedings of research conferences such as

GIS Research UK, which are published on the web (for example, Haklay *et al.*, 2010) give a feel for current research directions. Reading in other areas will also give you ideas for the development of GIS. One example is an interesting journal, *Personal Technologies*, which focuses on developments in hand-held and mobile information appliances.

Baumann J (2009) Roger Tomlinson on GIS history and future. *Geoconnexions: International Magazine*, February 2009. http://www.geoconnexion.com/uploads/roger_tomlinson_intv8i2.pdf

Berry J (2010) Topic 27: GIS Evolution and Future Trends. *Beyond Mapping III*. http://www.innovativegis.com/basis/MapAnalysis/Topic27/Topic27.htm

Goodchild M F (1992) Geographical Information Science, *International Journal of Geographical Information Systems* 6 (1): 31–45

Haklay M, Morley J, Rahemtulla, H (eds) (2010) *Proceedings of the GIS Research UK 18th Annual Conference GISRUK 2010*, University College London, London http://eprints.ucl.ac.uk/19284/

Maguire D J, Goodchild M F, Rhind D W (eds) (1991) *Geographical Information Systems: Principles and Applications*. Longman, London

Rhind D (1992) The next generation of Geographical Information Systems and the context in which they will operate. *Computers, Environment and Urban Systems*, 16: 256–68

Rhind D W, Goodchild M F, Maguire D J (1991) Epilogue. In: Maguire D J, Goodchild M F, Rhind D W (eds) *Geographical Information Systems: Principles and Applications*. Longman, London, vol. 2, pp. 313–27.

## WEB LINKS

### Online mapping services:

- Google maps UK: maps.google.co.uk/
- Multimap: www.multimap.com/
- Mapquest: www.mapquest.com/
- Streetmaps (UK): www.streetmap.co.uk/

### Mobile mapping services:

- Esri: www.esri.com/news/arcnews/summer10articles/arcgis-for-iphone.html
- British Geological Survey: www.bgs.ac.uk/iGeology/home.html?src=sfbFuture of GIS

### Geo-tagging photographs:

- Flickr: www.flickr.com/

### Internet GIS:

- Article on the present realities and the future of GIS: www.gisdevelopment.net/magazine/years/2004/july/29.shtml
- GIS and social media: www.slideshare.net/aGISGuy/gis-social-media-integration.
- GIS in natural resources: www.gis.wustl.edu/a-brief-history-and-probable-future.html
- Google Earth: earth.google.com/index.html
- NASA World Wind: worldwind.arc.nasa.gov/
- GIS blogs: wiki.gis.com/wiki/index.php/List_of_GIS-related_Blogs

## Professional bodies, training and certification:

- GISCI: www.gisci.org/
- AGI: www.agi.org.uk

Visit our website **www.pearsoned.co.uk/heywood** for further Web links, together with self-assessment questions, activities, data sets for practice opportunities and other interactive resources.

# References

**Abdul-Rahman A, Pilouk M** (2007) *Spatial Data Modelling for 3D GIS*. Springer-Verlag, Berlin

**Ackoff R L** (1971) Towards a system of systems concepts. *Management Science* **17** (11): 661–71

**Aho A V, Hopcroft J E, Ullmann J D** (1974) *The Design and Analysis of Computer Algorithms*. Addison-Wesley, Reading

**Allen, J.** (2006) Worry about crime in England and Wales. Home Office Online Report 15 June (http://rds.home-office.gov.uk/rds/pdfs06/rdsolr1506.pdf).

**Andersson S** (1987) The Swedish land data bank. *International Journal of Geographical Information Systems* **1** (3): 253–63

**Anon.** (1993) Survey of choice of system. *GIS World* **6**

**Anon.** (1994) Government is the largest GIS user. *GIS World* **7** (9): 12

**Anon.** (1995) *GIS World* interview: Ian McHarg reflects on the past, present and future of GIS. *GIS World* **8** (10): 46–9

**Anon.** (1996a) *GIS World* Interview: Roger Tomlinson the father of GIS. *GIS World* **9** (4): 56

**Anon.** (1996b) Platform preference varies with size of GIS user base. *GIS World* **9** (1): 16

**Anon.** (1997a) GIS software survey. *Mapping Awareness* **11** (2): 29–36

**Anon.** (1997b) Competitors unite to set standards. *Mapping Awareness* **11** (2): 29–36

**Anon.** (no date) Wigwam publicity flier, unpublished

**Applegate L N, Austin R D, Soule D L** (2009) *Corporate Information Strategy Management: Text and Cases*. 8th edn. McGrawHill-Irwin, Boston

**Aronoff S** (1991) *Geographic Information Systems: A Management Perspective*. WDL Publications, Ottawa

**Arrell K E** (2001) A fuzzy K-means classification of elevation derivatives to extract the natural landforms in Snowdonia, Wales. In: Kidner D B, Higgs G (eds) *Proceedings of GIS Research UK 9th Annual Conference: GISRUK 2001*. University of Glamorgan, Wales

**Arthur C** (1992) Ambulance computer system was too complicated. *New Scientist* **136** (1847): 7

**Arthur C** (1993) Pressurized managers blamed for ambulance failure. *New Scientist* **137** (1863): 5

**Arthur C, Cross M** (2006) Give us back our crown jewels. The *Guardian*, Thurs March 6, 2006, http://www.guardian.co.uk/technology/2006/mar/09/education.epublic

**Avison D E, Fitzgerald G** (1988) *Information Systems Development: Methodologies, Techniques and Tools*. Blackwell Scientific, Oxford

**Avison D E, Wood-Harper A T** (1991) Information systems development research: an exploration of ideas in practice. *Computer Journal* **34** (2): 98–112

**Balram S, Dragićević S** (2006) Modeling Collaborative GIS Processes Using Soft Systems Theory, UML and Object Oriented Design. *Transactions in GIS* **10** (2): 199–218

**Barnard M E** (1992) The Global Positioning System. *IEE Review* (March): 99–102

**Barr R** (1995a) Is GIS a Cheshire cat? *GIS Europe* **4** (10): 20–1

**Barr R** (1995b) Mind the generation gap. *GIS Europe* **4** (7): 18–19

**Barr R** (1995c) Addressing the issues. *GIS Europe* **4** (9): 18–19

**Barr R** (1996a) Data, information and knowledge in GIS. *GIS Europe* **5** (3): 14–15

**Barr R** (1996b) Look out! Someone is watching you. *GIS Europe* **5** (7): 12–13

**Barr R** (1996c) Desperately seeing solutions. *GIS Europe* **5** (8): 14–15

**Battista C** (1995) GIS fares well in utilities survey. *GIS World* **8** (12): 66–9

**Batty M, Xie Y** (1994a) Modelling inside GIS: Part 1. Model structures, exploratory spatial data analysis and aggregation. *International Journal of Geographical Information Systems* **8** (4): 291–307

**Batty M, Xie Y** (1994b) Modelling inside GIS: Part 2. Selecting and calibrating urban models using ARC/INFO. *International Journal of Geographical Information Systems* **8** (5): 451–70

**Batty M, Hudson-Smith A** (2005) Urban Simulacra: from real to virtual cities, back and beyond. *Architectural Design*. forthcoming

**Batty M, Dodge M, Doyle S, Smith A** (1998) *Visualizing Virtual Urban Environments*. CASA Working Paper 1. Centre for Advanced Spatial Analysis, University College London

**Batty M, Chapman D, Evans S, Haklay M, Kueppers S, Shiode N, Smith A, Torrens P** (2001) Visualizing the city: communicating urban design to planners and decision-makers. In: Brail R, Klosterman R (eds) *Planning Support Systems: Integrating Geographic Information Systems, Models, and Visualization Tools*. Esri Press, Redlands, pp. 405–43

**Batty P** (1990) Exploiting relational database technology in GIS. *Mapping Awareness* **4** (6): 25–32

**Baumann J** (2009) Roger Tomlinson on GIS history and future. *Geoconnexions: International Magazine* February 2009. http://www.geoconnexion.com/uploads/roger_tomlinson_intv8i2.pdf

**Beale H** (1987) The assessment of potentially suitable repository sites. In: *The Management and Disposal of Intermediate and Low Level Radioactive Waste.* Mechanical Engineering Publications, London, pp. 11–18

**Beck M B, Jakeman A J, McAleer M J** (1995) Construction and evaluation of models of environmental systems. In: Jakeman A J, Beck M B, McAleer M J (eds) *Modelling Change in Environmental Systems.* Wiley, Chichester, pp. 3–36

**Bedford M** (2004) *GIS for Water Management in Europe.* Esri Press, Redlands

**Bell S, Wood-Harper T** (1992) *Rapid Information Systems Development: A Non Specialist's Guide to Analysis and Design in an Imperfect World.* McGraw-Hill, London

**Berger P, Meysembourg P, Sales J, Johnston C** (1996) Toward a virtual reality interface for landscape visualization. *Proceedings of the 3rd International Conference/Workshop on Integrating GIS and Environmental Modeling* CD-ROM

**Bernhardsen T** (1992) *Geographic Information Systems.* VIAK IT, Norway

**Bernhardsen T** (1999) Choosing a GIS. In: Longley P A, Goodchild M F, Maguire D J and Rhind D W (eds) *Geographical Information Systems.* Wiley, New York, pp. 589–600

**Berry J K** (1987) Fundamental operations in computer-assisted map analysis. *International Journal of Geographical Information Systems* 1: 119–36

**Berry J K** (1993) *Beyond Mapping: Concepts, Algorithms and Issues in GIS.* GIS World Inc., Colorado

**Berry J** (2010) Topic 27: GIS Evolution and Future Trends. *Beyond Mapping III.* http://www.innovativegis.com/basis/MapAnalysis/Topic27/Topic27.htm

**Betak J, Vaidya A** (1993) GIS: a change agent in a transportation company. *Proceedings of the GIS in business '93,* conference, 7–10 March, Boston, Mass. GIS World Books, Fort Collins, Colorado, pp. 275–82

**Beven K J, Kirkby M J** (1979) A physically-based, variable contributing area model of basin hydrology. *Hydrological Sciences Bulletin* 24 (1): 43–69

**Bibby J S, Mackney D** (1969) *Land Use Capability Classification Soil Survey of Great Britain.* Technical Monograph no.1.

**Bickmore D P, Shaw M A** (1963) *Atlas of Great Britain and Northern Ireland.* Clarendon Press, Oxford

**Birkin M, Clarke G, Clarke M, Wilson A** (1996) *Intelligent GIS: Location Decisions and Strategic Planning.* GeoInformation International, Cambridge, UK

**Bishop I D** (2002) Determination of thresholds of visual impact: the case of wind turbines. *Environment and Planning B: Planning and Design* 29: 707–18

**Bishop I D, Miller D R** (2007) Visual assessment of off-shore wind turbines: the influence of distance, lighting, haze and movement. *Energy Policy* 32: 814–31

**Blakemore M J** (1984) Generalization and error in spatial databases. *Cartographica* 21: 131–9

**Bolstad P V, Gessler P, Lillesand T M** (1990) Positional uncertainty in manually digitized map data. *International Journal of Geographical Information Systems* 4 (4): 399–412

**Bonham-Carter G F** (1991) Integration of geoscientific data using GIS. In: Maguire D J, Goodchild M F, Rhind D W (eds) *Geographical Information Systems: Principles and Applications.* Longman, London, vol. 1, pp. 171–84

**Bonham-Carter G F** (1995) *Geographic Information Systems for Geoscientists: Modelling with GIS.* Pergamon Press, New York

**Boulding K E** (1956) On systems theory and analysis in urban planning. *Papers in Planning* 11. UWIST, Cardiff

**Brewer C** (2005) *Designing Better Maps: a Guide for GIS Users.* Esri Press, Redlands

**British Geological Survey** (2010) iGeology App: Britain's Rocks In Your Pocket. http://www.bgs.ac.uk/iGeology/home.html?src=sfb

**Brown A, Feringa W** (2003) *Colour Basics for GIS Users.* Pearson Education Limited, Harlow

**Brown C** (1989) *Implementing a GIS: Common Elements of Successful Sites.* Presented at Conference of URISA, Boston, Mass.

**Brown S J, Schreirer H E, Woods G, Hall S** (1994) A GIS analysis of forestry/caribou conflicts in the transboundary region of Mount Revelstoke and Glacier National Parks, Canada. In: Price M F, Heywood D I (eds) *Mountain Environments and Geographical Information Systems.* Taylor and Francis, London, pp. 235–48

**Brunsdon C, Carver S, Charlton M, Openshaw S** (1990) A review of methods for handling error propagation in GIS. *Proceedings of 1st European Conference on Geographical Information Systems,* Amsterdam, April, pp. 106–16

**Burgan R E, Rothermel R C** (1984) *BEHAVE: fire behaviour and prediction fuel modelling system – fuel subsystem.* General technical report INT–167, US Department of Agriculture, Ogden, Utah

**Burrough P A** (1986) *Principles of Geographical Information Systems for Land Resources Assessment.* Clarendon Press, Oxford

**Burrough P A, McDonnell R** (1998) *Principles of Geographical Information Systems.* Oxford University Press, Oxford

**Buttenfield B P, Mackaness W A** (1991) Visualization. In: Maguire D J, Goodchild M F, Rhind D W (eds) *Geographical Information Systems: Principles and Applications.* Longman, London, vol. 1, pp. 427–43

**Campbell H J** (1994) How effective are GIS in practice? A case study of British local government. *International Journal of Geographical Information Systems* 8 (3): 309–26

**Campbell H J** (1999) Institutional consequences of the use of GIS. In: Longley P A, Goodchild M F, Maguire D J, Rhind D W (eds) *Geographical Information Systems.* Wiley, New York, pp. 621–31

**Campbell H J, Masser I** (1992) GIS in local government: some findings from Great Britain. *International Journal of Geographical Information Systems* **6** (6): 529–46

**Campbell H J, Masser I** (1995) *GIS and Organizations: How Effective are GIS in Practice?* Taylor and Francis, London

**Campbell J B** (2007) *Introduction to Remote Sensing.* 3rd edn. Taylor and Francis, London

**Carver S J** (1991a) Error modelling in GIS: who cares? In: Cadoux-Hudson J, Heywood D I (eds) *The Association for Geographic Information Yearbook 1991.* Taylor and Francis/ Miles Arnold, London, pp. 229–34

**Carver S J** (1991b) Integrating multi-criteria evaluation with geographical information systems. *International Journal of Geographical Information Systems* **5** (3): 321–39

**Carver S J** (ed) (1998) *Innovations in GIS 5.* Taylor and Francis, London

**Carver S J, Heywood D I, Cornelius S C** (1995) Evaluating field-based GIS for environmental characterization, modelling and decision support. *International Journal of Geographical Information Systems* **9** (4): 475–86

**Carver S** (2001) Public participation using web-based GIS. *Environment and Planning B: Planning and Design* **28** (6): 803–4

**Carver S, Brunsdon C** (1994) Vector to raster conversion error and feature complexity: an empirical study using simulated data. *International Journal of Geographical Information Systems* **8** (3): 261–72

**Carver S, Evans A, Fritz S** (2002) Wilderness attribute mapping in the United Kingdom. *International Journal of Wilderness* **8** (1): 24–9

**Carver S, Evans A, Kingston R, Turton I** (2001) Public Participation, GIS, and Cyberdemocracy: Evaluating On-line Spatial Decision Support Systems. *Environment and Planning B:Planning and Design* **28**: 907–21

**Carver S, Blake M, Turton I, Duke-Williams O** (1997) Open spatial decision-making: evaluating the potential of the World Wide Web. In: Kemp Z (ed) *Innovations in GIS 4.* Taylor and Francis, London, pp. 267–78

**Cassettari S** (1993) *Introduction to Integrated Geo-information Management.* Chapman & Hall, London

**Charlesworth F R, Gronow W S** (1967) A summary of experience in the practical application of siting policy in the UK. In: *Containment and Siting of Nuclear Power Plants.* IAEA, Vienna, pp. 143–70

**Checkland P B** (1981) *Systems Thinking – Systems Practice.* John Wiley, Chichester

**Checkland P B** (1988) Information systems and systems thinking: time to unite? *International Journal of Information Management* **8**: 239–48

**Checkland P B, Howell S** (1998) *Information, Systems and Information Systems: Making Sense of the Field.* Wiley, London

**Checkland P B, Poulter J** (2006) *Learning for Action: A Short Definitive Account of Soft Systems Methodology and its use for Practitioners, Teachers and Students.* Wiley, Chichester

**Chen P P-S** (1976) The entity-relationship model – towards a unified view of data. *Association for Computing Machinery Transactions on Database Systems* **1** (1): 9–36

**Chen Z, Guevara J A** (1987) Systematic selection of very important points (VIP) from digital terrain models for construction of triangular networks. *Proceedings of Auto Carto 8.* ASPRS/ACSM, Falls Church, Virginia, pp. 50–6

**Chou Y-H** (1996) *Exploring Spatial Analysis in Geographic Information Systems.* OnWord Press, USA

**Chrisman N R** (1987) Efficient digitizing through the combination of appropriate hardware and software for error detection and editing. *International Journal of Geographical Information Systems* **1**: 265–77

**Chrisman N R** (1989) Modelling error in overlaid categorical maps. In: Goodchild M F, Gopal S (eds) *The Accuracy of Spatial Databases.* Taylor and Francis, London, pp. 21–34

**Chrisman N R** (2002) *Exploring Geographic Information Systems.* 2nd edn. Wiley, New York

**Churchman C W** (1968) *The Systems Approach.* Dell, New York

**Clarke K C** (1990) *Analytical and Computer Cartography.* Prentice-Hall, Englewood Cliffs, New Jersey

**Clayton K** (1995) The land from space. In: O'Riordan T (ed) *Environmental Science for Environmental Management.* 2nd edn.Longman, London, pp. 198–222

**Codd E** (1970) A relational model for large shared data banks. *Communications of the Association for Computing Machinery* **13** (6): 377–87

**Collins** (1981) *Pocket English Dictionary.* Collins, Glasgow

**Coppock J T** (1988) The analogue to digital revolution: a view from an unreconstructed geographer. *American Cartographer* **15** (3): 263–75

**Coppock J T, Rhind D W** (1991) The history of GIS. In: Maguire D J, Goodchild M F, Rhind D W (eds) *Geographical Information Systems: Principles and Applications.* Longman, London, vol. 1, pp. 21–43

**Cornelius S C, Medyckyj-Scott D** (1991) 'If only someone had said'. Human and organizational barriers to GIS success. *Mapping Awareness* **5** (7): 42–5

**Cornelius S C, Sear D A, Carver S J, Heywood D I** (1994) GPS, GIS and geomorphological field work. *Earth Surface Processes and Landforms* **19**: 777–87

**Craig W, Harris T, Weiner D** (1998) *NCGIA Specialist meeting: Empowerment, Marginalization and Public Participation GIS.* NCGIA, Santa Barbara

**Craig W, Harris T, Weiner D** (eds) (2002) *Community Empowerment, Public Participation and Geographic Information Science.* Taylor and Francis, London

**Crain I K** (1985) *Environmental Information Systems: An Overview.* Environment Canada, Ottawa

**Cresswell P** (1995) Customized and proprietary GIS: past, present and future. In: Longley P, Clarke G (eds) *GIS for Business and Service Planning*. Geoinformation International, Cambridge, UK, pp. 192–226

**Curran P** (1989) *Principles of Remote Sensing*. Longman, London

**Curry A, Flett P, Hollingsworth I** (2005) *Managing Information and Systems. The Business Perspective*. Routledge, Oxford

**Dale P F, McLaughlin J D** (1988) *Land Information Management: An Introduction with Special Reference to Cadastral Problems in Third World Countries*. Clarendon Press, Oxford

**Daniel L, Paula L, Whitener A** (1996) *Inside MapInfo Professional*. OnWord Press, USA

**Date C J** (1986) *An Introduction to Database systems*. 2nd edn. Addison-Wesley, Reading, Mass.

**Davies C, Medyckyj-Scott D** (1996) GIS users observed. *International Journal of Geographical Information Systems* **10** (4): 363–84

**Davis D** (1999) *GIS for Everyone*. Esri Press, Redlands

**Davis E B** (2001) *GIS A Visual Approach*. 2nd edn. Onward Press, Canada

**Davis J C** (1986) *Statistics and Data Analysis in Geology*. Wiley, New York

**Davis S** (2007) *GIS for Web developers: adding 'where' to your web applications*. Pragmatic Bookshelf, Raleigh, NC

**DeFloriani L, Magillo P** (1999) Intervisibility on terrains. In: Longley P A, Goodchild M F, Maguire D J, Rhind D W (eds) *Geographic Information Systems: Principles, Techniques, Management and Applications*. Wiley, London, pp. 543–56

**DeFloriani L, Fakidieno B, Nagy G, Pienovi C** (1984) A hierarchical structure for surface approximation. *Computer Graphics* **32**: 127–40

**DeMarco T, Lister, T** (2003) *Waltzing With Bears: Managing Risk on Software Projects*. Dorset House Publishing, London

**DeMers M N** (1997, 1999, 2005, 2008) *Fundamentals of Geographical Information Systems*. 4th edn. Wiley, New York

**DeMers M N** (2009) *GIS for Dummmies*. John Wiley and Sons, London

**Densham P J** (1991) Spatial decision support systems. In: Maguire D J, Goodchild M F, Rhind D W (eds) *Geographical Information Systems: Principles and Applications*. Longman, London, vol. 1, pp. 403–12

**Department of the Environment** (1985) *Disposal facilities on land for low and intermediate level radioactive wastes: principles for the protection of the human environment*. HMSO, London

**Department of the Environment** (1987) *Handling Geographic Information*. Report of the Committee of Enquiry chaired by Lord Chorley. HMSO, London

**DiBiase D** (2010) *Nature of Geographic Information*. Open Geospatial Textbook, https://www.e-education.psu.edu/natureofgeoinfo/node/1672

**Dominy N J, Duncan B** (2001) GPS and GIS methods in an African rain forest: applications to tropical ecology and conservation. *Conservation Ecology* **5** (2): 6 [online] http://www.consecol.org/vol5/iss2/art6/

**Dorling D** (1993) From computer cartography to spatial visualisation: a new cartogram algorithm. *AutoCarto* **11**: 208–17

**Dorling D** (1995) Visualizing changing social structure from a census. *Environment and Planning A* **26** (3): 353–78

**Dorling D, Openshaw S** (1992) Using computer animation to visualise space–time patterns. *Environment and Planning B* **19**: 639–50

**Douglas D H, Peucker T K** (1973) Algorithms for the reduction of the number of points required to represent a digitized line or its caricature. *Canadian Cartographer* **10** (4): 110–22

**Downey I, Heywood D I, Petch J R** (1991) Design and construction of a Geographical Information System (GIS) for Zdarske Vrchy landscape protected area, Czechoslovakia. In: Jackson M C, Mansell G J, Flood R L, Blackham R B, Probert S V E (eds) *Systems Thinking in Europe*. Plenum, New York, pp. 151–8

**Downey I, Pauknerova E, Petch J R, Brokes P, Corlyon A** (1992) Habitat analysis and modelling for endangered species. In: Willison et al. (eds) *Science and the Management of Protected Areas*. Elsevier, Amsterdam, pp. 271–7

**Dracos N** (1998) *The Shortest Path Algorithm*. Cornell University http://www.orie.cornell.edu/~or1154/handouts/ handout3/ handout3.html

**Drummond J, Billen R, Joao, E, Forrest D** (2007) (eds) *Dynamic and mobile GIS: investigating changes in space and time*. *Innovations in GIS*. CRC Press, Taylor and Francis Group, Boca Raton, FL

**Dunn R, Harrison A R, White J C** (1990) Positional accuracy and measurement error in digital databases of land use: an empirical study. *International Journal of Geographical Information Systems* **4** (4): 385–98

**Dyke J, MacEachren A M, Kraak M-J** (2005) *Exploring Geovisualization* Pergamon, Oxford

**Egenhofer M, Frank A** (1992) Object-Oriented Modeling for GIS. *Journal of the Urban and Regional Information Systems Association* **4**: 3-19

**Eason K D** (1988) *Information Technology and Organizational Change*. Taylor and Francis, London

**Eason K D** (1994) Planning for change: introducing a Geographical Information System. In: Medyckyj-Scott D J, Hearnshaw H M (eds) *Human Factors in Geographical Information Systems*. Belhaven, London, pp. 199–209

**Eastman R, Kyem P, Toledano J, Jin W** (1993) *GIS and Decision Making*. Explorations in Geographic Information Systems technology. Volume 4. United Nations Institute for Training and Research, Switzerland

**Elmasri R, Navathe S B** (1994) *Fundamentals of Database Systems*, 2nd edn. The Benjamin/Cummings, Calif.

**Esri** (1995) Esri builds national GIS for the real estate industry. *ARC News* **17** (4): 1–2

**Esri** (2001) Esri Timeline, http://www.esri.com/company/about/timeline/flash/index.html

**Esri** (2005) http://www.esri.com

**European Commission** (2005) *GALILEO European Satellite Navigation System*. http://europa.eu.int/comm/dgs/energy_transport/galileo/index_en.htm

**European Space Agency** (ESA) (2008) *Procurement of full GALILEO system begins*. http://www.esa.int/esaNA/SEM3ODSHKHF_galileo_0.html

**Evans I S** (1977) The selection of class intervals. *Transactions of the Institute of British Geographers* **2** (1): 98–124

**Ewert A W, Hollenhorst S J** (1997) Outdoor recreation and its implications for wilderness. *International Journal of Wilderness* **3** (2): 21–6

**Farhan B, Murray A T** (2005) A GIS-Based Approach for Delineating market Areas for Park and Ride Facilities. *Transactions in GIS* **9** (2): 91–108

**Faust N L** (1995) The virtual reality of GIS. *Environment and Planning B: Planning and Design* **22**: 257–68

**Fischer M M, Getis A** (2010) *Handbook of Applied Spatial Analysis: Software Tools, Methods and Applications*. Springer, Heidelberg

**Fisher P** (1991) Modelling soil map-unit inclusions by Monte Carlo simulation. *International Journal of Geographical Information Systems* **5** (2): 193–208

**Fisher P** (ed) (1995) *Innovations in GIS 2*. Taylor and Francis, London

**Fisher P, Dykes J, Moore K, Wood J, McCarthy T, Raper J, Unwin D, Williams N, Jenkins A** (1997) The virtual field course: an educational application of GIS. In: Carver S (ed) *Proceedings of GIS Research UK, 5th National Conference*, University of Leeds. pp. 37–40

**Fisher P, Wood J, Cheng T** (2004) Where is Helvellyn? Fuzziness of multi-scale landscape morphometry. *Transactions of the Institute of British Geographers* **29** (1): 106–28

**Fisher P F** (1993) Algorithm and implementation uncertainty in viewshed analysis. *International Journal of Geographical Information Systems* **7** (4): 331–47

**Fisher P F** (1994) Probable and fuzzy models of the view-shed operation. In: Worboys M (ed) *Innovations in GIS 1*. Taylor and Francis, London, pp. 161–75

**Fisher P F, Wood J** (1998) 'What is a mountain?' or the Englishman who went up a Boolean geographical concept but realised it was fuzzy. *Geography* **83** (3): 247–56

**Fishwick A, Clayson J** (1995) *GIS Development Project: Final Report to the Countryside Commission*. Lake District National Park Authority, Kendal, UK

**Foresman T** (ed) (1997) *The History of GIS*. Prentice Hall, New Jersey

**Fotheringham A S, O'Kelly M E** (1989) *Spatial Interaction Models: Formulations and Applications*. Kluwer Academic, Dordrecht

**Fotheringham A S, Rogerson P** (eds) (1994) *Spatial Analysis and GIS*. Taylor and Francis, London

**Fotheringam A S, Brunsdon C, Charlton M** (2000) *Quantitative Geography: Perspectives on Spatial Data Analysis*. Sage, London

**Fotheringham A S, Brunsdon C, Charlton M** (2002) *Geographically Weighted Regression: The Analysis of Spatially Varying Relationships*. Wiley, Chichester

**Fowler R J, Little, J J** (1979) Automatic extraction of irregular network digital terrain models. *Computer Graphics* **13**: 199–207

**Frank A U, Mark D M** (1991) Language issues for GIS. In: Maguire D J, Goodchild M F, Rhind D W (eds) *Geographical Information Systems: Principles and Applications*. Longman, London, vol. 1, pp. 147–63

**Fritz S, Carver S** (2000) Modelling remoteness in roadless areas using GIS. *Proceedings of 4th International Conference and Workshop on Integrating Geographical Information Systems and Environmental Modelling*. Banff, Canada, September 2000

**Gao J** (2002) Integration of GPS with remote sensing and GIS – reality and prospect. *Photogrammetric Engineering and Remote Sensing* **68** (5): 447–53

**Garcìa S G** (2004) GRASS GIS-embedded Decision Support Framework for Flood Simulation and Forecasting. *Transactions in GIS* **8** (2): 245–54

**Gatrell A C** (1991) Concepts of space and geographical data. In: Maguire D J, Goodchild M F, Rhind D W (eds) *Geographical Information Systems: Principles and Applications*. Longman, London, vol. 1, pp. 119–43

**Geertman S, Stillwell J** (eds) (2009) *Planning Support Systems Best Practice and New Methods*. Springer, Dordrecht

**Gelernter D** (1992) *Mirror World or the Day Software Puts the University in a Shoebox ... How it will Happen and What it will Mean*. Oxford University Press, New York

**Geoffrion A M** (1983) Can OR/MS evolve fast enough? *Interfaces* **13** (10)

**GeoInformation International** (1995) *GIS in Business: Discovering the Missing Piece in your Business Strategy*. GeoInformation International, Banbury, UK

**Gibson P, Power C H** (2000) *Introducing Remote Sensing: Digital Image Processing and Applications*. Routledge, London

**Gilfoyle I** (1997) The Chorley Report: philosophy and recommendations. What Chorley got right. Presented at: *The Future for Geographic Information: 10 Years after Chorley*, Association for Geographic Information, London, 1 May

**GISNews** (2001) *Guide to the new Ordnance Survey*. Supplement issued with vol. 2, no. 5

**Giusti de Perez R, Perez R** (2008) *Analyzing Urban Poverty: GIS for the developing world.* Esri Press, Redlands

**Goodchild M** (1978) Statistical aspects of the polygon overlay problem. In: Dutton G H (ed) *Harvard Papers on Geographical Information Systems*, vol. 6. Addison-Wesley, Reading, Mass., pp. 1–21

**Goodchild M F** (1988) Stepping over the line: the technological constraints and the new cartography. *American Cartographer* **15** (3): 277–89

**Goodchild M F** (1990) Spatial information science. *Proceedings of 4th International Symposium on Spatial Data Handling.* International Geographical Union, Columbus Ohio, vol. 1, pp. 3–12

**Goodchild M F** (1992) Geographical Information Science. *International Journal of Geographical Information Systems* **6** (1): 31–45

**Goodchild M F** (1995) GIS and geographic research. In: Pickles J (ed) *Ground Truth: The Social Implications of Geographic Information Systems.* Guilford Press, New York, pp. 31–50

**Goodchild M F** (1997) What is Geographic Information Science? NCGIACore Curriculum in GIScience. http// www.ncgia.ncsb.edu/giscc/units/u002/u002.html, posted 7 October 1997

**Goodchild M F, Gopal S** (eds) (1989) *The Accuracy of Spatial Databases.* Taylor and Francis, London

**Goodchild M F, Kemp K** (1990) Developing a curriculum in GIS: the NCGIA core curriculum project. In: Unwin D J (ed) *GIS Education and Training*, Collected Papers of a Conference, University of Leicester, 20–21 March 1990, Midlands Regional Research Laboratory, Leicester

**Goodchild M F, Rhind D W** (1990) The US National Center for Geographic Information and Analysis: some comparisons with the Regional Research Laboratories. In: Foster M F, Shand P J (eds) *The Association for Geographic Information Yearbook 1990.* Taylor and Francis, London, pp. 226–32

**Goodchild M F, Yang S** (1989) A hierarchical spatial data structure for global geographic information systems. *NCGIA Technical Paper 89–5.* National Center for Geographic Information and Analysis (NCGIA), University of California, Santa Barbara

**Goodchild M F, Parks B O, Steyaert L T** (eds) (1993) *Environmental Modelling with GIS.* Oxford University Press, Oxford and New York

**Goodwin T** (1995) GIS – getting started. In: *GIS in Business: Discovering the Missing Piece in your Business Strategy.* GeoInformation International, Banbury, UK, pp. 144–8

**Gould M I** (1992) The use of GIS and CAC by health authorities: results from a postal questionnaire. *Area* **24** (4): 391–401

**Gregory I, Ell P** (2007) *Historical GIS: Technologies, Methodologies and Scholarship.* Cambridge University Press, Cambridge

**Griffith D A, Amrhein C G** (1991) *Statistical Analysis for Geographers.* Prentice-Hall, Englewood Cliffs, New Jersey

**Grimshaw D J** (1999) *Bringing Geographical Information Systems into Business.* 2nd edn. Longman, London

**Gunn J, Hunting C, Cornelius S, Watson R** (1994) The proposed Cuilcagh National History Park, County Fermanagh: a locally based conservation initiative. In: O'Halloran D, Green C, Harley M, Stanley M, Knill J (eds) *Geological and Landscape Conservation.* Geological Society, London, pp. 337–41

**Haklay M, Morley J, Rahemtulla H** (eds) (2010) *Proceedings of the GIS Research UK 18th Annual Conference GISRUK 2010*, University College London, London. http://eprints.ucl.ac.uk/19284/

**Hadrian D R, Bishop I D, Mitcheltree R** (1988) Automated mapping of visual impacts in utility corridors. *Landscape and Urban Planning* **16**: 261–82.

**Haggett P, Chorley R J** (1967) Models, paradigms and the new geography. In: Chorley R J, Hagget P (eds) *Models in Geography.* Methuen, London, pp. 19–42

**Haines-Young R, Green R D, Cousins S H** (eds) (1994) *Landscape, Ecology and GIS.* Taylor and Francis, London

**Hale C** (1996) Fear of crime: a review of the literature *International Review of Victimology* **4** (2): 79–150

**Halpin P N** (1994) GIS analysis of the potential impacts of climate change in mountain ecosystems and protected areas. In: Price M F, Heywood D I (eds) *Mountain Environments and Geographical Information Systems.* Taylor and Francis, London, pp. 281–301

**Hanna K C** (1999) *GIS for Landscape Architects.* Esri Press, Redlands

**Hanold T** (1972), An executive view of MIS. *Datamation* **18** (11): 66

**Harder C** (1997) *ArcView GIS Means Business: Geographic Information System Solutions for Business.* Esri Press, Redlands

**Harder C** (1998) *Serving Maps on the Internet: Geographic Information and the World Wide Web.* Esri Press, Redlands

**Harding S M, Wilkinson G G** (1997) *A strategic view of GIS research and technology development for Europe.* Report of the expert panel convened at the Joint Research Centre, Ispra, 19–20 November 1996. EUR 17313 EN, European Commission, Brussels

**Hardisty J, Taylor D M, Metcalfe S E** (1993) *Computerized Environmental Modelling: a Practical Introduction Using Excel.* Wiley, New York

**Harley J B** (1975) *Ordnance Survey Maps: a Descriptive Manual.* OS, Southampton

**Healey P** (1997) *Collaborative Planning – Shaping Places in Fragmented Societies.* Macmillan, London

**Healey R G** (1991) Database management systems. In: Maguire D J, Goodchild M F, Rhind D W (eds) *Geographical Information Systems: Principles and Applications.* Longman, London, vol. 1, pp. 251–67

**Hearnshaw H M, Unwin D J** (1994) *Visualization in Geographical Information Systems.* Wiley, New York

**Heit M, Shortreid A** (1991) *GIS Applications in Natural Resources.* GIS World Inc., Colorado

**Hengl T, Reuter H I** (2009) *Geomorphometry: Concepts, Software, Applications.* Developments in Soil Science Volume 33. Elsevier, Amsterdam

**Henry B, Pugh J** (1997) CAD and GIS software find common ground: a review of the market leaders. *GIS World* **10** (5): 32–7

**Heritage G, Hetherington D** (2005) The use of high-resolution field laser scanning for mapping surface topography in fluvial systems. In: Walling D E and Horowitz A J (eds) *Sediment Budgets 1.* IAHS Publication 291: 269–77

**Heritage G, Hetherington D** (2007) Towards a protocol for laser scanning in fluvial geomorphology. *Earth Surface Processes and Landforms* **32**(1): 66–74

**Hernandez T, Bennison D** (1996) Selling GIS to retailers: organisational issues in the adoption and use of Geographical Information Systems. *Proceedings of the AGI/GIS 96 Conference.* 22–24 September, Birmingham, UK

**Hernandez T, Verrips A** (2000) Retail GIS: more than just pretty maps. *GeoEurope* **9** (4): 16–18

**Hernandez T, Scholten H J, Bennison D, Biasiotto M, Cornelius S, van der Beek M** (1999) Explaining retail GIS: the adoption, use and development of GIS by retail organizations in the Netherlands, the UK and Canada. *Netherlands Geographical Studies* **258**, Utrecht

**Hetherington D, Heritage G, Milan D** (2005) Daily fine sediment dynamics on an active Alpine glacier outwash plain. In: Walling D E, Horowitz A J (eds) *Sediment Budgets 1.* IAHS Publication, 291:278–284

**Heuvelink G B, Burrough P A, Stein A** (1990) Propagation of errors in spatial modelling with GIS. *International Journal of Geographical Information Systems* **3**: 303–22

**Hewitson B D, Crane R G** (1994) *Neural Nets: Applications in Geography.* Kluwer Academic, Dordrecht

**Heywood D I** (1997) *Beyond Chorley: Current Geographic Information Issues.* AGI, London

**Heywood D I, Blaschke T, Carlisle B** (1997) Integrating geographic information technology and multimedia: the educational opportunities. In: Hodgson S, Rumor M, Harts J J (eds) *Proceedings of the Joint European Conference and Exhibition on Geographical Information Systems,* Vienna, pp. 1281–90

**Heywood D I, Oliver J, Tomlinson S** (1995) Building an exploratory multi-criteria modelling environment for spatial decision support. In: Fisher P (ed) *Innovations in GIS 2.* Taylor and Francis, London, pp. 127–36

**Heywood I** (1990) Monitoring for change: a Canadian perspective on the environmental role for GIS. *Mapping Awareness* **4** (9): 24–6

**Heywood I, Carver S** (1994) Decision support or idea generation: the role for GIS in policy formulation. *Proceedings Symposium für Angewante Geographische Informationsverarbeitung (AGIT '94),* Salzburg, Austria, July, pp. 259–66

**Hobby J** (1992) This is an emergency. *Guardian,* 5 November

**Hoffman-Wellenhof B, Lichtenegger H, Collins J** (1994) *Global Positioning Systems: Theory and Practice.* Springer-Verlag

**Hohl P** (ed) (1998) *GIS Data Conversion: Strategies, Techniques, Management.* Onword Press, Santa Fe

**Holcroft C** (2003) A practical approach to OS MasterMap. *GI News,* Jan/Feb

**Holden J, Shotbolt L, Bonn A, Burt T P, Chapman P J, Dougill A J, Fraser E D G, Hubacek K, Irvine B, Kirkby M J, Reed M S, Prell C, Stagl S, Stringer L C, Turner A, Worrall F** (2007) Environmental change in moorland landscapes. *Earth-Science Reviews* **82**: 75–100

**Holland D** (2001) Delivering the digital national framework in GML. *GeoEurope* (August): 29–30.

**Hornsby K, Egenhofer M** (2000) Identity-based change: a foundation for spatio-temporal knowledge representation. *International Journal of Geographical Information Science* **14** (3): 207–24

**Hudson-Smith A, Evans S, Batty M** (2005) Building the virtual city: public participation through e-democracy. *Knowledge Technology and Policy* **18**: 62–85

**Huggett R** (1980) *Systems Analysis in Geography.* Clarendon Press, Oxford

**Hughes T** (1997) The new seekers. *Compuserve Magazine* **2**: 18–20

**Hull R B, Bishop I D** (1988) Scenic impacts of electricity transmission towers: the influence of landscape type and observer distance. *Journal of Environmental Management* **27**: 99–108.

**Huxhold W E** (1991) *An Introduction to Urban Geographic Information Systems.* Oxford University Press, New York

**IAEA** (1983) *Disposal of Low and Intermediate Level Solid Radioactive Wastes in Rock Cavities.* Safety Series No. 59, IAEA, Vienna

**Illiffe J, Lott R** (2008) *Datums and Map Projections for GIS, Remote Sensing and Surveying.* 2nd edn. Whittles Publishing, New York

**Intel** (2005) *Moore's Law.* ftp://download.intel.com/museum/Moores_Law/Printed_Materials/Moores_Law_2pg.pdf

**Jackson M J, Woodsford P A** (1991) GIS data capture hardware and software. In: Maguire D J, Goodchild M F, Rhind D W (eds) *Geographical Information Systems: Principles and Applications.* Longman, London, vol. 1, pp. 239–49

**Jacobsen F, Barr R, Theakstone W** (1993) Exploring glaciers with GIS: the Svartisen and Okstindan project. *GIS Europe* **2** (4): 34–6

**Jakeman A J, Beck M B, McAleer M J** (eds) (1995) *Modelling Change in Environmental Systems.* Wiley, Chichester

**Janssen R** (1992) *Multi-objective Decision Support for Environmental Management.* Kluwer Academic, Dordrecht

**Janssen R, Rievelt P** (1990) Multicriteria analysis and GIS: an application to agricultural land use in the Netherlands. In: Scholten H J, Stillwell J C H (eds) *Geographical Information Systems for Urban and Regional Planning.* Kluwer Academic, Dordrecht

**Jenks G F** (1981) Lines, computer and human frailties. *Annals of the Association of American Geographers* **71** (1): 142–7

**Karikari I, Stillwell J, Carver S** (2005) The application of GIS in the lands sector of a developing country: challenges facing land administrators in Ghana. *International Journal of Geographic Information Science* **19** (3): 343–62

**Karlsson A** (no date) GIS and Spatial Extensions with My SQL. http://dev.mysql.com/tech-resources/articles/4.1/gis-with-mysql.html

**Keates J S** (1982) *Understanding Maps.* Longman, London

**Keefer B K, Smith J L, Gregoire T G** (1988) Simulating manual digitizing error with statistical models. *Proceedings of GIS/LIS '88.* ACSM, ASPRS, AAG and URISA, San Antonio, Texas, pp. 475–83

**Kemp Z** (ed) (1997) *Innovations in GIS 4.* Taylor and Francis, London

**Kennedy M** (1996) *The Global Positioning System and GIS: An Introduction.* Ann Arbor Press, Ann Arbor, Michigan

**Kidner D B, Rallings, P J, Ware J A** (1997) Parallel Processing for Terrain Analysis in GIS: Visibility as a Case Study. *GeoInformatica* **1**(2): 183-207

**Kingston R, Carver S, Evans A, Turton I** (2000) Web-based public participation geographical information systems: an aid to local environmental decision-making. *Computers, Environment and Urban Systems,* **24** (2): 109–25

**Korte G B** (2000) *The GISBook.* 5th edn. OnWord Press, USA

**Kostblade J T** (1981) Mapping frequency distributions by the box and whisker. *Professional Geographer* **33** (4): 413–18

**Kraak M-J, Brown A** (eds) (2000) *Web Cartography: Developments and Prospects.* Taylor and Francis, London

**Kubo S** (1987) The development of GIS in Japan. *International Journal of Geographical Information Systems* **1** (3): 243–52

**Lam N S** (1983) Spatial interpolation methods: a review. *American Cartographer* **10**: 129–49

**Lane S N, Brookes C J, Kirkby M J, Holden J** (2004) A network-index-based version of TOPMODEL for use with high-resolution digital topographic data. *Hydrological Processes* **18**: 191–201

**Lang L** (1998) *Managing Natural Resources with GIS.* Esri Press, Redlands

**Langmuir E** (1984) *Mountaincraft and Leadership.* The Scottish Sports Council/MLTB

**Langran G** (1992) *Time in Geographical Information Systems.* Taylor and Francis, London

**Laurini R, Thompson D** (1992) *Fundamentals of Spatial Information Systems.* Academic Press, London

**Lee J** (1991) Comparison of existing methods for building triangular irregular network models of terrain from grid elevation models. *International Journal of Geographical Information Systems* **5** (3): 267–85

**Lee S, Choi J** (2004) Landslide susceptibility mapping using GIS and the weight-of-evidence model. *International Journal of Geographical Information Science* **18** (8): 789–814

**Lenhart D, Hinz S, Leitoff J, Stilla U** (2008) Automatic Traffic Monitoring Based on Aerial Image Sequencing. *Pattern Recognition and Image Analysis* **18** (3): 400–405

**Lewis O M, Ware J A** (1997) The use of census data in the appraisal of residential properties within the United Kingdom: a neural network approach. In: Hodgson S, Rumor M, Harts J J (eds) *Proceedings of the Joint European Conference and Exhibition on Geographical Information Systems,* Vienna, pp. 226–33

**Limp F** (1999) Mapping hits warp speed on the world wide web! *GeoWorld,* Sept 1999 available online at http://www.geoplace.com/ gw/1999/0999/999tec.asp

**Lindley S J, Crabbe H** (2004) What lies beneath? – Issues in the representation of air quality management data for public consumption. *Science of the Total Environment* **334–5**: 307–25

**Lindley S J, Walsh T** (2005) Inter-comparison of interpolated background nitrogen dioxide concentrations across Greater Manchester, UK. *Atmospheric Environment* **39**: 2709–2724.

**Lindley S J, Conlan D E, Raper D W, Watson A F R** (1999) Estimation of spatially resolved road transport emissions for air quality management applications in the North West region of England. *Science of the Total Environment* **235**: 119–32

**Lindley S J, Conlan D E, Raper D W, Watson A F** (2000) Uncertainties in the compilation of spatially resolved emission inventories – evidence from a comparative study. *Atmospheric Environment* **34** (3): 375–88

**Llobera M** (2003) Extending GIS based visual analysis: the concept of visualscapes. *International Journal of Geographical Information Science* **17** (1): 25–48

**Longley P, Batty M** (2003) *Advanced Spatial Analysis: The CASA Book of Spatial Analysis.* Esri Press, Redlands

**Longley P, Clarke G** (eds) (1995) *GIS for Business and Service Planning.* GeoInformation International, Cambridge, UK

Longley P A, Goodchild M F, Maguire D J, Rhind D W (eds) (1999) *Geographical Information Systems: Principles, Techniques, Management and Applications*. John Wiley, New York

Longley P A, Goodchild M F, Maguire D J, Rhind D W (eds) (2005) *Geographic Information Systems: Principles, Techniques, Management and Applications*. 2nd edn. Abridged version. John Wiley, New York

Longley P A, Goodchild M F, Maguire D J, Rhind D W (2010) *Geographic Information Systems and Science*. 3rd edn. Wiley, Chichester

MacEachren A M (1994) *Some Truth with Maps: a Primer on Symbolization and Design*. Association of American Geographers, Washington DC

MacEachren A M (1995) *How Maps Work: Representation, Visualization, and Design*. The Guilford Press, New York

MacEachren A M, Taylor D R F (1994) *Visualization in Modern Cartography*. Pergamon Press, New York

Maguire D J (1989) *Computers in Geography*. Longman, London

Maguire D J (1991) An overview and definition of GIS. In: Maguire D J, Goodchild M F, Rhind D W (eds) *Geographical Information Systems: Principles and Applications*. Longman, London, vol. 1, pp. 9–20

Maguire D J (1995) Implementing spatial analysis and GIS applications for business and service planning. In: Longley P, Clarke G (eds) *GIS for Business and Service Planning*. GeoInformation International, Cambridge, UK, pp. 171–270

Maguire D, Kouyoumjian V, Smith R (2008) *The business benefits of GIS: an ROI approach*. Redlands: Esri Press

Maguire D J, Goodchild M F, Rhind D W (eds) (1991) *Geographical Information Systems: Principles and Applications*. Longman, London

Maguire D J, Worboys M F, Hearnshaw H M (1990) An introduction to object oriented GIS. *Mapping Awareness* **4** (2): 36–9

Maher M (2010) *Lining up data in ArcGIS. A Guide to Map Projections*. Esri Press, Redlands

Mahoney R P (1991) GIS and utilities. In: Maguire D J, Goodchild M F, Rhind D W (eds) *Geographical Information Systems: Principles and Applications*. Longman, London, vol. 2, pp. 101–14

Malczewski J (1999) *GIS and Multi-criteria Decision Analysis*. Wiley, London

Mangold R (1997) Are the walls of GIS finally falling down? *Earth Observation Magazine* **6** (2): 4

Manning T, Dvorak M, Allen T, Poole A, Fear D, Stock G, Hussein A (2009) Reducing the Exploration Cycle with Effective Onboard Processing. *First Break*, 27 October: 93–96

Marble D (1997) Rebuilding the top of the pyramid: providing technical GIS education to support both GIS development and geographic research. Presentation to GIS in Higher Education 97 conference, 30 Oct–2 Nov, NCGIA and Towson University, Virginia, USA

Mark D M (2003) Geographic Information Science: Defining the Field. In Duckham M, Goodchild M F, Worboys M (eds) *Foundations of Geographic Information Science*. Taylor and Francis, London, pp. 1–17

Mark D M, Chrisman N, Frank A U, McHaffie P H, Pickels J (no date) The GIS History project: http://www.geog.buffalo.edu/ncgia/ gishist/bar_harbor.html

Martin D (1995) *Geographical Information Systems and thier Socio-economic Applications*. 2nd edn. Routledge, London

Martin D, Bracken I (1991) Techniques for modelling population related raster databases. *Environment and Planning A* **23**: 1065–79

Maskrey A (1997) The design of GIS for risk analysis in Latin America. Unpublished MSc thesis, Department of Environmental and Geographical Sciences, Manchester Metropolitan University

Masser I (1990) The Regional Research Laboratory initiative: an update. In: Foster M J, Shand P J (eds) *The Association for Geographic Information Yearbook 1990*. Taylor and Francis/Miles Arnold, London, pp. 259–63

Masser I, Campbell H, Craglia M (1996) *GIS Diffusion: The Adoption and Use of Geographical Information Systems in Local Government in Europe*. GISDATA 3, Taylor and Francis, London

Mather P M (1991) *Computer Applications in Geography*. Wiley, New York

Mather P M (ed) (1993) *Geographical Information Handling – Research and Applications*. Wiley, Chichester

McAlpine J R, Cook B G (1971) Data reliability from map overlay. *Proceedings of the Australian and New Zealand Association for the Advancement of Science 43rd Congress, Section 21 – Geographical Sciences*. Australian and New Zealand Association for the Advancement of Science, Brisbane, May 1971

McDonnell R, Kemp K (1998) *International GIS Dictionary*. 2nd edn. GeoInformation International, London

McHarg I L (1969) *Design with Nature*. Doubleday, New York

McLaren R A (1990) Establishing a corporate GIS from component datasets – the database issues. *Mapping Awareness* **4** (2): 52–8

McMaster R B, Shea K S (1992) *Generalization in Digital Cartography*. Association of American Geographers, Washington DC

Medyckyj-Scott D J (1989) Users and organisational acceptance of GIS: the route to success or failure. *Proceedings of 1st National Conference for Geographic Information Systems 'GIS – a corporate resource'*. AGI, London, pp. 4.3.1–4.3.8

**Medyckyj-Scott D J, Cornelius S C** (1991) A move to GIS: some empirical evidence of users' experiences. *Proceedings of Mapping Awareness 1991*, conference held in London, February, Blenheim Online, London

**Medyckyj-Scott D J, Cornelius S C** (1994) User viewpoint: a survey. In: Hart T, Tulip A (eds) *Geographic Information Systems Report*. Unicom Seminars, Uxbridge, UK

**Medyckyj-Scott D J, Hearnshaw H M** (1994) *Human Factors in Geographical Information Systems*. Belhaven, London

**Meehan B** (2007) *Empowering Electric and Gas Utilities with GIS*. Esri Press, Redlands

**Milan D J, Heritage G L, Hetherington, D** (2007) Application of 3D laser scanner in the assessment of erosion and deposition volumes and channel change in a proglacial river. *Earth Surface Processes and Landforms*, **32** (11): 1657–74

**Miller H J, Han J** (2001) *Geographic data mining and knowledge discovery*. Taylor and Francis, London

**Mitchell A** (1997) *Zeroing in: Geographic Information Systems at Work in the Community*. Esri Press, Redlands

**Mitchell A** (1999) *The Esri Guide to GIS Analysis. Volume 1: Geographic Patterns and Relationships*. Esri Press, Redlands

**Monmonier M** (1972) Contiguity-biased class-interval selection: a method for simplifying patterns on statistical maps. *Geographical Review* **62** (2): 203–28

**Monmonier M** (1993) *Mapping it Out*. University of Chicago Press, Chicago

**Monmonier M** (1995) *Drawing the Line*. Henry Holt, New York.

**Monmonier M** (1996) *How to Lie with Maps*. University of Chicago Press, Chicago

**Moore I D, Grayson R B, Ladson A R** (1991) Digital terrain modelling: a review of hydrological, geomorphological, and biological applications. *Hydro. Proc.* **5**: 3–30

**Mounsey H** (1991a) Eurodata – myth or reality? In Cadoux-Hudson J, Heywood D I (eds) *The Association for Geographic Information Yearbook 1991*. Taylor and Francis/ Miles Arnold, London, 59–64

**Mounsey H** (1991b) Multisource, multinational environmental GIS: lessons learnt from CORINE. In: Maguire D J, Goodchild M F, Rhind D W (eds) *Geographical Information Systems: Principles and Applications*. Longman, London, vol. 2, pp. 185–200

**Mounsey H** (1995) GIS: Justification in the face of organizational change. In: *GIS in Business: Discovering the Missing Piece in your Business Strategy*. GeoInformation International, Banbury, UK, pp. 25–8

**Muller J-C** (1993) Latest developments in GIS/LIS. *International Journal of Geographical Information Systems* **7** (4): 293–303

**Mullin J** (1992) Computer focus for ambulance inquiry. *Guardian*, 30 October

**Murray K J, Hart G, Allan P** (2004) The Digital National Framework – bridging information through geography. *Proceedings of the XXth International Society Photogrammety and Remote Sensing Congress*. www.isprs.org/ istanbul2004/ comm4/papers/359.pdf

**Nasirin S, Birks F D, Jones B** (2003) Re-examining Fundamental GIS Implementation Constructs Through Grounded Theory Approach. *Telematics and Informatics* **20** (4): 331–47

**National Research Council** (1997) *The Future of Spatial Data and Society*. National Academies Press, Washington DC

**NCGIA** (1989) The research plan of the National Center for Geographic Information and Analysis. *International Journal of Geographical Information Systems* **3** (2): 117–36

**Newcomer J A, Szajgin J** (1984) Accumulation of thematic map errors in digital overlay analysis. *American Cartographer* **11** (1): 58–62

**Nijkamp P** (1980) *Environmental Policy Analysis: Operational Methods and Models*. Wiley, New York

**Nijkamp P, Scholten H J** (1993) Spatial information systems: design, modelling and use in planning. *International Journal of Geographical Information Systems* **7** (1): 85–96

**NJUG** (1986) *Proposed data exchange format for utility map data. NJUG 11*, National Joint Utilities Group, 30 Millbank, London SW1P 4RD

**Obermeyer N J** (1999) Measuring the benefits and costs of GIS. In: Longley P A, Goodchild M F, Maguire D J, Rhind D W (eds) *Geographical Information Systems*. Wiley, New York, pp. 601–10

**O'Callaghan J F, Garner B J** (1991) Land and geographical information systems in Australia. In: Maguire D J, Goodchild M F, Rhind D W (eds) *Geographical Information Systems: Principles and Applications*. Longman, London, vol. 2, pp. 57–70

**O'Callaghan J F, Mark D M** (1984) The extraction of drainage networks from digital elevation data. *Computer Vision, Graphics and Image Processing* **28**: 328–44

**ODPM** (2004) *Community Involvement in Planning: The Government's Objectives*. HMSO, London

**Odum T H** (1971) *Environment, Power and Society*. Wiley, London

**OECD** (2001) *Citizens as Partners: Information, Consultation and Public Participation in Policy-Making*. OECD, Paris

**Onsrud H J, Pinto J K** (1991) Diffusion of geographic information innovations. *International Journal of Geographical Information Systems* **5** (4): 447–67

**Open Geospatial Consortium (OGC)** (2010a) OpenGIS Implementation Specification for Geographic Information – Simple feature access – Part 2: SQL option. http://www.opengeospatial.org/standards/sfs

**Open Geospatial Consortium (OGC)** (2010b) About OGC. http://www.opengeospatial.org/ogc

**Openshaw S** (1984) *The modifiable areal unit problem. Concepts and Techniques in Modern Geography*, vol. 38, GeoBooks, Norwich, UK

**Openshaw S** (1989) Learning to live with errors in spatial databases. In: Goodchild M F, Gopal S (eds) *The Accuracy of Spatial Databases*. Taylor and Francis, London, pp. 263–76

**Openshaw S** (1990) Spatial referencing for the user in the 1990s. *Mapping Awareness* **4** (2): 24

**Openshaw S** (1993) GIS crime and criminality *Environment and Planning A* **25** (4): 451–8

**Openshaw S** (ed) (1995) *The 1991 Census User's Handbook*. Longman GeoInformation, London

**Openshaw S, Goddard J** (1987) Some implications of the commodification of information and the emerging information economy for applied geographical analysis in the United Kingdom. *Environment and Planning A* **19**: 1423–39

**Openshaw S, Openshaw C O** (1997) *Artificial Intelligence in Geography*. John Wiley, Chichester

**Openshaw S, Perrée T** (1996) User-centred intelligent spatial analysis of point data. In: Parker D (ed) *Innovations in GIS 3*. Taylor and Francis, London, pp. 119–34

**Openshaw S, Rao L** (1995) Algorithms for reengineering 1991 census geography. *Environment and Planning A* **27**: 425–46

**Openshaw S, Carver S, Fernie J** (1989) *Britain's Nuclear Waste: Safety and Siting*. Belhaven, London

**Openshaw S, Charlton M, Carver S** (1991) Error propagation: a Monte Carlo simulation. In: Masser I, Blakemore M (eds) *Handling Geographic Information: Methodology and Potential Applications*. Longman, London, pp. 78–101

**Openshaw S, Charlton M, Wymer C, Craft A** (1987) A mark 1 geographical analysis machine for the automated analysis of point data sets. *International Journal of Geographical Information Systems* **1**: 335–58

**Openshaw S, Charlton M, Craft A W, Birch J M** (1988) An investigation of leukaemia clusters by use of a geographical analysis machine. *Lancet* **1**: 272–3

**Ordnance Survey** (2010) *Ordnance Survey Launches OS OpenData in Groundbreaking National Initiative*. April 2010. http://www.ordnancesurvey.co.uk/oswebsite/media/news/2010/April/OpenData.html

**Ormsby T, Alvi J** (1999) *Extending ArcView GIS*. Esri Press, Redlands

**Otawa T** (1987) Accuracy of digitizing: overlooked factor in GIS operations. *Proceedings of GIS '87*. ACSM, ASPRS, AAG and URISA, San Francisco, pp. 295–9

**Oxborrow E P** (1989) *Databases and Database Systems: Concepts and Issues*. 2nd edn. Chartwell Bratt, Sweden

**Parker D** (ed) (1996) *Innovations in GIS 3*. Taylor and Francis, London

**Perkal J** (1956) On the epsilon length. *Bulletin de L'Académie Polonaise des Sciences* **4** (7): 399–403

**Petch J R, Pauknerova E, Heywood D I** (1995) GIS in nature conservation: the Zdarske Vrchy project, Czech Republic. *ITC Journal* **2**: 133–42

**Peterson M P** (1995) *Interactive and Animated Cartography*. Prentice-Hall, Englewood Cliffs, NJ

**Peuquet D J** (1984) A conceptual framework and comparison of spatial data models. *Cartographica* **21** (4): 66–113

**Peuquet D J** (1990) A conceptual framework and comparison of spatial data models. In: Peuquet D J, Marble D F (eds) *Introductory Readings in Geographical Information Systems*. Taylor and Francis, London, pp. 209–14

**Peuquet D J** (1999) Time in GIS and geographical databases. In: Longley PA, Goodchild M F, Maguire D J, Rhind D W (eds) *Geographical Information Systems: Principles and Technical Issues*. Wiley, New York, vol. 1, pp. 91–103.

**Peuquet D J, Bacastow T** (1991) Organizational issues in the development of geographical information systems: a case study of US Army topographic information automation. *International Journal of Geographical Information Systems* **5** (3): 303–19

**Peuquet D J, Marble D F** (1990) *Introductory Readings in Geographical Information Systems*. Taylor and Francis, London

**Pick J B** (2008) *Geo-Business GIS in the Digital Organization*. Wiley, New Jersey

**Pickles J** (ed) (1995) *Ground Truth: the Social Implications of Geographical Information Systems*. Guilford Press, New York

**Plewe B** (1997) *GISOnline: Information Retrieval, Mapping, and the Internet*. OnWord Press, Santa Fe, NM

**Poiker T K** (1982) Looking at computer cartography. *GeoJournal* **6** (3): 241–9

**Price M F, Heywood D I** (eds) (1994) *Mountain Environments and Geographical Information Systems*. Taylor and Francis, London

**Ramasubramanian L** (1996) Building communities: GIS and participatory decision making. *Journal of Urban Technology* **3** (1): 67–79

**Rana S, Haklay M, Dodge M** (2001) GIS Timeline: www.casa.ucl.ac.uk/gistimeline

**Raper J** (1997) Experiences of integrating VRML and GIS. Paper presented at *ESIG97*, Lisbon

**Raper J F** (2000) *Multidimensional Geographic Information Science*. Taylor and Francis, London

**Raper J F, Kelk B** (1991) Three-dimensional GIS. In: Maguire D J, Goodchild M F, Rhind D W (eds)

*Geographical Information Systems: Principles and Applications.* Longman, London, vol. 1, pp. 299–317

**Raper J F, Rhind D W** (1990) UGIX(A): the design of a spatial language interface to a topological vector GIS. *Proceedings of the 4th International Conference on Spatial Data Handling.* International Geographical Union, Columbus, Ohio, pp. 405–12

**Raper J F, Rhind D W, Shepherd J W** (1992) *Postcodes: the New Geography.* Longman, Harlow, UK

**Raper J F, Livingstone D, Bristow C, Horn D** (1999) Developing process-response models for spits. In: Kraus D (ed) *Proceedings of Coastal Sediments 1999,* July 1999, Long Island, New York, pp. 1755–69

**Rees P, Thomas F, Duke-Williams O** (2002) Migration data from the census. In: Rees P, Martin D, Williamson P (eds) *The Census Data System.* Wiley, Chichester

**Reeve D E** (1996) *Module 4: Attribute data.* UNIGIS postgraduate diploma in GIS by distance learning. Course materials. 5th edn. Manchester Metropolitan University

**Reeve D E** (1997) *Module 10: GIS in organisations.* UNIGIS postgraduate diploma in GIS by distance learning. Course materials. 6th edn. Manchester Metropolitan University

**Reeve D E, Petch J R** (1998) *GIS Organizations and People: a Socio-technical Approach.* Taylor and Francis, London

**Reina P** (1997) At long last, AM/FM/GIS enters the IT mainstream. *Information Technologies for Utilities* (Spring): 29–34

**Rhind D** (1987) Recent developments in GIS in the UK. *International Journal of Geographical Information Systems* **1** (3): 229–41

**Rhind D** (1988) Personality as a factor in the development of a discipline: the example of computer-assisted cartography. *American Cartographer* **15** (3): 277–89

**Rhind D** (1992) The next generation of Geographical Information Systems and the context in which they will operate. *Computers, Environment and Urban Systems* **16:** 256–68

**Rhind D W** (1989) Why GIS? *ARC News* (Summer): pp. 28–9

**Rhind D W, Mounsey H W** (1989) Research policy and review 29: The Chorley Committee and 'Handling Geographic Information'. *Environment and Planning A* **21:** 571–85

**Rhind D W, Goodchild M F, Maguire D J** (1991) Epilogue. In: Maguire D J, Goodchild M F, Rhind D W (eds) *Geographical Information Systems: Principles and Applications.* Longman, London, vol. 2, pp. 313–27

**Rhind D W, Wyatt B K, Briggs D J, Wiggins J C** (1986) The creation of an environmental information system for the European Community. *Nachrichter aus dem Karten und Veressungsvesen Series 2,* **44:** 147–57

**Richards D R, Jones N L, Lin H C** (1993) Graphical innovations in surface water flow analysis. In: Goodchild M F, Parks B O, Steyaert L T (eds) *Environmental Modelling with GIS.* Oxford University Press, Oxford and New York, pp. 188–95

**Ripple W J** (ed) (1994) *The GIS Applications Book: Examples in Natural Resources: A Compendium.* American Society for Photogrammetry and Remote Sensing, Maryland

**Robertson B** (1988) *How to Draw Charts and Diagrams.* North Light, Cincinnati, Ohio

**Robinson A H, Morrison J L, Muehrecke P C, Kimerling A J, Guptill S C** (1995) *Elements of Cartography.* 6th edn. Wiley, New York

**Roche S, Caron C** (2009) *Organizational Facets of GIS.* Wiley, New Jersey

**Sahai R** (1996) *Inside MicroStation 95.* OnWord Press, USA

**Salge F** (1996) MEGRIN: a pragmatic approach towards a European Geographic Information infrastructure. In: Peckham R J (ed) *Proceedings of the 2nd EC-GIS Workshop, Genova, Italy, 26–28 June.* 16441 EN, European Commission, Brussels

**Samet H** (1989) *Applications of Spatial Data Structures: Computer Graphics Image Processing and Other Areas.* Addison-Wesley, London

**Sanderson M** (2005) personal communication

**Schimel D S, Burke I C** (1993) Spatial interactive models of atmosphere–ecosystem coupling. In: Goodchild M F, Parks B O, Steyaert L T (eds) *Environmental Modelling with GIS.* Oxford University Press, Oxford and New York, pp. 284–9

**Scholten H** (1997) *Towards a European spatial meta information system: ESMI.* Presentation at 3rd EC – GIS Workshop. 25–27 June 1997. Leuven, Belgium

**Seaborn D** (1995) *Database Management in GIS – Past, Present, Future: an Enterprise Perspective for Executives.* Seaborn Associates, 475 Cloverdale Road, Ottawa, Canada

**See L, Openshaw S** (1997) An introduction to the fuzzy logic modelling of spatial interaction. In: Hodgson S, Rumor M, Harts J J (eds) *Proceedings of the Joint European Conference and Exhibition on Geographical Information Systems,* Vienna, Austria, pp. 809–18

**Seegar H** (1999) Spatial referencing and coordinate systems. In: Longley PA, Goodchild M F, Maguire D J, Rhind D W (eds) *Geographical Information Systems.* Wiley, New York, vol. 1, pp. 427–36

**Shand P, Moore R** (eds) (1989) *AGI Yearbook 1989.* Taylor and Francis/Miles Arnold, London

**Shiffer M J** (1995a) Interactive multimedia planning support: moving from stand alone systems to the Web. *Environment and Planning B: Planning and Design,* **22:** 649–64.

**Shiffer M J** (1995b) Multimedia representational aids in urban planning support system. In: Marchese F (ed) *Understanding Images.* Springer-Verlag. New York, pp. 77–90

**Shupeng C** (1987) Geographical data handling and GIS in China. *International Journal of Geographical Information Systems* **1** (3): 219–28

**Sieber R E** (2000) GIS Implementation in the Grassroots. *Journal of the Urban and Regional Information Systems Association* **12** (1): 15–30

**Simmons J, Colleagues** (2002) *Crime in England and Wales 2001/2002*. HMSO, London. http://www.homeoffice. gov.uk/rds/pdfs2/hosb702.pdf

**Sinton D F** (1992) *Reflections on 25 Years of GIS*. Supplement to GIS World

**Skidmore S, Wroe B** (1988) *Introducing Systems Analysis*. Blackwell, Oxford

**Smith B, Mark D** (2003) Do mountains exist? Towards an ontology of landforms. *Environment and Planning B: Planning and Design* **30**: 411–27

**Smith D A, Tomlinson R F** (1992) Assessing costs and benefits of geographical information systems: methodological and implementation issues. *International Journal of Geographical Information Systems* **6** (3): 247–56

**Smith T R, Menon S, Star J L, Estes J E** (1987) Requirements and principles for the implementation of large-scale geographic information systems. *International Journal of Geographical Information Systems* **1** (1): 13–31

**Sobel J** (1990) Principal components of the Census Bureau's TIGER File. In: Peuquet D J, Marble D F (eds) *Introductory Readings in Geographic Information Systems*. Taylor and Francis, London, pp. 12–19

**Stair R, Reynolds G** (2010) *Principles of Information Systems, A Managerial Approach*. 9th edn. Course Technology, Boston.

**Steenberghen T, Dufays T, Thomas I, Flahaut B** (2004) Intra-urban location and clustering of road accidents using GIS: a Belgian example. *International Journal of Geographical Information Science* **18** (2): 169–81

**Stern N, Stern R A** (1993) *Computing in the Information Age*. Wiley, New York

**Stevens D, Dragicevic S, Rothley K** (2007) iCity: A GIS-CA modelling tool for urban planning and decision making. *Environmental Modelling and Software* **22** (6): 761–73

**Stewart I** (1989) *Does God Play Dice? The Mathematics of Chaos*. Basil Blackwell, Oxford

**Steyaert L T** (1993) A perspective on the state of environmental simulation modelling. In: Goodchild M F, Parks B O, Steyaert L T (eds) *Environmental Modelling with GIS*. Oxford University Press, Oxford and New York, pp. 16–30

**Stillwell J, Geertman S** (eds) (2003) *Planning Support Systems in Practice*. Springer, Heidelberg

**Stillwell J, Duke-Williams O, Feng Z, Boyle P** (2005) Delivering census interaction data to the user: data provision and software developments. *Working Paper 01/05*, School of Geography, University of Leeds, Leeds

**Stocks A M, Heywood D I** (1994) Terrain modelling for mountains. In: Price M F, Heywood D I (eds) *Mountain Environments and Geographical Information Systems*. Taylor and Francis, London, pp. 25–40

**Syed N, Birks F D, Jones B** (2003) Re-examining Fundamental GIS Implementation Constructs Through Grounded Theory Approach. *Telematics and Informatics* **20** (4): 331–47

**Symington D F** (1968) Land use in Canada: the Canadian land inventory. *Canadian Geographical Journal*

**Taylor D R F** (1991a) *Geographic Information Systems: the Microcomputer and Modern Cartography*. Pergamon Press, Oxford, London and New York

**Taylor D R F** (1991b) GIS and developing nations. In: Maguire D J, Goodchild M F, Rhind D W (eds) *Geographical Information Systems: Principles and Applications*. Longman, London, vol. 2, pp. 71–84

**Taylor P J** (1990) *GKS. Political Geography Quarterly* **9** (3): 211–12

**Thill J-C** (1999) *Spatial Multicriteria Decision Making and Analysis: a Geographic Information Systems Approach*. Ashgate, Aldershot

**Tobler W A** (1970) A computer movie simulating urban growth in the Detroit region. *Economic Geography* **46**: 234–40

**Tobler W R** (1976) Spatial interaction patterns. *Journal of Environmental Systems* **6**: 271–301

**Tomlin C D** (1983) Digital cartographic modelling techniques in environmental planning. Unpublished PhD dissertation, Yale University, Connecticut

**Tomlin C D** (1990) *Geographic Information Systems and Cartographic Modelling*. Prentice-Hall, Englewood Cliffs, New Jersey

**Tomlin C D** (1991) Cartographic modelling. In: Maguire D J, Goodchild M F, Rhind D W (eds) *Geographical Information Systems: Principles and Applications*. Longman, London, vol. 1, pp. 361–74

**Tomlinson R F** (1988) The impact of the transition from analogue to digital cartographic representation. *American Cartographer* **15** (3): 249–63

**Tomlinson R F** (1990) Geographic Information Systems – a new frontier. In: Peuquet D J, Marble D F (eds) *Introductory Readings in Geographic Information Systems*. Taylor and Francis, London, pp. 18–29

**Tomlinson R F** (2003) *Thinking About GIS: Geographic Information System Planning for Managers*. Esri Press, Redlands

**Tomlinson R F, Calkins H W, Marble D F** (1976) *Computer Handling of Geographic Data*. UNESCO Press, Paris

**Trotter A** (1998) Teachers find plenty of uses for software that covers the map. *Education Week* **17**, 15 April

**Trotter C M** (1991) Remotely sensed data as an information source for geographical information systems in natural resource management: a review. *International Journal of Geographical Information Systems* **5** (2): 225–39

**Tufte E** (1983) *The Visual Display of Quantitative Information.* Graphics Press, Cheshire, CT

**Tulloch D** (2003) What PPGIS Really Needs Is... *2nd Annual Public Participation GIS Conference,* Portland State University, Portland, Oregon

**Tyler C** (1989) Climatology: technology's answer to the British weather. *Geographical Magazine* **LXI** (7): 44–7

**UNECE** (1998) *Convention on Access to Information, Public Participation in Decision-making and Access to Justice in Environmental Matters.* UNECE, Geneva

**Unwin D** (1981) *Introductory Spatial Analysis.* Methuen, London

**Unwin D J** (1991) The academic setting of GIS. In: Maguire D J, Goodchild M F, Rhind D W (eds) *Geographical Information Systems: Principles and Applications.* Longman, London, vol. 1, pp. 81–90

**Unwin D J** et al. (1990) A syllabus for teaching geographical information systems. *International Journal of Geographical Information Systems* **4** (4): 457–65

**US Bureau of Census** (1990) Technical description of the DIME system. In: Peuquet D J, Marble D F (eds) *Introductory Readings in Geographic Information Systems.* Taylor and Francis, London, pp. 100–11

**Vasconcelos M J, Pereira J M C, Zeigler B P** (1994) Simulation of fire growth in mountain environments. In: Price M F, Heywood D I (eds) *Mountain Environments and Geographical Information Systems.* Taylor and Francis, London, pp. 167–85

**Vertex** (2005) http://www.vertex.com

**Vincent P, Daly R** (1990) GIS and large travelling salesman problems. *Mapping Awareness* **4** (1): 19–21

**Von Bertalanffy L** (1968) *General Systems Theory: Foundation, Developments and Applications.* Penguin, London

**Voogd H** (1983) *Multicriteria Evaluation for Urban and Regional Planning.* Pion, London

**Wainwright A** (2003) *A Pictorial Guide to the Lakeland Fells: Being an Illustrative Account of a Study and Exploration of the Mountains in the English Lake District: Northern Fells Book 5.* Frances Lincoln Publishers Ltd, London

**Walsby J** (1995) The causes and effects of manual digitizing on error creation in data input to GIS. In: Fisher P (ed) *Innovations in GIS 2.* Taylor and Francis, London, pp. 113–24

**Walsh S J, Lightfoot D R, Butler D R** (1987) Recognition and assessment of error in Geographic Information Systems. *Photogrammetric Engineering and Remote Sensing* **53** (10): 1423–30

**Ward R C** (1975) *Principles of Hydrology.* 2nd edn. McGraw-Hill, Maidenhead, UK

**Waters N M** (1989) *NCGIA Core Curriculum.* Lectures 40 and 41. University of California, Santa Barbara

**Weibel R, Heller M** (1991) Digital terrain modelling. In: Maguire D J, Goodchild M F, Rhind D W (eds) *Geographical Information Systems: Principles and Applications.* Longman, London, vol. 1, pp. 269–97

**Wheatley, D** (1995) Cumulative Viewshed Analysis: a GIS-based Method for Investigating Intervisibility, and its Archaeological Application. In G Lock and Z Stançic (eds) *Archaeology and Geographic Information Systems: a European Perspective.* Taylor and Francis, London, 171–85

**Wheatley D, Gillings M** (2000) Vision, Perception and GIS: Developing Enriched Approaches to the Study of Archaeological Visibility. In G R Lock (ed) *Beyond the Map: Archaeology and Spatial Technologies,* Amsterdam: NATO Science Series A: Life Sciences, **321**:1–27

**Wilson A G** (1975) Some new forms of spatial interaction model: a review. *Transportation Research* **9**: 167–79

**Wilson J D** (1997) Technology partnerships spark the industry. *GIS World* **10** (4): 36–42

**Wilson J P, Gallant J** (2000) *Terrain Analysis: Principles and Applications.* John Wiley, London

**Wise S** (2002) *GIS Basics.* Taylor and Francis, London

**Wise S, Craglia M** (eds) (2008) *GIS and Evidence-based policy making* (Innovations in GIS). CRC Press, Taylor and Francis Group, Boca Raton, FL

**Wood C H, Keller C P** (eds) (1996) *Cartographic Design.* Wiley, Chichester

**Wood D** (1993) *The Power of Maps.* Routledge, London

**Wood J** (1998) Modelling the continuity of surface form using digital elevation models. In: Poiker T, Chrisman N (eds) *Proceedings, 8th International Symposium on Spatial Data Handling.* IGU, Vancouver, pp. 725–36

**Woodcock C E, Sham C H, Shaw B** (1990) Comments on selecting a GIS for environmental management. *Environmental Management* **14** (3): 307–15

**Wood-Harper A T, Antill L, Avison D E** (1995) *Information Systems Definition: the Multiview Approach.* Blackwell Scientific, Oxford

**Worboys M F** (1995) *GIS: a Computing Perspective.* Taylor and Francis, London.

**Worboys M F** (1999) Relational databases and beyond. In: Longley P A, Goodchild M F, Maguire D J, Rhind D W (eds) *Geographical Information Systems.* Wiley, New York. vol. 1, pp. 373–84

**Worboys M F, Duckham M** (2004) *GIS: A Computing Perspective.* 2nd edn. Taylor and Francis, London

**Worboys M F, Hearnshaw H M, Maguire D J** (1991) Object-oriented data modelling for spatial databases. *International Journal of Geographical Information Systems* **4** (4): 369–83

**Zeiler M** (1994) *Inside ARC/INFO.* OnWord Press, USA

# Glossary

This glossary contains the definition of most of the non-standard words used in the text. It is not intended as a comprehensive GIS glossary. Links to several comprehensive GIS glossaries are provided on the book web page, www.pearsoned.co.uk/heywood.

*A posteriori* **models**: Models that are designed to explore an established theory.

*A priori* **models**: Models that are used to model processes for which a body of theory has yet to be established.

**Aerial photograph**: Photograph taken from an aerial platform (usually an aeroplane), either vertically or obliquely.

**Aggregation**: The process of combining smaller spatial units, and the data they contain, into larger spatial units by dissolving common boundaries and lumping the data together.

**AM/FM**: Automated mapping and facilities management.

**Analysis scheme**: The logical linking of spatial operations and procedures to solve a particular application problem.

**Analytical hill shading**: The process of calculating relief shadows and using these as a visualization technique to enhance the portrayal of relief on a map.

**Annotation**: Alphanumeric or symbolic information added to a map to improve communication.

**Arc**: An alternative term for a line feature formed from a connected series of points.

**ARC/INFO**: A leading GIS software package developed by the Environmental Systems Research Institute (Esri).

**Area**: The entity type used to represent enclosed parcels of land with a common description.

**Area cartogram**: A type of cartogram that is used to depict the size of areas relative to their importance according to some non-spatial variable such as population or per capita income.

**Artificial Intelligence (AI)**: Field of study concerned with producing computer programs capable of learning and processing their own 'thoughts'.

**Artificial Neural Network (ANN):** See 'Neural networks'.

**Aspatial query**: The action of questioning a GIS database on the basis of non-spatial attributes.

**Aspect**: The direction in which a unit of terrain faces. Aspect is usually expressed in degrees from north.

**Attributes**: Non-graphical descriptors relating to geographical features or entities in a GIS.

**Automated cartography**: See 'Computer cartography'.

**Azimuthal projections**: Map projections in which the surface of the globe is projected onto a flat plane.

**Basic Spatial Unit (BSU)**: The smallest spatial entity to which data are encoded.

**Block coding**: Extension of the run-length encoding method of coding raster data structures to two dimensions by using a series of square blocks to store the data.

**Boolean algebra**: Operations based on logical combinations of objects (using AND, NOT, OR and XOR).

**Boolean overlay**: A type of map overlay based on Boolean algebra.

**Buffering**: The creation of a zone of equal width around a point, line or area feature.

**Canadian Geographic Information System (CGIS):** An early GIS using data collected for the Land Inventory System that was developed as a result of the requirements of the Canadian Agriculture and Development Act.

**Cartesian co-ordinates**: The system for locating a point by reference to its distance from axes intersecting at right angles, often represented as a grid on a map.

**Cartograms**: Maps that show the location of objects relative to a non-Euclidean variable such as population density or relative distance.

**Cartographic modelling**: A generic methodology for structuring a GIS analysis scheme.

**Cartography**: The profession of map drawing; the study of maps.

**CASE tools**: Computer-assisted software engineering. Software tools used to help reduce program development time.

**CD-ROM**: Method of storing large volumes of data on optical discs.

**Census**: The collection of data about the entire population of a country or region.

**Central point linear cartogram**: A form of cartogram that shows space distorted from a central point on the basis of time taken or cost incurred to travel to a series of destination areas.

**Chain coding**: Method of raster data reduction that works by defining the boundary of the entity.

**Chaos theory**: Idea that seemingly minor events accumulate to have complex and massive effects on dynamic natural systems.

**Chart junk**: Unnecessary, often confusing, annotation added to maps and charts.

**Choice alternatives**: Range of feasible solutions to be evaluated in a decision problem in MCE.

**Choice set**: The set of choice alternatives in MCE.

**Choropleth map**: A thematic map that displays a quantitative attribute using ordinal classes. Areas are shaded according to their value and a range of shading classes.

**Cleaning**: See 'Data editing'.

**Computer-aided cartography**: See 'Computer cartography'.

**Computer-aided design (CAD)**: Software designed to assist in the process of designing and drafting. This is normally used for architectural and engineering applications, but can also be used for drafting maps as an input to GIS.

**Computer-assisted cartography (CAC)**: Software designed to assist in the process of designing and drawing maps.

**Computer-assisted drafting (CAD)**: See 'Computer-aided design (CAD)'.

**Computer cartography**: The generation, storage and editing of maps using a computer.

**Computer movies**: Animated maps showing how a chosen variable changes with time. These are used to spot temporal patterns in spatial data.

**Conceptual data model**: A model, usually expressed in verbal or graphical form, that attempts to describe in words or pictures quantitative and qualitative interactions between real-world features.

**Concordance–discordance analysis**: Method of MCE based on lengthy pairwise comparison of outranking and dominance relationships between each choice alternative in the choice set.

**Conic projections**: Map projections in which the surface of the globe is projected onto a cone (for example, Alber's Equal Area projection).

**Constraints**: Set of minimum requirements for a decision problem in MCE.

**Contour**: A line on a topographic map connecting points of equal height and used to represent the shape of the Earth's surface.

**Cookie cutting**: See 'IDENTITY overlay'.

**CORINE**: The Co-ordinated Information on the European Environment programme, initiated in 1985 by the European Union to create a database that would encourage the collection and co-ordination of consistent information to aid European Community policy.

**Criteria**: Attributes by which choice alternatives are evaluated in MCE.

**Cylindrical projections**: Map projections in which the surface of the globe is projected onto a cylinder (for example, the Mercator projection).

**Dangling arc**: A dead-end line or arc feature that is connected to other lines in the data layer at one end only.

**Data**: Observations made from monitoring the real world. Data are collected as facts or evidence, which may be processed to give them meaning and turn them into information.

**Data accuracy**: The extent to which an estimated data value approaches its true value.

**Data bias**: The systematic variation of data from reality.

**Data conversion**: The process of converting data from one format to another. With the many different GIS data formats available this can be a difficult task requiring specialized software.

**Data editing**: The process of correcting errors in data input into a GIS. This can be carried out manually or automatically.

**Data error**: The physical difference between the real world and a GIS facsimile.

**Data input**: The process of converting data into a form that can be used by a GIS.

**Data precision**: The recorded level of detail of the data.

**Data stream**: The process of progressing from raw data to integrated GIS database. This includes all stages of data input, transformation, re-projection, editing and manipulation.

**Database**: A collection of data, usually stored as single or multiple files, associated with a single general category.

**Database management system (DBMS)**: A set of computer programs for organizing information at the core of which will be a database.

**Decision support system (DSS)**: A system, usually computerized, dedicated to supporting decisions regarding a specific problem or set of problems.

**Delaunay triangulation**: Method of constructing a TIN model such that three points form the corners of a Delaunay triangle only when the circle that passes through them contains no other points.

**Deterministic model**: A model for which there is only one possible answer for a given set of inputs.

**Differential GPS**: The process of using two GPS receivers to obtain highly accurate and precise position fixes.

**Diffusion**: The spatio-temporal process concerned with the movement of objects from one area to another through time. Examples include forest fire development, the movement of pollutants through any medium, and the spread of a disease through a population.

**Digital elevation model (DEM):** A digital model of height (elevation or altitude) represented as regularly or irregularly spaced point height values.

**Digital mapping:** See 'Computer cartography'.

**Digital terrain model (DTM):** A digital model of a topographic surface using information on height, slope, aspect, breaks in slope and other topographic features.

**Digitizer:** A piece of computer hardware used to convert analogue data into digital format. See 'Digitizing table', 'Scanner' and 'Laser line follower'.

**Digitizing:** The process of converting data from analogue to digital format.

**Digitizing table:** A table underlain by a fine mesh of wires and connected to a computer that allows the user to record the location of features on a map using an electronic cursor. These are used to digitize maps and other graphical analogue data sources manually.

**Disaggregation:** The reverse of 'aggregation'.

**Distance decay:** A function that represents the way that some entity or its influence decays with distance from its geographical location.

**Douglas–Peucker algorithm:** A geometric algorithm used to thin out the number of points needed to represent the overall shape of a line feature.

**DXF:** Data eXchange Format. A file format used for the exchange of GIS data.

**Dynamic model:** A model in which time is the key variable whilst all other input variables remain constant. Outputs from the model vary as time progresses.

**Ecological fallacy:** A problem associated with the 'modifiable areal unit problem' that occurs when it is inferred that data for areas under study can be applied to individuals in those areas.

**Edge matching:** The process of joining data digitized from adjacent map sheets to ensure a seamless join.

**Electronic distance metering (EDM):** A theodolite combined with an optical rangefinder for accurate distance measurements.

**Enumeration district (ED):** The area over which one census enumerator delivers and collects census forms. The size of these units, though small, is enough to ensure anonymity of census respondents in data reported at this level of detail.

**Envelope polygon:** The external polygon of a vector map inside which all other features are contained.

**Epsilon modelling:** A method of estimating the effects of positional error in GIS overlay operations. Epsilon modelling is based on the use of buffer zones to account for digitizing error around point, line and area features.

**Equilibrium forecast model:** A model in which forecasts are made on the basis of change in one or more elements of the system or process whilst all other factors remain constant.

**Error propagation:** The generation of errors in a GIS database at various stages of the data stream and during subsequent analyses.

**Euclidean distance:** A straight-line distance measured as a function of Euclidean geometric space.

**Evaluation matrix:** A matrix containing criterion scores (rows) for each choice alternative (columns) in the choice set for any MCE problem.

**Evapotranspiration:** The process by which water vapour re-enters the atmosphere directly through evaporation and from plant transpiration.

**Eyeballing:** See 'Line threading'.

**Feasibility study:** An evaluation of the costs and benefits of adopting an IT solution to a problem.

**Feature codes:** Unique codes describing the feature to which they are attached in the GIS database.

**Features:** See 'Spatial entities'.

**Filter:** A cell matrix of varying shape and size used to modify the cell values in a raster map layer through a variety of mathematical procedures such as mean, sum, maximum and minimum. A filter is often used to smooth noisy data.

**GANTT charts:** Graphical time charts used to assist project management.

**Gazetteer:** A dictionary or index of geographical names.

**General systems theory:** A theory based on the idea that to understand the complexity of the real world, we must attempt to model this complexity.

**Generalization:** The process by which information is selectively removed from a map in order to simplify pattern without distortion of overall content.

**Geographical analysis:** Any form of analysis using geographical data.

**Geomorphology:** The study of the Earth's surface and the physical processes acting on it.

**Global Positioning System (GPS):** A system of orbiting satellites used for navigational purposes and capable of giving highly accurate geographic co-ordinates using hand-held receivers.

**GLONASS:** Global Orbiting Navigation Satellite System. The Russian equivalent of the US GPS system.

**Graphical User Interface (GUI):** Computer interface, such as MS Windows, that is used to provide a user-friendly interface between a user and the computer.

**Great circle:** A circle with the same diameter as the Earth. This is used to describe the shortest route between any two points on the Earth's surface.

**Groundwater percolation:** The process by which water from the soil enters the underlying geology by slow downward movement.

**Hard systems analysis:** A set of theory and methods for modelling the complexity of the real world.

**Hydrological cycle**: A dynamic natural system concerned with the circulation of water through the atmosphere, biosphere and geosphere.

**Ideal point analysis**: An MCE algorithm based on the evaluation of choice alternatives against a hypothetical ideal solution.

**IDENTITY overlay**: Polygon-on-polygon overlay corresponding to the Boolean OR and AND overlays. The output map will contain all those polygons from the first map layer and those which fall within these from the second map layer. Also referred to as 'cookie cutting'.

**Infiltration rate**: The rate at which surface runoff and precipitation soaks into the ground surface.

**Internet, The**: The global network of computers, originally set up as a means of secure communications for military and intelligence purposes.

**Internet protocols**: An established language used for the communication between systems on the Internet. HTTP and TCP/IP are examples.

**INTERSECT overlay**: Polygon-on-polygon overlay corresponding to the Boolean AND overlay. The output map will contain only those polygons that cover areas common to both sets of input polygons.

**Intranet**: A secure closed network that uses Internet protocols.

**Isoline**: A line joining areas of equal value.

**JAVA Applet**: A program that will run within a web page, written in the JAVA language.

**Keyboard entry**: A method of entering small volumes of data at the computer keyboard.

**Key-coding**: See 'Keyboard entry'.

**Landsat**: An orbiting satellite giving regular repeat coverage of the whole of the Earth's surface.

**Laser line follower**: An automatic digitizer that uses a laser beam to follow and digitize lines on a map.

**Latitude**: Angular measurement north and south of the equator. Represented on the globe as parallel lines circling the globe perpendicular to the lines of longitude.

**Layer-based approach**: An approach to organizing spatial data into thematic map layers, wherein each map layer contains information about a particular subject and is stored as a separate file (or series of files) for ease of management and use.

**Line**: Entity type used to represent linear features using an ordered set of (x,y) co-ordinate points or a chain of grid cells.

**Line-in-polygon overlay**: The process of overlaying a line map over a polygon map to determine which lines cross which polygons.

**Line of true scale**: Line on a map along which distances are not distorted due to the effects of the map projection method used.

**Line threading**: A method of spatial interpolation reliant on estimating isolines through a series of observed values by hand and eye.

**Lineage**: The record of a data set's origin and all the operations carried out on it.

**Linear weighted summation**: The simplest MCE algorithm based on the simple addition of weighted criterion scores for each alternative. Under this decision rule the choice alternative with the maximum score is the best.

**Link impedance**: The cost associated with traversing a network link, stopping, turning or visiting a centre.

**Linked display**: Method of dynamically linking map and non-map output such as charts and data plots such that changes in one are reflected by changes in the other. Such displays are used to aid exploratory data analysis.

**Local area network (LAN)**: A local network of computers, usually within the same building, which may be used, for instance, to connect a GIS fileserver to a number of GIS access terminals.

**Location-allocation modelling**: The use of network analysis to allocate the location of resources through the modelling of supply and demand through a network.

**Longitude**: Angular measurement east and west of the prime or Greenwich meridian. Represented on the globe as a series of great circles intersecting at both poles.

**Loose coupling**: A method of linking models to GIS in which the user may have to perform transformations on the data output from GIS, spatial analysis or process modelling software before they can be used in another software environment.

**Manhattan distance**: The method used to calculate distance between cells in a raster data layer by counting the distance along the sides of the raster cells.

**Manual digitizing**: The process of digitizing maps and aerial photographs using a table digitizer.

**Map algebra**: A method of combining raster data layers. Mathematical operations are performed on individual cell values from two or more input raster data layers to produce an output value.

**Map overlay**: The process of combining the data from two or more maps to generate a new data set or a new map. This is a fundamental capability of GIS.

**Map projection**: A mathematical and/or geometric method used to transfer the spherical surface of the Earth onto a flat surface such as a map. Many different map projections exist, each with its own advantages and disadvantages.

**Mapematics**: See 'Map algebra'.

**Mathematical model**: A model that uses one or more of a range of techniques, including deterministic, stochastic and optimization methods.

**METEOSAT**: A geostationary weather satellite monitoring the whole of the western hemisphere.

**Middleware**: Technology that breaks down information into packets for exchange between servers and clients.

**Modifiable areal unit problem (MAUP)**: A problem arising from the imposition of artificial units of spatial reporting on continuous geographic phenomena resulting in the generation of artificial spatial patterns.

**Monte Carlo simulation**: A statistical technique used to simulate the effects of random error in data analyses. Random 'noise' is added to the input data and analyses run repeatedly to estimate the effects on the output data.

**Multi-criteria evaluation (MCE)**: A method of combining several, possibly conflicting, criteria maps to derive suitability maps based on trade-off functions and user-specified criteria preference weights.

**Multi-criteria modelling**: See 'Multi-criteria evaluation (MCE)'.

**Multimedia**: Digital media presented in multiple forms, including text, pictures, sound and video.

**National Center for Geographic Information and Analysis (NCGIA)**: An organization set up with funding from the US National Science Foundation to lead US GIS research.

**National transfer format (NTF)**: Standard data transfer format developed and used by the Ordnance Survey in the UK.

**Natural analogue model**: A model that uses actual events or real-world objects as a basis for model construction.

**Natural language**: Spoken or written language.

**NAVSTAR**: American GPS satellite constellation consisting of 24 satellites used for location fixing.

**Network**: (1) A special type of line entity used to represent interconnected lines that allow for the flow of objects (for example, traffic) or information (for example, telephone calls). (2) Two or more interconnected computers used to facilitate the transfer of information.

**Neural networks**: Classification systems that look for pattern and order in complex multivariate data sets. A key component of Artificial Intelligence (AI) techniques.

**Node**: Point entity representing the start or end point of a line or its intersection with another line.

**Nuclear Industry Radioactive Waste Executive (NIREX)**: Organization set up to oversee the disposal of radioactive waste in the UK.

**Object-oriented approach**: Approach to organizing spatial data as discrete objects in a single map space.

**Octree**: Three-dimensional modification of the quadtree data structure.

**Optical character recognition (OCR)**: Software used to recognize the shape of alphanumeric characters in a scanned image and convert these to ASCII text files.

**Optimization model**: A model that is constructed to maximize or minimize some aspect of its output.

**Ordnance Survey**: UK mapping agency.

**PERT**: Program evaluation and review technique. A project management tool.

**Pilot project**: An example software application that shows the potential of a larger application. This is often referred to as a 'demonstrator'. See also 'Prototyping'.

**Pixel**: A single cell in a raster data model. Short for picture element.

**Planar co-ordinates**: See 'Cartesian co-ordinates'.

**Plotter**: Output device used to draw maps and other diagrams from digital data.

**Plug-ins**: Software that works automatically when particular types of files are encountered by a web browser.

**Point**: A single $(x,y)$ co-ordinate or pixel used to describe the location of an object too small to be represented as a polygon or area feature within the working scale of the GIS.

**Point dictionary**: Vector data structure in which all points are numbered sequentially and contain an explicit reference which records which co-ordinates are associated with which polygon.

**Point-in-polygon overlay (PIP)**: The process of laying a point map over a polygon map to determine which points fall within which polygon.

**Point mode digitizing**: A method of manual digitizing whereby the operator decides where to record each co-ordinate.

**Polygon**: A multifaceted vector graphic figure used to represent an area and formed from a closed chain of points.

**Polygon-on-polygon overlay**: Process of overlaying two polygon maps to determine how the two sets of polygons overlap. There are three main types of polygon overlay: UNION, INTERSECT and IDENTITY.

**Postal code**: Non-geographic co-ordinate system consisting of alphanumeric codes used to increase the efficiency of mail sorting and delivery. Linked to geographic co-ordinates these can be a useful form of spatial reference.

**Primary data**: Data collected through firsthand observation. Field notes, survey recordings and GPS readings are examples of primary data.

**Prototyping**: An IT software development and project management methodology that uses 'prototype' software applications to test ideas with potential users. See also 'Pilot project'.

**Public land survey system (PLSS)**: Recursive division of land into quarter sections which can be used as a non-co-ordinate referencing system. Used in the western USA.

**Quadtree**: A compact raster data structure in which geographical space is subdivided into variably sized homogeneous quarters.

**Quaternary triangular mesh (QTM)**: A referencing system that tries to deal with irregularities on the Earth's surface using a regular mesh of triangles. An alternative to the geographic (latitude and longitude) referencing system.

**Raster data model**: System of tessellating rectangular cells in which individual cells are used as building blocks to create images of point, line, area, network and surface entities.

**Rasterization**: The process of converting data from vector to raster format.

**Reclassification**: The process of reclassifying values in a map layer to produce a new map layer.

**Rectangular co-ordinates**: See 'Cartesian co-ordinates'.

**Regional Research Laboratories (RRL)**: A network of research laboratories set up by the Economic and Social Research Council to pump-prime GIS research in the UK.

**Relational database**: A computer database employing an ordered set of attribute values or records known as tuples grouped into two-dimensional tables called relations.

**Relations**: Two-dimensional tables of data in a relational database.

**Relief shadows**: Areas calculated to be in the shadow of hills using a DTM. The amount of shadow varies according to the relative position of the Sun as determined by the time of day and date.

**Remote sensing**: The science of observation without touching. Often used to refer to Earth observation from satellite platforms using electromagnetic sensors.

**Resolution**: The size of the smallest recording unit or the smallest feature that can be mapped and measured.

**Rich picture**: A graphical method for expressing the scope of a problem.

**Root definition**: A clear statement of purpose. In GIS this may be a statement of why the GIS is required and the purpose it is to serve.

**Route tracing**: Application of network analysis to trace the route of flows through a network from origin to destination. This is particularly useful where flows are unidirectional (as in hydrological networks).

**Routed line cartograms**: Maps in which routes are plotted as generalized lines showing the sequence of stops and connections rather than the actual route taken. An example is the map of the London Underground system.

**Rubber sheeting**: The method used to adjust the location of features in a digital map layer in a non-uniform or non-linear manner.

**Run-length encoding**: The method of reducing the volume of raster data on a row-by-row basis by storing a single value where there are a number of cells of a given type in a group, rather than storing a value for each individual cell.

**Satellite image**: Graphical image (usually in digital form) taken of the Earth's surface using electromagnetic sensors on board an orbiting satellite or spacecraft.

**Scale**: The size relationship or ratio between the map document and the area of the Earth's surface that it represents.

**Scale analogue model**: A model that is a scaled down and generalized replica of reality, such as a topographic map or aerial photograph.

**Scale-related generalization**: Cartographic simplification of mapped features, the level of which is directly related to the scale of the map being drawn.

**Scanner**: Raster input device similar to an office photocopier used to convert maps and other analogue data into digital data using the raster data model.

**Secondary data**: Data collected by another individual or organization for another primary purpose. Many data sources used in GIS, including maps, aerial photographs, census data and meteorological records, are secondary.

**Shortest path problem**: A classic problem in network analysis which involves identifying the shortest and/ or quickest route through a network, taking link and turn impedances as well as distance into account.

**Sieve mapping**: The consecutive overlay of various maps to find a set of feasible areas that satisfy a given set of criteria.

**Simple raster**: Simplest raster data structure in which the value of each pixel is stored as a separate value in a matrix.

**Sliver polygons**: Small, often long and narrow, polygons resulting from the overlay of polygons with a common but separately digitized boundary. Errors in the digitizing process mean two versions of the common boundary are slightly different.

**Slope**: The steepness or gradient of a unit of terrain, usually measured as an angle in degrees or as a percentage.

**Soft systems analysis**: A general-purpose methodology for investigating unstructured management problems.

**Spaghetti model**: Simplest form of vector data structure representing the geographic image as a series of independent (x,y) co-ordinate strings.

**Spatial**: Anything pertaining to the concepts of space, place and location.

**Spatial analysis**: See 'Geographical analysis'.

**Spatial autocorrelation**: Tobler's Law of Geography (1976), which states that points closer together in space are more likely to have similar characteristics than those that are further apart.

**Spatial data**: Data that have some form of spatial or geographical reference that enables them to be located in two- or three-dimensional space.

**Spatial data model**: A method by which geographical entities are represented in a computer. Two main methods exist: raster and vector data models.

**Spatial data retrieval**: The process of selectively retrieving items from a GIS database on the basis of spatial location.

**Spatial data structure**: Approach used to provide the information that the computer requires to reconstruct the spatial data model in digital form. There are many different data structures in use in GIS.

**Spatial decision support system (SDSS)**: A decision support system (DSS) with a strong spatial component and incorporating spatial data models and spatial analysis to assist the user in arriving at a solution.

**Spatial entities**: Discrete geographical features (points, lines and areas) represented in a digital data structure.

**Spatial interaction models**: Models that are used to help understand and predict the location of activities and the movement of materials, people and information.

**Spatial interpolation**: The procedure of estimating the values of properties at unsampled sites within an area covered by existing observations (Waters, 1989).

**Spatial model**: A model of the real world, incorporating spatial data and relationships, used to aid understanding of spatial form and process.

**Spatial process models**: Models that simulate real-world processes which have a significant spatial component. These are used to help evaluate and understand complex spatial systems.

**Spatial query**: Action of questioning a GIS database on the basis of spatial location. Spatial queries include 'What is here?' and 'Where is...?'

**Spatial referencing**: The method used to fix the location of geographical features or entities on the Earth's surface.

**Splines**: Mathematical method of smoothing linear features, often used to interpolate between points digitized along a curved feature.

**SPOT**: A French remote sensing satellite (Système Pour l'Observation de la Terre) yielding high-resolution (10 m) data.

**Spot height**: A single height value on a topographic map, usually representing the location and height of a prominent feature between contour lines.

**Standardization**: The process of transforming data onto a common scale of measurement. This is used in MCE to ensure comparability between criteria measured in different units and different scales of measurement.

**Stereoscopic aerial photography**: The use of overlapping vertical aerial photographs to determine height based on the parallax effect gained from viewing the same object from slightly different angles.

**Stochastic model**: A model that recognizes that there could be a range of possible outcomes for a given set of inputs, and expresses the likelihood of each one happening as a probability.

**Stream mode digitizing**: The method of manually digitizing lines whereby the digitizer automatically records a co-ordinate at a set time or distance interval.

**Structured Query Language (SQL)**: The computer language developed to facilitate the querying of relational databases. Also known as Standard Query Language.

**Structured systems analysis and design methodology (SSADM)**: A method used for the design of GIS projects.

**Surface**: The entity type used to describe the continuous variation in space of a third dimension, such as terrain.

**Surface drape**: The draping of an image on top of a 3D view of a terrain model for the purpose of landscape rendering or visualization.

**Surface significant points**: Points in a TIN model that cannot be closely interpolated from the height values of neighbouring points.

**SYMAP**: Synagraphic Mapping. An early commercial computer cartography package.

**SYMVU**: Early 3D computer mapping package developed from SYMAP.

**System life cycle**: An IT software development and methodology that uses a linear approach to project management.

**Temporal data**: Data that can be linked to a certain time or period between two moments in time.

**Terrain model**: Surface model of terrain. See 'digital terrain model (DTM)', 'digital elevation model (DEM)' and 'triangulated irregular network (TIN)'.

**Thematic data**: Data that relate to a specific theme or subject.

**Thematic maps**: Maps pertaining to one particular theme or subject.

**Theodolite**: A survey instrument for measuring horizontal and vertical angles using a rotating and tilting telescope.

**Thiessen polygons**: An exact method of interpolation that assumes the values of unsampled locations are equal to the value of the nearest sampled point to produce a mesh of irregular convex polygons.

**Tight coupling**: A method of linking models to GIS in which the link between the GIS and the model is hidden from the user by an application interface and GIS and model share the same database.

**Topographic maps**: Maps whose primary purpose is to indicate the general lie of the land. These maps generally show terrain, basic land use, transport networks, administrative boundaries, settlements and other man-made features.

**Topology**: The geometric relationship between objects located in space. Adjacency, containment and connectivity can describe this.

**Total station**: A theodolite or EDM combined with data logger and automated mapping software.

**Transformation**: The process of converting data from one co-ordinate system to another.

**Travelling salesperson problem**: A problem in network analysis where the best route between a series of locations to be visited in one journey must be identified.

**Trend surface analysis**: A routine that interpolates a complex surface or series of data points to produce a much-simplified surface showing the overall trend in the data.

**Triangulated irregular network (TIN)**: An irregular set of height observations in a vector data model. Lines connect these to produce an irregular mesh of triangles. The faces represent the terrain surface and the vertices represent the terrain features.

**Tuples**: Individual records (rows) in a relational database.

**UNION overlay**: Polygon-on-polygon overlay corresponding to the Boolean OR overlay. The output map will contain a composite of all the polygons in both of the input map layers.

**Universal soil loss equation (USLE)**: Method of predicting soil loss over large areas based on relationships between soil loss and determining factors measured empirically using small erosion plots.

**Vector data model**: A spatial data model using two-dimensional Cartesian (x,y) co-ordinates to store the shape of spatial entities.

**Vectorization**: The process of converting data from raster to vector format.

**Viewshed**: A polygon map resulting from a visibility analysis showing all the locations visible from a specified viewpoint.

**Virtual reality (VR)**: The production of realistic-looking computer-generated worlds using advanced computer graphics and simulation models.

**Virtual Reality Modelling Language (VRML)**: A computer language developed to assist VR development on the World Wide Web.

**Visibility analysis**: An analysis of visible features on a terrain surface using a DTM.

**Voronoi polygons**: See 'Thiessen polygons'.

**Weights**: Priorities or preferences attached to criteria in MCE. Usually specified by the user in order to indicate the relative importance of each criterion.

**Weird polygons**: See 'Sliver polygons'.

**World Wide Web (the Web)**: Interconnected global network of computers and the software interface used to access and exchange digital information and multimedia.

**ZIP code**: See 'Postal code'.

# Index